$\bar{z}$	conjugate of z	(6.2)
ρ or $\lvert z \rvert$	rho, absolute value or modulus of z	(6.3)
θ	theta, argument of a complex number	(6.3)
z^0	the complex number $1 + 0i$ or $(1, 0)$	(6.5)
$z^{-n}, n \in N$	$1/z^n$	(6.5)
$z^{1/n}, n \in N$	nth root of z	(6.5)
$\{a, b\}$	the set whose elements (or members) are a and b	(A.1)
$\in$	is a member of, is an element of ($\notin$, is not a member of, is not an element of)	(A.1)
$\{x \mid \ldots\}$	the set of all x such that ...	(A.1)
$\cup$	the union of	(A.1)
$\cap$	the intersection of	(A.1)
$\varnothing$	the null set, the empty set	(A.1)
H	the set of irrational numbers	(A.2)
J	the set of integers	(A.2)
N	the set of natural numbers	(A.2)
Q	the set of rational numbers	(A.2)
R	the set of real numbers	(A.2)
$\sqrt{a}, a \geq 0$	the nonnegative square root of a	(A.2)
$\approx$	approximately equal to	(A.2)
$\lvert a \rvert$	the absolute value of a	(A.2)
$<$	is less than	(A.2)
$\leq$	is less than or equal to	(A.2)
$>$	is greater than	(A.2)
$\geq$	is greater than or equal to	(A.2)
(a, b)	the ordered pair of numbers whose first component is a and whose second component is b	(A.3)
f, F, g, etc.	names a function or a relation	(A.3)
$f(x)$	f of x, or the value of f at x	(A.3)
$\overline{P_1 P_2}$	the line segment whose endpoints are P_1 and P_2	(A.4)
$P_1 P_2$	the length of line segment $\overline{P_1 P_2}$	(A.4)
$\mathscr{R}$	a relation	(A.6)
$\mathscr{R}^{-1}$	the inverse of relation $\mathscr{R}$	(A.6)
AB	ray with endpoint A and including B	(A.7)
$\angle ABC$	angle ABC (the vertex is at B)	(A.7)
$\perp$	is perpendicular to	(A.7)
$\triangle ABC$	a triangle with vertices A, B, C	(A.8)
$\cong$	is congruent to	(A.9)
$\sim$	is similar to	(A.9)

TRIGONOMETRY
An Analytic Approach

Fifth Edition

TRIGONOMETRY

AN ANALYTIC APPROACH

Irving Drooyan
Walter Hadel
Charles C. Carico
Emeriti, Los Angeles Pierce College

Macmillan Publishing Company
New York

Collier Macmillan Publishers
London

Printed in the United States of America.

Macmillan Publishing Company
866 Third Avenue, New York, New York 10022

Collier Macmillan Canada, Inc.

Library of Congress Cataloging in Publication Data

Drooyan, Irving.
Trigonometry: an analytic approach.

Includes index.
1. Trigonometry, Plane. I. Hadel, Walter.
II. Carico, Charles C. III. Title.
QA533.D75 1987 516.2′4 85-23172
ISBN 0-02-330650-5

Printing: 2 3 4 5 6 7 8 Year: 7 8 9 0 1 2 3 4 5 6

ISBN 0-02-330650-5

PREFACE

This fifth edition of *Trigonometry: An Analytic Approach* reflects twenty years of classroom experience with the earlier editions. As in the previous editions, the circular functions, which are pairings of real numbers with real numbers, and their properties are introduced in Chapter 1 and are the basis for a mathematical characterization of periodic phenomena. This approach gives early importance to those aspects of trigonometry that students who plan to continue their study of mathematics will need in calculus.

This edition, like the previous editions, is designed for use in a one-semester or one-quarter course in trigonometry. The topics included are shown in the detailed Contents.

Prerequisites

Although it is assumed that students undertaking the study of trigonometry have had prerequisite work in algebra and geometry, Appendix A presents a review of topics in these subjects that are needed as a background for the course. Students are referred to a section of this review when the material can be helpful in understanding a particular trigonometric concept.

Changes in This Edition

- Approximations for circular (and trigonometric) function values that were obtained in the previous editions by using tables are obtained by using a calculator. We have continued our emphasis on obtaining exact function values of real numbers $\pi/6$ and $\pi/4$ and angles with measures of 30° and 45° (and integral multiples thereof) by always specifying in the instructions when *exact values* are expected.
- Section 2.6 of the previous edition has been split into two sections to allow a more detailed consideration of inverse functions.

- The new Section 4.1 in this edition, providing work in simplifying expressions involving trigonometric function values, has been added to motivate "proving identities" in the following sections.
- Section 4.7 of the previous edition has been split into two sections to provide an expanded approach to conditional trigonometric equations.
- Some textual material has been rewritten to improve the presentation, and some exercise sets have been revised and expanded.
- A discussion of significant digits, rounding-off procedures, and scientific notation is included in Appendix B.
- A discussion of how tables can be used instead of a calculator is now included in Appendix C. These tables also give students a visual representation of how values of the trigonometric functions increase/decrease over their ranges and provide insight into the cofunction relationships between pairs of these functions.

Pedagogical Features

As in the previous editions, the subject matter of the text is continually reviewed through the use of Chapter Summaries and Review Exercises. Items in both the Summaries and Reviews are cross-referenced to the appropriate sections of the text, as are the symbols listed inside the front cover. The detailed summary of the important topics in the text that appears inside the back cover can serve as a convenient reference for students completing assignments or studying for examinations.

All Exercise Sets include basic exercises designated A; many also include more challenging exercises designated B, which provide the instructor with flexibility in making assignments, depending on the time available and the objectives of the course.

Answers, including graphs, for odd-numbered exercises in each Exercise Set and for all of the Review Exercises follow the Appendixes.

Ancillary materials are available to instructors.

Acknowledgments

We gratefully acknowledge the encouraging and helpful comments from the reviewers of this fifth edition: Kent Aeschliman, Oakland Community College; John M. Carman, San Diego Mesa College; Steven J. Davidson, San Jacinto College Central; Katherine Franklin, Los Angeles Pierce College; and Richard K. Matthei, Henry Ford Community College.

A special thanks is extended to our production supervisor Elisabeth Belfer for her outstanding professional help in the preparation of each edition of this book.

Woodland Hills, California

Irving Drooyan
Walter Hadel
Charles C. Carico

CONTENTS

APPENDIXES

A ALGEBRA AND GEOMETRY REVIEW 297

B ROUNDING-OFF PROCEDURES 325

C USING TABLES 329

ANSWERS TO ODD-NUMBERED EXERCISES 347

INDEX 389

TO THE STUDENT

Comment on Calculators

If you are not already familiar with the calculator that you are using, you should read the appropriate sections of the instruction manual that pertain to the functions or operations that are being considered in the text at a particular time.

As you use your calculator, you will eventually be able to perform many complicated combined operations involving trigonometric function values by using a single chain of key strokes. Until you become proficient in performing combined operations, however, you can simply use your calculator in a sequence of steps:

1. Obtain approximations for trigonometric function values instead of using tables.
2. Write these values in expressions that involve one or more computations.
3. Complete the computations using the simple function keys for the basic operations.

1

1 CIRCULAR FUNCTIONS

Trigonometry (from the Greek, meaning "triangle measurement") was used as early as the second century B.C.E. to find the lengths of sides of triangles and the measures of angles in such fields as astronomy, surveying, and navigation. With the development of the sciences, including physics and engineering, the study of mathematical models, which can be used to describe periodic phenomena such as alternating current, pendulum oscillations, and wave motion, has become an important part of trigonometry. These models appear as functions that are called *periodic functions*. The most common periodic functions are the *circular functions,* which we consider in this chapter, and the *trigonometric functions,* which we consider in Chapter 3.

A summary of prerequisite topics of algebra and geometry is to be found in Appendix A. You may want to refer to particular topics as necessary during your work in this course. The review of simple algebraic functions and their graphs in Sections A.3 and A.4 would be particularly helpful at this time.

1.1 The Cosine and Sine Functions

In this section we will define two functions in terms of the coordinates of a point on a circle of radius one, a **unit circle.**

Lengths of Arcs on a Unit Circle

Consider the graph of $x^2 + y^2 = 1$, a unit circle with center at the origin. Each arc with initial point (1, 0) on this circle has a certain length,

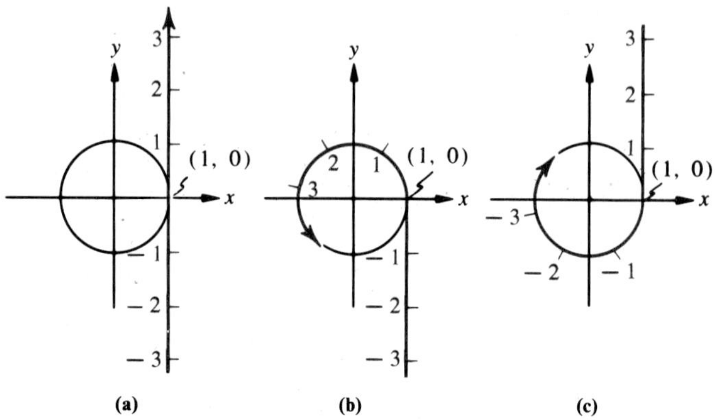

Figure 1.1

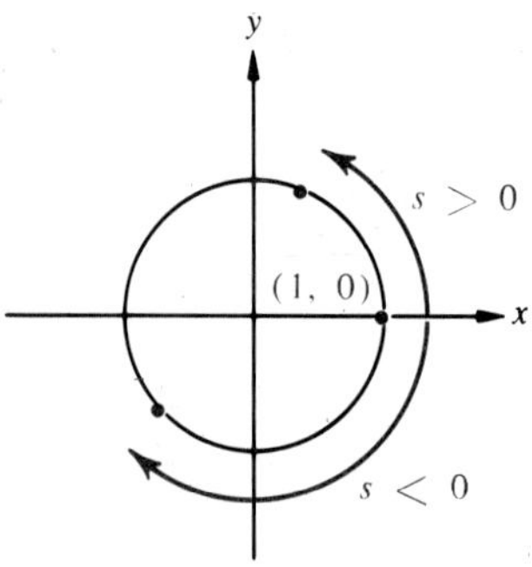

Figure 1.2

which is a real number. Hence we can establish a one-to-one correspondence between all such arcs on the circle and the set R of real numbers. We can visualize the correspondence as shown in Figure 1.1, where the real line is "wrapped" about the circle, matching points from $(1, 0)$, which is placed at the origin of the line, to the terminal point of the arc, which corresponds to a point on the real line representing the length of the arc. If the line is wrapped in a counterclockwise direction from $(1, 0)$ as in Figure 1.1b, a positive real number (its length) is assigned to the arc; if the line is wrapped in the clockwise direction, as in Figure 1.1c, a negative real number is assigned, the negative of the arc's length. We will use the variable s to represent the lengths of such arcs, as shown in Figure 1.2, where $s \in R$.

The radius of a unit circle is 1. Hence the circumference, given by $2\pi r$, is equal to 2π. The distance halfway around is π, one-fourth of the way around

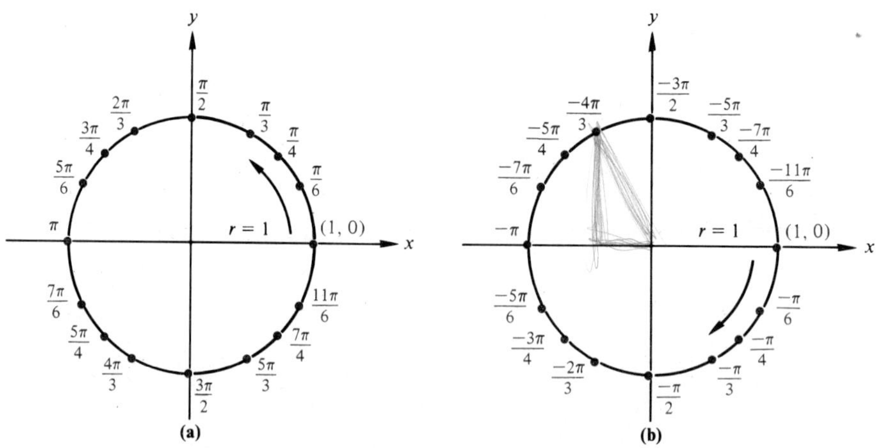

Figure 1.3

is $\pi/2$, and so on. End points of selected arcs (viewed in a *counterclockwise direction* and their corresponding arc lengths in units of π are shown in Figure 1.3a. End points of selected arcs (viewed in a *clockwise direction*) and their corresponding arc lengths in units of π are shown in Figure 1.3b.

In our work we will often need to know the quadrant in which the terminal point of an arc lies. It is usually more convenient to use multiples of the irrational number π as the basic unit of measurement, unless otherwise noted. However, when we use decimals we will use $\pi \approx 3.14$, so that the points where the unit circle intersects the axes, namely (0, 1), (−1, 0), (0, −1), and (1, 0) will correspond to arc lengths of

$$\frac{\pi}{2} \approx 1.57, \quad \pi \approx 3.14, \quad \frac{3\pi}{2} \approx 4.71, \quad \text{and} \quad 2\pi \approx 6.28,$$

respectively.*

Examples

a. The terminal point of the arc with length $11\pi/12$ lies in the second quadrant because $\pi/2 < 11\pi/12 < \pi$.

b. The terminal point of the arc with length 4.2 lies in the third quadrant because $3.14 < 4.2 < 4.71$.

c. The terminal point of the arc with length -5.21 lies in the first quadrant because $-6.28 < -5.21 < -4.71$.

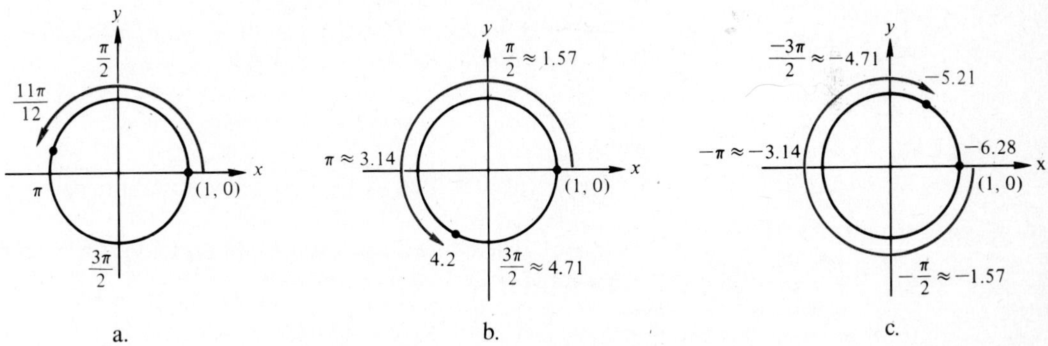

As noted in Figure 1.1, the number line can be thought of as "wrapped" about the circle. For directed lengths on the number line greater than 2π or less than -2π, the wrapping will exceed the circumference of the circle, as shown in the following examples. To find the quadrant in which the terminal point of such an arc lies, we must add or subtract multiples of 2π in order to obtain a real number between 0 and 2π.

* The symbol ≈ is read "is approximately equal to."

Examples

a. The terminal point of the arc with length $11\pi/4$ lies in the second quadrant because $(11\pi/4) - 2\pi = 3\pi/4$ and $\pi/2 < 3\pi/4 < \pi$.

b. The terminal point of the arc with length -7.31 lies in the fourth quadrant because $-7.31 - (-6.28) = -1.03$ and $-1.57 < -1.03 < 0$.

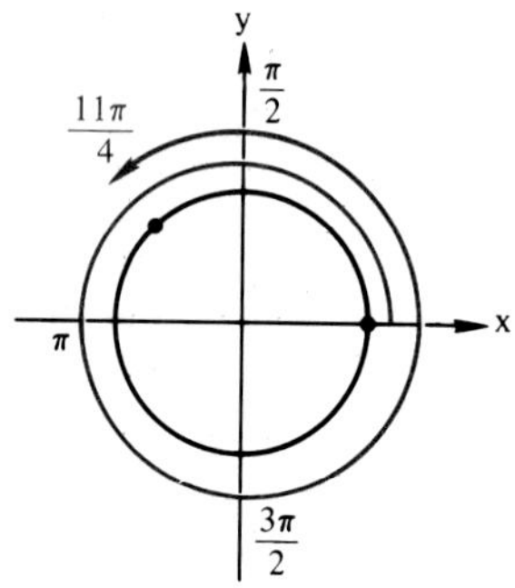

a.

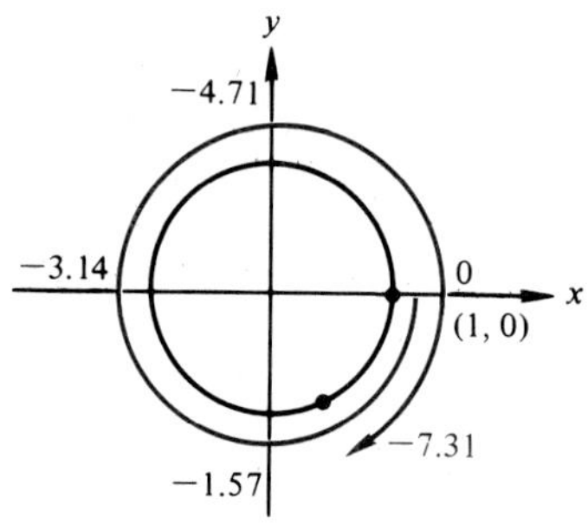

b.

Coordinates of Points on a Unit Circle

The coordinates of the terminal point of an arc on the unit circle $x^2 + y^2 = 1$ with initial point (1,0) are given special names in terms of the arc length s.

Definition 1.1 *For each arc of length s on a unit circle, with initial point* (1, 0) *and terminal point* (x, y), $x, y \in R$,

$$\textbf{cosine } s = x \qquad \textit{and} \qquad \textbf{sine } s = y.$$

Cosine s is abbreviated as **cos s,** and sine s is abbreviated as **sin s.**

Example The figure illustrates Definition 1.1 for the case of an arc having its terminal point in Quadrant I.

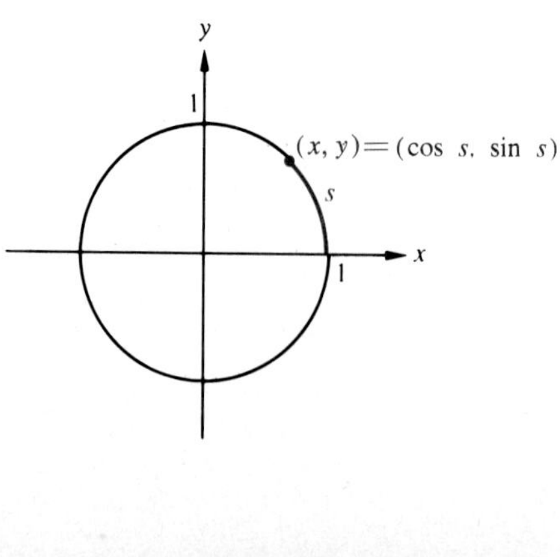

Circular Functions, Cosine and Sine

The set of pairings of each arc length s with x or $\cos s$ is called the **cosine function,** and the set of pairings of each arc length s with y or $\sin s$ is called the **sine function.*** Thus,

$$\text{cosine} = \{(s, \cos s)\} \qquad \text{and} \qquad \text{sine} = \{(s, \sin s)\},$$

where the domain of each function is R. This notation is similar to notations such as $(x, f(x))$ and $(x, g(x))$ that is used for algebraic functions. However, by convention we write $\cos s$ and $\sin s$ instead of $\cos(s)$ and $\sin(s)$.

The ranges of the functions are determined by the fact that $\cos s$ and $\sin s$ are the coordinates of points on the unit circle. Hence,

$$-1 \leq \cos s \leq 1 \qquad \text{or} \qquad |\cos s| \leq 1,$$

since $-1 \leq x \leq 1$ for points on the unit circle; and

$$-1 \leq \sin s \leq 1 \qquad \text{or} \qquad |\sin s| \leq 1,$$

since $-1 \leq y \leq 1$ for points on the unit circle.

Observe that both the cosine function and the sine function *pair real numbers with real numbers.* These functions are called **circular functions** because $\cos s$ and $\sin s$ are the coordinates of points on a circle.

Signs of cos *s* and sin *s*

Notice that $\cos s$ and $\sin s$, the coordinates of a point on the unit circle, are either positive or negative depending on the quadrant in which the point lies, as shown in Figure 1.4.

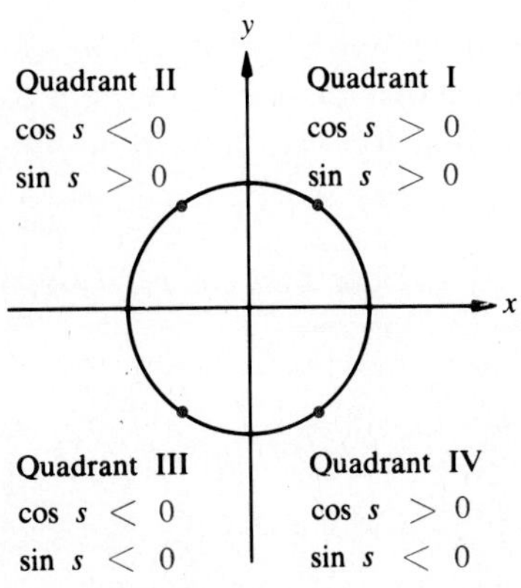

Figure 1.4

These results are summarized in Table 1.1. The quadrant in each case is that in which the terminal point of the arc with length s lies.

TABLE 1.1

quadrant	**cos s**	**sin s**
I	+	+
II	−	+
III	−	−
IV	+	−

Examples

a. If $s = 3\pi/4$, the terminal point of the corresponding arc is in Quadrant II. Thus, $\cos(3\pi/4)$ is negative and $\sin(3\pi/4)$ is positive.

* As we have suggested, you may find it helpful to review the notion of a function introduced in Sections A.3 and A.4.

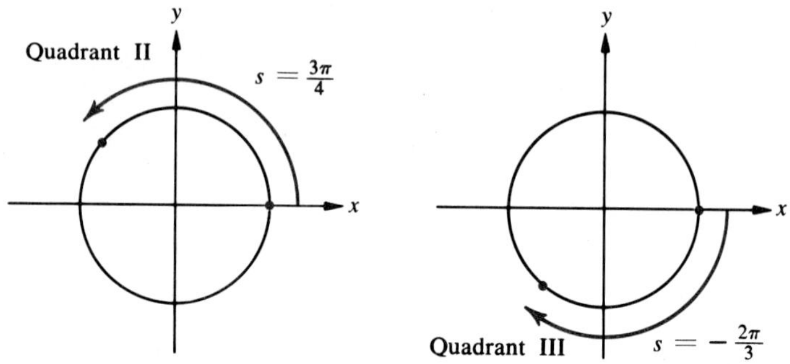

a. b.

b. If $s = -2\pi/3$, the terminal point of the corresponding arc is in Quadrant III. Thus, $\cos(-2\pi/3)$ is negative and $\sin(-2\pi/3)$ is negative.

Relationship Between cos *s* and sin *s*

An important relationship exists between $\cos s$ and $\sin s$. Because an equation of a unit circle is $x^2 + y^2 = 1$, $x = \cos s$ and $y = \sin s$, we have the following basic identity, which is sometimes called the **Pythagorean identity.**

Theorem 1.1 *If* $s \in R$, *then*

$$\cos^2 s + \sin^2 s = 1.$$

Notice that we have used the more convenient symbols $\cos^2 s$ and $\sin^2 s$ for $(\cos s)^2$ and $(\sin s)^2$, respectively. We can use this relationship to find the function values for $\cos s$ if we know the value of $\sin s$, and to find $\sin s$ for a given value of $\cos s$.

Example If $\sin s = -3/4$ and $\cos s < 0$, we can find $\cos s$ by substituting $-3/4$ for $\sin s$ in $\cos^2 s + \sin^2 s = 1$ to obtain

$$\cos^2 s + \left(-\frac{3}{4}\right)^2 = 1 \quad \leftrightarrow \quad \cos^2 s + \frac{9}{16} = 1 \quad \leftrightarrow \quad \cos^2 s = \frac{7}{16},^*$$

from which

$$\cos s = \frac{\sqrt{7}}{4} \quad \text{or} \quad \cos s = -\frac{\sqrt{7}}{4}.$$

Since $\cos s < 0$, we take the negative value, $-\sqrt{7}/4$.

* The symbol ↔ is read "is equivalent to."

EXERCISE SET 1.1

A

Name the quadrant that contains the terminal point of each arc (of a unit circle) with the given length s and initial point (1, 0).

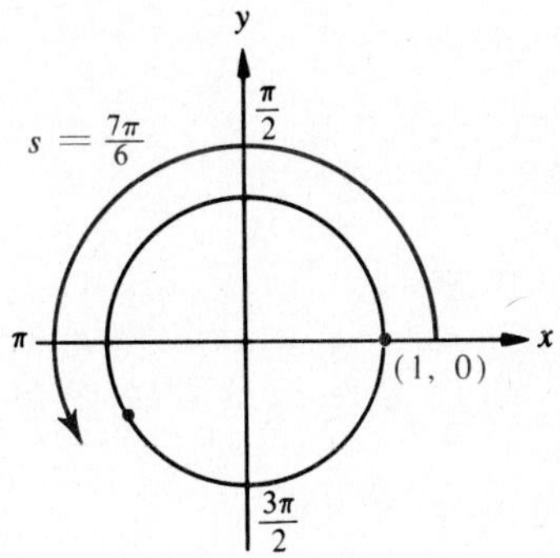

Example $\frac{7\pi}{6}$

Solution The terminal point of the arc lies in the third quadrant because $\pi < 7\pi/6 < 3\pi/2$.

1. $\frac{2\pi}{3}$ **2.** $\frac{3\pi}{4}$ **3.** $\frac{11\pi}{6}$ **4.** $\frac{7\pi}{4}$

5. $\frac{13\pi}{6}$ **6.** $\frac{10\pi}{3}$ **7.** $-\frac{\pi}{3}$ **8.** $-\frac{5\pi}{6}$

9. $-\frac{5\pi}{4}$ **10.** $-\frac{11\pi}{6}$ **11.** $-\frac{13\pi}{3}$ **12.** $-\frac{9\pi}{4}$

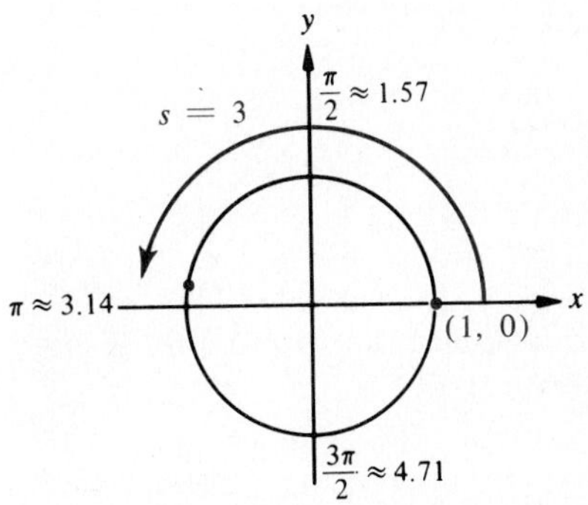

Example 3

Solution The terminal point of the arc lies in the second quadrant because $1.57 < 3 < 3.14$.

13. 2 **14.** 4 **15.** 5 **16.** 6

17. 9 **18.** 10 **19.** -1 **20.** -2

21. -3 **22.** -6 **23.** -8 **24.** -9

Determine the quadrant in which the terminal point of each arc with the length s lies.

25. $\sin s > 0$ and $\cos s > 0$ **26.** $\sin s < 0$ and $\cos s < 0$

27. $\sin s < 0$ and $\cos s > 0$ **28.** $\sin s > 0$ and $\cos s < 0$

Example If $\cos s = 4/5$ and $\sin s < 0$, find $\sin s$.

Solution Using Theorem 1.1 and substituting 4/5 for $\cos s$ in

$$\cos^2 s + \sin^2 s = 1,$$

(continued)

we obtain

$$\left(\frac{4}{5}\right)^2 + \sin^2 s = 1 \quad \leftrightarrow \quad \frac{16}{25} + \sin^2 s = 1 \quad \leftrightarrow \quad \sin^2 s = \frac{9}{25},$$

from which

$$\sin s = \frac{3}{5} \quad \text{or} \quad \sin s = -\frac{3}{5}.$$

For $\sin s < 0$, we take $\sin s = -3/5$.

29. If $\sin s = 3/5$ and $\cos s > 0$, find $\cos s$.

30. If $\cos s = 8/17$ and $\sin s > 0$, find $\sin s$.

31. If $\cos s = 1/2$ and $\sin < 0$, find $\sin s$.

32. If $\sin s = -5/13$ and $\cos s > 0$, find $\cos s$.

33. If $\sin s = 1/2$ and $\cos s > 0$, find

a. $\cos s$ **b.** $\dfrac{1}{\cos s}$ **c.** $\dfrac{1}{\sin s}$

34. If $\cos s = 40/41$ and $\sin s < 0$, find

a. $\dfrac{\sin s}{\cos s}$ **b.** $\dfrac{\cos s}{\sin s}$

B

Name the quadrant that contains the terminal point of each arc (of a unit circle) with the given length s and initial point (1, 0).

Example 54.3

Solution Because $54.3 \div 6.28 \approx 8.65$, the arc with length 54.3 would wind around a unit circle 8 times and $0.65(6.28) \approx 4.08$ more. Hence the arc terminates in Quadrant III.

35. 29 **36.** 36 **37.** 49.7

38. 39.3 **39.** -23.4 **40.** -28.2

1.2 Special Values for cos *s* and sin *s*

In order to use the circular functions in applications, we first have to find the function values cos *s* and sin *s* (elements in the ranges) for all real numbers s (elements in the domains). Some elementary notions of geometry can be used to find such function values, for selected values of s, that will be useful in our work.

Function Values for s Equal to 0, $\pi/2$, π, and $3\pi/2$

Arcs with lengths $\pi/2$, π, and $3\pi/2$ are shown in Figure 1.5 with their corresponding ordered pairs $(0, 1)$, $(-1, 0)$, and $(0, -1)$. An arc with length 0 is considered to correspond to the ordered pair $(1, 0)$. Thus,

$$\cos 0 = 1 \quad \text{and} \quad \sin 0 = 0,$$

$$\cos \frac{\pi}{2} = 0 \quad \text{and} \quad \sin \frac{\pi}{2} = 1,$$

$$\cos \pi = -1 \quad \text{and} \quad \sin \pi = 0,$$

$$\cos \frac{3\pi}{2} = 0 \quad \text{and} \quad \sin \frac{3\pi}{2} = -1.$$

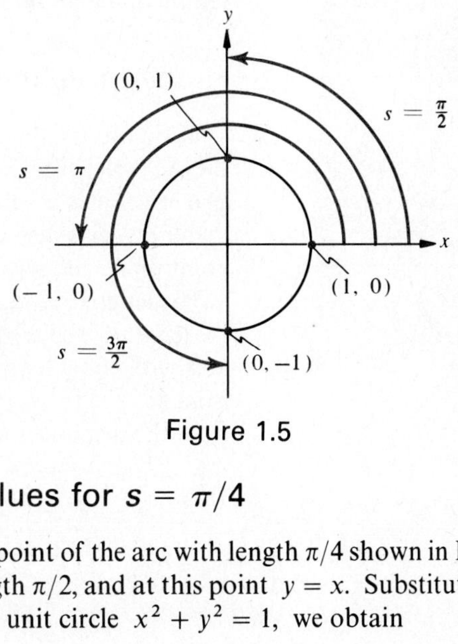

Figure 1.5

Function Values for $s = \pi/4$

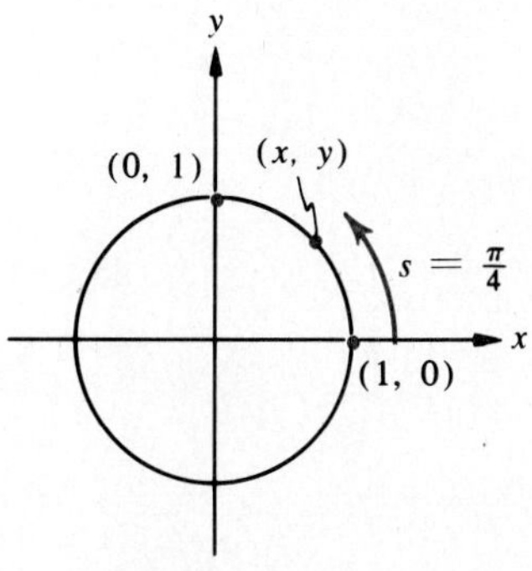

Figure 1.6

The terminal point of the arc with length $\pi/4$ shown in Figure 1.6 bisects the arc with length $\pi/2$, and at this point $y = x$. Substituting x for y in the equation for the unit circle $x^2 + y^2 = 1$, we obtain

$$x^2 + x^2 = 1,$$

$$2x^2 = 1,$$

from which $x^2 = 1/2$, and since the point is in Quadrant I $(x, y > 0)$,

$$x = \frac{1}{\sqrt{2}}.$$

Then, since $y = x$,

$$y = \frac{1}{\sqrt{2}}.$$

Thus,

$$\cos\frac{\pi}{4} = \frac{1}{\sqrt{2}} \quad \text{and} \quad \sin\frac{\pi}{4} = \frac{1}{\sqrt{2}}.$$

Note that the fraction $1/\sqrt{2}$ is shown with a radical expression as the denominator. Of course, we could rationalize the denominator as follows:

$$\frac{1}{\sqrt{2}} = \frac{1\cdot\sqrt{2}}{\sqrt{2}\cdot\sqrt{2}} = \frac{\sqrt{2}}{2}.$$

However, for our work it is convenient to use fractions in which radicals are retained in denominators.

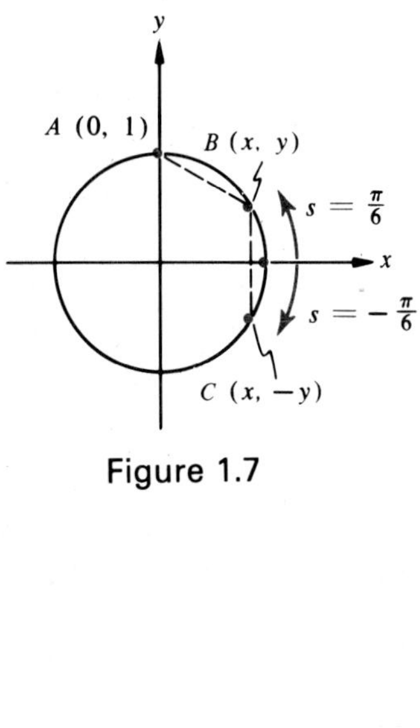

Figure 1.7

Function Values for $s = \pi/6$

In Figure 1.7, the ordered pair (x, y) corresponds to the terminal point of the arc with length $\pi/6$ and initial point $(1, 0)$ on a unit circle. Observe again that if (x, y) is a solution of the equation of the unit circle $x^2 + y^2 = 1$, then $(x, -y)$ is also a solution. Hence $(x, -y)$ corresponds to the terminal point of the arc with length $-\pi/6$.

The length of the arc from $(x, -y)$ to (x, y) equals $\pi/6 + \pi/6 = \pi/3$, and the length of the arc from (x, y) to $(0, 1)$ equals $\pi/2 - \pi/6 = \pi/3$. In a circle, arcs with equal lengths subtend chords with equal lengths, so the distance from $(x, -y)$ to (x, y) is the same as the distance from (x, y) to $(0, 1)$. Thus, from the distance formula (see Appendix A.4),

$$d^2 = (x_2 - x_1)^2 + (y_2 - y_1)^2,$$

we have

$$BC = BA$$

from which

$$(x - x)^2 + (-y - y)^2 = (x - 0)^2 + (y - 1)^2$$

or

$$4y^2 = x^2 + y^2 - 2y + 1.$$

The equation of the unit circle is $x^2 + y^2 = 1$. Hence, substituting 1 for

$x^2 + y^2$ in the right-hand member, we have

$$4y^2 = 1 - 2y + 1,$$

from which

$$4y^2 + 2y - 2 = 0,$$
$$2(2y - 1)(y + 1) = 0.$$

Hence,

$$y = \frac{1}{2} \quad \text{or} \quad y = -1.$$

Because (x, y) is in Quadrant I $(x, y > 0)$, we select $1/2$ as the y-coordinate. Substituting $1/2$ for y in $x^2 + y^2 = 1$, we have

$$x^2 + \left(\frac{1}{2}\right)^2 = 1,$$
$$x = \pm\frac{\sqrt{3}}{2}.$$

Again selecting the positive value, we have

$$\cos\frac{\pi}{6} = \frac{\sqrt{3}}{2} \quad \text{and} \quad \sin\frac{\pi}{6} = \frac{1}{2}.$$

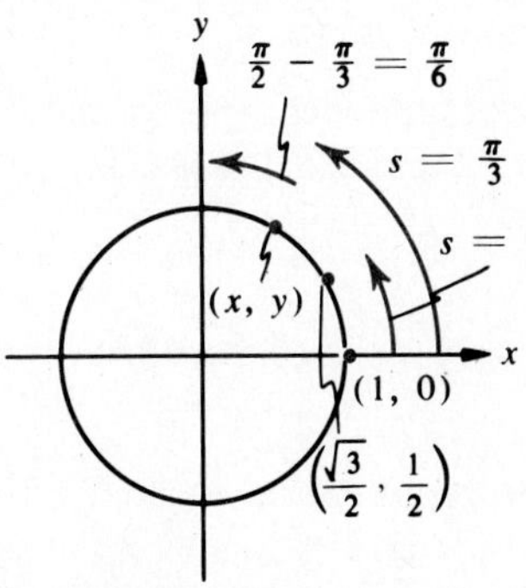

Figure 1.8

Function Values for $s = \pi/3$

In Figure 1.8, we have the ordered pair $(\sqrt{3}/2, 1/2)$ on a unit circle corresponding to the terminal point of the arc with length $\pi/6$ whose initial point is $(1, 0)$. Also, the ordered pair corresponding to the terminal point of the arc with length $\pi/3$ is shown as (x, y). An application of the distance formula, as above, enables us to find x and y. In this case,

$$(x, y) = \left(\frac{1}{2}, \frac{\sqrt{3}}{2}\right)$$

and

$$\cos\frac{\pi}{3} = \frac{1}{2} \quad \text{and} \quad \sin\frac{\pi}{3} = \frac{\sqrt{3}}{2}.$$

The results that we have obtained for cos s and sin s for

$$s \in \left\{0, \frac{\pi}{6}, \frac{\pi}{4}, \frac{\pi}{3}, \frac{\pi}{2}, \pi, \frac{3\pi}{2}\right\}$$

are summarized in Table 1.2. The entries cos s and sin s are called **function values, or values of the function.** For example, $\sin(\pi/3)$, or $\sqrt{3}/2$, is the function value or simply the value of the sine function for $s = \pi/3$. Similarly, $\cos(\pi/4)$, or $1/\sqrt{2}$, is the function value or the value of the cosine function for $s = \pi/4$.

Note that the values $\sqrt{3}/2$ and $1/\sqrt{2}$, which involve the radicals $\sqrt{3}$ and $\sqrt{2}$, are irrational numbers and are *exact values* for $\sin(\pi/3)$ and $\cos(\pi/4)$, respectively. As we shall see, we can obtain decimal approximations for these function values, and the other values in Table 1.2 that are irrational, by using a calculator or special tables. However, since the exact values shown in the table arise in advanced work in mathematics, *we will use these exact values rather than decimal approximations when working with the domain elements* $\pi/6$, $\pi/4$, and $\pi/3$.

TABLE 1.2

s	x or cos s	y or sin s
0	1	0
$\frac{\pi}{6}$	$\frac{\sqrt{3}}{2}$	$\frac{1}{2}$
$\frac{\pi}{4}$	$\frac{1}{\sqrt{2}}$	$\frac{1}{\sqrt{2}}$
$\frac{\pi}{3}$	$\frac{1}{2}$	$\frac{\sqrt{3}}{2}$
$\frac{\pi}{2}$	0	1
π	-1	0
$\frac{3\pi}{2}$	0	-1

Approximations for Other Function Values

The selected function values in Table 1.2 were obtained by using geometric methods. Such methods cannot be used for all real numbers s. However, approximations for function values cos s and sin s can be obtained for all $s \in R$ by using formulas that are considered in the study of the calculus.

Two such formulas are introduced in the exercises on pages 15–16. Approximate function values obtained by using these formulas (and other circular function values that we will consider) are listed in a set of tables in the appendixes. Approximations for function values can also be obtained by using a scientific calculator.

In our work we will use a calculator to obtain such approximations whenever values of s are not listed in Table 1.2 and exact values for cos s and sin s cannot be obtained.

Approximations for cos x and sin x, $0 < x < \pi/2$, Using a Calculator

We have been using the variable $s \in R$ to represent an element in the domain of cosine and sine functions, that is the length of an arc on the unit circle. Other variables can also be used; in particular, x is frequently used for this purpose. Because of this general practice, we use x in this way, even though the variable is now used in two different ways.

Finding function values for circular functions by using a hand-held scientific calculator is a simple process. However, because the steps to be taken vary somewhat for different types of calculators, we will not consider specific procedures. The instruction booklet for your particular calculator will provide this information.

Most calculators can be set to accept entries in either degrees (DEG) or radians (RAD). These measures will be discussed in Chapter 3. For our purpose at this time the circular function values we have defined can be obtained by using the radian setting. Calculators vary in the number of digits that are shown in the display; however, we shall round off all readings to four places.*

Examples Set the calculator for radians.

a. $\sin 0.714 \approx 0.6548620$ b. $\cos 0.393 \approx 0.9237643$

Rounding off the result to four decimal places gives

a. $\sin 0.714 \approx 0.6549$ b. $\cos 0.393 \approx 0.9238$

Using Tables

Before the advent of scientific calculators, approximations for function values were obtained from extensive tables prepared for this purpose. Although such tables have lost their primary purpose, you may wish to examine Table II in Appendix C, which explicitly lists function values for cosine and sine, and other functions that we consider in Section 1.4, to get an overview of their behavior.

Note that this table readily suggests that the range of both the cosine and sine functions is the set of real numbers from -1 to 1, inclusive.

EXERCISE SET 1.2

A

1. Complete the entries for (cos *s*, sin *s*) on the unit circle.

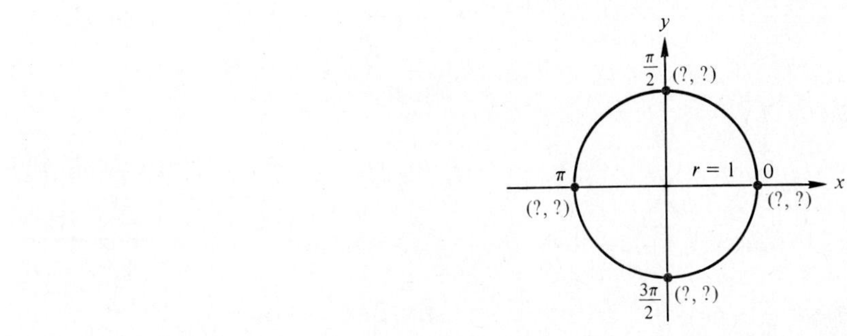

2. Complete the entries for (cos *s*, sin *s*) on the unit circle.

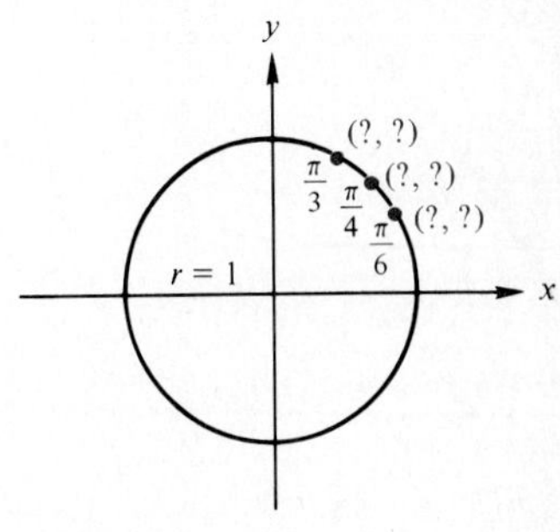

* See Appendix B for a brief review of rounding-off procedures.

Compute each expression if the expression represents a real number. Find exact values.

Example $\sin \frac{\pi}{2} + \left(\cos^2 \frac{\pi}{6}\right)\left(\cos \frac{\pi}{3}\right)$

Solution The function values can be obtained from Table 1.2, or by referring to appropriate coordinates on a unit circle.

$$\sin \frac{\pi}{2} = 1, \quad \cos \frac{\pi}{6} = \frac{\sqrt{3}}{2}, \quad \text{and} \quad \cos \frac{\pi}{3} = \frac{1}{2}.$$

Substituting these values in the given expression, we obtain

$$\begin{aligned} 1 + \left(\frac{\sqrt{3}}{2}\right)^2 \left(\frac{1}{2}\right) &= 1 + \frac{3}{4} \cdot \frac{1}{2} \\ &= 1 + \frac{3}{8} = \frac{11}{8}. \end{aligned}$$

3. $\sin^2 \frac{\pi}{6}$

4. $\cos^2 \frac{\pi}{6}$

5. $2\left(\sin \frac{\pi}{2}\right)\left(\cos^2 \frac{\pi}{4}\right)$

6. $3(\cos \pi)\left(\sin^2 \frac{\pi}{3}\right)$

7. $3 \sin \frac{\pi}{6} - \cos^2 \frac{\pi}{6}$

8. $2 \cos \frac{\pi}{3} - \sin^2 \frac{\pi}{4}$

9. $\left(\cos^2 \frac{\pi}{2}\right)\left(\sin \frac{3\pi}{2}\right)(\cos 0)$

10. $\left(\sin \frac{\pi}{4}\right)\left(\cos \frac{3\pi}{2}\right)\left(\sin^2 \frac{\pi}{3}\right)$

11. $2(\sin^2 \pi)\left(\cos^2 \frac{\pi}{4}\right) + \sin \frac{\pi}{4}$

12. $\cos^2 \frac{\pi}{3} - \left(\sin \frac{3\pi}{2}\right)\left(\cos \frac{\pi}{4}\right)$

13. $\dfrac{2 \cos \frac{\pi}{3}}{\left(\sin \frac{\pi}{3}\right)(\cos^2 0)}$

14. $\dfrac{4\left(\sin^2 \frac{\pi}{6}\right)(\cos \pi)}{\sin \frac{\pi}{2}}$

15. $\dfrac{3 \sin 0 + 2 \sin \pi}{\cos \pi}$

16. $\dfrac{2 \cos \pi - \sin 0}{\sin \pi}$

17. $\sin^2 \frac{\pi}{6} + \cos^2 \frac{\pi}{6}$

18. $\sin^2 \frac{\pi}{4} + \cos^2 \frac{\pi}{4}$

Find a decimal approximation to four places for each function value.

Examples a. cos 0.23 b. sin 1.345

Solutions Set the calculator for radians.

a. $\cos 0.23 \approx 0.9736664$ b. $\sin 1.345 \approx 0.9746161$

Rounding off to four decimal places yields

a. $\cos 0.23 \approx 0.9737$ b. $\sin 1.345 \approx 0.9746$

19. cos 0.08 **20.** sin 0.15 **21.** sin 0.35

22. cos 0.57 **23.** cos 0.72 **24.** sin 0.96

25. sin 0.255 **26.** cos 0.536 **27.** cos 1.283

28. sin 1.097 **29.** sin 0.393 **30.** cos 1.178

Find a decimal approximation to four places for each expression.

31. $\sin 0.42 + \cos 1.23$ **32.** $\cos 0.45 + \sin 1.04$

33. $2 \cos 0.47 - \sin 0.21$ **34.** $3 \sin 1.24 - \cos 0.42$

35. $\dfrac{3.4 \sin 1.21}{4.7}$ **36.** $\dfrac{12.1 \cos 0.45}{6.2}$

37. $\dfrac{1}{\sin 0.47}$ **38.** $\dfrac{1}{\cos 1.42}$

39. $\dfrac{4}{\cos 1.27}$ **40.** $\dfrac{3}{\sin 0.57}$

41. Verify that each value of s listed in Table 1.2 satisfies the relationship $\cos^2 s + \sin^2 s = 1$.

42. Using the relationship in Exercise 41, show that

$$\sin s = \pm\sqrt{1 - \cos^2 s} \quad \text{and} \quad \cos s = \pm\sqrt{1 - \sin^2 s}.$$

B

In the calculus it is shown that for all $x \in R$ *and* $n \in N$,

$$\cos x = 1 - \frac{x^2}{2!} + \frac{x^4}{4!} + \cdots + (-1)^{n-1} \frac{x^{2n-2}}{(2n-2)!} + \cdots \tag{1}$$

and

$$\sin x = x - \frac{x^3}{3!} + \frac{x^5}{5!} + \cdots + (-1)^{n-1}\frac{x^{2n-1}}{(2n-1)!} + \cdots, \qquad (2)$$

where $2! = 1 \cdot 2$, $3! = 1 \cdot 2 \cdot 3$, $4! = 1 \cdot 2 \cdot 3 \cdot 4$, *and so on.*

The expression in the right member of each equation is called an **infinite power series in *x*.** *By taking any finite number of terms in the power series to obtain an approximation for the function value, the error that occurs is less than the absolute value of the next term in the series.*

Example Find an approximation for cos 0.3 correct to four decimal places.

Solution Replacing x in Equation (1) by 0.3 yields

$$\begin{aligned}\cos 0.3 &= 1 - \frac{(0.3)^2}{2!} + \frac{(0.3)^4}{4!} - \frac{(0.3)^6}{6!} + \cdots \\ &= 1 - \frac{0.09}{2} + \frac{0.0081}{24} - \frac{0.000729}{720} + \cdots \\ &= 1 - 0.045 + 0.00034 - 0.000001 + \cdots.\end{aligned}$$

Since we desire accuracy to four decimal places, and since the error, by taking the first three terms, is less than the absolute value of the fourth term (0.000001), then

$$\cos 0.3 \approx 1 - 0.045 + 0.00034 = 0.9553$$

is correct to at least four decimal places.

Note that this is the approximation for cos 0.3 that is obtained by using a calculator.

Use Equation (1) or (2) above to find each function value correct to four decimal places.

43. sin 0.01 **44.** cos 0.02 **45.** cos 0.50

46. sin 0.60 **47.** sin 1.00 **48.** cos 1.10

1.3 Function Values cos x and sin x for $x \in R$

In Section 1.2 we obtained values for cos x and sin x in the interval 0 to $\pi/2$. In this section we shall consider ways to find values for these functions for all $x \in R$.

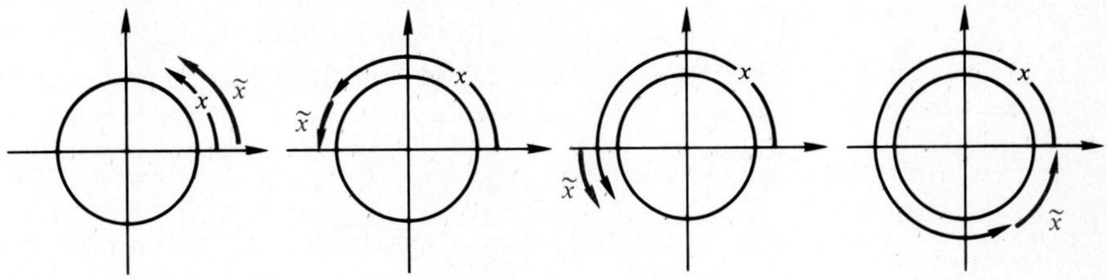

Figure 1.9

In order to find function values for integral multiples of $\pi/6$ and $\pi/4$ we shall use a sketch employing a *reference arc,* $\tilde{x}$, where $0 \le \tilde{x} < \pi/2$. The **reference arc** is the positive arc between the terminal point of the arc and the nearest point at which the unit circle meets the horizontal axis. Thus, for values of x such that $0 \le x \le 2\pi$, as shown in Figure 1.9,

in Quadrant I, $\tilde{x} = x$;

in Quadrant II, $\tilde{x} = \pi - x$;

in Quadrant III, $\tilde{x} = x - \pi$; and

in Quadrant IV, $\tilde{x} = 2\pi - x$.

However, if $x > 2\pi$, these rules must be modified. As an example, see Example d below.

Examples

a. The arc with length $3\pi/4$ terminates in Quadrant II. Hence,

$$\tilde{x} = \pi - \frac{3\pi}{4} = \frac{\pi}{4}.$$

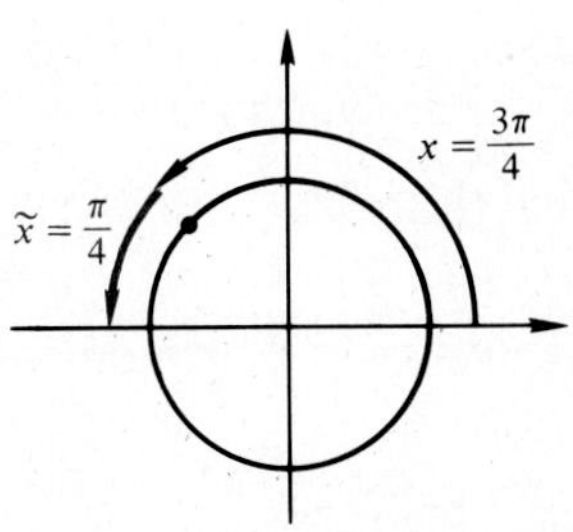

b. The arc with length $7\pi/6$ terminates in Quadrant III. Hence,

$$\tilde{x} = \frac{7\pi}{6} - \pi = \frac{\pi}{6}.$$

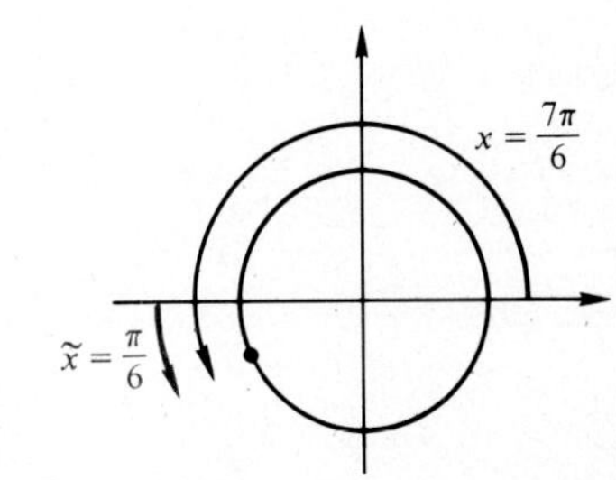

(*continued*)

c. The arc with length 5.01 terminates in Quadrant IV. Hence, $\tilde{x} \approx 2(3.14) - 5.01 = 1.27$

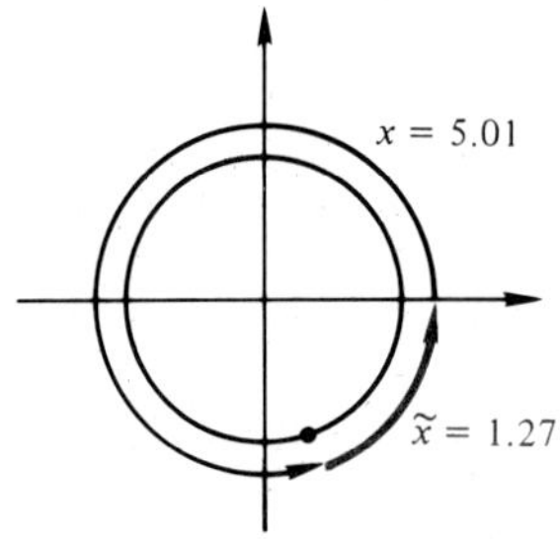

d. The arc with length 8.26 terminates in Quadrant II. Hence, $\tilde{x} \approx 3(3.14) - 8.26 = 1.16$

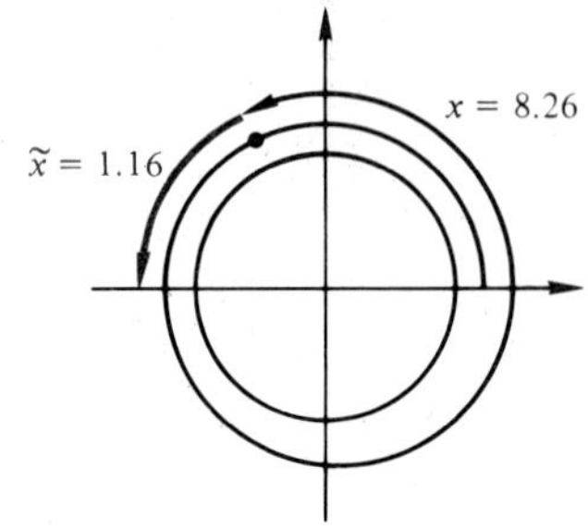

The Identities $|\cos x| = \cos \tilde{x}$ and $|\sin x| = \sin \tilde{x}$

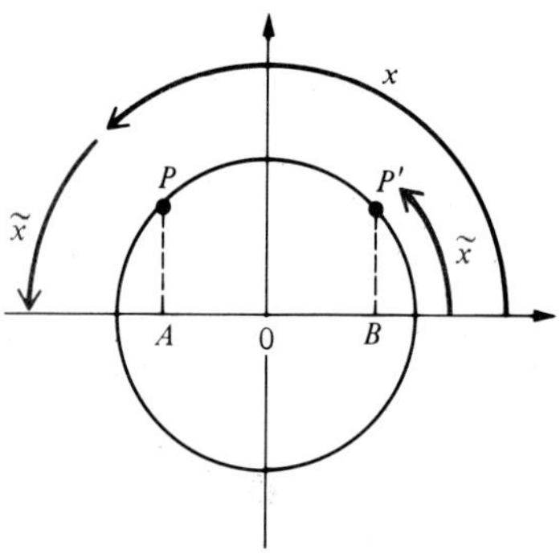

Figure 1.10

Consider an arc of length x on a unit circle, as shown in Figure 1.10, having a reference arc of length $\tilde{x}$. If we construct an arc of length $\tilde{x}$ on the circle in Quadrant I, the points P and P' are symmetric with respect to the vertical axis and the absolute value of the x-coordinate OA of P is equal to the x-coordinate OB of P'.

Since

$$|OA| = |\cos x| \qquad \text{and} \qquad OB = \cos \tilde{x},$$

$$|\cos x| = \cos \tilde{x}.$$

Now, while the arc with length x in Figure 1.10 terminates in the second quadrant, the discussion above can be generalized to include an arc terminating in any quadrant and can be used to prove the following theorem.

Theorem 1.2 *For* $x \in R$,

$$|\cos x| = \cos \tilde{x} \qquad \textit{and} \qquad |\sin x| = \sin \tilde{x},$$

where $\tilde{x}$ *is the length of the reference arc corresponding to* x.

We can now find function values $\cos x$ or $\sin x$ for all integral multiples of $\pi/6$ and $\pi/4$ by simply performing the following steps.

1. Find the reference arc length $\tilde{x}$.
2. Use Theorem 1.2 and Table 1.2 to find $\cos \tilde{x}$ or $\sin \tilde{x}$.
3. Use Table 1.1 to determine the algebraic sign for $\cos x$ or $\sin x$.

Example Find $\cos \dfrac{7\pi}{6}$.

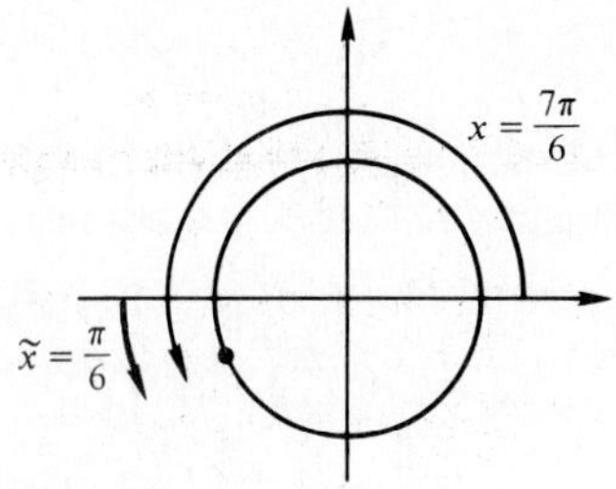

Solution Using the notion of a reference arc, we observe from the figure that $\tilde{x} = (7\pi/6) - \pi = \pi/6$. Hence, because the arc with length $7\pi/6$ terminates in Quadrant III as shown, $\cos(7\pi/6)$ is negative, and

$$\cos\frac{7\pi}{6} = -\cos\tilde{x} = -\cos\frac{\pi}{6}.$$

Using Table 1.2, we have

$$\cos\frac{7\pi}{6} = -\cos\frac{\pi}{6} = -\frac{\sqrt{3}}{2}.$$

In the above example, the element in the domain, $7\pi/6$, is in the interval between 0 and 2π. However, the procedure is perfectly general and is applicable for all $x \in R$.

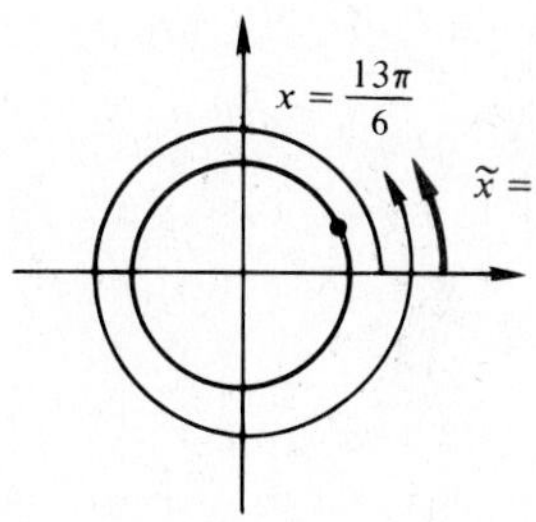

Examples a. Find $\sin\frac{13\pi}{6}$. b. Find $\sin\frac{25\pi}{6}$.

Solutions

a. In this case, $\tilde{x} = (13\pi/6) - 2\pi = \pi/6$. Hence, because the arc with length $13\pi/6$ terminates in Quadrant I,

$$\sin\frac{13\pi}{6} = \sin\frac{\pi}{6} = \frac{1}{2}.$$

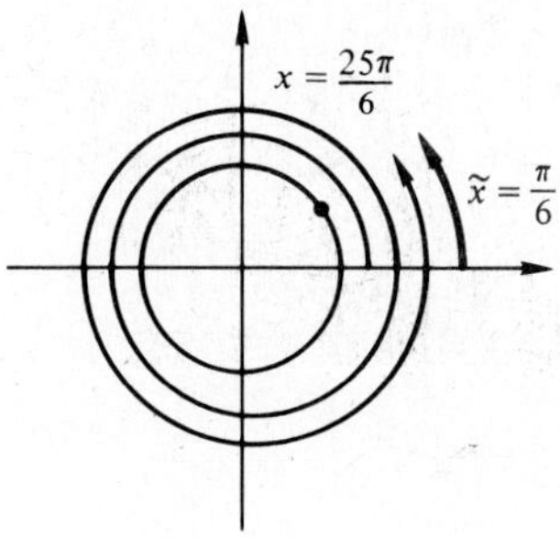

b. In this case, $\tilde{x} = (25\pi/6) - 4\pi = \pi/6$. Hence,

$$\sin\frac{25\pi}{6} = \sin\frac{\pi}{6} = \frac{1}{2}.$$

In the examples above we found that

$$\sin\frac{\pi}{6} = \sin\left(\frac{\pi}{6} + 2\pi\right) = \sin\left(\frac{\pi}{6} + 2\cdot 2\pi\right).$$

This suggests that for all $x \in R$,

$$\sin(x + k\cdot 2\pi) = \sin x,$$

where k is an integer. Such a property is referred to as a "periodic" property, and the sine function is called a "periodic" function. This important property is discussed in more detail in Section 1.5, where we will see that all circular functions are periodic.

Values for cos x and sin x for $x < 0$

Arcs that are measured in a clockwise direction, which have negative directed length, also have reference arcs. Hence, we can use $\tilde{x}$, as required to find cos x and sin x.

Example Find $\cos\left(-\frac{15\pi}{4}\right)$.

Solution Observe from the figure that

$$\tilde{x} = 4\pi - \frac{15}{4}\pi = \frac{\pi}{4}.$$

Hence, because the arc with length x terminates in Quadrant I as shown,

$$\cos\left(-\frac{15\pi}{4}\right) = \cos\tilde{x} = \cos\frac{\pi}{4}.$$

Using Table 1.2, we have

$$\cos\left(-\frac{15\pi}{4}\right) = \cos\frac{\pi}{4} = \frac{1}{\sqrt{2}}.$$

Approximations for cos x and sin x

Calculators are programmed to give an approximation for a function value for each element in the domain of the function.

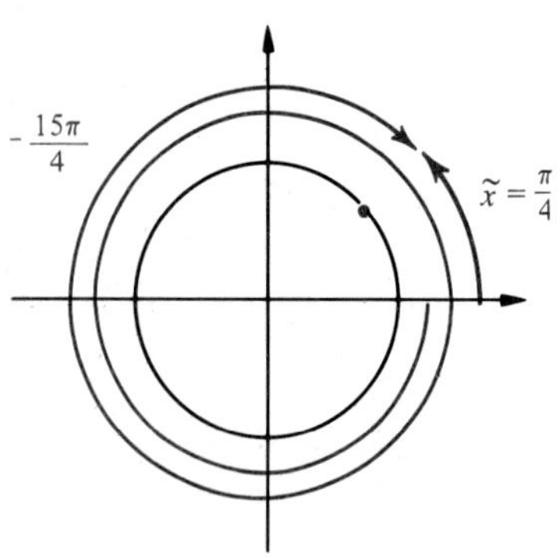

Examples To find a decimal approximation for each function value to four places, set the calculator for radians.

a. $\sin 5.84 \approx -0.4288$
b. $\cos 5.84 \approx 0.9034$
c. $\sin 8.32 \approx 0.8934$
d. $\cos 8.32 \approx -0.4493$
e. $\sin(-4.2) \approx 0.8716$
f. $\cos(-4.2) \approx -0.4903$

Although we can obtain approximations for function values of all angles using calculators, *we shall continue to obtain exact function values, which require the use of reference arcs, for elements in the domains that are integral multiples of $\pi/6$ and $\pi/4$.*

Using Tables

The use of tables to obtain function values outside the interval 0 to $\pi/2$ is discussed in Appendix C. Reference arcs in conjunction with tables are used for such values, and 3.14 is used as an approximation for π. The approximations obtained by using the tables are less accurate than those obtained by using a calculator partly because the calculator uses a closer approximation to π in its computations.

EXERCISE SET 1.3

A

1. Complete the entries for (cos x, sin x) on the unit circle.

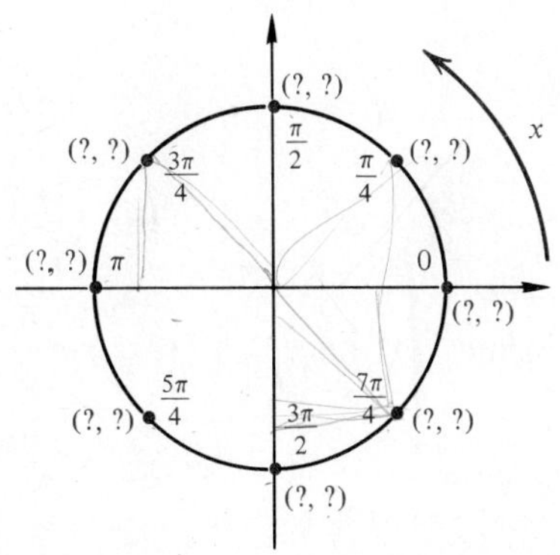

2. Complete the entries for (cos x, sin x) on the unit circle.

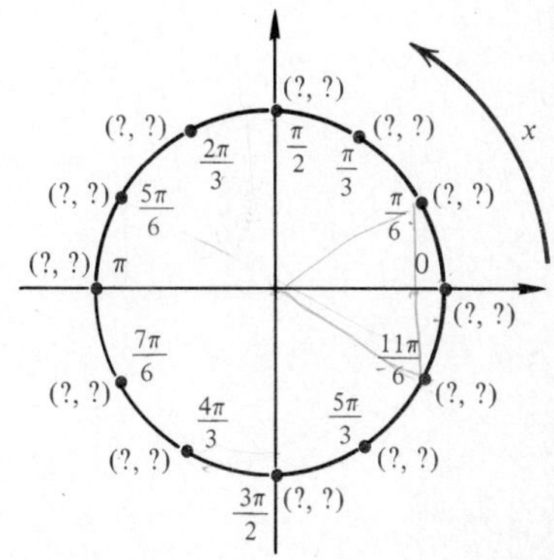

Sketch each arc with the given length, and its reference arc.

Example $\frac{5\pi}{6}$

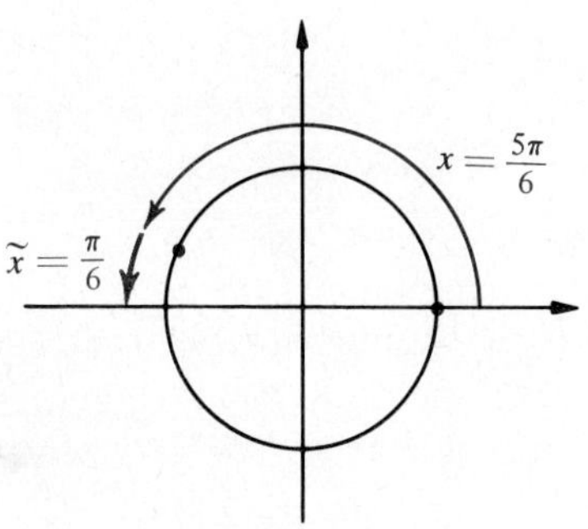

Solution An arc with length $5\pi/6$ terminates in Quadrant II because $\pi/2 < 5\pi/6 < \pi$.

3. $\frac{7\pi}{6}$ **4.** $\frac{11\pi}{6}$ **5.** $-\frac{5\pi}{3}$ **6.** $-\frac{5\pi}{4}$

7. $\frac{17\pi}{3}$ **8.** $\frac{21\pi}{4}$ **9.** $-\frac{19\pi}{6}$ **10.** $-\frac{11\pi}{3}$

11. 6 **12.** -7 **13.** -4 **14.** 3

Find each function value (exact value).

Example $\cos\frac{5\pi}{4}$

(continued)

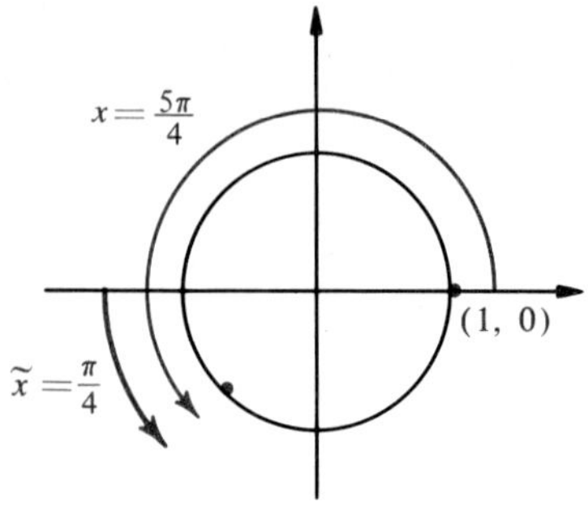

Solution The arc with length $5\pi/4$ and the corresponding reference arc are shown in the figure. Because the arc with length $5\pi/4$ terminates in Quadrant III, $\cos(5\pi/4)$ is negative, and we have

$$\cos\frac{5\pi}{4} = -\cos\tilde{x} = -\cos\frac{\pi}{4}.$$

From Table 1.2, $\cos(\pi/4) = 1/\sqrt{2}$. Thus,

$$\cos\frac{5\pi}{4} = -\cos\frac{\pi}{4} = -\frac{1}{\sqrt{2}}.$$

15. $\cos\frac{2\pi}{3}$ **16.** $\cos\frac{3\pi}{4}$ **17.** $\sin\frac{5\pi}{6}$

18. $\sin\frac{7\pi}{4}$ **19.** $\cos\frac{7\pi}{6}$ **20.** $\cos\frac{4\pi}{3}$

21. $\sin\frac{11\pi}{6}$ **22.** $\sin\frac{5\pi}{3}$ **23.** $\cos\frac{9\pi}{4}$

24. $\cos\frac{7\pi}{3}$ **25.** $\sin\frac{11\pi}{4}$ **26.** $\sin\frac{13\pi}{6}$

27. $\cos\left(-\frac{3\pi}{4}\right)$ **28.** $\cos\left(-\frac{7\pi}{6}\right)$ **29.** $\sin\left(-\frac{4\pi}{3}\right)$

30. $\sin\left(-\frac{13\pi}{6}\right)$

Find approximate function values to four places for the following.

Examples a. $\sin 5.28$ b. $\cos(-3.74)$

Solutions Set the calculator for radians.

a. $\sin 5.28 \approx -0.8432$ b. $\cos(-3.74) \approx -0.8262$

31. $\sin 1.76$ **32.** $\cos 3.81$ **33.** $\cos 4.91$

34. $\sin 4.23$ **35.** $\sin 6.06$ **36.** $\cos 2.91$

37. $\cos(-0.26)$ **38.** $\sin(-1.27)$ **39.** $\sin(-2.55)$

40. $\cos(-3.57)$ **41.** $\cos 10.13$ **42.** $\sin 9.30$

43. $3\sin 2.3 + \cos 4.2$ **44.** $2\cos 5.1 - \sin 3.8$

45. $2 \cos 5.32 - 3 \sin 1.92$

46. $4 \sin 2.01 + 2 \cos 4.4$

47. $\dfrac{3.44 \cos 1.83}{2.14}$

48. $\dfrac{4.61 \sin 2.13}{3.04}$

49. $\dfrac{2}{\sin 5.62}$

50. $\dfrac{3}{\cos 4.02}$

1.4 Tangent, Secant, Cosecant, and Cotangent Functions

Four additional circular functions can be defined in terms of $\cos x$ and $\sin x$.

Tangent Function

The ratio of the sine function value to the cosine function value is used so often that it is given a special name.

Definition 1.2 *For all* $x \in R \left(x \neq \frac{\pi}{2} + k\pi,\ k \in J \right)$,

$$\textbf{tangent } x = \frac{\sin x}{\cos x}.$$

Tangent x is abbreviated as **tan** $\boldsymbol{x}$. Pairings of the form $(x, \tan x)$ make up the **tangent function.** Note that the *domain* of the tangent function contains all real numbers x such that

$$x \neq \frac{\pi}{2} + k\pi,\ k \in J,$$

for which $\cos x = 0$.

The range of the tangent function is the set of real numbers R. This can be seen by taking values of x close to $\pi/2$, and considering the value of the quotient

$$\frac{\sin x}{\cos x} = \tan x. \qquad \left(x \neq \frac{\pi}{2} + k\pi,\ k \in J \right)$$

Since $\sin x$ is close to 1 for these values, and $\cos x$ is close to zero, the value of $\tan x$ is large in absolute value. On the other hand, for x values close to zero, $\tan x$ will also be close to zero.

Special Values for tan x

We can find some special values for tan x by referring to the cosine and sine function values in Table 1.2. Thus,

$$\tan 0 = \frac{\sin 0}{\cos 0} = \frac{0}{1} = 0,$$

$$\tan \pi = \frac{\sin \pi}{\cos \pi} = \frac{0}{-1} = 0;$$

$\tan(\pi/2)$ and $\tan(3\pi/2)$ are undefined since, from Definition 1.2, $\pi/2$ and $3\pi/2$ are not in the domain of the tangent function. Also,

$$\tan\frac{\pi}{6} = \frac{\sin\frac{\pi}{6}}{\cos\frac{\pi}{6}} = \frac{\frac{1}{2}}{\frac{\sqrt{3}}{2}} = \frac{1}{\sqrt{3}},$$

$$\tan\frac{\pi}{4} = \frac{\sin\frac{\pi}{4}}{\cos\frac{\pi}{4}} = \frac{\frac{1}{\sqrt{2}}}{\frac{1}{\sqrt{2}}} = 1,$$

$$\tan\frac{\pi}{3} = \frac{\sin\frac{\pi}{3}}{\cos\frac{\pi}{3}} = \frac{\frac{\sqrt{3}}{2}}{\frac{1}{2}} = \sqrt{3}.$$

TABLE 1.3

x	$\cos x$	$\sin x$	$\tan x$
0	1	0	0
$\frac{\pi}{6}$	$\frac{\sqrt{3}}{2}$	$\frac{1}{2}$	$\frac{1}{\sqrt{3}}$
$\frac{\pi}{4}$	$\frac{1}{\sqrt{2}}$	$\frac{1}{\sqrt{2}}$	1
$\frac{\pi}{3}$	$\frac{1}{2}$	$\frac{\sqrt{3}}{2}$	$\sqrt{3}$
$\frac{\pi}{2}$	0	1	undefined
π	-1	0	0
$\frac{3\pi}{2}$	0	-1	undefined

These special function values for the tangent function, together with those for the cosine and sine, are shown in Table 1.3.

By Definition 1.2,

$$\tan x = \frac{\sin x}{\cos x}.$$

Thus, $\tan x > 0$ if the arc of length x terminates in Quadrant I or III, since $\sin x$ and $\cos x$ are either both positive or both negative (see Table 1.1). Tan $x < 0$ if the arc corresponding to x terminates in Quadrant II or IV, since $\sin x$ is positive and $\cos x$ is negative, or $\sin x$ is negative and $\cos x$ is positive, respectively.

These results along with the data from Table 1.1 are summarized in Table 1.4. The quadrant in each case is the quadrant in which the terminal point of the arc with length x lies.

TABLE 1.4

quadrant	$\cos x$	$\sin x$	$\tan x$
I	+	+	+
II	−	+	−
III	−	−	+
IV	+	−	−

Secant, Cosecant, and Cotangent Functions

The reciprocals of $\sin x$ and $\cos x$ and the ratio $\cos x/\sin x$ are commonly used in mathematics and are also given special names.

Definition 1.3 *For all* $x \in R$,

$$\text{I.}\quad \textbf{secant } x = \frac{1}{\cos x} \quad \left(x \neq \frac{\pi}{2} + k\pi,\ k \in J\right),$$

$$\text{II.}\quad \textbf{cosecant } x = \frac{1}{\sin x} \quad (x \neq k\pi,\ k \in J),$$

$$\text{III.}\quad \textbf{cotangent } x = \frac{\cos x}{\sin x} \quad (x \neq k\pi,\ k \in J).$$

Secant x, cosecant x, and cotangent x are abbreviated as **sec *x*, csc *x*,** and **cot *x*,** respectively.

Properties of Secant

Ordered pairs of the form $(x, \sec x)$ make up the **secant function.** Since $\sec x$ is the reciprocal of $\cos x$, the domain of the secant function is the set of real numbers, except those for which $\cos x = 0$. Hence the domain contains all elements $x \in R$ such that

$$x \neq \frac{\pi}{2} + k\pi, \quad k \in J.$$

Since $|\cos x| \leq 1$ for all $x \in R$, its reciprocal satisfies $|\sec x| \geq 1$, which determines the range of the secant function.

Properties of Cosecant

Ordered pairs of the form $(x, \csc x)$ make up the **cosecant function.** Since $\csc x$ is the reciprocal of $\sin x$, the domain of the cosecant function is the set of real numbers, except those values of x for which $\sin x = 0$. Hence the domain contains all elements $x \in R$ such that

$$x \neq k\pi, \quad k \in J.$$

Since $|\sin x| \leq 1$ for all $x \in R$, the range of the cosecant function is determined by the fact that $|\csc x| \geq 1$ for all x in the domain.

Properties of Cotangent

Ordered pairs of the form $(x, \cot x)$ make up the **contangent function.** Since $\cot x = \cos x/\sin x$, the domain of the cotangent function is the set of real numbers, except those values of x for which $\sin x = 0$. Hence the domain contains all elements $x \in R$ such that

$$x \neq k\pi, \quad k \in J.$$

Since

$$\frac{\cos x}{\sin x} = \frac{1}{\sin x/\cos x} \quad (\sin x, \cos x \neq 0),$$

$$\mathbf{\cos x = \frac{1}{\tan x}}, \quad x \neq k \cdot \frac{\pi}{2}, \quad k \in J.$$

We shall view this statement as an alternate definition of cotangent x. As with the tangent function, the range of the cotangent function is R, the set of real numbers.

Special Values for sec x, csc x, and cot x

We can find exact function values for secant, cosecant, and cotangent for some values of x in the set

$$\left\{0, \frac{\pi}{6}, \frac{\pi}{4}, \frac{\pi}{3}, \frac{\pi}{2}, \pi, \frac{3\pi}{2}\right\}$$

by using Definition 1.3 and the function values for the sine, cosine, and tangent functions shown in Table 1.3.

Examples

a. $\sec 0 = \dfrac{1}{\cos 0} = \dfrac{1}{1} = 1$

b. $\csc \dfrac{\pi}{6} = \dfrac{1}{\sin \dfrac{\pi}{6}} = \dfrac{1}{\dfrac{1}{2}} = 2$

Whether particular function values of $\sec x$, $\csc x$, and $\cot x$ are negative or positive real numbers can be readily determined from their definitions and the data in Table 1.4 applicable to $\sin x$, $\cos x$, and $\tan x$.

Using Reference Arcs

A geometric argument similar to the one used to suggest Theorem 1.2 leads to the following result.

Theorem 1.3 *For $x \in R$,*

$$|\tan x| = \tan \tilde{x}, \qquad |\sec x| = \sec \tilde{x},$$

$$|\csc x| = \csc \tilde{x}, \qquad \textit{and} \qquad |\cot x| = \cot \tilde{x},$$

where $\tilde{x}$ is the length of the reference arc corresponding to x.

Some additional exact function values for these functions can now be found directly from Definition 1.3 and Table 1.3 by using references arcs.

Examples

a. $\sec \frac{3\pi}{4} = \frac{1}{\cos \frac{3\pi}{4}} = \frac{1}{-\cos \tilde{x}} = \frac{1}{-\cos \frac{\pi}{4}} = \frac{1}{-\frac{1}{\sqrt{2}}} = -\sqrt{2}.$

b. $\cot \frac{7\pi}{4} = \frac{1}{\tan \frac{7\pi}{4}} = \frac{1}{-\tan \tilde{x}} = \frac{1}{-\tan \frac{\pi}{4}} = \frac{1}{-1} = -1.$

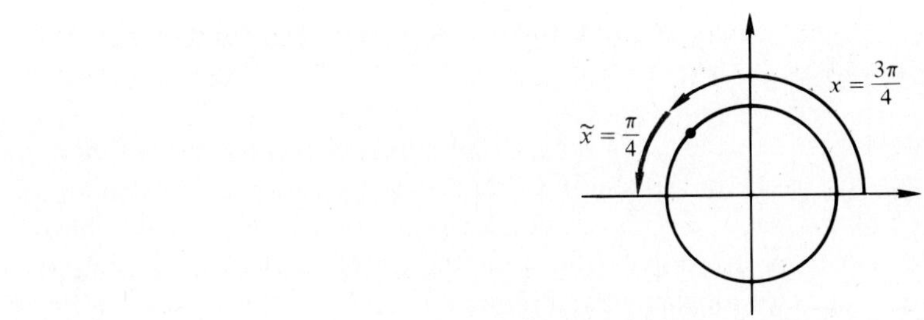

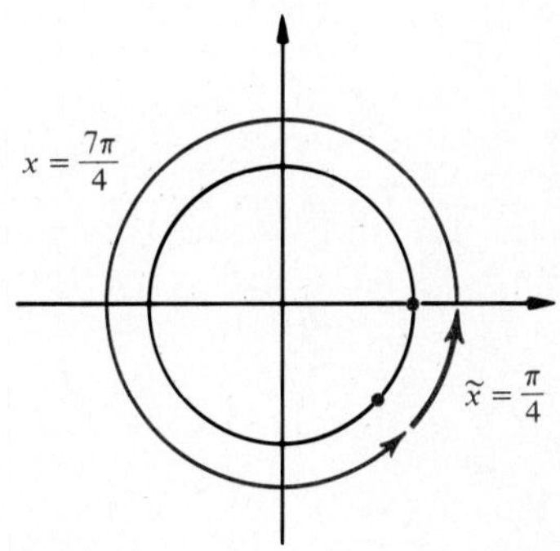

Using a Calculator

We can use a calculator to find decimal approximations for values of the functions introduced in this section in a manner similar to the way that we used a calculator to find approximations for $\cos x$ and $\sin x$. Because scientific calculators have a TAN key, the procedure used to find values for $\tan x$ is exactly the same.

Examples Find decimal approximations to four places.

a. $\tan 2.48$

b. $\tan(-2.84)$

Solutions Set the calculator for radians.

a. $\tan 2.48 \approx -0.7787$

b. $\tan(-2.84) \approx 0.3111$

Values for sec x, csc x, and cot x can be obtained by using the reciprocal relationships specified by Definition 1.3, and the alternate definition of cot x.

Examples Find decimal approximations to four places.

a. sec 2.14

b. cot(−1.32)

Solutions

a. From Definition 1.3, sec 2.14 = 1/cos 2.14. Thus,

$$\sec 2.14 = \frac{1}{\cos 2.14} \approx -1.8554.$$

b. From the alternate definition for cot x (page 26),

$$\cot(-1.32) = \frac{1}{\tan(-1.32)}.$$

Thus,

$$\cot(-1.32) = \frac{1}{\tan(-1.32)} \approx -0.2562.$$

Using Tables

As we have noted, the use of tables to obtain approximations for function values is discussed in Appendix C. Table II, Appendix C, lists function values for tan x, sec x, csc x, and cot x for values of x from 0 to 1.57. The use of reference arcs in conjunction with this table enables us to find the function values for all elements in their respective domains.

EXERCISE SET 1.4

A

Find each function value (exact value).

Example $\tan \dfrac{7\pi}{6}$

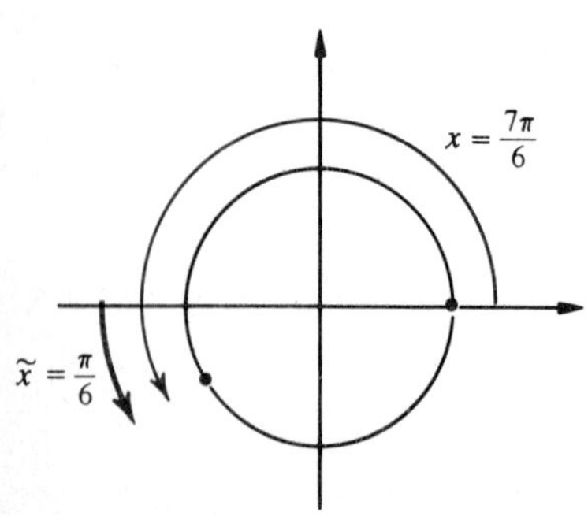

Solution Because the arc with length $7\pi/6$ terminates in Quadrant III, $\tan(7\pi/6)$ is positive, and we have

$$\tan \frac{7\pi}{6} = \tan \tilde{x} = \tan \frac{\pi}{6}.$$

From Table 1.3, $\tan(\pi/6) = 1/\sqrt{3}$. Thus,

$$\tan\frac{7\pi}{6} = \tan\frac{\pi}{6} = \frac{1}{\sqrt{3}}.$$

1. $\tan\frac{5\pi}{4}$ **2.** $\tan\frac{5\pi}{6}$ **3.** $\tan\frac{7\pi}{3}$

4. $\tan\frac{9\pi}{4}$ **5.** $\tan\left(-\frac{7\pi}{6}\right)$ **6.** $\tan\left(-\frac{5\pi}{3}\right)$

7. $\csc\frac{5\pi}{6}$ **8.** $\csc\frac{2\pi}{3}$ **9.** $\csc\frac{9\pi}{4}$

10. $\csc\frac{13\pi}{6}$ **11.** $\csc\left(-\frac{9\pi}{4}\right)$ **12.** $\csc\left(-\frac{11\pi}{6}\right)$

13. $\sec\frac{5\pi}{4}$ **14.** $\sec\frac{7\pi}{6}$ **15.** $\sec\frac{7\pi}{3}$

16. $\sec\frac{11\pi}{4}$ **17.** $\sec\left(-\frac{5\pi}{3}\right)$ **18.** $\sec\left(-\frac{11\pi}{6}\right)$

19. $\cot\frac{4\pi}{3}$ **20.** $\cot\frac{5\pi}{4}$ **21.** $\cot\frac{7\pi}{3}$

22. $\cot\frac{9\pi}{4}$ **23.** $\cot\left(-\frac{13\pi}{6}\right)$ **24.** $\cot\left(-\frac{7\pi}{4}\right)$

Find the exact value of each expression.

25. $\cot\frac{\pi}{6} + \csc\frac{\pi}{4}$ **26.** $\csc\frac{\pi}{3} + \sec\pi$

27. $2\sec 0 - 3\cot\frac{\pi}{4}$ **28.** $3\cos\frac{3\pi}{2} - \cot\frac{\pi}{4}$

29. $\dfrac{2\csc\frac{\pi}{4} + \cot\frac{\pi}{6}}{\tan\frac{\pi}{3}}$ **30.** $\dfrac{\sec\frac{\pi}{4} + \tan\frac{\pi}{3}}{\cot\frac{\pi}{4}}$

Find an approximation for each function value.

Examples a. cot 1.93 b. sec(−4.20)

(continued)

Solutions

a. $\cot 1.93 = 1/\tan 1.93$. Hence,

$$\cot 1.93 = \frac{1}{\tan 1.93} \approx -0.3755.$$

b. $\sec(-4.20) = 1/\cos(-4.20)$. Hence,

$$\sec(-4.20) = \frac{1}{\cos(-4.20)} \approx -2.0397.$$

31. $\tan 0.48$ **32.** $\sec 0.32$ **33.** $\csc 0.47$

34. $\cot 0.51$ **35.** $\tan 4.91$ **36.** $\sec 4.23$

37. $\csc 6.06$ **38.** $\cot 2.91$ **39.** $\cot(-0.26)$

40. $\csc(-1.27)$ **41.** $\sec(-2.55)$ **42.** $\tan(-3.57)$

Compute each expression.

43. $\csc 0.44 + \sec 0.10$ **44.** $\cot 1.38 + \csc 1.93$

45. $3 \csc 1.17 - \cot 0.33$ **46.** $3 \sec 2.07 - 3 \cot 1.51$

47. $\dfrac{2 \csc 3.89 + \tan 0.65}{\cot 0.60}$ **48.** $\dfrac{\sec 3.84 + \cot 1.78}{\tan 1.22}$

Determine the quadrant in which the terminal point of each arc with the length x lies.

49. $\sin x > 0$ and $\tan x < 0$ **50.** $\cos x < 0$ and $\tan x > 0$

51. $\tan x > 0$ and $\csc x < 0$ **52.** $\tan x < 0$ and $\sec x > 0$

53. $\cot x < 0$ and $\csc x > 0$ **54.** $\cot x > 0$ and $\sec x < 0$

Find each function value (exact value).

Example If $\cos x = 4/5$ and $\tan x < 0$, find $\tan x$.

Solution Since $\cos x > 0$ and $\tan x < 0$, from Table 1.4, the terminal point of the arc x lies in Quadrant IV and $\sin x < 0$. Substituting 4/5 for $\cos x$ in $\cos^2 x + \sin^2 x = 1$, we have

$$\left(\frac{4}{5}\right)^2 + \sin^2 x = 1 \quad \leftrightarrow \quad \sin^2 x = \frac{9}{15},$$

from which

$$\sin x = \frac{3}{5} \quad \text{or} \quad \sin x = -\frac{3}{5}.$$

Because $\sin x < 0$, we take $\sin x = -3/5$. By Definition 1.2, and substituting $-3/5$ for $\sin x$ and $4/5$ for $\cos x$, we have

$$\tan x = \frac{-\frac{3}{5}}{\frac{4}{5}} = -\frac{3}{4}.$$

55. If $\cos x = 1/\sqrt{2}$ and $\sin x < 0$, find $\tan x$.

56. If $\sin x = \sqrt{3}/2$ and $\cos x > 0$, find $\tan x$.

57. If $\cos x = 1/2$ and $3\pi/2 < x < 2\pi$, find $\csc x$.

58. If $\sin x = -\sqrt{3}/2$ and $\pi < x < 3\pi/2$, find $\sec x$.

59. If $\cos x = 3/5$ and $\sin x < 0$, find $\cot x$.

60. If $\sin x = 1/\sqrt{2}$ and $\cos x < 0$, find $\cot x$.

Use the relationships of Definitions 1.2 and 1.3 and $\sin^2 x + \cos^2 x = 1$ *for all* $x \in R$ *to write each expression in terms of* $\sin x$ *and/or* $\cos x$ *and simplify.*

Examples a. $\tan x \cdot \csc x$ b. $\dfrac{\cot^2 x}{\cos x}$

Solutions

a. Since $\tan x = \sin x/\cos x$ and $\csc x = 1/\sin x$, we have that

$$\tan x \cdot \csc x = \frac{\sin x}{\cos x} \cdot \frac{1}{\sin x} = \frac{1}{\cos x}.$$

b. Since $\cot x = \cos x/\sin x$, we have that

$$\frac{\cot^2 x}{\cos x} = \frac{\left(\frac{\cos x}{\sin x}\right)^2}{\cos x} = \frac{\cos^2 x}{\sin^2 x} \cdot \frac{1}{\cos x} = \frac{\cos x}{\sin^2 x}.$$

61. $\tan x \cdot \csc^2 x$

62. $\cot^2 x \cdot \sec x$

63. $\sin^2 x \cdot \cot^2 x$

64. $\tan^2 x \cdot \cos^2 x$

65. $\dfrac{\cot^2 x}{\sec x}$

66. $\dfrac{\tan^2 x}{\csc x}$

67. $\dfrac{\tan^2 x + 1}{\sec x}$

68. $\dfrac{\cot^2 x + 1}{\csc x}$

69. $\dfrac{\csc^2 x - 1}{\cot x}$

70. $\dfrac{\sec^2 x - 1}{\tan x}$

B

71. Show that $\sin\left(\dfrac{\pi}{6} + k \cdot 2\pi\right) = \dfrac{1}{2}$ for $k \in \{-2, -1, 0, 1, 2\}$.

72. Show that $\tan\left(\dfrac{\pi}{4} + k\pi\right) = 1$ for $k \in \{-2, -1, 0, 1, 2\}$.

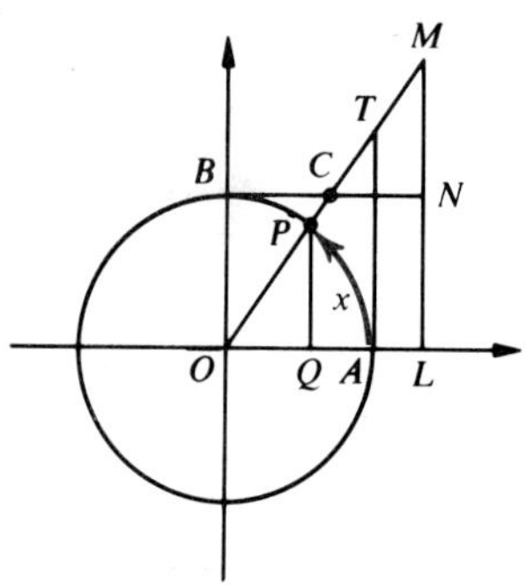

Circular function values can be directly associated with the lengths of certain line segments. For example, in the figure, $\overline{OA}$ *and* $\overline{OB}$ *are radii of a unit circle,* $\overline{BN}$ *is perpendicular to* $\overline{AT}$ *and* $\overline{LM}$*, and* $\overline{BN}$ *and* $\overline{AT}$ *are tangents to the circle. Since the coordinates of the point P are* (cos x, sin x), *the length OQ equals* cos x *and the length QP equals* sin x.

73. Show that AT equals $\tan x$.

74. Show that OT equals $\sec x$.

75. Show that OC equals $\csc x$.

76. Show that BC equals $\cot x$.

1.5 Periodic Functions; Applications

In Section 1.3 we noted that an element in the range of the sine function is paired with a number of different elements in its domain. For example,

$$\sin \frac{\pi}{6} = \frac{1}{2}, \quad \sin\left(\frac{\pi}{6} + 2\pi\right) = \frac{1}{2}, \quad \text{and} \quad \sin\left(\frac{\pi}{6} + 4\pi\right) = \frac{1}{2},$$

where $1/2$ is paired with several different elements in the domain. These examples suggest that

$$x = \frac{\pi}{6} + k \cdot 2\pi, \quad k \in J,$$

are solutions of $\sin x = 1/2$. These examples further suggest a periodic property common to a variety of functions including each of the circular functions.

Definition 1.4 *If there exists a real number $a \neq 0$ such that*

$$f(x + a) = f(x),$$

for all x in the domain of f, then f is a **periodic function** *and* **a** *is a* **period.**

Periodic Properties

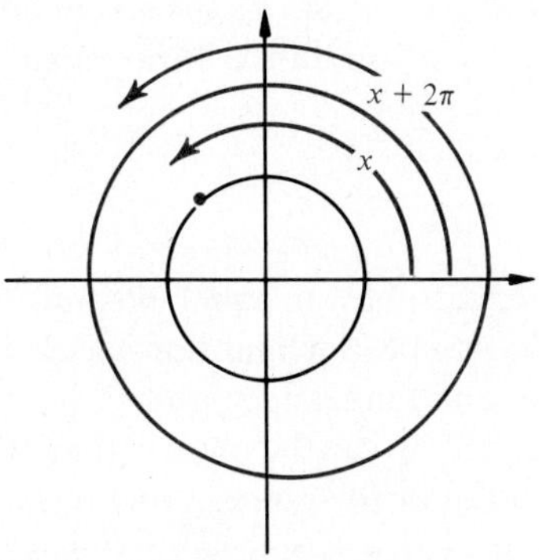

Figure 1.11

The circumference of a unit circle is 2π. Hence, the cosine and sine functions are periodic with period 2π as suggested in Figure 1.11.

Theorem 1.4 *If $x \in R$ and k is an integer, then*

$$\text{I.}\quad \cos(x + k \cdot 2\pi) = \cos x,$$

$$\text{II.}\quad \sin(x + k \cdot 2\pi) = \sin x.$$

Using Theorem 1.4, we can readily obtain some additional function values that we previously obtained by using geometric considerations.

Examples

a. $\sin\dfrac{5\pi}{2} = \sin\left(\dfrac{\pi}{2} + 2\pi\right)$

$= \sin\dfrac{\pi}{2} = 1$

b. $\cos 5\pi = \cos(\pi + 4\pi)$

$= \cos \pi = -1$

The smallest positive period of a periodic function is known as the **fundamental period.** Since the terminal point of an arc on a unit circle must travel completely around the circle, a distance of 2π, before the coordinates of the point start to repeat, it seems plausible that the fundamental period of both the sine and cosine function is 2π.

It can be shown that the points with coordinates $(\cos x, \sin x)$ and $(-\cos x, -\sin x)$ are the end points of a diameter on a unit circle. (See Figure 1.12). Hence, the arc joining them has a length of π. Furthermore, since for each of these points,

$$\tan x = \frac{\sin x}{\cos x} = \frac{-\sin x}{-\cos x},$$

it follows that the fundamental period of the tangent function is π.

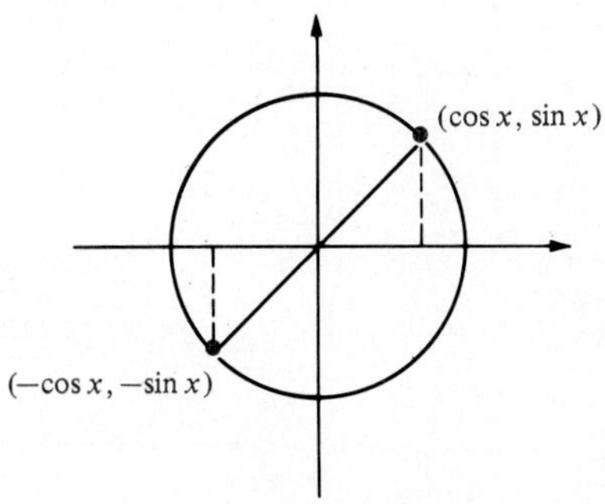

Figure 1.12

Theorem 1.5 *If $x \in R$ $(x \neq \pi/2 + k\pi)$ and k is an integer, then*

$$\tan(x + k\pi) = \tan x.$$

Examples

a. $\tan \dfrac{7\pi}{6} = \tan\left(\dfrac{\pi}{6} + \pi\right) = \tan \dfrac{\pi}{6} = \dfrac{1}{\sqrt{3}}$

b. $\tan \dfrac{13\pi}{4} = \tan\left(\dfrac{\pi}{4} + 3\pi\right) = \tan \dfrac{\pi}{4} = 1$

Note that because $\sec x = 1/\cos x$ and $\csc x = 1/\sin x$, the secant and cosecant functions have period 2π; because $\cot x = 1/\tan x$, the cotangent function has period π.

Applications

If a point is moving so that the motion repeats itself in equal intervals of time, the motion is said to be **periodic,** or **harmonic.** Such motions occur in the vibration of a guitar string, a swinging pendulum, and sound waves, to name a few examples. Using laws of physics, it can be shown that the mathematical models determining the position of the point at any instant involve the sine and cosine functions. For *simple harmonic motion,* the models take the form

$$d = A \sin Bt \qquad \text{or} \qquad d = A \cos Bt,$$

where d is a measure of displacement, A and B are constants for the particular motion, and t is a measure of time.

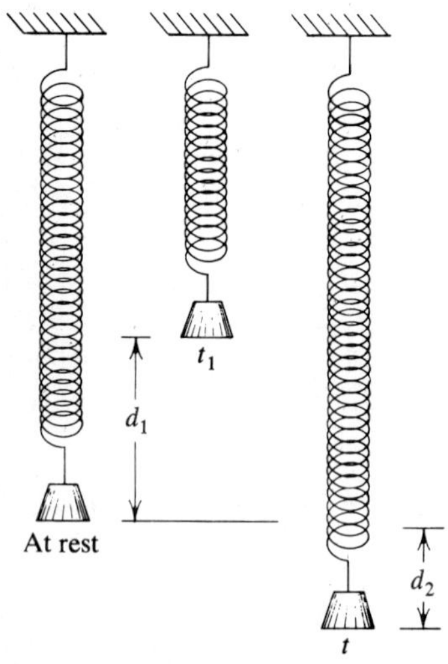

Example A spring is displaced from its resting position as shown in the figure. The displacement d is given by the equation $d = A \cos \omega t$, where A, the initial displacement, and d are expressed in centimeters, t is expressed in seconds, and $\omega = 6\pi$. If the initial displacement is 12 centimeters, find d when $t = 2$.

Solution Since $A = 12$, $\omega = 6\pi$, and $t = 2$,

$$\begin{aligned} d &= A \cos \omega t \\ &= 12 \cos 6\pi(2) = 12 \cos 12\pi. \end{aligned}$$

Since $\cos 12\pi = \cos(0 + 6 \cdot 2\pi) = 1$, we have

$$d = 12 \cos 0 = 12(1) = 12.$$

Hence, the displacement when $t = 2$ is 12 centimeters.

Besides mechanical systems such as the example above, certain electromagnetic systems can be analyzed using the same mathematical models.

Example The voltage e in an electrical circuit can be expressed as $e = E \sin t$, where E is the maximum voltage and t is the time in seconds. If $E = 100$, find e to three significant digits when $t = 2$.

Solution Since $E = 100$ and $t = 2$,

$$e = E \sin t = 100 \sin 2 \approx 90.9$$

Hence, e is approximately 90.9 volts when $t = 2$.

EXERCISE SET 1.5

A

Use Theorem 1.4 and Theorem 1.5 to write each of the following expressions in the form $\cos(x + k \cdot 2\pi)$, $\sin(x + k \cdot 2\pi)$, *or* $\tan(x + k\pi)$, *where* $0 \le x \le 2\pi$, $k \in J$.

Examples

a. $\cos \dfrac{9\pi}{4} = \cos\left(\dfrac{\pi}{4} + \dfrac{8\pi}{4}\right) = \cos\left(\dfrac{\pi}{4} + 2\pi\right)$

b. $\tan \dfrac{10\pi}{3} = \tan\left(\dfrac{\pi}{3} + \dfrac{9\pi}{3}\right) = \tan\left(\dfrac{\pi}{3} + 3\pi\right)$

1. $\cos \dfrac{11\pi}{4}$ **2.** $\sin \dfrac{8\pi}{3}$ **3.** $\tan \dfrac{5\pi}{4}$

4. $\tan \dfrac{7\pi}{3}$ **5.** $\sin \dfrac{13\pi}{4}$ **6.** $\cos \dfrac{17\pi}{6}$

7. $\tan \dfrac{15\pi}{4}$ **8.** $\cos \dfrac{11\pi}{3}$ **9.** $\sin \dfrac{19\pi}{6}$

10. $\cos \dfrac{17\pi}{4}$ **11.** $\sin \dfrac{13\pi}{3}$ **12.** $\tan \dfrac{13\pi}{6}$

Solve Exercises 13–16 for the conditions for the spring in the example on page 34.

13. If the initial displacement is 10, find d when $t = 4$ and when $t = 4\frac{1}{2}$.

14. If the initial displacement is -10, find d when $t = 4$ and when $t = 4\frac{1}{2}$.

15. If the initial displacement is 8, find d when $t = \frac{1}{2}$ and when $t = \frac{1}{4}$.

16. If the initial displacement is -8, find d when $t = \frac{1}{2}$ and when $t = \frac{1}{4}$.

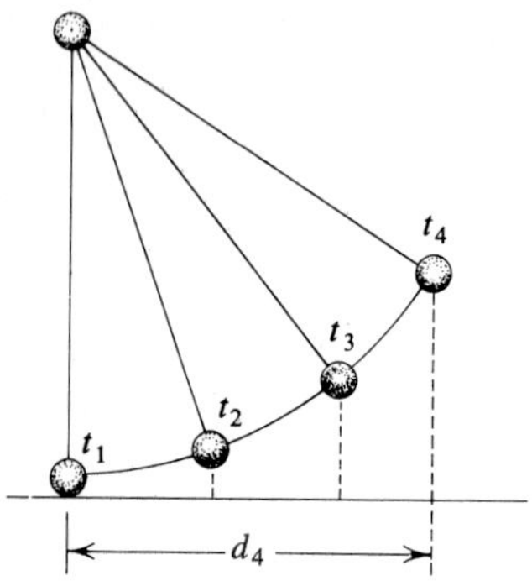

The horizontal displacement d of the bob on an oscillating pendulum is described by the equation $d = K \sin 2\pi t$, where K and d are expressed in centimeters and t is expressed in seconds.

17. Find d when $t = \frac{1}{4}$ and $K = 22$.

18. Find d when $t = \frac{1}{6}$ and $K = 20$.

19. Find K when $d = 20$ and $t = \frac{1}{6}$.

20. Find K when $d = 10$ and $t = \frac{1}{12}$.

The instantaneous voltage e in a circuit at time t is given by the equation $e = 200 \sin 3700t$ where e is expressed in volts and t in seconds. Find the voltage to three significant digits at each of the following times.

Example $t = 0.004$

Solution Evaluate formula for $t = 0.004$.

$$e = 200 \sin 3700t = 200 \sin 3700(0.004)$$
$$= 200 \sin 14.8 \approx 158.$$

Hence, the voltage is approximately 158 volts when $t = 0.004$.

21. $t = 0.001$ **22.** $t = 0.002$ **23.** $t = 0.005$

24. $t = 0.008$ **25.** $t = 0.012$ **26.** $t = 0.020$

The instantaneous current i in an alternating system is given by the equation $i = I \sin \omega t$, where i is expressed in amperes and t is expressed in seconds. Find the current to the nearest tenth at each of the following times if $I = 0.05$ and $\omega = 377$.

27. $t = 0.01$ **28.** $t = 0.03$ **29.** $t = 0.2$

30. $t = 0.4$ **31.** $t = 1.0$ **32.** $t = 3.0$

The pressure P in a traveling sound wave is described by the equation $P = 10 \sin 200\pi[t - (x/1000)]$, where x is the distance from source of the sound in centimeters, t is the time expressed in seconds, and P is expressed in dynes per square centimeter. Find P to three significant digits for each of the following conditions.

33. $x = 200$ and $t = 0.06$ **34.** $x = 330$ and $t = 0.01$

35. $x = 660$ and $t = 0.02$ **36.** $x = 1650$ and $t = 0.05$

B

37. The displacement d of a point on a vibrating spring under certain conditions is given by $d = (1/50) \sin 50t - t \cos 50t$, where d is in feet and t is in seconds. Find the displacement when $t = 0.3$ second.

38. It is shown in calculus that an object having simple harmonic motion (described on page 34) has a velocity v given by $v = AB \cos Bt$ and an acceleration given by $a = -AB^2 \sin Bt$. The motion of a certain spring is expressed by $d = 3 \sin(\pi t/2)$, where d is a displacement in centimeters and t is in seconds.

a. Express the velocity and the acceleration of a point on the free end of the spring in terms of t.

b. Find the velocity and acceleration when t equals 2, 3, and 4 seconds.

Chapter Summary

1.1 If $P(x, y)$ is the terminal point of an arc on the unit circle with initial point $(1, 0)$, then the coordinates of P are functions of the length s of the arc.

$$x = \text{cosine } s \quad (\cos s),$$

$$y = \text{sine } s \quad (\sin s).$$

For all $s \in R$ the ranges of these functions are given by $|\cos s| \leq 1$ and $|\sin s| \leq 1$.

The following important relationship follows from the definitions for the sine and cosine functions.

$$\cos^2 s + \sin^2 s = 1.$$

1.2 Function values can be obtained for the special elements $0, \pi/2, \pi, 3\pi/2, \pi/6, \pi/4$, and $\pi/3$ in the domain of the cosine and sine functions by geometric methods.

The variable x is now used instead of s to represent an element in the domain of the cosine and sine functions.

A calculator or Table II, Appendix C can be used to find approximations for $\cos x$ and $\sin x$.

1.3 Values for $\cos x$ and $\sin x$ for $x > \pi/2$ and $x < 0$ can be obtained by using a reference arc, whose length is $\tilde{x}$, in conjunction with Table 1.2 or a calculator or Table II in Appendix C.

1.4 Four additional circular functions, **tangent, secant, cosecant,** and **cotangent** are defined in terms of $\cos x$ and $\sin x$.

$$\tan x = \frac{\sin x}{\cos x}, \quad x \neq \frac{\pi}{2} + k\pi, \ k \in J,$$

$$\sec x = \frac{1}{\cos x}, \quad x \neq \frac{\pi}{2} + k\pi, \ k \in J,$$

$$\csc x = \frac{1}{\sin x}, \quad x \neq k\pi, \ k \in J,$$

$$\cot x = \frac{\cos x}{\sin x}, \quad x \neq k\pi, \ k \in J.$$

Values for tan x, sec x, csc x, and cot x for specified values of x can be obtained by first expressing them in terms of cos x and sin x.

1.5 Functions with the property that $f(x + a) = f(x)$ for all x in the domain of f and for some $a \in R$ $(a \neq 0)$, are said to be **periodic.** The smallest positive number a for which a function is periodic is called its **fundamental period.**

The fundamental period of the sine, cosine, cosecant, and secant functions is 2π. The fundamental period of the tangent and cotangent functions is π.

The equations $d = A \sin Bt$ and $d = A \cos Bt$ are often used as mathematical models to describe periodic phenomena.

A list of the symbols introduced in this chapter is shown inside the front cover, and important properties are shown inside the back cover.

Review Exercises

1.1 *Name the quadrant in which the terminal point of each arc with the given length s lies [initial point at* (1, 0)].

1. $s = \frac{5\pi}{6}$ **2.** $s = \frac{9\pi}{4}$ **3.** $s = -\frac{7\pi}{3}$

4. $s = -\frac{19\pi}{6}$ **5.** $s = 5$ **6.** $s = -4$

Determine the quadrant in which the terminal point of each arc with length s lies [initial point at (1, 0)].

7. $\sin s < 0$ and $\cos s < 0$ **8.** $\cos s < 0$ and $\sin s > 0$

If $\sin s = -4/5$ *and* $\cos s < 0$, *find the value of each of the following.*

9. $\cos s$ **10.** $\dfrac{\sin s}{\cos s}$ **11.** $\dfrac{1}{\sin s}$ **12.** $\dfrac{1}{\cos s}$

1.2 *Compute each expression (exact value).*

13. $\dfrac{\sin^2 \dfrac{\pi}{6} - \cos \dfrac{\pi}{3}}{\cos \dfrac{\pi}{4}}$ **14.** $\dfrac{2\left(\sin^2 \dfrac{\pi}{3}\right) + \cos^2 \dfrac{\pi}{4}}{\sin^2 \dfrac{\pi}{2}}$

Find a decimal approximation to four places for each expression.

15. $\cos 0.96$ **16.** $\sin 0.28$

17. $\sin 1.08$ **18.** $\cos 1.36$

19. $\sin 1.32 + 3 \cos 0.41$ **20.** $2 \cos 1.32 - \sin 1.14$

21. $2 \sin 0.17 - \cos 1.44$ **22.** $\dfrac{3.4 \sin 0.2}{4.7}$

1.3 *Sketch each arc with the given length and its reference arc,* $\tilde{x}$.

23. $\dfrac{5\pi}{6}$ **24.** $-\dfrac{7\pi}{4}$ **25.** 5 **26.** -3

Find each function value (exact value).

27. $\sin \dfrac{3\pi}{4}$ **28.** $\cos \dfrac{7\pi}{6}$

29. $\cos\left(-\dfrac{5\pi}{3}\right)$ **30.** $\sin\left(-\dfrac{11\pi}{4}\right)$

Find approximate function values to four places for the following.

31. $\cos 4.58$ **32.** $\sin(-2.89)$

1.4 *Find each function value (exact value).*

33. $\tan \dfrac{5\pi}{6}$ **34.** $\tan\left(-\dfrac{7\pi}{6}\right)$ **35.** $\csc \dfrac{5\pi}{4}$

36. $\csc\left(-\frac{5\pi}{3}\right)$ **37.** $\sec\frac{4\pi}{3}$ **38.** $\sec\left(-\frac{13\pi}{6}\right)$

39. $\cot\frac{5\pi}{4}$ **40.** $\cot\left(-\frac{7\pi}{6}\right)$

Compute each expression (exact value).

41. $3\sec\frac{\pi}{4} - 2\cos\pi$ **42.** $\dfrac{\cos\frac{\pi}{3} + \cot\frac{\pi}{6}}{\tan\frac{\pi}{4}}$

Find an approximation for each expression.

43. $\tan 0.74$ **44.** $\sec 0.98$

45. $\csc(-2.21)$ **46.** $\cot(-3.23)$

47. $2\cot 1.09 + \sec 3.04$ **48.** $\dfrac{2\cot 1.51 - \tan 0.61}{\csc 0.57}$

Determine the quadrant in which the terminal point of each arc with the length x lies.

49. $\cos x > 0$ and $\cot x > 0$ **50.** $\sec x < 0$ and $\tan x > 0$

Find each function value (exact value).

51. If $\sin x = -\frac{1}{2}$ and $\frac{3\pi}{2} < x < 2\pi$, find $\sec x$.

52. If $\cos x = \frac{\sqrt{3}}{2}$ and $\sin x < 0$, find $\tan x$.

Write each expression in terms of $\sin x$ *and/or* $\cos x$ *and simplify.*

53. $\csc x \cdot \tan^2 x$ **54.** $\dfrac{\cot^2 x}{\cos x}$

1.5 *Write each of the following expressions in the form of* $\cos(x + k\cdot 2\pi)$ *or* $\sin(x + k\cdot 2\pi)$, $k \in J$.

55. $\cos\frac{8\pi}{3}$ **56.** $\sin\frac{17\pi}{4}$

57. The displacement of a spring is given by $d = 4 \cos 8\pi t$, where d is expressed in centimeters and t is expressed in seconds. Find d when $t = 2\frac{1}{2}$ seconds.

58. The horizontal displacement d of the bob on an oscillating pendulum is given by the equation $d = 18 \sin 2\pi t$, where d is expressed in centimeters and t is expressed in seconds. Find d when $t = 2\frac{1}{2}$ seconds.

59. The instantaneous voltage e in a circuit at time t is given by the equation $e = 1.3 \sin 980\pi t$, where e is expressed in volts and t is expressed in seconds. Find the voltage when $t = 3$.

60. The instantaneous current i in an alternating current system is given by the equation $i = I \sin \omega t$, where i and I are expressed in amperes, ω in radians per second, and t in seconds. Find the current to three significant digits at $t = 0.002$ if $I = 0.04$ and $\omega = 352$.

2 GRAPHS; INVERSE CIRCULAR FUNCTIONS

The periodic characteristics of the trigonometric and circular functions can be exhibited very effectively by graphing these functions in a Cartesian coordinate system. You may want to review the material on graphing in Appendix A.4 before starting this chapter.

In graphing the circular functions, it is convenient to view line segments on the x-axis of a Cartesian coordinate system as obtained by unwinding the arc on a unit circle with the same length, as shown in Figure 2.1. A similar association could be made for the lengths of arcs in a clockwise direction and line segments on the negative axis.

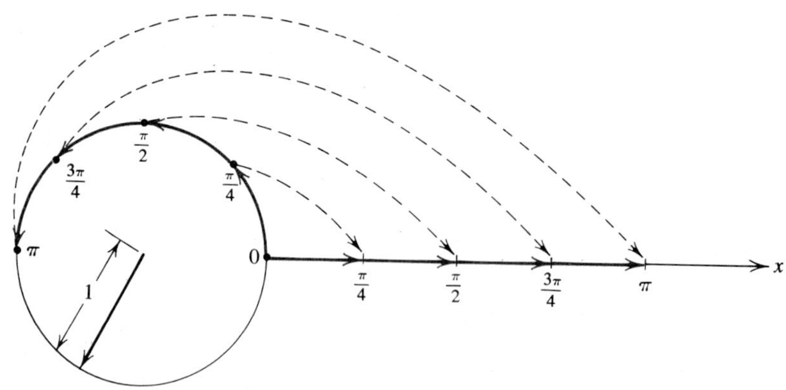

Figure 2.1

Sometimes it is convenient to use integers instead of rational multiples of π as scale numbers on the x-axis. Both of these scale numbers are shown in a Cartesian system in Figure 2.2. The y-axis is scaled in integers.

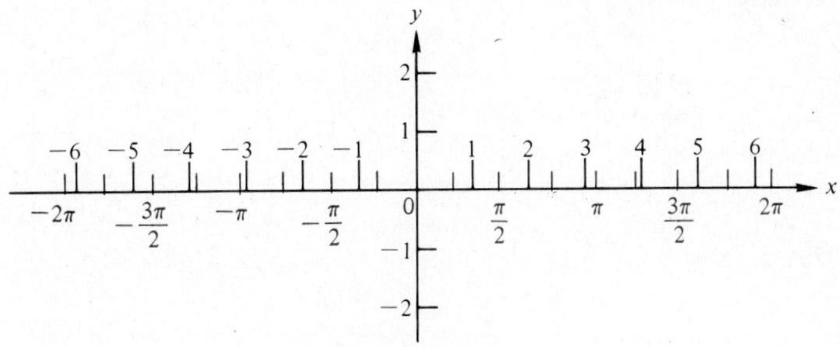

Figure 2.2

2.1 Sine and Cosine Functions

The graphs of the sine and cosine functions, defined by the two equations $y = \sin x$ and $y = \cos x$, consist of the set of points corresponding to all ordered pairs (x, y) in the respective functions.

Although a calculator can be used to find good approximations for the values of $\sin x$ and $\cos x$, function values from Table 1.3 (page 24) with which you are probably now familiar and the notion of a reference arc, are usually sufficient to obtain good approximations to the graphs of these functions for $0 \le x \le 2\pi$.

The Sine Graph

We first consider the graph of

$$y = \sin x. \tag{1}$$

In order not to distort the form of the graph of (1) and the following graphs, we shall use the same units of measurement on the x-axis and y-axis. To obtain selected ordered pairs $(x, \sin x)$ in the function shown in Figure 2.3a on page 44, we first note from Table 1.3 (or memory) that

$$\sin \frac{\pi}{4} = \frac{1}{\sqrt{2}} \approx 0.7;$$

then using a reference arc, we obtain

$$\sin \frac{3\pi}{4} = \sin \frac{\pi}{4} \approx 0.7,$$

$$\sin \frac{5\pi}{4} = -\sin \frac{\pi}{4} \approx -0.7,$$

$$\sin \frac{7\pi}{4} = -\sin \frac{\pi}{4} \approx -0.7.$$

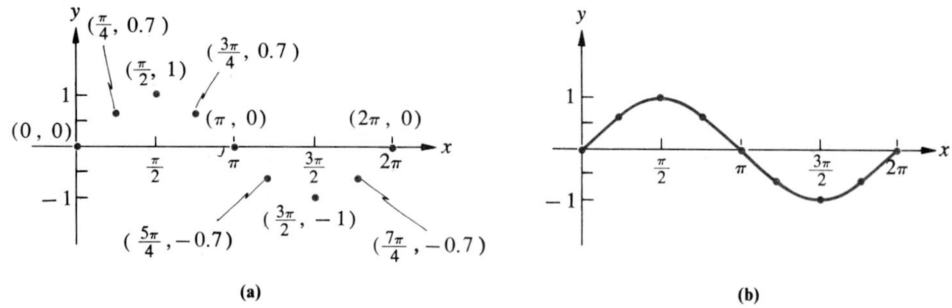

Figure 2.3

In calculus it is shown that the trigonometric functions are continuous over the intervals for which they are defined. For us this means that since the sine function is defined for all real numbers, its graph contains no "breaks." Furthermore, since sine function values increase as x increases from 0 to $\pi/2$ and then decrease as x increases from $\pi/2$ to $3\pi/2$, and so on, we connect the points in Figure 2.3a with a smooth curve to produce Figure 2.3b. Of course, a calculator can be used to obtain additional points to assist us in graphing the function in the interval 0 to 2π.

Now, because the sine function is periodic with period 2π,

$$\sin(x + k \cdot 2\pi) = \sin x, \ k \in J.$$

Hence we repeat the graph in Figure 2.3b in both directions and obtain Figure 2.4 which is part of the graph of $y = \sin x$ for all real numbers x. The graph is called a **sine wave** or **sinusoidal wave. The graph** over one period is called a **cycle** of the wave.

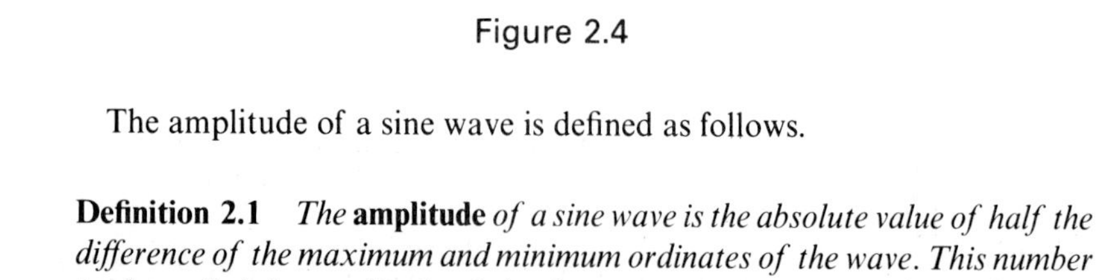

Figure 2.4

The amplitude of a sine wave is defined as follows.

Definition 2.1 *The* **amplitude** *of a sine wave is the absolute value of half the difference of the maximum and minimum ordinates of the wave. This number is also called the amplitude of the function.*

Thus, the amplitude of the wave in Figure 2.4 is

$$\left|\frac{1}{2}[1-(-1)]\right| = 1.$$

Also observe that $\sin x = 0$ when $x = k\pi$, $k \in J$, so the zeros of the function (See Appendix A.5) are the values $k\pi$, $k \in J$. At these values the curve intersects the x-axis. Furthermore, note that the range of the function is

$$\{y \mid -1 \leq y \leq 1\}.$$

The Cosine Graph

We obtain the graph of the cosine function,

$$y = \cos x,$$

in the same way that we obtained the graph of the sine function. Some ordered pairs in the function for $0 \leq x \leq 2\pi$ can be obtained using Table 1.3. First noting that

$$\cos\frac{\pi}{4} = \frac{1}{\sqrt{2}} \approx 0.7,$$

we then obtain

$$\cos\frac{3\pi}{4} = -\cos\frac{\pi}{4} \approx -0.7,$$

$$\cos\frac{5\pi}{4} = -\cos\frac{\pi}{4} \approx -0.7,$$

$$\cos\frac{7\pi}{4} = \cos\frac{\pi}{4} \approx 0.7.$$

The ordered pairs are shown in Figure 2.5a on page 46. In this case, function values decrease as x increases from zero to π and they increase as x increases from π to 2π. We connect the points in Figure 2.5a with a smooth curve to produce Figure 2.5b. Again, of course, a calculator can be used to obtain additional points for graphing.

Since the cosine function is also periodic with period 2π,

$$\cos(x + k \cdot 2\pi) = \cos x, \; k \in J.$$

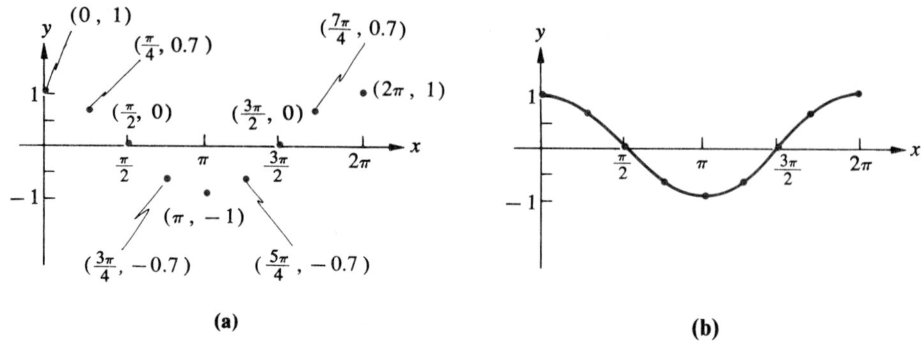

Figure 2.5

Hence we repeat the pattern in Figure 2.5b and obtain Figure 2.6, which is part of the graph of $y = \cos x$ for all real numbers x. We observe that, like the sine function, the cosine function has the range

$$\{y \mid -1 \leq y \leq 1\}.$$

Also like the sine function, the amplitude of the wave is 1. Furthermore also observe that $\cos x = 0$ where $x = (\pi/2) + k\pi$ so that the zeros of the function are the values $(\pi/2) + k\pi$, $k \in J$. At these values the curve intersects the x-axis.

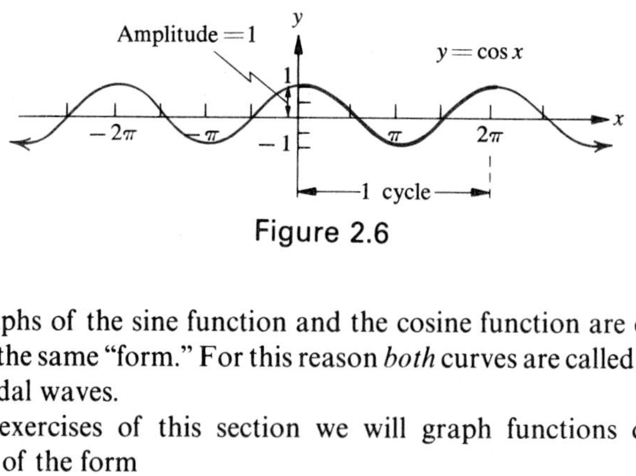

Figure 2.6

The graphs of the sine function and the cosine function are congruent; they have the same "form." For this reason *both* curves are called sine waves or sinusoidal waves.

In the exercises of this section we will graph functions defined by equations of the form

$$y = A \sin B(x - C) \qquad \text{or} \qquad y = A \cos B(x - C)$$

by first obtaining a number of ordered pairs in the function. This will give us some insight regarding the properties of these graphs.

EXERCISE SET 2.1

A

1. Graph the elements of

$$\left\{-2\pi, \frac{-3\pi}{2}, -\pi, \frac{-\pi}{2}, 0, \frac{\pi}{2}, \pi, \frac{3\pi}{2}, 2\pi\right\}$$

on the number line below scaled in integral units. Use $\pi \approx 3.14$.

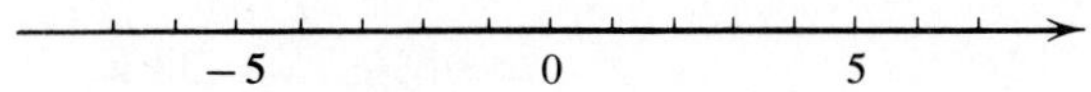

2. Graph the elements of

$$\left\{0, \frac{\pi}{4}, \frac{\pi}{2}, \frac{3\pi}{4}, \pi, \frac{5\pi}{4}, \frac{3\pi}{2}, \frac{7\pi}{4}, 2\pi\right\}$$

on a number line scaled in integral units as shown in Exercise 1. Use $\pi \approx 3.14$.

3. Graph the elements of

$$\left\{0, \frac{\pi}{4}, \frac{\pi}{2}, \frac{3\pi}{4}, \pi, \frac{5\pi}{4}, \frac{3\pi}{2}, \frac{7\pi}{4}, 2\pi\right\}$$

on the number line below scaled in multiples of π.

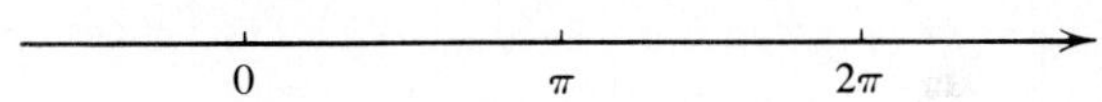

4. Graph the elements of

$$\left\{-2\pi, \frac{-7\pi}{4}, \frac{-3\pi}{2}, \frac{-5\pi}{4}, -\pi, \frac{-3\pi}{4}, \frac{-\pi}{2}, \frac{-\pi}{4}, 0\right\}$$

on a number line scaled in multiples of π as shown in Exercise 3.

5. Graph the elements of

$$\left\{-2\pi, \frac{-5\pi}{3}, \frac{-4\pi}{3}, -\pi, \frac{-2\pi}{3}, \frac{-\pi}{3}, 0, \frac{\pi}{3}, \frac{2\pi}{3}, \pi, \frac{4\pi}{3}, \frac{5\pi}{3}, 2\pi\right\}$$

on a number line scaled in multiples of π.

6. Graph the elements of

$$\left\{0, \frac{\pi}{6}, \frac{\pi}{3}, \frac{\pi}{2}, \frac{2\pi}{3}, \frac{5\pi}{6}, \pi, \frac{7\pi}{6}, \frac{4\pi}{3}, \frac{3\pi}{2}, \frac{5\pi}{3}, \frac{11\pi}{6}, 2\pi\right\}$$

on a number line scaled in multiples of π.

In Exercises 7–18, (**a**) *graph the function defined by each equation by graphing one cycle and then repeating the pattern, and* (**b**) *specify the zeros and the amplitude of the function.*

Example $y = 2 \sin x$

x	$\sin x$	$2 \sin x$ *or* y
0	0	0
$\frac{\pi}{2}$	1	2
π	0	0
$\frac{3\pi}{2}$	-1	-2
2π	0	0

Solution

a. Some ordered pairs in the function defined by the equation are obtained in order to graph one cycle. The systematic tabulation of the data shown is helpful. The ordered pairs (x, y) are then graphed as shown, connected with a smooth curve, and the pattern of the wave duplicated. A calculator can be used to obtain additional points for graphing.

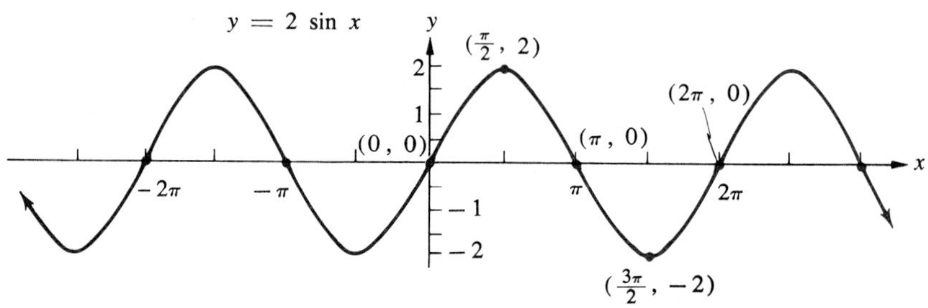

b. Zeros: $\{x \mid x = k\pi,\ k \in J\}$; amplitude: $\left|\frac{1}{2}[2 - (-2)]\right| = 2.$

7. $y = 3 \sin x$ **8.** $y = 2 \cos x$ **9.** $y = \frac{1}{2} \cos x$

10. $y = \frac{1}{3} \sin x$ **11.** $y = -\sin x$ **12.** $y = -\cos x$

Example $y = \cos 2x$

Solution

a. Some ordered pairs in the function defined by the equation are obtained in order to graph one cycle. The ordered pairs (x, y) obtained from

x	$2x$	$\cos 2x$ or y
0	0	1
$\frac{\pi}{4}$	$\frac{\pi}{2}$	0
$\frac{\pi}{2}$	π	-1
$\frac{3\pi}{4}$	$\frac{3\pi}{2}$	0
π	2π	1

the table are then graphed as shown, connected with a smooth curve, and the pattern of the wave duplicated.

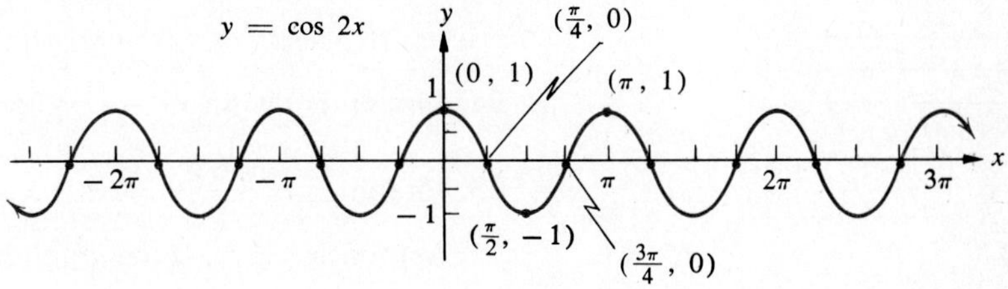

b. Zeros: $\left\{x \mid x = \frac{\pi}{4} + k \cdot \frac{\pi}{2},\ k \in J\right\}$; amplitude: $\left|\frac{1}{2}[1 - (-1)]\right| = 1.$

13. $y = \sin 2x$ **14.** $y = \sin \frac{1}{2} x$ **15.** $y = \cos \frac{1}{2} x$

16. $y = \cos \frac{1}{3} x$ **17.** $y = \sin\left(x - \frac{\pi}{2}\right)$ **18.** $y = \cos\left(x + \frac{\pi}{4}\right)$

Graph two cycles of each of the following functions.

Example $y = \cos \frac{\pi}{3} x$

x	$\frac{\pi}{3} x$	$\cos \frac{\pi}{3} x$ or y
0	0	1
1	$\frac{\pi}{3}$	$\frac{1}{2}$
2	$\frac{2\pi}{3}$	$-\frac{1}{2}$
3	π	-1
4	$\frac{4\pi}{3}$	$-\frac{1}{2}$
5	$\frac{5\pi}{3}$	$\frac{1}{2}$
6	2π	1

Solution In such cases where the coefficient of x is a multiple of π, it is helpful to label the x-axis in *rational units* 1, 2, 3, . . . , and to select *rational numbers* for x, as shown in the table. Some ordered pairs in the function defined by the equation are graphed as shown and connected with a smooth curve. Then the pattern of the wave is duplicated.

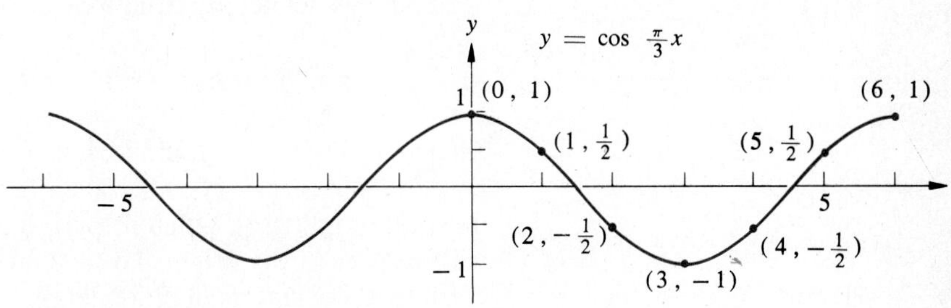

19. $y = \cos \frac{\pi}{2} x$ **20.** $y = \sin \frac{\pi}{2} x$

21. $y = \sin \pi x$ **22.** $y = \cos \pi x$

B

23. Graph $y = \frac{\sin x}{x}$, $-\pi < x < \pi$, and describe the graph as x approaches zero. Is the graph periodic? Why or why not?

24. Graph $y = x \sin x$, $-4\pi \le x \le 4\pi$.

25. Graph $y = \frac{\cos x}{x}$, $-\pi < x < \pi$, and describe the graph as x approaches zero. Is the graph periodic? Why or why not?

26. Graph $y = x \cos x$, $-4\pi \le x \le 4\pi$.

Examples of noncircular periodic functions are given below. The symbol $[x]$ *denotes the number that is the greatest integer less than or equal to* x *for each* $x \in R$.

Examples

a. $[3] = 3$

b. $[5.91] = [5 + 0.91]$
$= 5$

c. $[-1.21] = [-2 + 0.79]$
$= -2$

Graph the function defined by each equation. Specify the period of the function.

27. $y = x - [x]$, $-4 \le x < 4$

28. $y = x + [x]$, $-4 \le x < 4$

29. $y = 2x - [2x]$, $-3 \le x < 3$

30. $y = -2x + [2x]$, $-3 \le x < 3$

2.2 General Sine Waves

Functions defined by equations of the form

$$y = A \sin B(x - C) \quad \text{or} \quad y = A \cos B(x - C), \tag{1}$$

where $A, B, C, x, y \in R$ and $A, B \neq 0$, always have sine waves for their graphs. Depending upon the values for A, B, and C, these sine waves have different amplitudes, different periods, and different horizontal positions with respect to the origin. To facilitate graphing such functions, we now consider these factors in more detail.

Amplitude

First consider how the value of A in the defining equation affects the graph of the functions in (1) above. For each value of x, each ordinate of the graph of

$$y = A \sin x$$

is A times the ordinate of the graph of $y = \sin x$. The effect of A, then, is to multiply the amplitude of the graph of $y = \sin x$ by $|A|$. The graph of $y = A \cos x$ is related to the graph of $y = \cos x$ in a similar way.

Example Graph $y = 2 \cos x$ over the interval $-2\pi \leq x \leq 4\pi$.

Solution

1. Sketch the graph of $y = \cos x$, over the interval $0 \leq x \leq 2\pi$, as a reference.
2. The amplitude of the graph of $y = 2 \cos x$ is 2. Therefore, sketch the desired graph on the same coordinate system over the same interval by making each ordinate 2 times the ordinate of the graph of $y = \cos x$.
3. Repeat the cycle over the interval $-2\pi \leq x \leq 4\pi$.

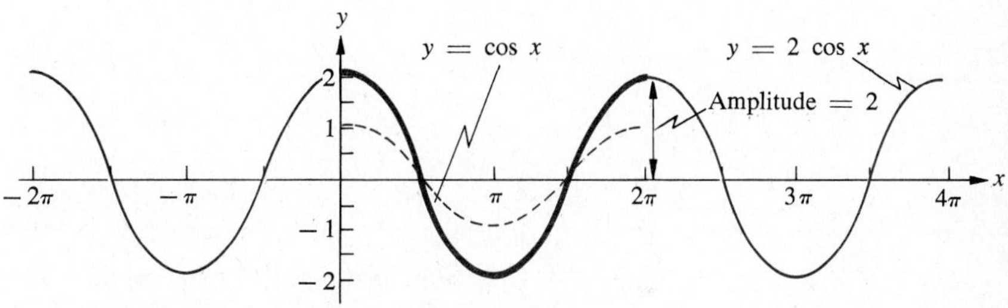

The first cycle is shown sketched with a heavier line for emphasis.

In the above example, the coefficient of $\cos x$ is a positive number. If the coefficient were a negative number, say -2, each ordinate of the graph of $y = -2 \cos x$ would be the negative of the corresponding ordinate of $y = 2 \cos x$. The graph of $y = -2 \cos x$, over the interval $-2\pi \leq x \leq 4\pi$, is shown in Figure 2.7.

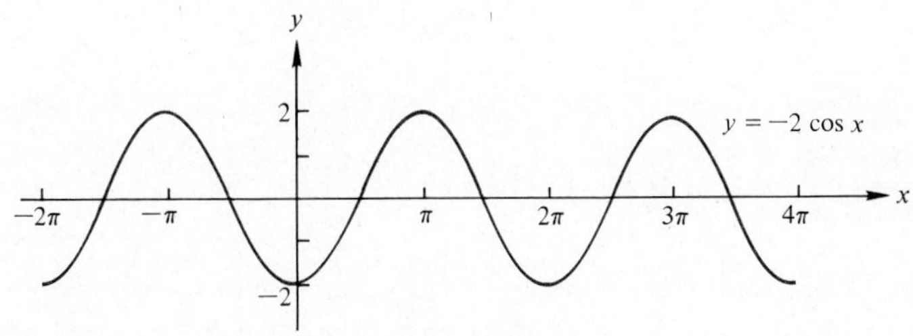

Figure 2.7

Period

Now consider how different values for B affect the graph of

$$y = \sin Bx \qquad \text{or} \qquad y = \cos Bx.$$

A complete cycle of the graph of $y = \sin x$ (or $y = \cos x$) occurs over the period 2π. Therefore, a complete cycle of the graph of $y = \sin Bx$ (or of $y = \cos Bx$) is obtained as Bx increases from 0 to 2π. Thus, the period of either the sine function or the cosine function is the solution of the equation

$$Bx = 2\pi \qquad \text{or} \qquad x = \frac{2\pi}{B}.$$

In order to specify the period by a positive number, we use the absolute value of B. This argument establishes the following theorem.

Theorem 2.1 *The period P of a periodic function defined by an equation of the form* $y = \sin Bx$ or $y = \cos Bx$ is

$$P = \frac{2\pi}{|B|}.$$

Example Graph $y = \sin 3x$, over the interval $0 \leq x \leq 2\pi$.

Solution

1. Sketch the graph of $y = \sin x$, over the interval $0 \leq x \leq 2\pi$, as a reference.

2. The period: $P = \dfrac{2\pi}{|B|} = \dfrac{2\pi}{3}$.

   Therefore, sketch a complete cycle of the desired graph on the same coordinate system over the interval $0 \leq x \leq 2\pi/3$.

3. Repeat the cycle obtained in step 2 over the interval $0 \leq x \leq 2\pi$.

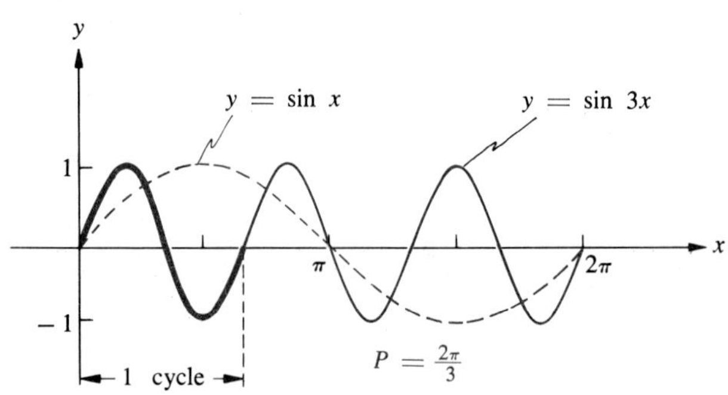

Phase Shift

Finally, consider how different values for C affect the graph of

$$y = \sin(x - C).$$

As shown in Figure 2.8, the graph of this equation is the same as that of $y = \sin x$, except that it is shifted (translated) horizontally. This is so because a cycle begins at the point at which

$$x - C = 0 \qquad \text{or} \qquad x = C$$

and starting at that point, the graph will have one full cycle as x changes by one period.

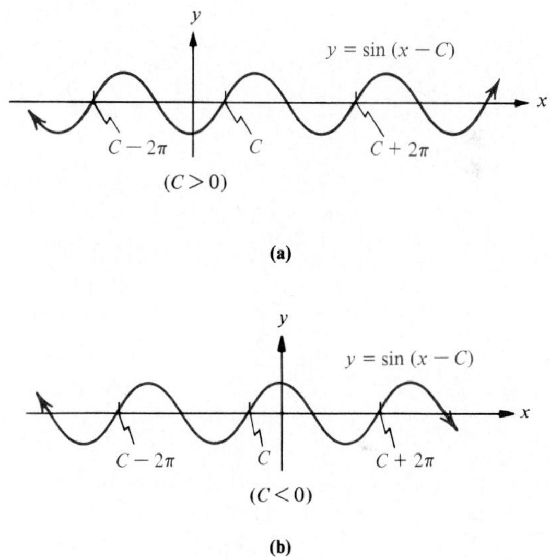

Figure 2.8

The graphs of $y = \cos(x - C)$ and $y = \cos x$ are related in a similar way.

Definition 2.2 *The number* $|C|$ *is the* **phase shift** *of the graph of periodic functions defined by equations of the form*

$$y = \sin(x - C) \qquad \textit{or} \qquad y = \cos(x - C).$$

The number $|C|$ *is also the phase shift of the corresponding function.*

Example Graph $y = \sin\left(x - \dfrac{\pi}{3}\right)$ over the interval $-2\pi \le x \le 4\pi$.

(continued)

Solution

1. Sketch the graph of $y = \sin x$, over the interval $0 \leq x \leq 2\pi$, as a reference.
2. Since $C = \pi/3$, the graph of $y = \sin x$ is shifted $\pi/3$ units to the right to obtain the desired graph.
3. Sketch the graph over the interval $-2\pi \leq x \leq 4\pi$.

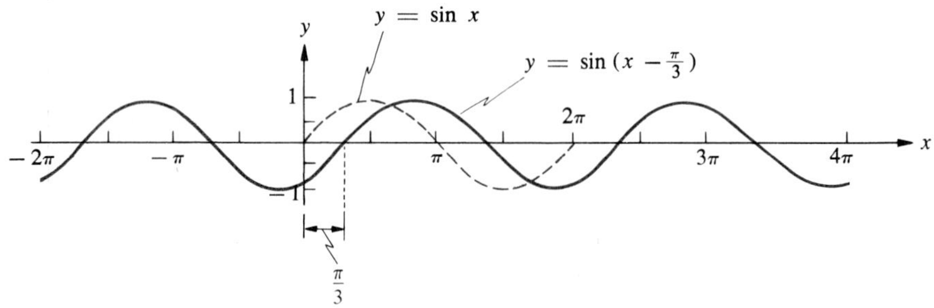

Note that in the above example the graph of $y = \sin(x - \pi/3)$ was shifted $\pi/3$ *units to the right* of the graph of $y = \sin x$ because $\pi/3$, the *value of* C, is *positive*. The graph of $y = \sin(x + \pi/3)$, which is equivalent to $y = \sin[x - (-\pi/3]$, would be shifted *to the left* of the graph of $y = \sin x$ because $-\pi/3$, the *value of* C, is *negative*. An example is shown in the exercise set on page 56.

We can now use all information about the effects of A, B, and C on the graphs of functions defined by the equation $y = A \sin B(x - C)$ or by the equation $y = A \cos B(x - C)$ to facilitate making such graphs.

If an equation were written in the form $y = A \cos(Bx - C)$, it would be necessary to first factor B from $Bx - C$ in order to make a judgment about the period and phase shift of the graph. For example, we would write

$$y = 2\cos\left(3x + \frac{\pi}{2}\right) = 2\cos 3\left[x - \left(-\frac{\pi}{6}\right)\right],$$

from which we could observe that the period is $2\pi/3$ and the phase shift is $\pi/6$ unit to the left of the graph of $y = 2 \cos 3x$. The graph is shown in the example in the exercise set on page 57.

The Period P, a Positive Rational Number

If the period P of a graph is a positive rational number, instead of a multiple of π such as we encountered in the examples above, it is helpful to use selected rational values as elements of the domain and to label the x-axis in rational units to facilitate sketching the graph, as we suggested on page 49.

Example Graph $y = \frac{1}{2} \cos \pi x$ over the interval $-5 \le x \le 5$.

Solution

1. The phase shift $|C|$ is zero. Therefore compare the equation with

$$y = A \cos Bx.$$

It is evident that the graph of $y = \frac{1}{2} \cos \pi x$ has

$$\text{amplitude: } |A| = \frac{1}{2}; \quad \text{period: } P = \frac{2\pi}{|B|} = \frac{2\pi}{\pi} = 2.$$

2. Sketch the graph, marking the x-axis in rational units.

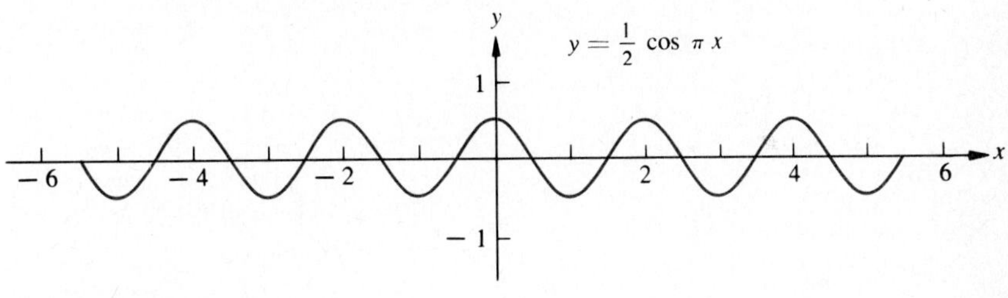

EXERCISE SET 2.2

A

Sketch the graph of each equation over the interval $-2\pi \le x \le 2\pi$.

Example $y = \frac{1}{2} \cos 2x$

Solution

1. Compare the equation with $y = A \cos Bx$. It is evident that the graph of $y = \frac{1}{2} \cos 2x$ has

$$\text{amplitude: } |A| = \frac{1}{2}; \quad \text{period: } P = \frac{2\pi}{|B|} = \frac{2\pi}{2} = \pi.$$

2. Sketch the graph.

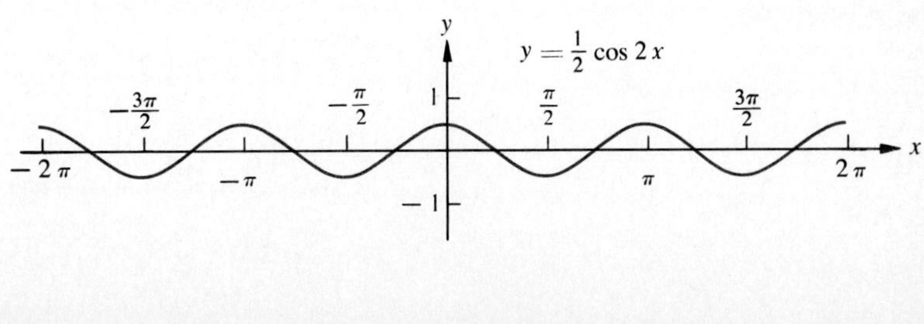

1. $y = 3 \sin x$ **2.** $y = 3 \cos x$ **3.** $y = -2 \cos x$

4. $y = -5 \sin x$ **5.** $y = \frac{1}{2} \sin x$ **6.** $y = \frac{1}{3} \cos x$

7. $y = \cos 2x$ **8.** $y = \sin 3x$ **9.** $y = \sin \frac{1}{2} x$

10. $y = \cos \frac{1}{3} x$ **11.** $y = 2 \sin 2x$ **12.** $y = \frac{1}{3} \cos \frac{1}{2} x$

Example $y = 3 \sin\left(x + \frac{\pi}{2}\right)$

Solution

1. Write the equation as

$$y = 3 \sin\left[x - \left(-\frac{\pi}{2}\right)\right].$$

2. Compare this equation with $y = A \sin B(x - C)$. It is evident that the graph has

$$\text{amplitude: } |A| = 3; \quad \text{period: } P = \frac{2\pi}{|B|} = 2\pi;$$

phase shift: $\pi/2$ units to the left of the graph of $y = 3 \sin x$.

3. Sketch the graph.

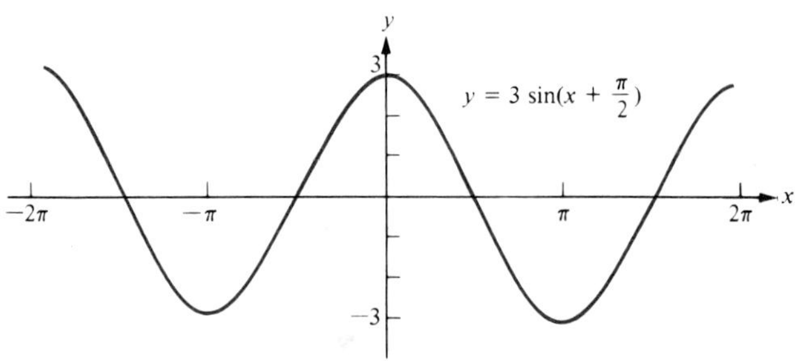

13. $y = \cos(x + \pi)$ **14.** $y = \sin\left(x - \frac{\pi}{2}\right)$

15. $y = 2 \sin\left(x - \frac{\pi}{4}\right)$ **16.** $y = \frac{1}{2} \cos\left(x + \frac{\pi}{6}\right)$

17. $y = \frac{1}{2}\sin\left(x + \frac{\pi}{3}\right)$

18. $y = \cos\left(x - \frac{\pi}{4}\right)$

Example $y = 2\cos\left(3x + \frac{\pi}{2}\right)$

Solution

1. Rewrite the equation by factoring 3 from the expression in parentheses and then as

$$y = 2\cos 3\left[x - \left(-\frac{\pi}{6}\right)\right].$$

2. Compare this equation with $y = A\cos B(x - C)$. It is evident that the graph has

$$\text{amplitude: } |A| = 2; \quad \text{period: } P = \frac{2\pi}{|B|} = \frac{2\pi}{3};$$

phase shift: $\pi/6$ unit to the left of the graph of $y = 2\cos 3x$.

3. Sketch the graph.

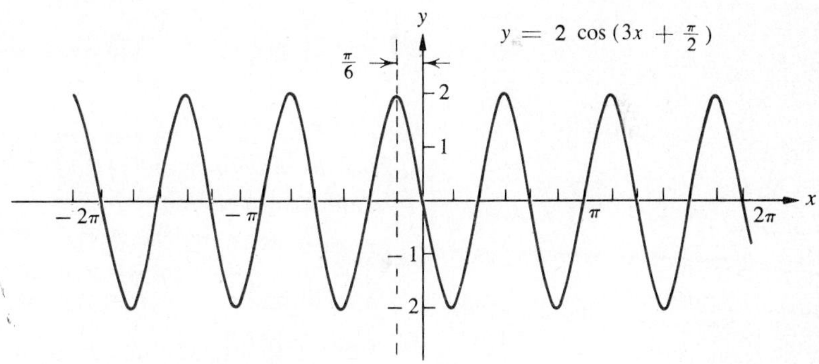

19. $y = 2\cos\left(2x + \frac{\pi}{2}\right)$

20. $y = \frac{1}{2}\sin\left(\frac{1}{2}x - \frac{\pi}{4}\right)$

21. $y = \frac{1}{2}\sin\left(\frac{1}{2}x + \frac{\pi}{2}\right)$

22. $y = 2\cos(2x - \pi)$

23. $y = 3\cos\left(\frac{1}{3}x - \frac{\pi}{3}\right)$

24. $y = 4\sin\left(3x + \frac{3\pi}{4}\right)$

Graph two cycles of each of the following with positive elements in the domain. Hint: *Scale the x-axis in rational units* 1, 2, 3,...

Example $y = \sin \pi x$

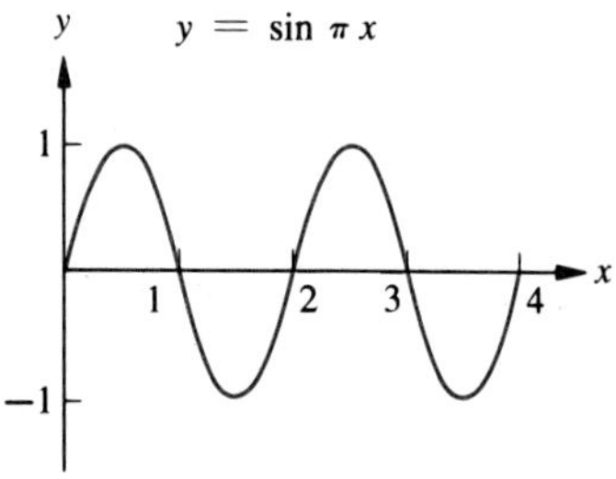

Solution

1. Compare the equation with $y = \sin Bx$. It is evident that the graph of $y = \sin \pi x$ has period

$$P = \frac{2\pi}{|B|} = \frac{2\pi}{\pi} = 2.$$

Sketch the graph on a coordinate system in which the x-axis is scaled in rational units.

25. $y = \cos \pi x$

26. $y = \sin \frac{\pi}{2} x$

27. $y = \frac{1}{2} \sin \frac{\pi}{3} x$

28. $y = 2 \cos \frac{\pi}{4} x$

B

Graph two cycles of each function.

29. $y = |\sin x|$

30. $y = |\cos x|$

31. $y = 3|\cos 2x|$

32. $y = 2\left|\sin \frac{1}{2} x\right|$

33. $y = |\sin \pi x|$

34. $y = |\cos \pi x|$

In each of the following problems
a. *Find the amplitude and period of the function.*
b. *Graph the function over one period.*

35. The horizontal displacement, d, of the bob on a pendulum is given by $d = 20 \sin 2\pi t$, where d is expressed in centimeters and t in seconds.

36. For another pendulum, as in Exercise 35, the displacement is given by $d = 24 \sin 2\pi t$.

37. The voltage E in an electric circuit is given by $E = 4.2 \cos 80\pi t$, where t is time measured in seconds.

38. The voltage in another circuit is given by $E = 2.6 \cos 60\pi t$.

39. The pressure, P, in a traveling sound wave is given by $P = 10 \sin 200\pi(t - 0.2)$, where P is expressed in dynes per square centimeter and t in seconds.

40. For another sound wave, as in Exercise 39, the pressure is given by $P = 12 \sin 200\pi(t - 0.5)$.

2.3 Graphical Addition

Functions defined by equations such as

$$y = 2 \sin x + \cos x \qquad \text{and} \qquad y = x + \cos x$$

can be graphed by methods similar to those we used in the preceding sections. We can find some elements in the range for several elements in the domain by first finding for $\cos x$ and $\sin x$ and then performing the indicated operations. Then we can graph the set of ordered pairs obtained and complete the graph by interpolating for other elements in the range as we connect these points.

An alternative, and somewhat simpler, method involves the **graphical addition of ordinates.** Two examples illustrate this method.

Example Graph $y = 2 \sin x + \cos x$ over the interval $0 \leq x \leq 2\pi$.

Solution Sketch the graphs of $y_1 = 2 \sin x$ and $y_2 = \cos x$ on the same coordinate system over the given interval, as shown. Note that the ordinate of the graph of $y = 2 \sin x + \cos x$ for each x is the algebraic sum of the ordinates of the graph of $y_1 = 2 \sin x$ and the graph of $y_2 = \cos x$. For example, for $x = \pi/3$,

$$2 \sin x = 2\left(\frac{\sqrt{3}}{2}\right) \approx 1.7 \qquad \text{and} \qquad \cos x = 0.5,$$

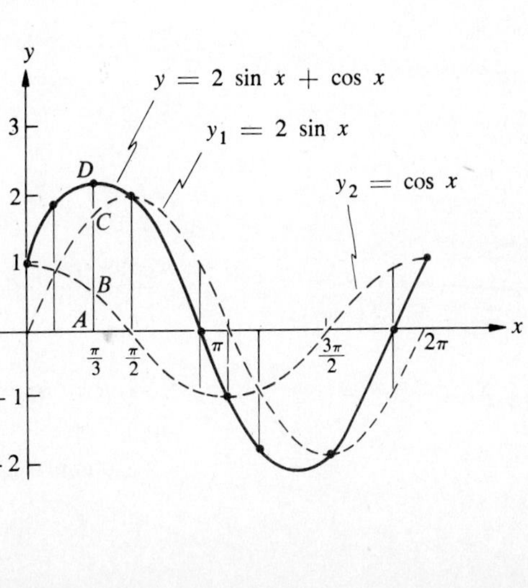

$AB + AC = AD$

from which

$$y = 2 \sin x + \cos x$$

$$\approx 1.7 + 0.5 = 2.2.$$

The ordinate 2.2 can be approximated graphically by "adding" vertical directed segments, $\overrightarrow{AC}$ and $\overrightarrow{AB}$, from the x-axis to the respective curves at $x = \pi/3$. Other ordinates can be found graphically in a similar way. The points in the graph of

$$y = 2 \sin x + \cos x$$

for which $2 \sin x$ or $\cos x$ equals zero are particularly easy to locate. Connect the points with a smooth curve to obtain the graph shown.

Example Graph $y = x + \cos x$ over the interval $-\pi \leq x \leq 2\pi$.

Solution Sketch the graphs of $y_1 = x$ and $y_2 = \cos x$ on the same set of axes over the given interval. Then "add" directed segments for arbitrary values of x. The points in the graph of $y = x + \cos x$ for which $\cos x = 0$ are easy to locate. Connect the points with a smooth curve to obtain the graph shown.

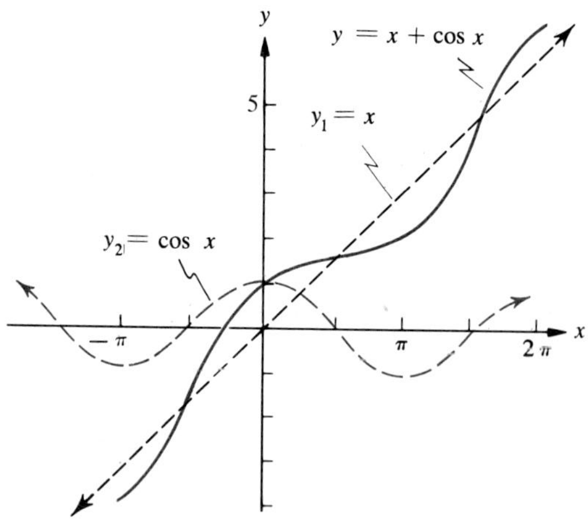

EXERCISE SET 2.3

A

Sketch the graph of each equation over the interval $-\pi \leq x \leq 2\pi$.

Example $y = 3 + 2 \sin x$

Solution

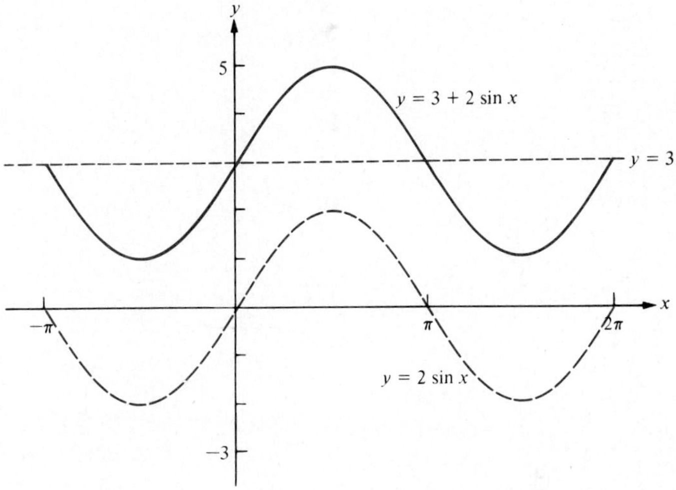

1. $y = 1 + 3 \sin x$

2. $y = 2 + 2 \cos x$

3. $y = -2 + 3 \cos x$

4. $y = -3 + 4 \sin x$

Sketch the graph of each equation over the interval $-\pi \leq x \leq 2\pi$.

Example $y = \sin x + \cos x$

Solution

1. Graph $y_1 = \sin x$ and $y_2 = \cos x$ on the same set of axes.
2. Add the ordinates of several pairs of points for several values of x and draw a smooth curve.

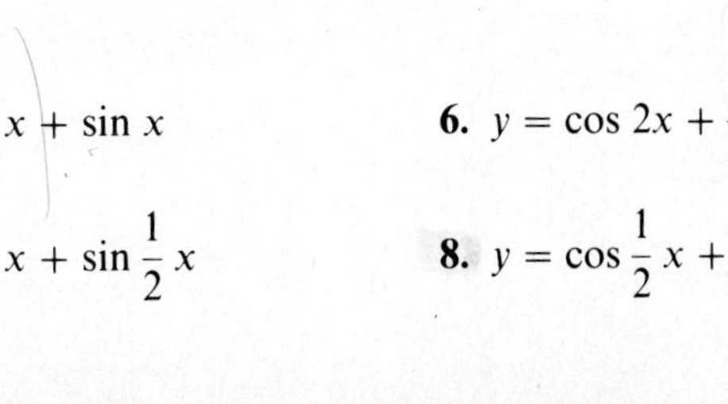

5. $y = 2 \cos x + \sin x$

6. $y = \cos 2x + \frac{1}{2} \sin x$

7. $y = \frac{1}{2} \cos x + \sin \frac{1}{2} x$

8. $y = \cos \frac{1}{2} x + \frac{1}{3} \sin x$

9. $y = \sin 2x + \frac{1}{2}\cos x$

10. $y = \sin\frac{1}{2}x + \cos x$

11. $y = 2\sin x + \cos\frac{1}{2}x$

12. $y = \sin\frac{1}{2}x + 2\cos\frac{1}{2}x$

13. $y = 2\sin x - \cos x$

14. $y = \cos x - \frac{1}{2}\sin x$

15. $y = \cos\frac{1}{2}x - 2\sin x$

16. $y = 2\cos x - \frac{1}{3}\sin x$

17. $y = x + \sin x$

18. $y = \frac{x}{2} - \cos x$

B

Sketch the graph of each equation over the specified interval.

19. $y = x - 2 + \cos x, \quad 0 \le x \le 2\pi.$
Hint: First graph $y_1 = x - 2$ and $y_2 = \cos x$.

20. $y = 3 - x + \sin x, \quad 0 \le x \le 2\pi.$
Hint: First graph $y_1 = 3 - x$ and $y_2 = \sin x$.

21. $y = 2\cos x + \frac{x^2}{12}, \quad 0 \le x \le 2\pi$

22. $y = 2\sin x + \frac{1}{x}, \quad 0 \le x \le 2\pi$

23. $y = \sin \pi x - x + 4, \quad 0 \le x \le 4$

24. $y = \cos\frac{\pi}{2}x - \frac{x^2}{4}, \quad 0 \le x \le 4$

Sketch one cycle of the graphs of each pair of equations and verify that the graphs are identical.

25. $y = \cos x$

$y = \sin\left(x + \frac{\pi}{2}\right)$

26. $y = \sin x$

$y = \cos\left(x - \frac{\pi}{2}\right)$

27. $y = -\sin x$

$y = \cos\left(x + \frac{\pi}{2}\right)$

28. $y = -\cos x$

$y = \sin\left(x - \frac{\pi}{2}\right)$

29. $y = \sin x + \cos x$

$$y = \sqrt{2}\sin\left(x + \frac{\pi}{4}\right)$$

30. $y = \sin x - \cos x$

$$y = \sqrt{2}\sin\left(x - \frac{\pi}{4}\right)$$

31. $y = \cos x - \sin x$

$$y = \sqrt{2}\cos\left(x + \frac{\pi}{4}\right)$$

32. $y = \sin x + \cos x$

$$y = \sqrt{2}\cos\left(x - \frac{\pi}{4}\right)$$

2.4 Graphs of Other Circular Functions

We can graph the tangent, cotangent, secant, and cosecant functions on a Cartesian coordinate system in the same way that we graphed the sine and cosine functions in Sections 2.1 and 2.2. Although the graphs of these functions are not sine waves, the graphs do exhibit the periodic nature and other important features of the functions.

Tangent Function

The period of the tangent function is π. Thus, to graph $y = \tan x$ we need some values of x only in the interval 0 to π. Using Table 1.3, we obtain some ordered pairs in the function and graph these ordered pairs as shown in Figure 2.9. Observe that

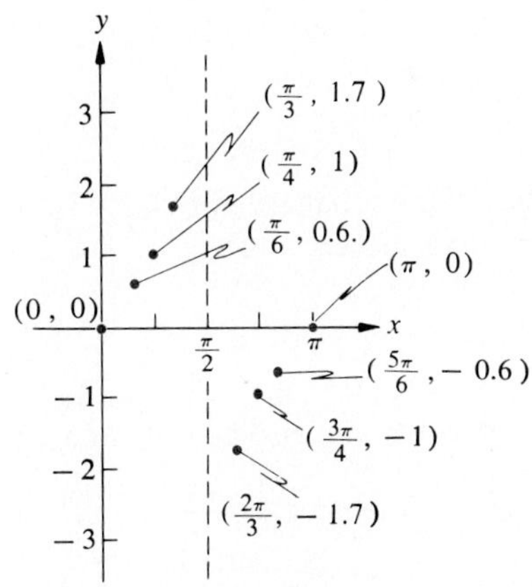

Figure 2.9

$$\tan\frac{\pi}{6} = \frac{1}{\sqrt{3}} \approx 0.6,$$

and

$$\tan\frac{\pi}{3} = \sqrt{3} \approx 1.7.$$

It can be shown that the function defined by $y = \tan x$ is continuous in the domain

$$0 \le x < \frac{\pi}{2} \quad \text{and} \quad \frac{\pi}{2} < x \le \pi$$

($\cos x \neq 0$ in these intervals). Thus we connect the points in Figure 2.9 with a smooth curve to produce one cycle (heavy line), and then we repeat this pattern in both directions to obtain Figure 2.10. Of course, a calculator can be used to obtain additional points to assist in graphing the entire function.

Because $\tan x$ is undefined for $x = \pi/2$, there is no point in the graph of the function when $x = \pi/2$. Notice, however, that $|\tan x|$ increases indefinitely as x is taken closer and closer to $\pi/2$. The vertical dashed lines in

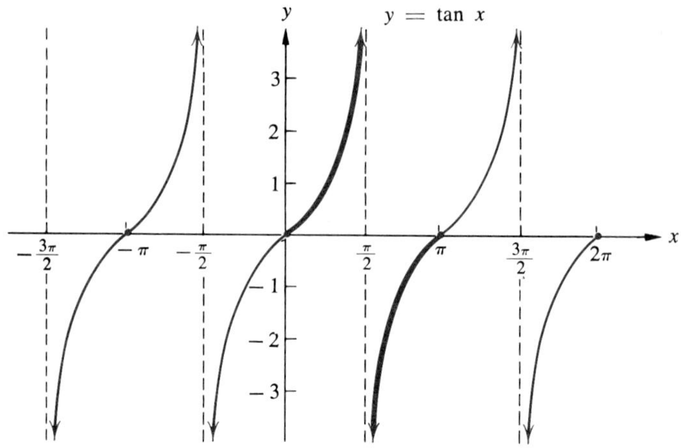

Figure 2.10

Figures 2.9 and 2.10 are called **asymptotes.** The entire set of asymptotes to the curve is the set of lines which are the graphs of

$$x = \frac{\pi}{2} + k\pi, \quad k \in J.$$

Furthermore, observe that the zeros of the function that are associated with the points where the curve intersects the x-axis are $k\pi$, $k \in J$.

Cotangent Function

The graph of the function defined by $y = \cot x$ can also be obtained by plotting points. However, for $x \neq k(\pi/2)$, $k \in J$,

$$\cot x = \frac{1}{\tan x}.$$

Thus, each ordinate of the graph of $y = \cot x$, where $x \neq k(\pi/2)$ is the reciprocal of the corresponding ordinate of the graph of $y = \tan x$. With this information we can quickly sketch the graph of $y = \cot x$ if we first sketch the graph of $y = \tan x$ on the same coordinate system. The graphs appear in Figure 2.11. Notice that the zeros of the function defined by $y = \cot x$ are

$$\frac{\pi}{2} + k\pi, \quad k \in J,$$

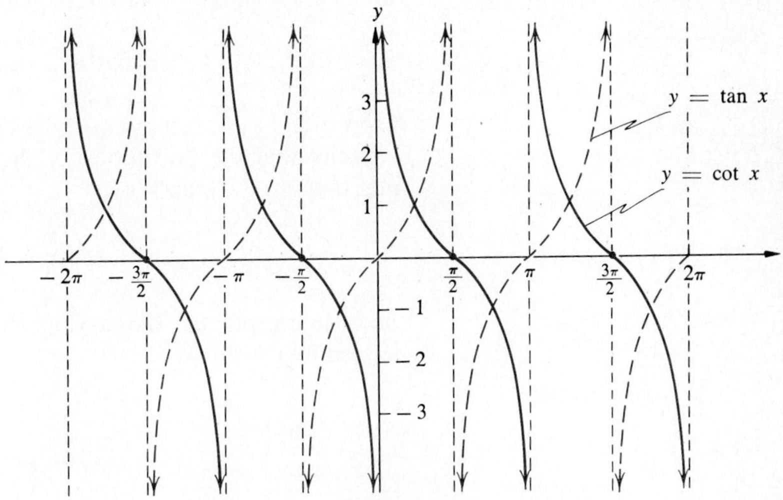

Figure 2.11

and that the asymptotes to the curve are given by

$$x = k\pi, \ k \in J.$$

Also note that the range of either the tangent or cotangent function is $\{y \mid y \in R\}$.

Secant and Cosecant Functions

The graphs of the functions defined by $y = \sec x$ and $y = \csc x$ can also be obtained by using the facts that

$$\sec x = \frac{1}{\cos x}, \quad x \neq \frac{\pi}{2} + k\pi, \ k \in J,$$

and

$$\csc x = \frac{1}{\sin x}, \quad x \neq k\pi, \ k \in J.$$

Hence, each ordinate of the graph of $y = \sec x$ is the reciprocal of the corresponding ordinate of the graph of $y = \cos x$, and each ordinate of the graph of $y = \csc x$ is the reciprocal of the corresponding ordinate of the graph of $y = \sin x$, except for the noted restrictions. Thus the graphs of

$$y = \sec x \qquad \text{and} \qquad y = \csc x$$

can be obtained directly from the graphs of

$$y = \cos x \qquad \text{and} \qquad y = \sin x,$$

respectively, as shown in Figures 2.12 and 2.13 (see exercises 7 and 8).

Notice that the asymptotes of the graph of $y = \sec x$ are the vertical lines that are the graphs of

$$x = \frac{\pi}{2} + k\pi, \ k \in J;$$

the x-intercepts of the asymptotes are the zeros of the function defined by $y = \cos x$.

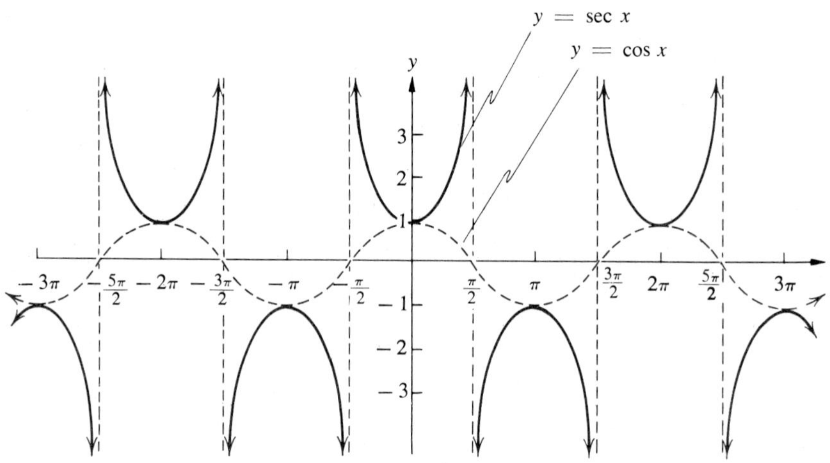

Figure 2.12

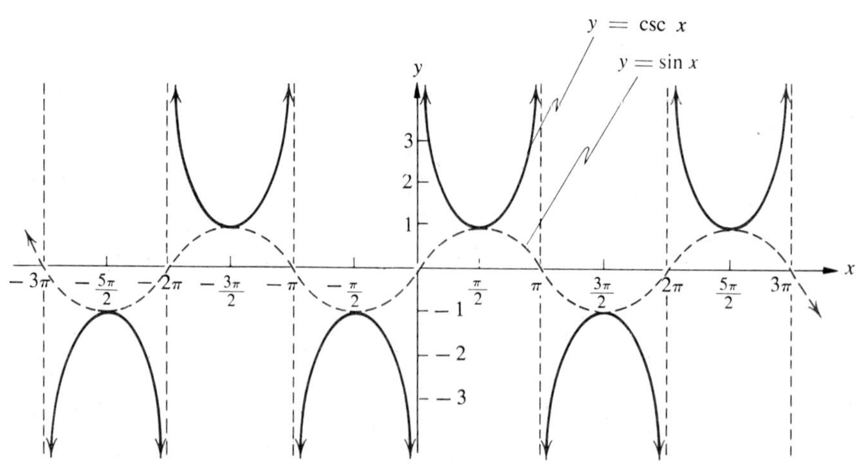

Figure 2.13

The asymptotes of the graph of $y = \csc x$ are the vertical lines that are the graphs of

$$x = k\pi, \ k \in J;$$

the x-intercepts of the asymptotes are the zeros of the function defined by $y = \sin x$. The range of either the secant or the cosecant function is $\{y \mid y \leq -1 \text{ or } y \geq 1\}$. Thus, there are no zeros of these functions.

Variation of Constants

Variations in the constants A, B, and C in equations such as

$$y = A \tan B(x - C) \tag{1}$$

affect their graphs in much the same way that these constants affect the graphs of $y = A \sin B(x - C)$ and $y = A \cos B(x - C)$. However, in the graphs of equations of the form of Equation (1), the coefficient A is not referred to as the amplitude.

Domains and Ranges of the Circular Functions

The domains and ranges of the six circular functions have been briefly considered in Chapter 1, and in this chapter. For convenience we summarize this information in Table 2.1.

TABLE 2.1

Function	Domain	Range
$\{(x, \sin x)\}$	$x \in R$	$-1 \leq \sin x \leq 1$
$\{(x, \cos x)\}$	$x \in R$	$-1 \leq \cos x \leq 1$
$\{(x, \tan x)\}$	$x \in R \ \left(x \neq \frac{\pi}{2} + k\pi\right)$	$\tan x \in R$
$\{(x, \cot x)\}$	$x \in R \ (x \neq k\pi)$	$\cot x \in R$
$\{(x, \sec x)\}$	$x \in R \ \left(x \neq \frac{\pi}{2} + k\pi\right)$	$\lvert\sec x\rvert \geq 1$
$\{(x, \csc x)\}$	$x \in R \ (x \neq k\pi)$	$\lvert\csc x\rvert \geq 1$

EXERCISE SET 2.4

A

Sketch the graph of each equation over the interval $0 \leq x \leq 2\pi$ *and specify* **(a)** *the range,* **(b)** *the equations of the asymptotes, and* **(c)** *the zeros of the function (if they exist).*

Example $y = \tan 2x$

Solution Since the period of $y = \tan x$ is π, the period of $y = \tan 2x$ is $\pi/2$. The graph is sketched as shown.

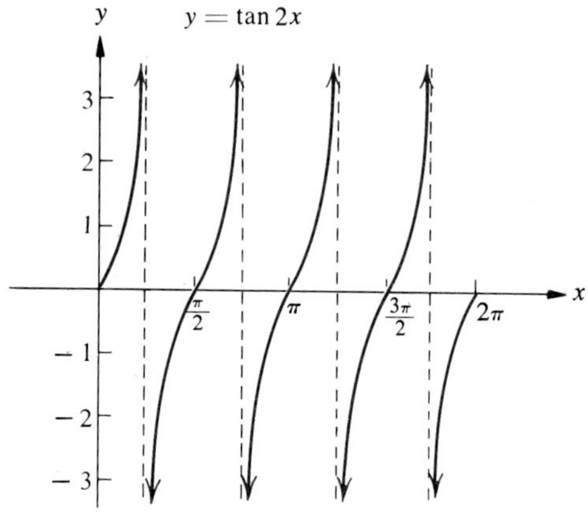

a. Range: $\{y \mid y \in R\}$

b. Since the equations of the asymptotes of $y = \tan x$ are

$$x = \frac{\pi}{2} + k\pi, \; k \in J,$$

the equations of the asymptotes of $y = \tan 2x$ are

$$x = \frac{1}{2}\left(\frac{\pi}{2} + k\pi\right) = \frac{\pi}{4} + k \cdot \frac{\pi}{2}, \; k \in J.$$

c. Zeros: $\left\{x \mid x = k \cdot \frac{\pi}{2}, \; k \in J\right\}$

1. $y = \tan 3x$

2. $y = \tan \frac{1}{2} x$

3. $y = \tan\left[x - \left(-\frac{\pi}{6}\right)\right]$

4. $y = \tan\left(x - \frac{\pi}{4}\right)$

5. $y = \cot \frac{1}{2} x$

6. $y = \cot 2x$

7. Without referring to Figure 2.11, graph $y = \sec x$, by using the fact that $\sec x = \frac{1}{\cos x}$, $x \neq \frac{\pi}{2} + k\pi$, $k \in J$.

8. Without referring to Figure 2.12, graph $y = \csc x$ by using the fact that $\csc x = \dfrac{1}{\sin x}$, $x \neq k\pi,\ k \in J$.

9. Use the graph obtained in Exercise 7 to write the equations of the asymptotes of the graph of $y = \sec x$.

10. Use the graph obtained in Exercise 8 to write the equations of the asymptotes of the graph of $y = \csc x$.

B

Graph each of the following over the interval $-\pi \leq x \leq 2\pi$.

11. $y = \dfrac{1}{2} \sec x$

12. $y = 2 \csc x$

13. $y = \csc 2x$

14. $y = \sec \dfrac{1}{2} x$

15. $y = \sec\left(x + \dfrac{\pi}{4}\right)$

16. $y = \csc\left(x - \dfrac{\pi}{6}\right)$

2.5 Graphing Systems of Equations

It is sometimes convenient to use systems of equations involving trigonometric function values to specify conditions on variables. As in algebraic equations, the solution of a system of equations is the intersection of the solution sets of the equations in the system.

In this section we consider graphical methods of *approximating solutions* of such systems. In Chapter 4 we will consider algebraic methods to find such solutions.

Example Solve the system

$$y = 2 \sin x$$

$$y = \frac{3}{2}$$

by graphical methods in the interval $0 \leq x \leq 2\pi$.

Solution Graph the solution set of each equation and approximate the coordinates of the points of intersection.

(continued)

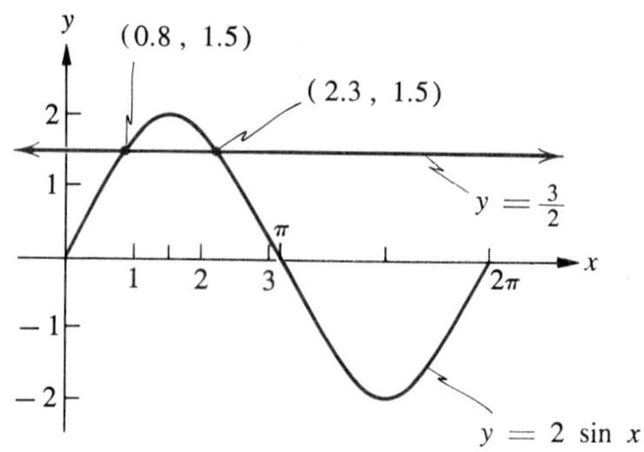

An approximation to the solution set is {(0.8, 1,5), (2.3, 1.5)}.

Example Solve the system

$$y = 2 \sin x$$
$$y = \cos 2x$$

by graphical methods in the interval $0 \leq x \leq 2\pi$.

Solution Graph each equation and approximate the coordinates of the points of intersection.

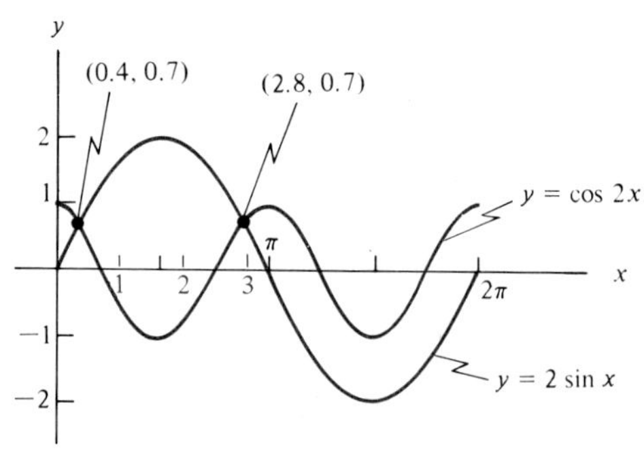

An approximation to the solution set is {(0.4, 0.7), (2.8, 0.7)}.

Parametric Equations

Sometimes it is convenient to define a relation (or function) in which the variable representing an element in the domain and the variable representing an element in the range are each related to a third variable, frequently a variable representing time or an angle. For example, the pair of equations

$$x = f(t) = t - 3 \qquad \text{and} \qquad y = g(t) = t + 1$$

express x in terms of t and y in terms of t. Such equations are called **parametric equations;** t is the **parameter.** To graph the relation between x and y as shown in the example below, we can choose arbitrary real number replacement values for t, and then obtain the corresponding values for x and y. Alternatively, by solving the system of equations, we can eliminate the parameter t to obtain an equation in the variables x and y. Then we can graph the equation in the usual way.

Example

a. Graph the function defined by the parametric equations

$$x = t - 3 \qquad \text{and} \qquad y = t + 1 \tag{1}$$

by assigning arbitrary values for t to obtain some ordered pairs (x, y) in the function

b. Eliminate t and find a Cartesian (or rectangular) equation.

Solution

a. Choosing 0, 1, 2, and 3 as replacements for t, the corresponding ordered pairs (x, y) shown in the figure can be obtained. The construction of a table is sometimes helpful to obtain these ordered pairs.

t	x	y
0	−3	1
1	−2	2
2	−1	3
3	0	4

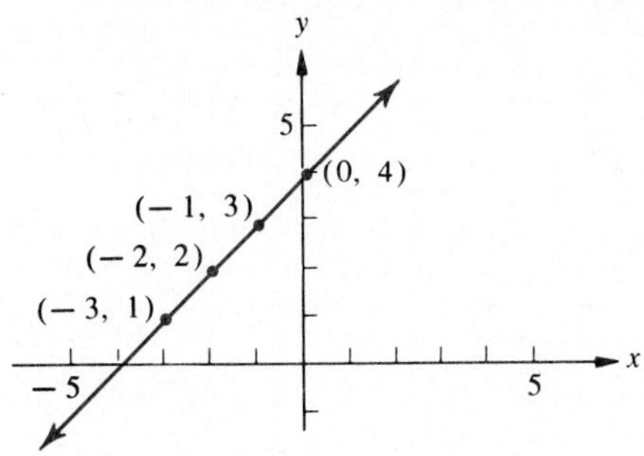

b. Equations (1) can be written equivalently as

$$t = x + 3 \qquad \text{and} \qquad t = y - 1,$$

from which

$$x + 3 = y - 1 \quad \leftrightarrow \quad x - y = -4,^*$$

the Cartesian equation of a straight line.

Example

a. Graph the function defined by the parametric equations

$$x = 3\cos t \qquad \text{and} \qquad y = 4\sin t$$

by assigning arbitrary values for t to obtain some ordered pairs (x, y) in the function.

b. Eliminate t and find the Cartesian equation.

Solution

a. Choosing selected elements of $\{t \mid 0 \le t \le 2\pi\}$ as replacements for t, we can obtain the corresponding ordered pairs (x, y) as shown in the figure. The tabulation of values is helpful.

t	x $3\cos t$	y $4\sin t$
0	3	0
$\frac{\pi}{6}$	$\frac{3\sqrt{3}}{2} \approx 2.6$	2
$\frac{\pi}{4}$	$\frac{3}{\sqrt{2}} \approx 2.1$	$\frac{4}{\sqrt{2}} \approx 2.8$
$\frac{\pi}{2}$	0	4
$\frac{5\pi}{6}$	$-\frac{3\sqrt{3}}{2} \approx -2.6$	2
⋮	⋮	⋮
2π	3	0

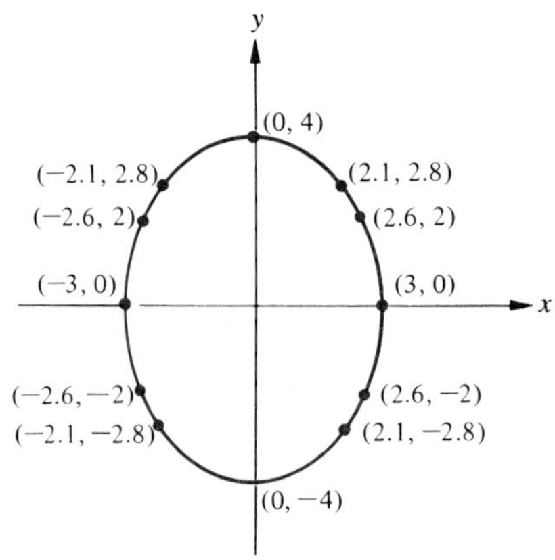

b. Motivated by the fact that $\cos^2 t + \sin^2 t = 1$, we first solve the given equations for $\cos t$ and $\sin t$.

$$x = 3\cos t \quad \leftrightarrow \quad \cos t = \frac{x}{3};$$

$$y = 4\sin t \quad \leftrightarrow \quad \sin t = \frac{y}{4}.$$

Hence,

$$\cos^2 t + \sin^2 t = \frac{x^2}{9} + \frac{y^2}{16}.$$

Because $\cos^2 t + \sin^2 t = 1$,

$$\frac{x^2}{9} + \frac{y^2}{16} = 1$$

from which

$$16x^2 + 9y^2 = 144.$$

EXERCISE SET 2.5

A

Solve each system by graphical methods over the interval $0 \le x \le 2\pi$.

Example $y = \sin x$

$y = 2 \cos x$

Solution

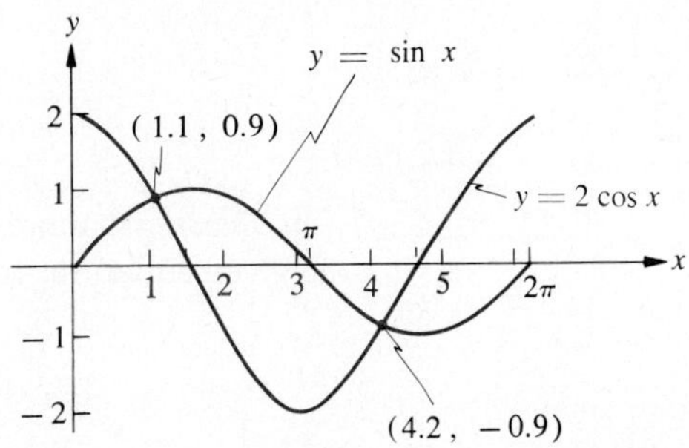

An approximation to the solution set is $\{(1.1, 0.9), (4.2, -0.9)\}$.

1. $y = \cos x$
$y = 1$

2. $y = \sin 2x$
$y = -1$

3. $y = \cos \frac{1}{2}x$
$y = \frac{1}{2}$

4. $y = 2 \sin x$
$y = -\frac{1}{2}$

5. $y = \tan x$
$y = -1$

6. $y = \cot x$
$y = \sqrt{3}$

7. $y = \sin 2x$
$y = \frac{1}{2} \sin x$

8. $y = \cos 2x$
$y = 2 \sin x$

9. $y = 2 \cos 2x$
$y = \sin \frac{1}{2}x$

10. $y = \cos x$
$y = 2 \cos \frac{1}{2}x$

In Exercises 11–18 (**a**) *graph the curve represented by the parametric equations by assigning values to $t \in R$, as specified, to obtain some ordered pairs (x, y), and* (**b**) *eliminate the parameter to find the Cartesian equation.*

Example $x = 4 \cos t,\ y = 3 \sin t;\quad 0 \le t \le 2\pi.$

Solution

a. Choosing selected values as replacements for t, we obtain the corresponding ordered pairs (x, y) as shown in the figure. Tabulation of the values is also shown.

t	x $4 \cos t$	y $3 \sin t$
0	4	0
$\frac{\pi}{6}$	$\frac{4\sqrt{3}}{2} \approx 3.5$	1.5
$\frac{\pi}{4}$	$\frac{4}{\sqrt{2}} \approx 2.8$	$\frac{3}{\sqrt{2}} \approx 2.1$
$\frac{\pi}{2}$	0	3
$\frac{5\pi}{6}$	$-\frac{4\sqrt{3}}{2} \approx -3.5$	1.5
⋮	⋮	⋮
2π	4	0

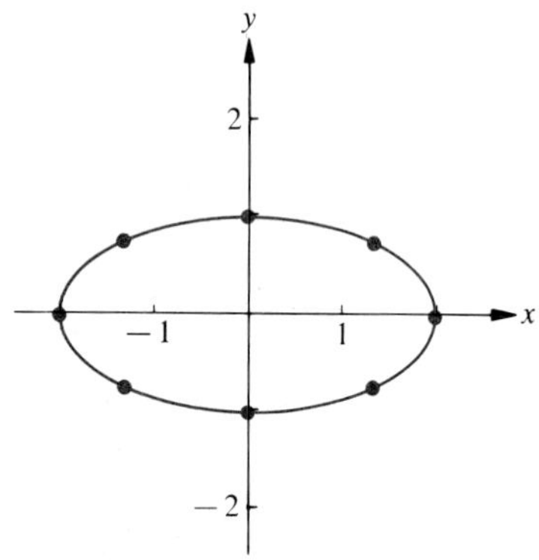

b. Bearing in mind that $\cos^2 t + \sin^2 t = 1$ (Theorem 1.1), we first solve the given equations for $\cos t$ and $\sin t$.

$$x = 4 \cos t \quad \text{or} \quad \cos t = \frac{x}{4}$$

$$y = 3 \sin t \quad \text{or} \quad \sin t = \frac{y}{3}$$

Hence,

$$\cos^2 t + \sin^2 t = \frac{x^2}{16} + \frac{y^2}{9}.$$

Because $\cos^2 t + \sin^2 t = 1$,

$$9x^2 + 16y^2 = 144.$$

11. $x = 3t + 3,\ y = 2t - 5;\quad -2 \le t \le 3$

12. $x = t - 3,\ y = t^2;\quad -3 \le t \le 3$

13. $x = 2 \cos t,\ y = 2 \sin t;\quad 0 \le t \le 2\pi$

14. $x = 2 \cos t,\ y = 5 \sin t;\quad 0 \le t \le 2\pi$

15. $x = \cos t,\ y = \sin^2 t;\quad 0 \le t \le \pi$

16. $x = \cos^2 t,\ y = 2 \sin t;\quad -\frac{\pi}{2} \le t \le \frac{\pi}{2}$

17. $x = \sin 2t, \ y = \cos 2t; \quad 0 \le t \le \dfrac{\pi}{2}$

18. $x = \sin \dfrac{1}{2} t, \ y = \cos \dfrac{1}{2} t; \quad -\pi \le t \le \pi$

B

In Exercises 19 and 20 obtain one equation in x and y.

19. $\begin{cases} x = 2 \cot t \\ y = 2 \sin^2 t \end{cases}$

20. $\begin{cases} x = \sin t \\ y = \cos^2 (2 + \sin t)t \end{cases}$

21. A curve called the **cycloid** is determined by the equations

$$x = a(\alpha - \sin \alpha),$$
$$y = a(1 - \cos \alpha).$$

Graph these equations in an x, y coordinate system for $a = 1$ and $0 \le \alpha \le 2\pi$.

22. Graph the equations in Exercise 21 for $a = 2$ and for $a = 3$.

In Exercises 23 and 24, obtain one equation in x and y and sketch the graph over the interval $-\pi \le t \le \pi$.

23. $\begin{cases} x = \sec t \\ y = \cos t \end{cases}$

24. $\begin{cases} x = \sec t \\ y = \tan t \end{cases}$

In Exercises 25 and 26, obtain one equation that does not contain the parameter t.

25. $\begin{cases} x = a \cos t + h \\ y = b \sin t + k \end{cases}$

26. $\begin{cases} x = a \sec t + h \\ y = b \csc t + k \end{cases}$

2.6 Inverse Circular Functions

In Chapter 1 we obtained function values (elements in the range) for each of the circular functions, for given elements in their respective domains. We observed that *for each element in a domain there is exactly one element in the respective range*. However, since the circular functions are periodic, *for each element in a range, there are an infinite number of elements in their respective domains*. For example, consider the graph of $y = \cos x$ in

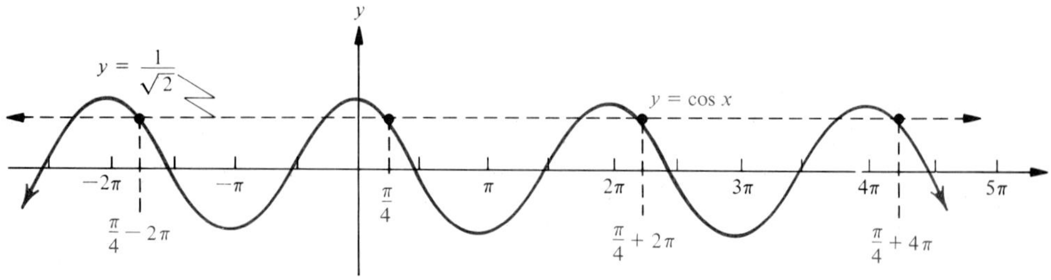

Figure 2.14

Figure 2.14, and note that the *least positive* x for which

$$\cos x = \frac{1}{\sqrt{2}}$$

(approximately equal to 0.7) is $\pi/4$. However, each value of x such that

$$x = \frac{\pi}{4} + k \cdot 2\pi, \ k \in J,$$

is also a solution of $\cos x = 1/\sqrt{2}$ because the cosine function has a period 2π.

Similar considerations are also applicable to other circular functions. In this chapter we are interested in obtaining elements in the domains of the circular functions for given elements in their ranges. Furthermore, we will restrict the domains in such a way that there will be *one and only one* element in the domain for a given element in its range. Before studying the inverses of the circular functions in this chapter, it would be very helpful to review the notions relating to the inverses of algebraic functions considered in Appendix A.6.

As noted in Section A.6, the inverse of a function is the relation obtained by interchanging the components of every ordered pair in the function. Hence, the defining equation of the inverse relation of a function can be obtained by interchanging the variables in the defining equation of the original function. Furthermore, the graph of the inverse of a function is the reflection of the graph of the function with respect to the graph of the equation $y = x$.

Inverse of the Sine Function

Consider the graph of the function

$$y = \sin x,$$

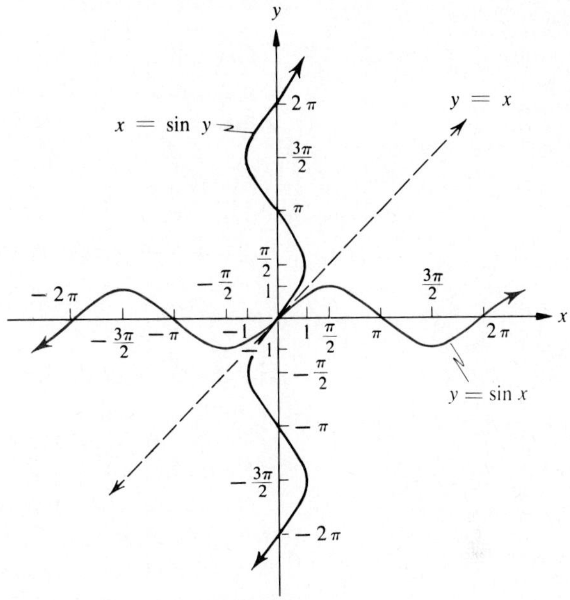

Figure 2.15

with domain R and range $\{y \mid -1 \leq y \leq 1\}$, and its inverse

$$x = \sin y,$$

with domain $\{x \mid -1 \leq x \leq 1\}$ and range $\boldsymbol{R}$, in Figure 2.15. Notice that the graph of $x = \sin y$ is the reflection of the graph of $y = \sin x$ about the graph of the equation $y = x$. Further, notice that

$$x = \sin y$$

defines a relation that is not a function, because for each element x in its domain, there are an infinite number of elements y_1, y_2, etc., in its range as shown in Figure 2.16.

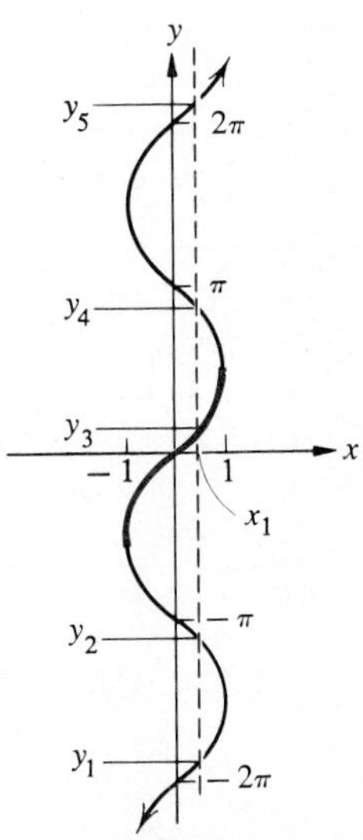

Figure 2.16

Other Inverse Circular Functions

Each circular function has an inverse relation. By suitably restricting the domain of the function or the range of the respective inverse relation, we can define an *inverse function* for each circular function. If we restrict the domain of the sine function to $\{x \mid -\pi/2 \leq x \leq \pi/2\}$, then

$$x = \sin y, \quad -\frac{\pi}{2} \leq y \leq \frac{\pi}{2}, \tag{1}$$

defines an inverse sine function (there is only one y for each x in the domain). Its graph is shown in color in Figure 2.16.

Inverse Notation

Generally, we wish to write the defining equation (1) in a form in which y is expressed explicitly in terms of x. This is done by using the **inverse notation**

$$y = \text{Arcsin}\ x,$$

which is read "y equals arcsin x," or

$$y = \text{Sin}^{-1}\ x,$$

with the same meaning and which is read "y equals inverse sin x." It should be understood that $\text{Sin}^{-1}\ x$ does not mean $1/\text{Sin}\ x$.

Similar notation is used for the other inverse circular functions, as shown in Definition 2.3.

Notice that the ranges of the inverse functions are specified in the definition rather than following the usual procedure of stating the domain. Furthermore, the choices for the ranges are somewhat arbitrary. Those that have been chosen here involve small values of y and include values between 0 and $\pi/2$, have relatively simple graphs, and yield a one-to-one correspondence between elements in the domain and range. They are also the values that can be obtained directly with most calculators.

Definition 2.3 *The inverse notation corresponding to the six inverse circular functions is*

$$y = \text{Sin}^{-1}\ x \quad \leftrightarrow \quad x = \sin y, \quad -\frac{\pi}{2} \le y \le \frac{\pi}{2},$$

$$y = \text{Cos}^{-1}\ x \quad \leftrightarrow \quad x = \cos y, \quad 0 \le y \le \pi,$$

$$y = \text{Tan}^{-1}\ x \quad \leftrightarrow \quad x = \tan y, \quad -\frac{\pi}{2} < y < \frac{\pi}{2},$$

$$y = \text{Csc}^{-1}\ x \quad \leftrightarrow \quad x = \csc y, \quad -\frac{\pi}{2} \le y < 0 \quad \text{or} \quad 0 < y \le \frac{\pi}{2},$$

$$y = \text{Sec}^{-1}\ x \quad \leftrightarrow \quad x = \sec y, \quad 0 \le y < \frac{\pi}{2} \quad \text{or} \quad \frac{\pi}{2} < y \le \pi,$$

$$y = \text{Cot}^{-1}\ x \quad \leftrightarrow \quad x = \cot y, \quad -\frac{\pi}{2} < y < 0 \quad \text{or} \quad 0 < y \le \frac{\pi}{2},$$

The function values in $\text{Sin}^{-1}\ x$, $\text{Cos}^{-1}\ x$, etc., are sometimes called the **principal values** of the corresponding relation.

Function Values

It is sometimes easier to interpret a relationship expressed in inverse notation if it is first rewritten as an equation without using such notation. For example, the equation $y = \text{Arcsin}(1/2)$ can be rewritten equivalently without inverse notation as

$$\sin y = \frac{1}{2}, \quad -\frac{\pi}{2} \le y \le \frac{\pi}{2}.$$

This equation is satisfied for $y = \pi/6$. Hence,

$$y = \text{Arcsin}\,\frac{1}{2} = \frac{\pi}{6}.$$

With a little practice you will be able to find an inverse function value directly from the expression involving the inverse notation.

Examples Find the value of each of the following by using Table 1.3.

a. $\text{Arcos}\,\dfrac{1}{\sqrt{2}}$ b. $\text{Tan}^{-1}\,\dfrac{1}{\sqrt{3}}$ c. $\text{Csc}^{-1}\,\dfrac{2}{\sqrt{3}}$

Solutions From Definition 2.3,

a. $\pi/4$ is the number between 0 and $\pi/2$ such that $\cos(\pi/4) = 1/\sqrt{2}$. Hence,

$$\text{Arccos}\,\frac{1}{\sqrt{2}} = \frac{\pi}{4}.$$

b. $\pi/6$ is the number between 0 and $\pi/2$ such that $\tan(\pi/6) = 1/\sqrt{3}$. Hence,

$$\text{Tan}^{-1}\,\frac{1}{\sqrt{3}} = \frac{\pi}{6}.$$

c. Because $\csc x = 1/\sin x$, we want a number x between 0 and $\pi/2$ such that $\csc x = 2/\sqrt{3}$ or $\sin x = \sqrt{3}/2$. From Table 1.3, $\pi/3$ is that number. Hence

$$\text{Csc}^{-1}\,\frac{2}{\sqrt{3}} = \frac{\pi}{3}.$$

Careful attention should be given to the ranges specified in Definition 2.3 when finding function values that have *negative* domains.

Example Find the value of each of the following using Table 1.3.

a. $\text{Arccos}\left(-\frac{1}{\sqrt{2}}\right)$ b. $\text{Tan}^{-1}\left(-\frac{1}{\sqrt{3}}\right)$ c. $\text{Csc}^{-1}\left(-\frac{2}{\sqrt{3}}\right)$

Solutions From Definition 2.3,

a. Note that in the preceding example, Arccos $(1/\sqrt{2}) = \pi/4$. For $\text{Arccos}(-1/\sqrt{2})$, the number between $\pi/2$ and π is $\pi - (\pi/4)$, or $3\pi/4$, such that $\cos(3\pi/4) = -1/\sqrt{2}$. Hence,

$$\text{Arccos}\left(-\frac{1}{\sqrt{2}}\right) = \frac{3\pi}{4}.$$

b. Since $-\pi/6$ is the number between 0 and $-\pi/2$ such that $\tan(-\pi/6) = -1/\sqrt{3}$,

$$\text{Tan}^{-1}\left(-\frac{1}{\sqrt{3}}\right) = -\frac{\pi}{6}.$$

c. The number x between 0 and $-\pi/2$ such that $\csc x = -2/\sqrt{3}$ or $\sin x = -\sqrt{3}/2$ is $-\pi/3$. Hence,

$$\text{Csc}^{-1}\left(-\frac{2}{\sqrt{3}}\right) = -\frac{\pi}{3}.$$

Using a Calculator

A calculator (or the tables in Appendix C) can be used to find approximations for values of the six trigonometric inverse functions for elements in their respective domains that are not listed in Table 1.3. Values for Arcsin x, Arccos x, and Arctan x can be found directly.

Examples Find an approximation for each of the following to the nearest hundredth. Use a calculator in radian mode.

a. Arcsin 0.2377 b. $\text{Tan}^{-1}(-0.6989)$

Solutions a. Arcsin $0.2377 \approx 0.24$. b. $\text{Tan}^{-1}(-0.6989) \approx -0.61$.

Values for Arcsec x, Arccsc x, and Arccot x cannot be obtained directly when using a calculator. However, we can obtain them by using the reciprocal relationships

$$\cos x = \frac{1}{\sec x}, \quad \sin x = \frac{1}{\csc x}, \quad \text{and} \quad \tan x = \frac{1}{\cot x}.$$

For example, if $y = \text{Arcsec } x$, then $x = \sec y$, so

$$\cos y = \frac{1}{x} \quad \text{or} \quad y = \text{Arccos}\left(\frac{1}{x}\right).$$

Thus

$$\text{Arcsec } x = \text{Arccos } \frac{1}{x}.$$

Examples Find an approximation for each of the following to the nearest hundredth.

a. Arcsec 1.016 b. Cot^{-1} 1.163

Solutions

a. $\text{Arcsec } 1.016 = \text{Arccos } \dfrac{1}{1.016} \approx 0.18.$

b. $\text{Cot}^{-1} 1.163 = \text{Tan}^{-1} \dfrac{1}{1.163} \approx 0.71$

EXERCISE SET 2.6

A

Rewrite each expression without using inverse notation and then specify the value of y for which the statement is true.

Example $y = \text{Arccot } 1$

Solution From Definition 2.3,

$$y = \text{Arccot } 1 \quad \leftrightarrow \quad \cot y = 1, \quad 0 < y \leq \frac{\pi}{2}.$$

Hence, $y = \pi/4$.

1. $y = \text{Arccos } \dfrac{\sqrt{3}}{2}$

2. $y = \text{Arctan } 1$

3. $y = \text{Sin}^{-1} 1$

4. $y = \text{Csc}^{-1} 2$

5. $y = \text{Cot}^{-1} \dfrac{1}{\sqrt{3}}$

6. $y = \text{Sec}^{-1} \sqrt{2}$

Find the exact value of each of the following.

Example $\text{Cos}^{-1}\left(-\frac{\sqrt{3}}{2}\right)$

Solution Let $y = \text{Cos}^{-1}(-\sqrt{3}/2)$. Then, from Definition 2.3,

$$\cos y = -\frac{\sqrt{3}}{2}, \quad 0 \le y \le \pi,$$

from which

$$y = \text{Cos}^{-1}\left(-\frac{\sqrt{3}}{2}\right) = \frac{5\pi}{6}.$$

7. $\text{Arcsin}\ \frac{1}{\sqrt{2}}$ **8.** $\text{Arccos}\ \frac{\sqrt{3}}{2}$ **9.** $\text{Arctan}\ 1$

10. $\text{Arcsin}\ 0$ **11.** $\text{Tan}^{-1}\sqrt{3}$ **12.** $\text{Cos}^{-1}\ 0$

13. $\text{Cos}^{-1}\ 1$ **14.** $\text{Tan}^{-1}\ 0$ **15.** $\text{Sec}^{-1}\ \frac{2}{\sqrt{3}}$

16. $\text{Csc}^{-1}\ 1$ **17.** $\text{Cot}^{-1}\ 1$ **18.** $\text{Sec}^{-1}\ 1$

19. $\text{Sin}^{-1}\left(-\frac{1}{2}\right)$ **20.** $\text{Cos}^{-1}(-1)$ **21.** $\text{Tan}^{-1}(-\sqrt{3})$

22. $\text{Tan}^{-1}(-1)$ **23.** $\text{Cos}^{-1}\left(-\frac{1}{2}\right)$ **24.** $\text{Sin}^{-1}(-1)$

Find an approximation to the nearest hundredth for each of the following by using a calculator in radian mode.

Examples a. $\text{Sin}^{-1}\ 0.8772$ b. $\text{Arcsec}\ 1.238$

Solutions a. $\text{Sin}^{-1}\ 0.8772 \approx 1.07.$ b. $\text{Arcsec}\ 1.238 = \text{Arccos}\ \frac{1}{1.238}$

$$\approx 0.63.$$

25. $\text{Arctan}\ 0.5726$ **26.** $\text{Arcsin}\ 0.6442$ **27.** $\text{Arccos}\ 0.8961$

28. $\text{Arctan}\ 0.1104$ **29.** $\text{Sin}^{-1}\ 0.3523$ **30.** $\text{Cos}^{-1}\ 0.6600$

31. $\text{Tan}^{-1}\ 2.912$ **32.** $\text{Sin}^{-1}\ 0.8724$ **33.** $\text{Sec}^{-1}\ 8.299$

34. $\text{Cot}^{-1}\ 0.6563$ **35.** $\text{Csc}^{-1}\ 1.077$ **36.** $\text{Sec}^{-1}\ 1.004$

37. $\text{Tan}^{-1}(-0.5463)$ **38.** $\text{Sin}^{-1}(-0.6518)$ **39.** $\text{Cos}^{-1}(-0.9759)$

40. $\text{Cos}^{-1}(-0.3153)$ **41.** $\text{Sin}^{-1}(-0.9927)$ **42.** $\text{Tan}^{-1}(-1.827)$

2.7 More About Inverse Circular Functions

A circular function value is sometimes combined with an inverse function value. Such "composite" function values can be evaluated in the same way that we evaluated each value in earlier sections.

Examples Find the value of each of the following.

a. $\mathrm{Cos}^{-1}\left(\sin\frac{\pi}{2}\right)$ b. $\mathrm{Tan}^{-1}(\cos 0)$ c. $\sin(\mathrm{Tan}^{-1}\sqrt{3})$

Solutions Using Definition 2.3 and Table 1.3:

a. Since $\sin \pi/2 = 1$,

$$\mathrm{Cos}^{-1}\left(\sin\frac{\pi}{2}\right) = \mathrm{Cos}^{-1}(1) = 0.$$

b. Since $\cos 0 = 1$,

$$\mathrm{Tan}^{-1}(\cos 0) = \mathrm{Tan}^{-1}(1) = \frac{\pi}{4}.$$

c. Since $\mathrm{Tan}^{-1}\sqrt{3} = \pi/3$,

$$\sin(\mathrm{Tan}^{-1}\sqrt{3}) = \sin\frac{\pi}{3} = \frac{\sqrt{3}}{2}.$$

Examples Find approximations for each of the following to the nearest hundredth.

a. $\tan(\mathrm{Cos}^{-1} 0.4321)$ b. $\mathrm{Sin}^{-1}(\cos 0.52)$

Solutions Use Definition 2.3 and a calculator:

a. Since $\mathrm{Cos}^{-1} 0.4321 \approx 1.1240$,

$$\tan(\mathrm{Cos}^{-1} 0.4321) \approx \tan(1.1240) \approx 2.09.$$

b. Since $\cos 0.52 \approx 0.8678$,

$$\mathrm{Sin}^{-1}(\cos 0.52) \approx \mathrm{Sin}^{-1} 0.8678 \approx 1.05.$$

Writing Equivalent Equations

Inverse notation for the circular functions enables us to solve equations for one variable in terms of the other variables and constants.

Example Solve $4 \cos 3x = 1$ for x using inverse notation. Assume restrictions specified in Definition 2.3.

Solution We can multiply each member of

$$4 \cos 3x = 1$$

by 1/4 to obtain

$$\cos 3x = \frac{1}{4}.$$

By Definition 2.3,

$$\cos 3x = \frac{1}{4} \quad \leftrightarrow \quad 3x = \text{Cos}^{-1} \frac{1}{4}.$$

Then, multiplying each member by 1/3, we obtain

$$x = \frac{1}{3} \text{Cos}^{-1} \frac{1}{4}.$$

Of course, if we want to, we can now obtain a decimal approximation for x by first finding an approximation for $\text{Cos}^{-1}(1/4)$.

Example Solve $y = \frac{1}{3} \tan \frac{x}{2}$ explicitly for x in terms of y using inverse notation. Assume restrictions specified in Definition 2.3.

Solution We can multiply each member of

$$y = \frac{1}{3} \tan \frac{x}{2}$$

by 3 to obtain

$$3y = \tan \frac{x}{2} \quad \leftrightarrow \quad \frac{x}{2} = \text{Tan}^{-1} 3y.$$

Then multiplying each member by 2, we obtain

$$x = 2 \text{Tan}^{-1} 3y.$$

Graphs of the Inverse Circular Functions

The graph of the Arcsine function shown in color in Figure 2.16 on page 77 and the graphs of the other inverse circular functions, which can be obtained by graphing ordered pairs or by reflecting the graphs of the respective functions over the appropriate domain, are shown in Figure 2.17.

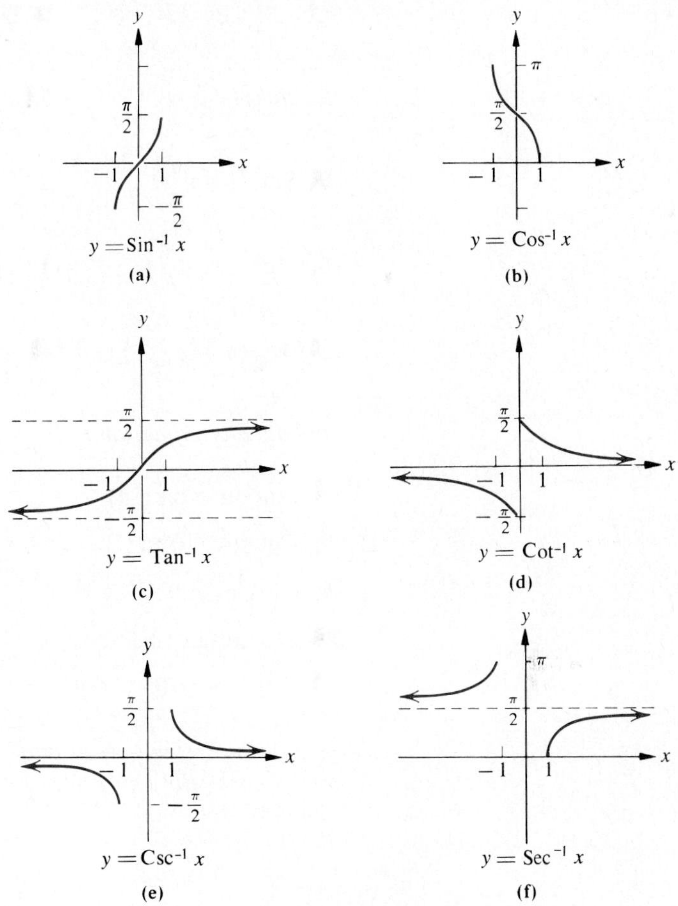

Figure 2.17

EXERCISE SET 2.7

A

Find the exact value of each of the following.

Example $\cos\left(\text{Sin}^{-1}\frac{1}{2}\right)$

(continued)

Solution $\text{Sin}^{-1}\left(\frac{1}{2}\right) = \frac{\pi}{6}$; therefore

$$\cos\left(\text{Sin}^{-1}\frac{1}{2}\right) = \cos\frac{\pi}{6} = \frac{\sqrt{3}}{2}.$$

1. $\cos\left(\text{Cos}^{-1}\frac{\sqrt{3}}{2}\right)$ **2.** $\tan(\text{Tan}^{-1} 1)$ **3.** $\tan(\text{Sin}^{-1} 0)$

4. $\cos(\text{Sin}^{-1} 1)$ **5.** $\sin[\text{Arctan}(-1)]$ **6.** $\sin\left[\text{Cos}^{-1}\left(-\frac{1}{2}\right)\right]$

7. $\text{Sin}^{-1}\left(\sin\frac{\pi}{4}\right)$ **8.** $\text{Cos}^{-1}\left(\cos\frac{\pi}{3}\right)$ **9.** $\text{Cos}^{-1}\left(\sin\frac{\pi}{3}\right)$

10. $\text{Sin}^{-1}\left(\cos\frac{\pi}{6}\right)$ **11.** $\text{Tan}^{-1}\left(\cos\frac{\pi}{2}\right)$ **12.** $\text{Csc}^{-1}(\cos \pi)$

13. $\text{Arcsec}\left(\tan\frac{3\pi}{4}\right)$ **14.** $\text{Arccot}\left(\sin\frac{3\pi}{2}\right)$

Find an approximation to the nearest hundredth for each of the following.

15. $\tan(\text{Sin}^{-1} 0.8134)$ **16.** $\cos(\text{Tan}^{-1} 1.432)$

17. $\sin(\text{Cos}^{-1} 0.4321)$ **18.** $\sin(\text{Tan}^{-1} 0.632)$

19. $\cot(\text{Cos}^{-1} 0.5321)$ **20.** $\sec(\text{Sin}^{-1} 0.6432)$

21. $\csc(\text{Arctan } 1.632)$ **22.** $\cot(\text{Arccos } 0.8214)$

23. $\cot[\text{Arccos}(0.3421)]$ **24.** $\csc[\text{Arcsin}(0.6321)]$

Rewrite each equation so that x is expressed explicitly in terms of y and/or the constants using inverse notation. Assume restrictions specified in Definition 2.3.

Example $3 \sin 2x = 1$

Solution Multiplying each member by 1/3 yields $\sin 2x = 1/3$. By Definition 2.3,

$$\sin 2x = \frac{1}{3} \quad \leftrightarrow \quad 2x = \text{Sin}^{-1}\frac{1}{3}.$$

Then, multiplying each member by 1/2, we obtain

$$x = \frac{1}{2}\text{Sin}^{-1}\frac{1}{3}.$$

25. $4 \cos x = 3$

26. $2 \tan x = 3$

27. $\sin 3x = \frac{1}{4}$

28. $\cos 2x = \frac{1}{3}$

29. $2 \tan 2x = 5$

30. $3 \sin 4x = 2$

31. $2 \cos 3x = 1.24$

32. $5 \tan 2x = 2.25$

33. $y = 2 \tan 3x$

34. $y = 3 \cos 2x$

35. $y = 4 \sin \frac{x}{2}$

36. $y = 2 \tan \frac{x}{3}$

B

Simplify each expression.

37. $\sin(\text{Sin}^{-1} x)$

38. $\cos(\text{Cos}^{-1} x)$

39. $\tan(\text{Tan}^{-1} 3x)$

40. $\sec(\text{Sec}^{-1} 3x)$

41. $\cot\left(\text{Tan}^{-1} \frac{1}{2x}\right)$

42. $\csc\left(\text{Sin}^{-1} \frac{1}{2x}\right)$

Graph each function. Hint: Use the graphs shown in Figure 2.17.

43. $y = \text{Arcsin}\ x + \frac{\pi}{2}$

44. $y = \text{Arccos}\ x + \pi$

45. $y = \text{Arctan}\ x + \frac{\pi}{2}$

46. $y = \text{Arctan}\ x + \pi$

Chapter Summary

2.1 Graphs of sine or cosine functions are called **sine waves** or **sinusoidal waves.** The part of the graph over one period is called a **cycle** of the wave. The absolute value of half the difference of the maximum and the minimum ordinates is called the **amplitude** of the wave.

Zeros of the sine function, that is, values of x for which $\sin x = 0$, are $k\pi$, $k \in J$. Zeros of the cosine function are $(\pi/2) + k\pi$, $k \in J$. The zeros of a function are the x-coordinates of the points where the graph intersects the x-axis (see Figures 2.4 and 2.6).

2.2 Functions defined by equations of the form

$$y = A \sin B(x - C) \qquad \text{or} \qquad y = A \cos B(x - C),$$

where $A, B, C, x, y \in R$ and $A, B \neq 0$, always have sine waves for their graphs. The **amplitude** of the wave or function is $|A|$, the **period** is $2\pi/|B|$, and the **phase shift** is $|C|$ (see Figure 2.8).

2.3 Functions defined by equations of the form

$$y = A_1 \sin B_1 x + A_2 \cos B_2 x$$

can be graphed by first graphing $y_1 = A_1 \sin B_1 x$ and $y_2 = A_2 \cos B_2 x$, and then "adding" their ordinates graphically for arbitrary values of x.

2.4 The graphs of the circular functions defined by $y = \tan x$, $y = \cot x$, $y = \sec x$, and $y = \csc x$ exhibit the periodic nature of the functions (see Figures 2.10–2.13). The following table shows the zeros and/or the equations for the asymptotes of these circular functions.

Function	**Zeros**	**Equations for Asymptotes**
tangent	$k\pi,\ k \in J$	$x = \frac{\pi}{2} + k\pi,\ k \in J$
cotangent	$\frac{\pi}{2} + k\pi,\ k \in J$	$x = k\pi,\ k \in J$
secant	none	$x = \frac{\pi}{2} + k\pi,\ k \in J$
cosecant	none	$x = k\pi,\ k \in J$

2.5 Solutions of systems of equations involving circular function values can be approximated by graphical methods.

A pair of equations of the form

$$x = f(t) \quad \text{and} \quad y = g(t),$$

in which two variables, in this case x and y, are each expressed in terms of a third variable, are called **parametric equations.** The relation between x and y can be graphed by arbitrarily choosing real number replacements for the third variable, in this case t, and then obtaining the corresponding values for x and y.

2.6–2.7 The **inverse of a function** is the relation obtained by interchanging the components of every ordered pair in the function.

By suitably restricting the domains of the circular functions, or the ranges of the respective inverse relations, an inverse function is defined for each circular function (see Definition 2.3).

The function values in Arcsine, Arccosine, etc., are sometimes called the principal values in the corresponding relation.

A calculator or Table II, Appendix C, can be used to find approximations for function values of the six inverse circular functions for elements in their respective domains. Table 1.3 can be used to find exact values for special elements in their domains.

A list of symbols introduced in this chapter is shown inside the front cover, and important properties are shown inside the back cover.

Review Exercises

2.1–2.2 *Sketch the graph of the function defined by each equation over the interval* $-2\pi \le x \le 2\pi$. *Specify* **(a)** *the amplitude,* **(b)** *the period, and* **(c)** *the phase shift (if applicable).*

1. $y = 4 \sin x$

2. $y = \dfrac{1}{2} \cos 2x$

3. $y = \sin\left(x + \dfrac{\pi}{2}\right)$

4. $y = 2 \cos\left(\dfrac{1}{2}x - \dfrac{\pi}{6}\right)$

Graph two cycles of each of the following.

5. $y = \cos \dfrac{\pi}{4} x$

6. $y = 2 \sin \dfrac{\pi}{3} x$

2.3 *Using the method of graphical addition of ordinates, graph the equations over the interval* $0 \le x \le 2\pi$.

7. $y = \sin 2x - \dfrac{1}{2} \cos x$

8. $y = \dfrac{1}{3} x - \sin x$

2.4 *Sketch the graph of the function defined by each equation over the interval* $0 \le x \le 2\pi$. *Specify* **(a)** *the range,* **(b)** *the equations of the asymptotes, and* **(c)** *the set of all zeros of the function (if applicable).*

9. $y = \tan\left(x - \dfrac{\pi}{3}\right)$

10. $y = \cot 3x$

11. $y = \csc \dfrac{1}{2} x$

12. $y = 2 \sec x$

2.5 *Find approximations to solutions of each system over the interval* $0 \le x \le 2\pi$ *by graphical methods.*

13. $y = \cos x$
$y = \frac{1}{3}$

14. $y = 2 \sin x$
$y = \frac{3}{4}$

15. $y = \sin 2x$
$y = \frac{1}{2}\cos x$

16. $y = \cos 2x$
$y = \frac{1}{2}\sin x$

17. Graph $x = 4t - 1$ and $y = 2t^2$ in an xy system.

18. Graph $x = 3\cos t$ and $y = 4\sin t$ in an xy system.

19. Eliminate the parameter in the equations in Exercise 17 to find an equation in x and y.

20. Eliminate the parameter in the equations in Exercise 18 to find an equation in x and y.

2.6 *Rewrite each expression without using inverse notation and then specify the value of y for which the statement is true.*

21. $y = \text{Arccos}\,\dfrac{1}{\sqrt{2}}$

22. $y = \text{Arccot}(-1)$

23. $y = \text{Csc}^{-1}(-2)$

24. $y = \text{Sec}^{-1}\,\dfrac{2}{\sqrt{3}}$

Find the exact value of each expression.

25. Arcsin 1

26. $\text{Arccot}(-\sqrt{3})$

27. $\text{Cos}^{-1}\left(-\dfrac{1}{\sqrt{2}}\right)$

28. $\text{Tan}^{-1}\,\dfrac{1}{\sqrt{3}}$

Find an approximation to the nearest hundredth for each of the following.

29. Arccos 0.1205

30. Arctan 1.398

31. $\text{Sin}^{-1}\,0.4529$

32. $\text{Cot}^{-1}(-0.6142)$

2.7 *Find the exact value of each of the following.*

33. $\tan\left(\text{Cos}^{-1}\,\dfrac{1}{2}\right)$

34. $\cos(\text{Arccot}\,\sqrt{3})$

35. $\text{Sin}^{-1}\left(\tan\dfrac{\pi}{4}\right)$

36. $\text{Tan}^{-1}(\cos 0)$

Find an approximation to the nearest hundredth for each of the following.

37. $\sin(\text{Cos}^{-1}\,0.6072)$

38. $\tan(\text{Sin}^{-1}\,0.3021)$

39. $\cot(\text{Tan}^{-1}\,1.832)$

40. $\sec(\text{Cos}^{-1}\,0.2932)$

Rewrite each equation using inverse notation so that x is expressed explicitly in terms of y and/or the constants. Assume restrictions specified in Definition 2.3.

41. $5 \cos x = 3$

42. $\sin 2x = \dfrac{1}{5}$

43. $y = 3 \tan 2x$

44. $y = 2 \cos \dfrac{x}{2}$

3 TRIGONOMETRIC FUNCTIONS; SOLUTION OF TRIANGLES

The circular functions that we considered in Chapters 1 and 2 are pairings in which the domain and range are sets of real numbers. In this chapter we discuss functions in which the domain is a set of *angles* and the range is a set of real numbers. As you will see, a very close relationship exists between these functions and the circular functions.

3.1 Angles and Their Measures

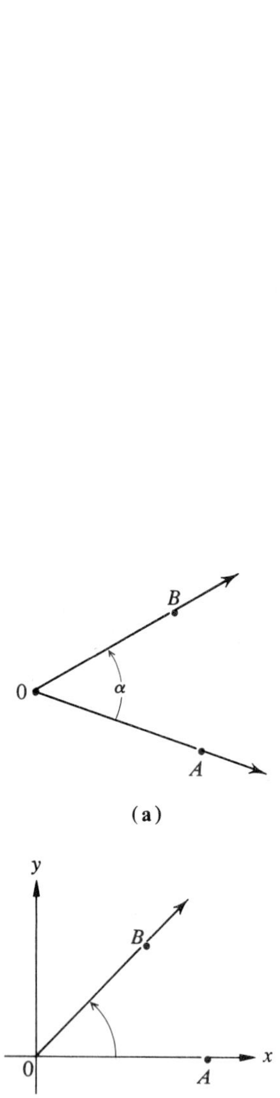

Figure 3.1

In geometry an angle is defined as the union of two rays having a common endpoint (see Appendix A.7). In trigonometry we also consider the rotation associated with an angle. For example, consider $\angle AOB$ as shown in Figure 3.1a and visualize the angle as being formed by rotating ray $\overrightarrow{OA}$, with O as the pivotal point, to the ray $\overrightarrow{OB}$. Viewed in this manner, ray $\overrightarrow{OA}$ is called the **initial side,** and ray $\overrightarrow{OB}$ is called the **terminal side** of the angle. Angles for which rotation is considered are called **directed angles.**

Standard Position of an Angle

Each angle in the plane can be made to coincide with (or is congruent to) an angle that has one of its sides along the positive half of the x-axis and its vertex at the origin, as shown in Figure 3.1b. Such an angle is said to be in **standard position.** The terminal side of an angle in standard position may lie in any quadrant or on any axis. The angle is said to be in that quadrant in which the terminal side lies. If the terminal side lies on an axis, the angle is called a **quadrantal angle.** For example, in Figure 3.2, α is in Quadrant I,

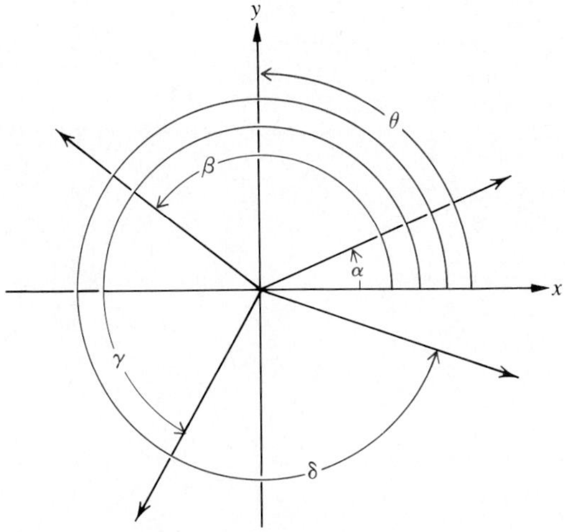

Figure 3.2

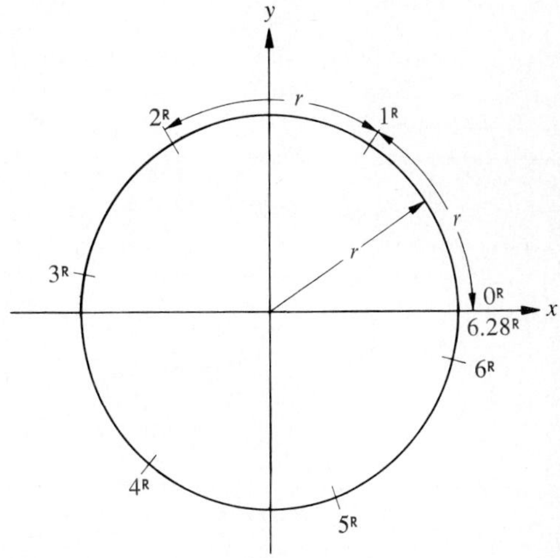

Figure 3.3

β (beta) is in Quadrant II, γ (gamma) is in Quadrant III, δ (delta) is in Quadrant IV, and θ (theta) is a quadrantal angle.

Measure of an Angle

Just as different real numbers can be specified as the length of a particular line segment, say 36, 3, or 1, depending on whether the unit of length is an inch, a foot, or a yard, different real numbers can be specified as the measure of a particular angle depending on the unit of measure. Two kinds of angle measure most commonly used are called **degree measure** and **radian measure.** Such measures can be assigned to an angle by using a circle whose center is at the vertex of the angle.

If the circle is divided into 360 arcs of equal length, the number of such unit arcs intercepted by a central angle, α, is called the measure of the angle in *degree units* (see Appendix A.7) and is designated by $\boldsymbol{m^\circ(\alpha)}$.

In radian measure, the circle is divided into unit arcs, each having a length equal to the length of the circle's radius, as shown in Figure 3.3. The circumference of a circle whose radius has length r is $2\pi r$; hence the circle will include $2\pi r/r$ or 2π (approximately 6.28) of these arcs. The number of such arcs intercepted by a central angle is called the measure of the angle in *radian units* and is designated by $\boldsymbol{m^R(\alpha)}$.

Positive numbers are assigned for the measures of angles in which the rotation from the initial to the terminal side is in a counterclockwise direction, and negative numbers are assigned for the measure if the rotation is in a clockwise direction. In Figure 3.4 on page 94, a circle is scaled in degree units and radian units for some positive measures. In Figure 3.5, a circle is scaled for similar negative units. The correspondence between degree

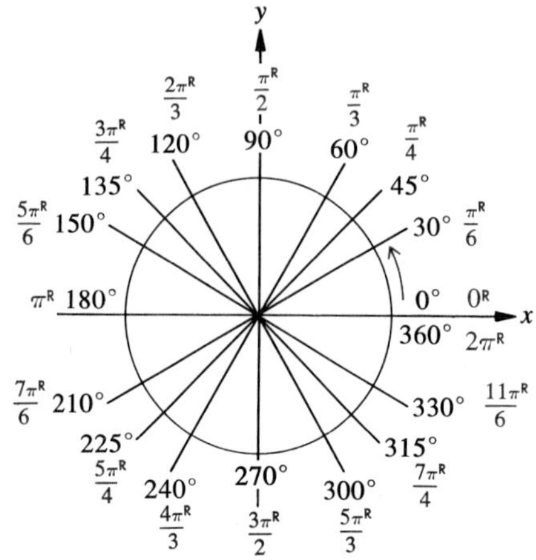

Figure 3.4

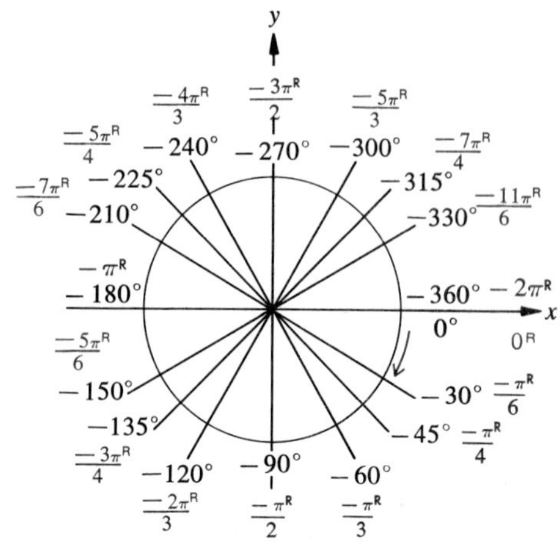

Figure 3.5

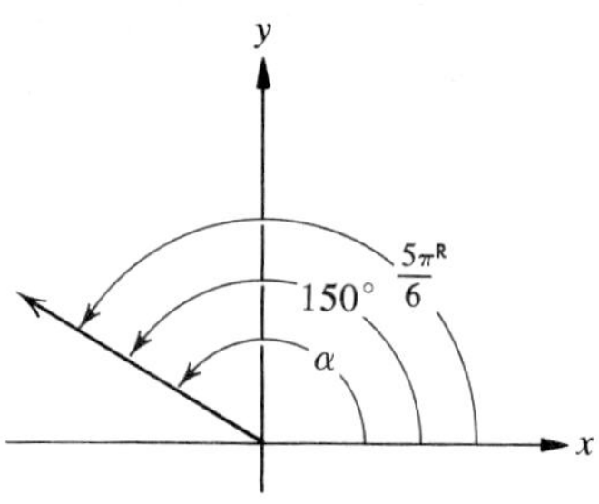

(a)

Figure 3.6

units and radian units is verified by elementary algebraic operations. If $2\pi^{\mathrm{R}}$ and 360° are the measures of the same angle, then so are π^{R} and 180°, $(\pi/2)^{\mathrm{R}}$ and 90°, etc.

We shall use the usual conventions concerning symbolism for angles and their measures. For example, we shall express the measures of the angle in Figure 3.6 as

$$m^\circ(\alpha) = 150 \qquad \text{or} \qquad m^{\mathrm{R}}(\alpha) = \frac{5\pi}{6},$$

or alternatively as

$$m^\circ(\alpha) = 150^\circ \qquad \text{or} \qquad m^{\mathrm{R}}(\alpha) = \frac{5\pi^{\mathrm{R}}}{6},$$

where the use of the degree or radian symbol in the right-hand member of the equations should be interpreted as a redundancy for emphasis. We shall also simply write

$$\alpha = 150^\circ \qquad \text{and} \qquad \alpha = \frac{5\pi^{\mathrm{R}}}{6}.$$

Furthermore, we shall also refer to a "least positive angle" as meaning the member of a set of angles that has the least positive measure, a "negative angle" as meaning an angle with a negative measure, etc.

Notice that for a given angle the number of units in one measure generally does not equal the number in another measure. For example,

$150 \neq 5\pi/6$. However, an angle whose measure in degree units is 150 is congruent to (has the same amount of rotation as) an angle whose measure in radian units is $5\pi/6$. Symbolically, this is written as

$$150° = \frac{5\pi}{6}^{\mathrm{R}},$$

where 150 and $5\pi/6$ are simply measures in degree units and radian units, respectively, for the same or congruent angles.

Degree–Radian Conversion Formulas

The relations between the radian and degree measures of some angles are shown in Figures 3.4 and 3.5. The measure of any angle in radians can be related to the measure of the angle in degrees. Since the ratio of the measure of a given central angle to the measure of the central angle intercepting the entire circle is the same for any unit of measurement, we have the following relationship between degree measure and radian measure.

Theorem 3.1 *For all angles α,*

$$\frac{m°(\alpha)}{360} = \frac{m^{\mathrm{R}}(\alpha)}{2\pi}.$$

The equation in this theorem is sometimes written in either of the following equivalent forms.

$$m°(\alpha) = \frac{180}{\pi}\, m^{\mathrm{R}}(\alpha), \tag{1}$$

$$m°(\alpha) = \frac{\pi}{180}\, m°(\alpha). \tag{2}$$

Examples Find the degree measure (to the nearest tenth of a degree) of each angle α with the given radian measure.

a. $\alpha = \dfrac{3\pi}{5}^{\mathrm{R}}$ b. $\alpha = 0.81^{\mathrm{R}}$

Solutions

a. $m°(\alpha) = \left(\dfrac{180}{\pi} \cdot \dfrac{3\pi}{5}\right)^{\circ}$

$= 108.0°$

b. $m°(\alpha) = \left(\dfrac{180}{\pi} \cdot 0.81\right)^{\circ}$

$\approx 46.4°$

In Example b above, we used a calculator approximation for π equal to 3.1415926 to obtain the approximation shown. Slightly different values may be expected for different approximations.

Examples Find the radian measure (to the nearest hundredth of a radian) of each angle α with the given degree measure.

a. $\alpha = 60^\circ$

b. $\alpha = 330^\circ$

Solutions

a. $m^R(\alpha) = \left(\dfrac{\pi}{180} \cdot 60\right)^R = \dfrac{\pi}{3}^R \approx 1.05^R$

b. $m^R(\alpha) = \left(\dfrac{\pi}{180} \cdot 330\right)^R = \dfrac{11\pi}{6}^R \approx 5.76^R$

From conversion formulas (1) and (2) on page 95, we see that the measure in degrees of an angle α with a measure of 1^R is given by

$$m^\circ(\alpha) = \left(\frac{180}{\pi} \cdot 1\right)^\circ \approx 57.30^\circ$$

and that the measure in radians of an angle α with a measure of 1° is

$$m^R(\alpha) = \left(\frac{\pi}{180} \cdot 1\right)^R \approx 0.01745^R.$$

These results can also be used to make conversions from radian units to degree units and from degree units to radian units, respectively.

Examples

a. The degree measure of an angle $\alpha = 0.37^R$ is given by

$$m^\circ(\alpha) \approx (57.30 \cdot 0.37)^\circ \approx 21.2^\circ.$$

b. The radian measure of an angle $\alpha = 200^\circ$ is given by

$$m^R(\alpha) \approx (0.01745 \cdot 200)^R = 3.49^R.$$

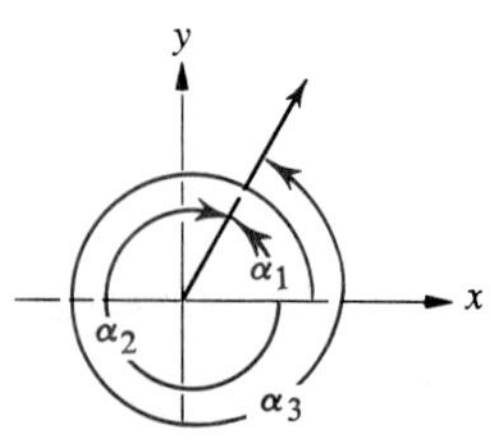

Figure 3.7

Coterminal Angles

If an angle is viewed as being formed by rotation of the terminal side from the initial side to its final position, then the terminal side may be rotated through more than 360°. Thus, as indicated in Figure 3.7, there are several conditions under which angles may have the same initial side and the same terminal side but have different measures.

Definition 3.1 **Coterminal angles** *are two or more angles that have the same initial side and the same terminal side.*

In Figure 3.7, α_1, α_2, and α_3 are coterminal.

Examples Find the least positive angle that is coterminal with each of the following.

a. 6.59^R b. $520°$

Solutions

a. Sketch the angle, showing its initial and terminal sides. Since the angle generated in one revolution is $2\pi^R \approx 6.28^R$, the least positive angle having the same sides is approximately $(6.59 - 6.28)^R = 0.31^R$.

b. Sketch the angle. In this case, the least positive angle having the same initial and terminal sides is $(520 - 360)° = 160°$.

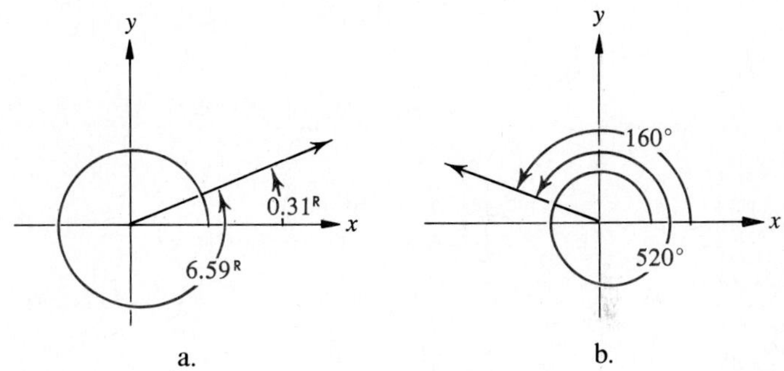

Sketches can also be helpful in determining the measures of angles coterminal with angles having negative measures.

Examples Find the least positive angle that is coterminal with each of the following.

a. $-\dfrac{3\pi}{4}^R$ b. $-210°$

Solutions

a. Sketch the angle, showing the initial and terminal sides. The least positive angle is $\left(2\pi - \dfrac{3\pi}{4}\right)^R = \dfrac{5\pi}{4}^R$.

(continued)

b. Sketch the angle. The least positive angle having the same initial and terminal sides is $(360 - 210)^\circ = 150^\circ$.

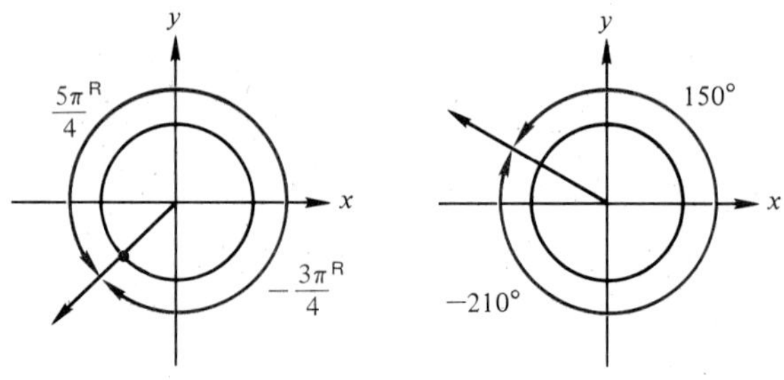

One method of specifying the measures of the infinite number of angles coterminal with a given angle is shown in the following example.

Example Write an expression for all angles coterminal with 45°.

Solution Since one revolution from a terminal side of any angle brings us back to that terminal side, then all angles coterminal with an angle of 45°, or $(\pi/4)^{\mathrm{R}}$, are given by

$$(45 + k \cdot 360)^\circ,\ k \in J \qquad \text{or} \qquad \left(\frac{\pi}{4} + k \cdot 2\pi\right)^{\mathrm{R}},\ k \in J.$$

Note that

$$\text{for } k = 0, \quad \alpha_1 = (45 + 0)^\circ = 45^\circ \quad \text{or} \quad \left(\frac{\pi}{4} + 0\right)^{\mathrm{R}} = \frac{\pi}{4}^{\mathrm{R}};$$

$$\text{for } k = 1, \quad \alpha_2 = (45 + 360)^\circ = 405^\circ \quad \text{or} \quad \left(\frac{\pi}{4} + 2\pi\right)^{\mathrm{R}} = \frac{9\pi}{4}^{\mathrm{R}};$$

$$\text{for } k = -1, \quad \alpha_3 = (45 - 360)^\circ = -315^\circ \quad \text{or} \quad \left(\frac{\pi}{4} - 2\pi\right)^{\mathrm{R}} = -\frac{7\pi}{4}^{\mathrm{R}}.$$

EXERCISE SET 3.1

A

1. State the conversion formulas relating the degree measure and the radian measure of an angle α.
2. Complete the degree–radian conversion wheels for special angles without referring to Figures 3.4 and 3.5. (These wheels are a handy reference.)

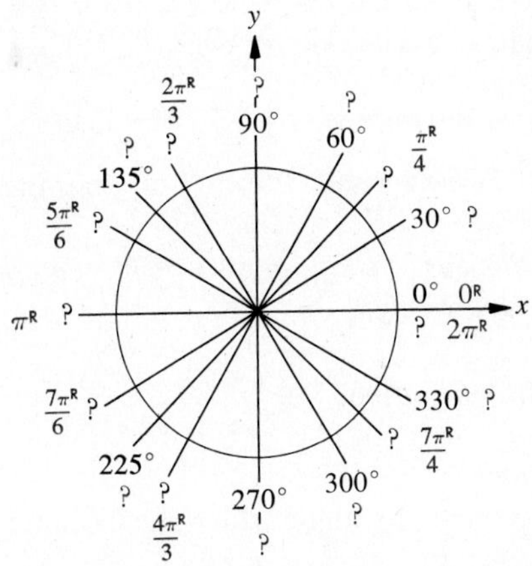

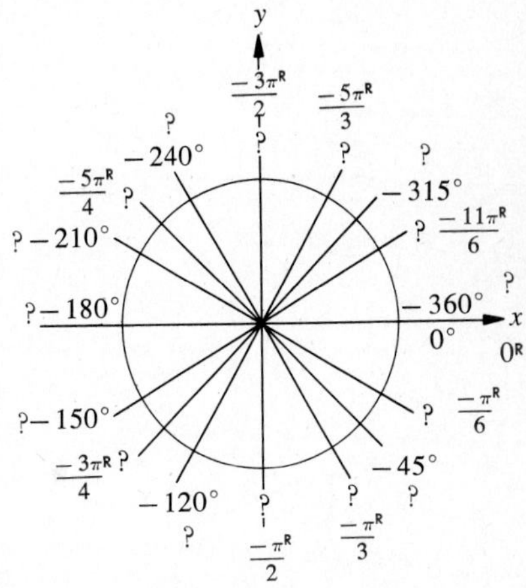

Find the degree measure to the nearest tenth of a degree of an angle α with the given radian measure.

Examples a. $\dfrac{6\pi}{5}^{\mathsf{R}}$ b. -1.24^{R}

Solutions

a. $m^\circ(\alpha) = \left(\dfrac{180}{\pi} \cdot \dfrac{6\pi}{5}\right)^\circ$

$= 216.0^\circ$

b. $m^\circ(\alpha) \approx \left[\left(\dfrac{180}{\pi}\right)(-1.24)\right]^\circ$

$\approx -71.0^\circ$

3. $\dfrac{\pi}{3}^{\mathsf{R}}$ **4.** $\dfrac{2\pi}{3}^{\mathsf{R}}$ **5.** $\dfrac{4\pi}{3}^{\mathsf{R}}$ **6.** $\dfrac{5\pi}{4}^{\mathsf{R}}$

7. $\dfrac{8\pi}{5}^{\mathsf{R}}$ **8.** $\dfrac{11\pi}{8}^{\mathsf{R}}$ **9.** $\dfrac{-13\pi}{8}^{\mathsf{R}}$ **10.** $\dfrac{-\pi}{5}^{\mathsf{R}}$

11. 0.18^{R} **12.** 0.94^{R} **13.** -1.54^{R} **14.** -4.55^{R}

Find the radian measure to the nearest hundredth of a radian of an angle α with the given degree measure.

Examples a. 118° b. -140°

Solutions

a. $m^{\mathsf{R}}(\alpha) = \left(\dfrac{\pi}{180} \cdot 118\right)^{\mathsf{R}}$

$\approx 2.06^{\mathsf{R}}$

b. $m^{\mathsf{R}}(\alpha) = \left[\dfrac{\pi}{180} \cdot (-140)\right]^{\mathsf{R}}$

$\approx -2.44^{\mathsf{R}}$

15. 30° **16.** 45° **17.** 75° **18.** 125°

19. 43° **20.** 137° **21.** -202° **22.** -342°

23. $1'$ [One minute $(1') = \frac{1}{60}$ of a degree.]

24. $1''$ [One second $(1'') = \frac{1}{60}$ of a minute.]

Find the least positive angle which is coterminal with each of the following. Sketch each angle.

Examples a. 6.65^R b. -242°

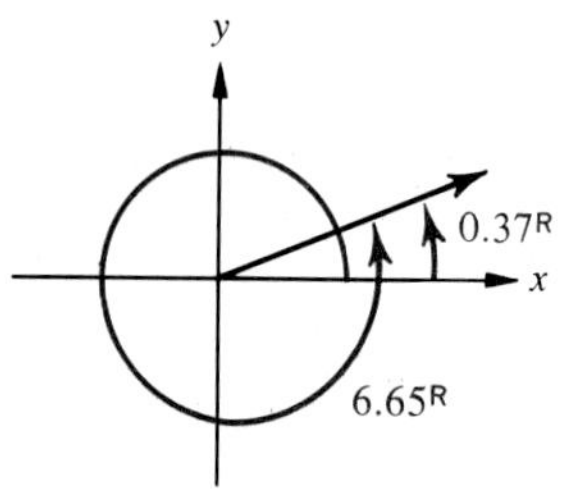

a.

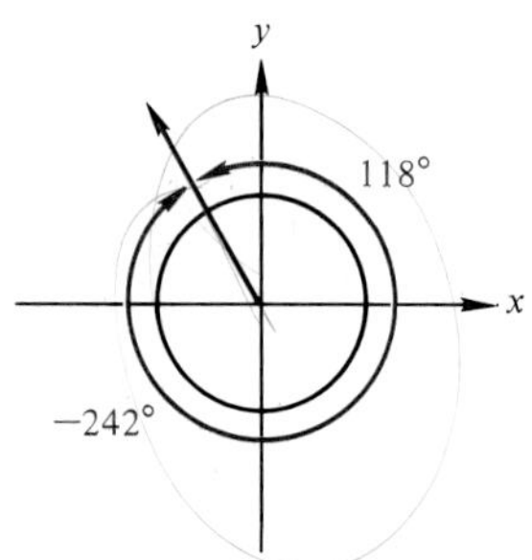

b.

Solutions

a. Sketch the angle, showing its initial and terminal sides. The angle generated in one revolution is $2\pi^R \approx 6.28^R$. Hence the angle of least positive measure having the same initial and terminal sides is approximately equal to

$$(6.65 - 6.28)^R = 0.37^R.$$

b. Sketch the angle. The angle of least positive measure having the same initial and terminal sides is

$$(360 - 242)^\circ = 118^\circ.$$

25. 6.72^R **26.** 8.91^R **27.** 379° **28.** 423°

29. 11.11^R **30.** 13.48^R **31.** -792° **32.** -863°

Write a general expression for all angles coterminal with an angle having the given measure. Give each expression in terms of the least positive measure.

Examples a. 60° b. $-\dfrac{2\pi}{3}^R$

Solutions

a. One revolution from a terminal side of any angle brings us back to that terminal side. Therefore all angles coterminal with an angle 60° are given by

$$(60 + k \cdot 360)^\circ, \quad k \in J$$

b. All angles coterminal with an angle $(-2\pi/3)^R$ are given by

$$\left(-\frac{2\pi}{3} + k \cdot 2\pi\right)^R, \quad k \in J$$

33. $21°$ **34.** $128°$ **35.** $-\frac{5\pi}{6}^{R}$

36. $\frac{5\pi}{3}^{R}$ **37.** $242°$ **38.** $-310°$

B

If a wheel is rotating at the given number of revolutions per minute (rpm), find the exact radian measure and degree measure of the angle through which a spoke rotates in the given time.

39. 3 rpm, 1 min. **40.** 4 rpm, 2 min.

41. 12 rpm, 30 sec. **42.** 60 rpm, 2 sec.

43. 6000 rpm, 0.1 sec. **44.** 12,000 rpm, 0.01 sec.

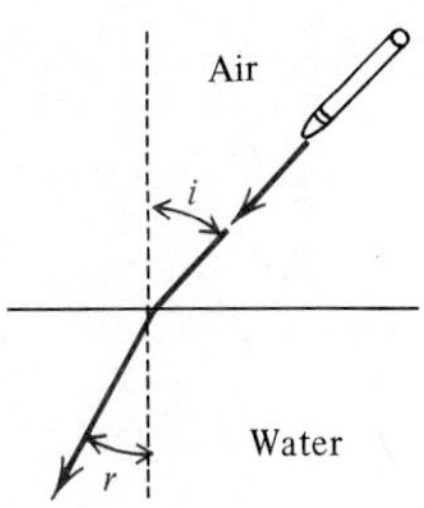

In Exercises 45–48 use the fact that the path of a light ray changes direction as it passes from one medium to another in accordance with the following law.

$$n_i \sin i = n_r \sin r,$$

where i and r are angles with the vertical of the incident (initial) and refracted ray, and n_i and n_r are the respective constants for specific media (see figure). In particular, $n_i = 1.00$ for air and $n_r = 1.33$ for water. Express answers to the nearest tenth.

45. Find the angle of refraction if the angle of incidence is $28.1°$.

46. Find the angle of incidence if the angle of refraction is $14.7°$.

47. Find the angle of incidence if the angle of refraction is $42°\ 24'$.

48. Find the angle of refraction if the angle of incidence is $36°\ 18'$.

3.2 Uniform Circular Motion

In this section we will consider a point moving along a circle at a constant (uniform) speed. First, however, we discuss a relationship between the measure of a central angle of a circle, the length of the intercepted arc, and the length of the radius.

Arc Length

The measure of an angle in radians, $m^R(\alpha)$, indicates how many times the length r of the radius of the circle is used as a unit length along the

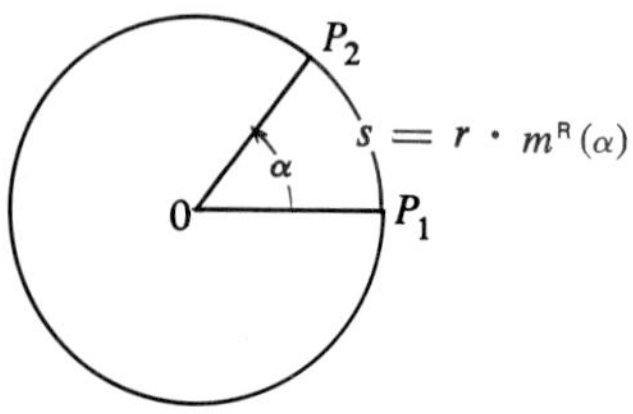

Figure 3.8

intercepted arc of length s (see Figure 3.8). We have the following theorem as a direct consequence.

Theorem 3.2 *If r is the length of the radius of a circle, then the length s of the arc intercepted by a central angle α is*

$$s = r \cdot m^R(\alpha). \tag{1}$$

Notice that the length s is a directed length; it is a positive or a negative real number as $m^R(\alpha)$ is a positive or a negative real number. Also notice that the central angle α in (1) is measured in radians. Thus, the measure of an angle that is given in degrees must first be changed to radians before Equation (1) can be used.

Examples On each circle with the given radius, find the length of the arc intercepted by the given angle.

a. $r = 10$ centimeters; $\alpha = 2^R$ b. $r = 2$ meters; $\alpha = -150°$

Solutions

a. From Theorem 3.2,

$$s = r \cdot m^R(\alpha) = 10 \cdot 2 = 20.$$

Therefore the arc is 20 centimeters in length.

b. From Theorem 3.1, the radian measure of α is given by

$$m^R(\alpha) = \left[\frac{\pi}{180}(-150)\right]^R = -\frac{5\pi}{6}^R.$$

From Theorem 3.2,

$$s = 2\left(-\frac{5\pi}{6}\right) = -\frac{5}{3}\pi \approx -5.24.$$

Therefore the arc length (directed) is approximately -5.24 meters.

Linear and Angular Velocity

Now consider a point moving along a circle at a constant speed. If the point moves from P_1 to P_2 in time t, the distance traveled equals the length of the arc shown in Figure 3.8.

Definition 3.2 *The* **linear velocity** *v of a point moving along an arc of a* *circle at a constant speed is*

$$v = \frac{\textit{length of the arc}}{\textit{time}} = \frac{r \cdot m^{R}(\alpha)}{t}. \tag{2}$$

Definition 3.3 *The* **angular velocity** ω (omega) *of a point moving along an arc of a circle at a constant speed is*

$$\omega = \frac{\textit{measure of the angle of rotation}}{\textit{time}}.$$

If the measure of the angle is in radians,

$$\omega = \frac{m^{R}(\alpha)}{t}, \tag{3}$$

where ω *is in radians per unit time.*

Notice that the linear velocity and angular velocity are both positive if the motion is in a counterclockwise direction because the length of the arc and the measure of the angle of rotation are both positive. The linear velocity and angular velocity are both negative if the motion is in a clockwise direction.

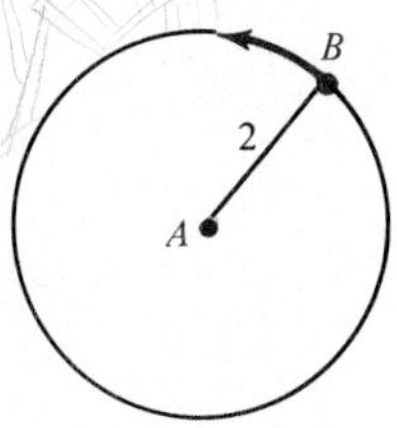

Example In the figure, point B revolves in a circular path about A making 10 complete revolutions in 4 seconds. Find the linear and the angular velocities of point B if the radius is 2 centimeters.

Solution One revolution corresponds to an angle of rotation of 2π radians. From Definition 3.2, with $r = 2$,

$$v = \frac{2 \cdot 10(2\pi)}{4}$$
$$= 10\pi \approx 31.4.$$

Hence the linear velocity is approximately 31.4 centimeters per second. From Definition 3.3,

$$\omega = \frac{10 \cdot 2\pi}{4}$$
$$= 5\pi \approx 15.7$$

Therefore, the angular velocity is approximately 15.7 radians per second.

Substituting ω, the left-hand member of Equation (3), for $m^R(\alpha)/t$ in the right-hand member of Equation (2) leads to the following theorem.

Theorem 3.3 *The linear velocity v of a point moving along the arc of a circle at a constant speed is*

$$\mathbf{v = r\omega,} \tag{4}$$

where r is the length of the radius of the circle and ω is the angular velocity of the moving point in radians per unit time.

Example A turbine with a radius of length 0.74 meter turns at the rate of 10,000 revolutions per minute. What is the linear velocity of a point on the rim of the turbine?

Solution One revolution corresponds to 2π radians. Therefore, 10,000 revolutions per minute corresponds to an angular velocity of $20{,}000\pi$ radians per minute. From Theorem 3.3, with $r = 0.74$, we have

$$\begin{aligned} v = r\omega &= 0.74 \cdot 20{,}000\pi \\ &= 46{,}496. \end{aligned}$$

Hence, the linear velocity is approximately 46,496 meters per minute.

EXERCISE SET 3.2

A

On each circle with the given radius, find the length of the arc (to the nearest tenth of a unit) intercepted by each angle with given measure.

Example $r = 9$ centimeters; $240°$

Solution From Theorem 3.1, the related radian measure of $240°$ is

$$m^R(\alpha) = \left(\frac{\pi}{180} \cdot 240\right)^R = \frac{4\pi}{3}^R.$$

Thus,

$$\begin{aligned} s = r \cdot m^R(\alpha) &= 9 \cdot \frac{4\pi}{3} \\ &= 12\pi \approx 37.7, \end{aligned}$$

and the length of the intercepted arc is approximately 37.7 centimeters.

1. $r = 2$ meters; 3.1^{R}

2. $r = 30$ centimeters; 1.5^{R}

3. $r = \pi$ decimeters; $150°$

4. $r = 2\pi$ meters; $210°$

5. $r = 3$ feet; $-135°$

6. $r = 20$ inches; $-240°$

Find the measure (to the nearest tenth of a degree) of the central angle that intercepts each arc on a circle with given radius.

Example $s = 15$ centimeters; $r = 5$ centimeters

Solution Substituting 5 for r and 15 for s in $s = r \cdot m^{R}(\alpha)$ yields

$$15 = 5 \cdot m^{R}(\alpha).$$

Thus

$$m^{R}(\alpha) = \frac{15^{R}}{5} = 3^{R}.$$

From Theorem 3.1,

$$m°(\alpha) = \left(\frac{180}{\pi} \cdot 3\right)° \approx 171.9°.$$

7. $s = 31$ centimeters; $r = 14$ centimeters

8. $s = 2.5$ meters; $r = 2.5$ meters

9. $s = 4.9$ centimeters; $r = 1$ centimeter

10. $s = 14.20$ inches; $r = 1.27$ inches

In Exercises 11–18, consider a flywheel with uniform circular motion, where v is the linear velocity of a point on the rim and ω is the angular velocity. Find v, r, or ω to the nearest tenth of a unit.

Example Find ω in degrees per hour if $v = 0.5$ meter per minute and $r = 0.63$ meter.

Solution 0.5 meter per minute is 60(0.5) or 30 meters per hour. From Theorem 3.3, we have that

$$30 = 0.63\omega,$$

from which ω equals approximately 47.62 radians per hour. Therefore, in

(continued)

degrees per hour,

$$\omega = 47.62\left(\frac{180}{\pi}\right) \approx 2728.4.$$

11. Find v if $r = 0.3$ meter and $\omega = 12$ radians per second.

12. Find r if $v = 10$ centimeters per second and $\omega = 10$ radians per second.

13. Find ω in radians per minute if $v = 3$ meters per second and $r = 6$ centimeters.

14. Find ω in degrees per minute if $v = 4$ meters per second and $r = 0.2$ meter.

15. Find v if $r = 9$ decimeters and $\omega = -3$ revolutions per second.

16. Find v if $r = 4$ decimeters and $\omega = -120$ degrees per second.

17. Find ω in degrees per second if $r = 2.5$ feet and $v = -10$ feet per minute.

18. Find r in inches if $\omega = 1/12$ revolution per second and $v = 8.2$ feet per minute.

19. The center line of a freeway curve is laid out as an arc of a circle of radius 300 meters. What is the length of the center line of the curve if it subtends an angle of 20° at the center of the circle?

20. A pendulum 24 decimeters long oscillates 4.2° on each side of its vertical position. Find the length of arc through which the end of the pendulum swings.

21. Two points on the earth's surface (assuming the earth to be a perfect sphere) are located so that the radii from the center of the earth to each point form an angle of 25°. Assuming the earth's radius to be 4000 miles, find the distance between the points measured along the earth's surface.

22. A wheel 10 inches in diameter is turning at a rate of 200 revolutions per minute. If it is rolling on the ground without slipping, how far in *feet* will it travel in 5 minutes?

23. A belt moving at a rate of 100 centimeters per minute is turning a pulley at an angular velocity of 8 revolutions per minute. Find the pulley diameter.

24. The rear gear of a bicycle has a radius of 10 centimeters and the front gear has a radius of 25 centimeters. Through what angle (in radians) does the front gear turn if the back gear turns through 5 radians?

25. A propeller, 2 meters from tip to tip, is rotating at a rate of 800 revolutions per minute. Find the linear velocity of a tip of the propeller in meters per minute.

26. Find the linear velocity, in centimeters per minute, of the tip of an hour hand of a clock, if the hand has a length of 1.5 decimeters.

27. A racing driver moves around a curved track at a speed of 160 miles per hour. If the radius of the circle, of which the driver's path is an arc, is 1500 feet, what is the driver's angular speed in radians per minute around the curve?

28. Using 8000 miles for the diameter of the earth, find the linear velocity, in miles per hour, of a point on the equator.

29. Assuming that the earth's orbit about the sun is circular, what is the angular velocity of the earth in radians per hour (assuming 365 days per year and 24 hours per day)?

30. Using the information in Exercise 29, what is the linear velocity in miles per hour of the earth in its orbit around the sun if the circular path of the earth has a radius of 9.3×10^7 miles?

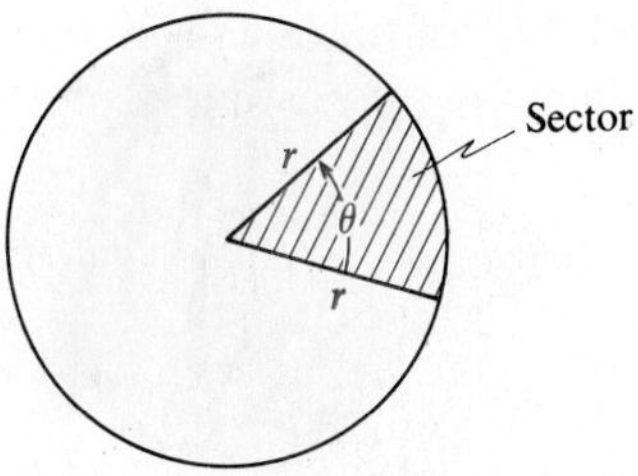

Ex. 33

31. Assuming that the moon's orbit about the earth is circular with a radius of 2.4×10^5 miles and that it takes 28 days for one orbit, what is the linear velocity of the moon in miles per hour?

32. The planet Mars moves around the sun in 686 earth-days. If the radius of Mars' orbit (assumed to be circular) around the sun is approximately 227,000,000 kilometers, find the linear velocity of Mars in its orbit in kilometers per hour. Assume 1 earth-day equals 24 hours.

B

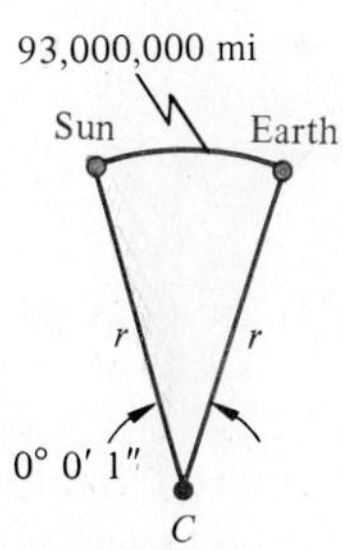

Ex. 35

33. Show that the area $\mathscr{A}$ of the sector of a circle (see figure) with central angle θ is given by $\mathscr{A} = \frac{1}{2}r^2 \theta^R$.

34. Use the formula in Exercise 33 to find the area of a sector of a circle whose radius has a length of 1 meter, where the sector has a central angle with measure of 148°.

35. A unit of distance used in astronomy is the **parsec.** It is the distance r shown in the drawing. Assuming point C is the center of a circle through the sun and the earth on which the indicated arc is 93,000,000 miles long, find r, or 1 parsec, in miles. Give the answer to three significant figures.

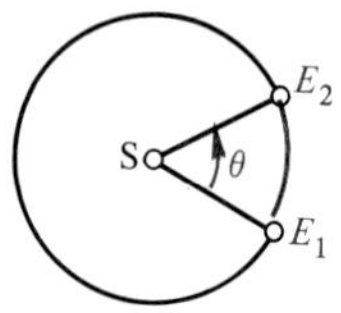

Ex. 36

36. Assuming that the earth travels in a circular orbit about the sun as in the figure, what would be the measure of θ during one week's travel? If the earth is 93,000,000 miles from the sun, how far does it travel in its orbit in one week?

37. If a bicycle wheel has a diameter of 27 inches, find the radian measure of the angle through which the wheel turns as the bicycle travels a distance of 1 mile.

38. If a pulley has a diameter of 20 centimeters, find the radian measure of the angle through which it turns as 4 meters of pulley belt is pulled over it. Assume the belt does not slip on the pulley.

3.3 Trigonometric Ratios and Functions

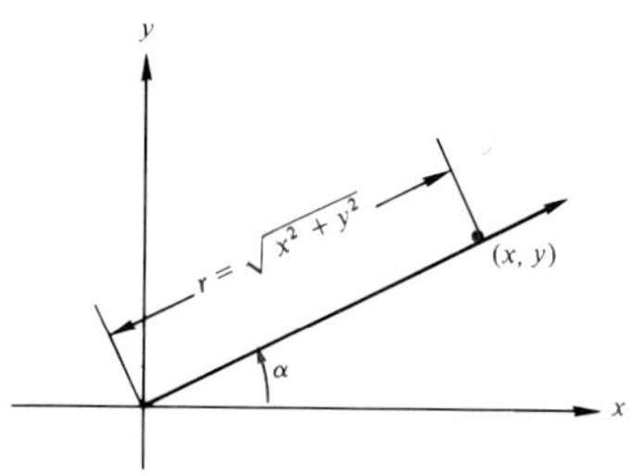

Figure 3.9

Consider the three real numbers x, y and $r = \sqrt{x^2 + y^2}$ that can be associated with any point on the terminal side of an angle in standard position, as shown in Figure 3.9. The components of the ordered pair (x, y) are the coordinates of the point and r is the length of the line segment from the origin to the point. For each angle α, there are six possible ratios that may be formed using these three numbers.

Trigonometric Ratios

The names given to the ratios formed by the three numbers x, y, and r are the same as the ones that are used to name elements in the ranges of the circular functions because, as we shall see, they are closely related. These ratios are defined as follows.

Definition 3.4 *For all angles α in standard position, if x and y (not both 0) are the coordinates of any point in the terminal side of α and $r = \sqrt{x^2 + y^2}$, then*

$$\textbf{cosine } \alpha = \frac{x}{r}, \qquad \textbf{secant } \alpha = \frac{r}{x} \quad (x \neq 0),$$

$$\textbf{sine } \alpha = \frac{y}{r}, \qquad \textbf{cosecant } \alpha = \frac{r}{y} \quad (y \neq 0),$$

$$\textbf{tangent } \alpha = \frac{y}{x} \quad (x \neq 0), \qquad \textbf{cotangent } \alpha = \frac{x}{y} \quad (y \neq 0).$$

These ratios are sometimes called **trigonometric ratios.** Cosine, sine, etc., are abbreviated in the same way as they were for the circular functions. Note that in each ratio where r occurs, only the positive square root of $x^2 + y^2$ is used.

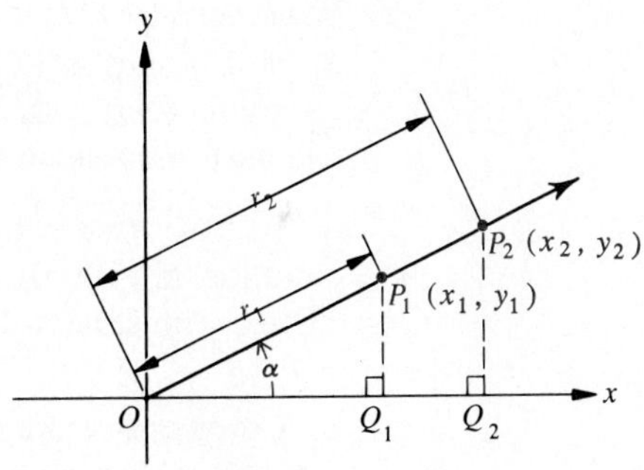

Figure 3.10

In Figure 3.10, P_1 and P_2 are two points on the terminal side of angle α. Hence, triangles OQ_1P_1 and OQ_2P_2 are similar. From this it follows that

$$\frac{x_1}{r_1} = \frac{x_2}{r_2}, \quad \frac{y_1}{r_1} = \frac{y_2}{r_2}, \quad \frac{y_1}{x_1} = \frac{y_2}{x_2},$$

$$\frac{r_1}{x_1} = \frac{r_2}{x_2}, \quad \frac{r_1}{y_1} = \frac{r_2}{y_2}, \quad \text{and} \quad \frac{x_1}{y_1} = \frac{x_2}{y_2}.$$

Thus:

The six trigonometric ratios of an angle in standard position are determined only by the position of the terminal side of the angle.

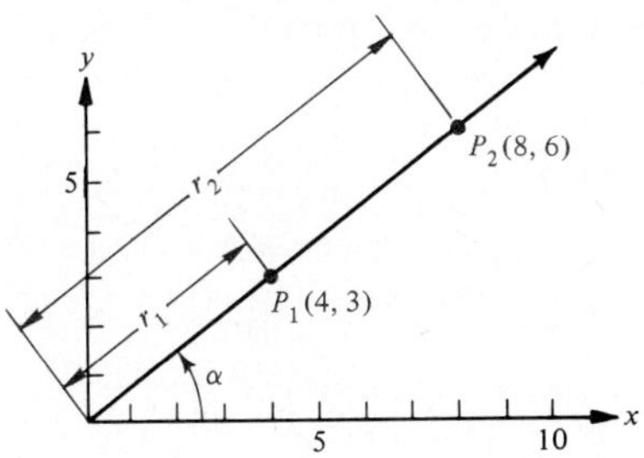

Example Consider P_1 (4, 3) and P_2 (8, 6), both points being in the terminal side of angle α. Using P_1 (4, 3), where $r_1 = \sqrt{4^2 + 3^2} = 5$, we have

$$\cos \alpha = \frac{4}{5}.$$

Using P_2 (8, 6), where $r_2 = \sqrt{8^2 + 6^2} = 10$,

$$\cos \alpha = \frac{8}{10} = \frac{4}{5}.$$

Because r, equal to $\sqrt{x^2 + y^2}$, is nonnegative for all (x, y), whether a trigonometric ratio is less than zero, equal to zero, or greater than zero depends on the values of x and y, which in turn depend on the quadrant in which the angle lies. Therefore, Table 1.4 (page 24), which pertains to circular function values, is also applicable to trigonometric function values.

Examples Determine the quadrant in which α lies.

a. $\cos \alpha > 0$ and $\cot \alpha < 0$. b. $\cos \alpha < 0$ and $\tan \alpha > 0$.

(continued)

Solutions

a. For $\cos\alpha > 0$, the terminal side of α lies in either Quadrant I or IV and for $\cot\alpha < 0$, the terminal side of α lies in either Quadrant II or IV. Thus, α must lie in Quadrant IV, the intersection of $\{I, IV\}$ and $(II, IV\}$.

b. For $\cos\alpha < 0$, the terminal side of α lies in either Quadrant II or III and for $\tan\alpha > 0$, the terminal side of α lies in either Quadrant I or III. Thus, α must lie in Quadrant III, the intersection of $\{II, III\}$ and $\{I, III\}$.

Given one of the six trigonometric ratios of an angle α and the quadrant in which α lies, we can find the other five ratios.

Example If $\sec\alpha = -5/3$ and α is in Quadrant II, find the other five ratios.

Solution By Definition 3.4, $\sec\alpha = r/x$. Because $\sec\alpha = -5/3$ and r is always positive, $x = -3$ and $r = 5$. Then using $r = \sqrt{x^2 + y^2}$, we have

$$5 = \sqrt{(-3)^2 + y^2},$$
$$25 = 9 + y^2,$$

from which

$$y = \pm 4.$$

Since α is in Quadrant II, $y > 0$ and we take $y = 4$; then

$$\cos\alpha = \frac{y}{r} = \frac{-3}{5}, \qquad \left(\sec\alpha = \frac{-5}{3}\right)$$

$$\sin\alpha = \frac{y}{r} = \frac{4}{5}, \qquad \csc\alpha = \frac{r}{y} = \frac{5}{4},$$

$$\tan\alpha = \frac{y}{x} = \frac{-4}{3}, \qquad \cot\alpha = \frac{x}{y} = \frac{-3}{4}.$$

Trigonometric Functions

Because the six trigonometric ratios of an angle in standard position are determined only by the position of its terminal side, each angle α in standard position can be paired with a unique real number for each ratio. Hence, the equations in Definition 3.4 determine six functions. Each of these **trigonometric functions** is named by the name of the trigonometric ratio in its range. The domains of these functions are determined by their definitions. For $A = \{\text{all angles}\}$, the domains are as follows:

cosine and sine: $\{\alpha \mid \alpha \in A\}$;

tangent and secant: $\left\{\alpha \mid \alpha \in A, \quad \alpha \neq \left(\frac{\pi}{2} + k\pi\right)^{\mathsf{R}}, \quad k \in J\right\}$;

cotangent and cosecant: $\{\alpha \mid \alpha \in A, \quad \alpha \neq k\pi^{\mathsf{R}}, \ k \in J\}$.

EXERCISE SET 3.3

A

Find the value of each of the six trigonometric functions of α, if the terminal side of α (in standard position) contains the given point. Sketch each angle.

Example $(-5, 12)$

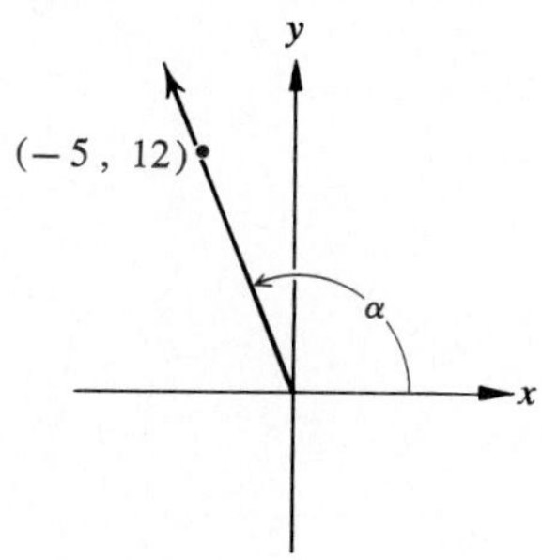

Solution $r = \sqrt{x^2 + y^2} = \sqrt{(-5)^2 + 12^2} = 13.$ Hence, by Definition 3.4,

$$\cos\alpha = \frac{x}{r} = \frac{-5}{13}, \qquad \sec\alpha = \frac{r}{x} = \frac{13}{-5},$$

$$\sin\alpha = \frac{y}{r} = \frac{12}{13}, \qquad \csc\alpha = \frac{r}{y} = \frac{13}{12},$$

$$\tan\alpha = \frac{y}{x} = \frac{12}{-5}, \qquad \cot\alpha = \frac{x}{y} = \frac{-5}{12}.$$

1. $(3, 4)$ **2.** $(-15, 8)$ **3.** $(-12, -5)$

4. $(1, -1)$ **5.** $(2, 3)$ **6.** $(-3, 5)$

7. $(-\sqrt{3}, -1)$ **8.** $(1, -\sqrt{3})$ **9.** $(3\sqrt{3}, 3)$

10. $(-5, -5)$ **11.** $(3, 0)$ **12.** $(0, -4)$

Determine the quadrant in which each angle lies.

Examples

a. $\sin\alpha < 0$ and $\tan\alpha > 0$ b. $\sin\alpha > 0$ and $\cos\alpha < 0$

Solutions

a. For $\sin\alpha < 0$, the terminal side of α lies in either Quadrant III or IV and for $\tan\alpha > 0$, the terminal side of α lies in Quadrant I or III. Thus, α must lie in Quadrant III, the intersection of {III, IV} and {I, III}.

(continued)

b. For $\sin\alpha > 0$, the terminal side of α lies in either Quadrant I or II, and for $\cos\alpha < 0$, the terminal side of α lies in Quadrant II or III. Thus, α must lie in Quadrant II, the intersection of $\{I, II\}$ and $\{II, III\}$.

13. $\sin\alpha < 0$ and $\cos\alpha < 0$

14. $\sin\alpha < 0$ and $\tan\alpha < 0$

15. $\sin\alpha > 0$ and $\sec\alpha > 0$

16. $\sin\alpha > 0$ and $\cot\alpha < 0$

17. $\cos\alpha < 0$ and $\tan\alpha > 0$

18. $\cos\alpha < 0$ and $\csc\alpha > 0$

19. $\cos\alpha > 0$ and $\cot\alpha > 0$

20. $\tan\alpha < 0$ and $\csc\alpha < 0$

Find the other five trigonometric ratios for each angle α.

Example $\tan\alpha = -2$; α in Quadrant IV.

Solution By Definition 3.4, $\tan\alpha = y/x$. Because $\tan x = -2 = -2/1$, and α is in Quadrant IV, we can choose $y = -2$ and $x = 1$. Then using $r = \sqrt{x^2 + y^2}$,

$$r = \sqrt{(1)^2 + (-2)^2} = \sqrt{5}.$$

Finally,

$$\cos\alpha = \frac{x}{r} = \frac{1}{\sqrt{5}}, \qquad \sec\alpha = \frac{r}{x} = \sqrt{5},$$

$$\sin\alpha = \frac{y}{r} = \frac{-2}{\sqrt{5}}, \qquad \csc\alpha = \frac{r}{y} = \frac{-\sqrt{5}}{2},$$

$$(\tan\alpha = -2), \qquad \cot\alpha = \frac{x}{y} = \frac{1}{-2}.$$

21. $\tan\alpha = -\frac{2}{3}$; α in Quad II

22. $\cos\alpha = \frac{2}{3}$; α in Quad IV.

23. $\sin\alpha = -\frac{3}{5}$; α in Quad III.

24. $\sec\alpha = -\frac{3}{2}$; α in Quad III.

25. $\csc\alpha = -4$; α in Quad IV.

26. $\cot\alpha = -4$; α in Quad II.

Specify the measures of angles for which the given function is (**a**) *not defined and* (**b**) *equal to* 0.

27. Tangent

28. Cotangent

29. Secant

30. Cosecant

B

Express each of the six trigonometric functions of α in terms of the given variable, if the terminal side of α contains the point with the given coordinates.

31. $(x, \sqrt{4 - x^2}),\ 0 < x < 2$

32. $(x, \sqrt{9 - x^2}),\ 0 < x < 3$

33. $(\sqrt{16 - y^2}, y),\ 0 < y < 4$

34. $(\sqrt{25 - y^2}, y),\ 0 < y < 5$

35. $(x, \sqrt{x^2 - 4}),\ x > 2$

36. $(x, \sqrt{x^2 - 9}),\ x > 3$

3.4 Function Values

The circular functions were defined in Chapter 1 by making use of a unit circle. Since the unit circle can also be used to assign measures to angles, we can relate the circular and trigonometric functions.

Basic Relationship

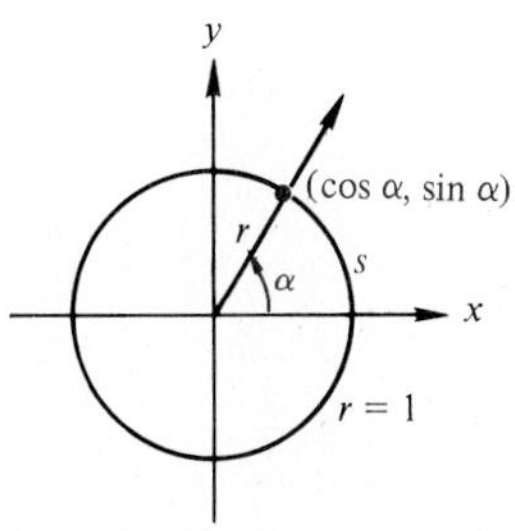

Figure 3.11

The trigonometric functions are defined in terms of the coordinates of any point, except (0, 0), on the terminal side of an angle in standard position. If a point is taken such that $r = 1$ (see Figure 3.11), then, since the arc length $s = r \cdot m^{\mathsf{R}}(\alpha)$, it follows that $s = m^{\mathsf{R}}(\alpha)$. That is, the length of the arc on a unit circle intercepted by angle α is equal to the measure of angle α in radians. Thus, if an element in the range of any one of the six trigonometric functions is denoted by $\text{Trig}(\alpha^{\mathsf{R}})$ or $\text{Trig}(\alpha^{\circ})$, and $\text{Circ}(s)$ is an element in the range of the analogous circular function, then

$$\text{Circ}(s) = \text{Trig}(\alpha^{\mathsf{R}}) = \text{Trig}(\alpha^{\circ}), \tag{1}$$

where s is the length of the arc intercepted by α on the unit circle.

This demonstrates the close relationship between the circular functions, whose domains are sets of real numbers, and the trigonometric functions, whose domains are sets of angles. The elements in the ranges of both circular and trigonometric functions are real numbers. For example,

$$\sin \frac{\pi^{\mathsf{R}}}{4} = \sin \frac{\pi}{4} \qquad \text{and} \qquad \sin 45^{\circ} = \sin \frac{\pi}{4},$$

where the left-hand member in each equation is an element in the range of the trigonometric function and the right-hand member is an element in the range of the corresponding circular function.

As a consequence of Equation (1), we can find trigonometric function values from the tables used for circular functions. We can use Table 1.3 or a calculator or Table II (for angles in radian measure) and Table III (for angles in degree measure) in Appendix C.

In general, we shall use the relationships in Table 1.3 to obtain exact function values when we work with the angles whose measures are 0°, 30°, 45°, 60°, 90°, 180°, and 270°, or integral multiples thereof. We shall use a calculator to obtain decimal approximations for function values when we work with other angles.

Since the relationships given in Table 1.3 are often employed when working with trigonometric functions, the entries in this table are shown again in Table 3.1 for convenient reference. In this table, however, the entries are shown in relation to elements (angles) in the domain of the trigonometric functions.

TABLE 3.1

$m^R(\alpha)$	$m^\circ(\alpha)$	$\cos\alpha$	$\sin\alpha$	$\tan\alpha$
0	0	1	0	0
$\frac{\pi}{6}$	30	$\frac{\sqrt{3}}{2}$	$\frac{1}{2}$	$\frac{1}{\sqrt{3}}$
$\frac{\pi}{4}$	45	$\frac{1}{\sqrt{2}}$	$\frac{1}{\sqrt{2}}$	1
$\frac{\pi}{3}$	60	$\frac{1}{2}$	$\frac{\sqrt{3}}{2}$	$\sqrt{3}$
$\frac{\pi}{2}$	90	0	1	undef.
π	180	-1	0	0
$\frac{3\pi}{2}$	270	0	-1	undef.

Examples

a. From Table 3.1,

$$\sin\frac{\pi^R}{3} = \frac{\sqrt{3}}{2}.$$

b. From Definition 1.3 and Table 3.1,

$$\cot\frac{\pi^R}{6} = \frac{1}{\tan(\pi/6)^R} = \frac{1}{1/\sqrt{3}} = \sqrt{3}.$$

Using Angles to Obtain Function Values

It is sometimes preferable to obtain a function value directly from a sketch of the angle rather than by using Table 3.1. Trigonometric function values for the quadrantal angles 0°, 90°, 180°, and 270° can be obtained directly by using the coordinates of a point on the unit circle, just as we found the corresponding values for the circular functions in Chapter 1. See Figure 3.12.

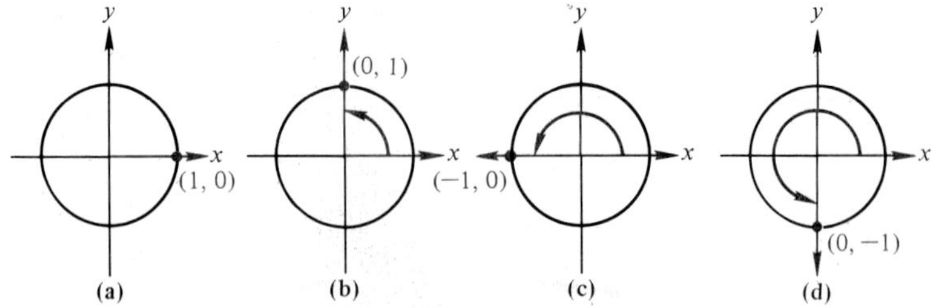

Figure 3.12

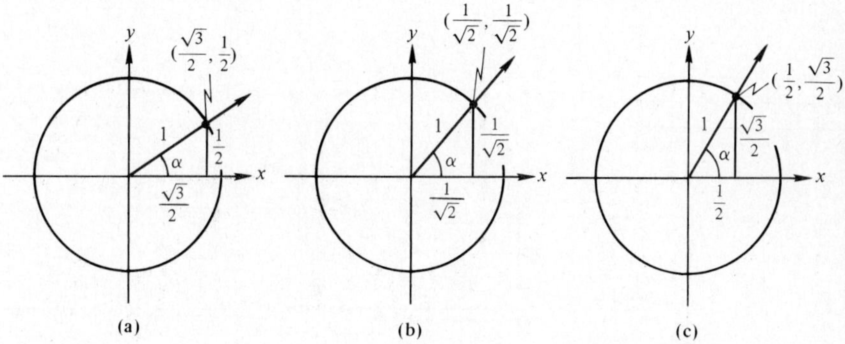

$$\alpha = \frac{\pi}{6}^R = 30^\circ \qquad \alpha = \frac{\pi}{4}^R = 45^\circ \qquad \alpha = \frac{\pi}{3}^R = 60^\circ$$

Figure 3.13

Trigonometric function values for 30°, 45°, and 60° can also be obtained by sketching angles in a unit circle as shown in Figure 3.13.

Alternatively, it is often convenient to select a point on the terminal side of the angle whose coordinates are not fractions. The right triangles ABC and $AB'C'$ in Figure 3.14 on page 116 are similar. Hence, the lengths of their sides are proportional (see Appendix A.9) and the coordinates at B with the corresponding value for r can now be used to obtain the trigonometric ratios of 30°, 45°, and 60°. Thus,

$$\sin 30^\circ = \frac{y}{r} = \frac{1}{2}, \qquad \csc 30^\circ = \frac{r}{y} = 2,$$

$$\cos 30^\circ = \frac{x}{r} = \frac{\sqrt{3}}{2}, \qquad \sec 30^\circ = \frac{r}{x} = \frac{2}{\sqrt{3}},$$

$$\tan 30^\circ = \frac{y}{x} = \frac{1}{\sqrt{3}}, \qquad \cot 30^\circ = \frac{x}{y} = \sqrt{3},$$

and

$$\sin 45^\circ = \frac{y}{r} = \frac{1}{\sqrt{2}}, \qquad \csc 45^\circ = \frac{r}{y} = \sqrt{2},$$

$$\cos 45^\circ = \frac{x}{r} = \frac{1}{\sqrt{2}}, \qquad \sec 45^\circ = \frac{r}{x} = \sqrt{2},$$

$$\tan 45^\circ = \frac{y}{x} = 1, \qquad \cot 45^\circ = \frac{x}{y} = 1,$$

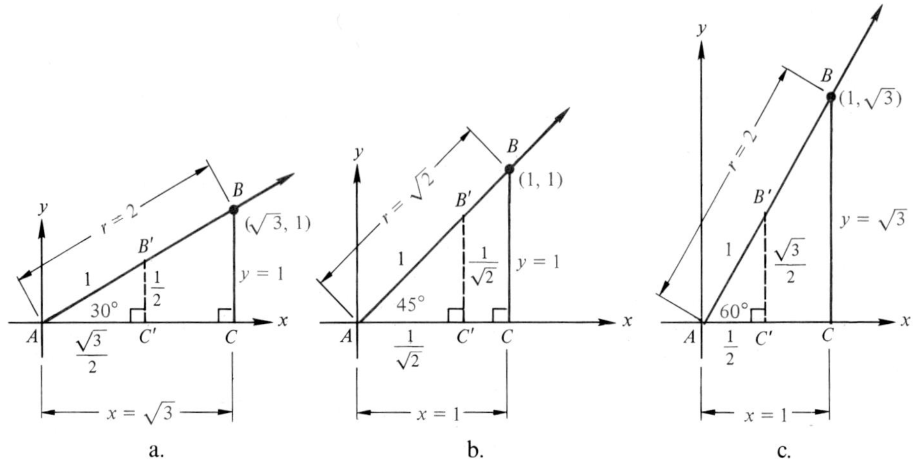

Figure 3.14

and

$$\sin 60^\circ = \frac{y}{r} = \frac{\sqrt{3}}{2}, \qquad \csc 60^\circ = \frac{r}{y} = \frac{2}{\sqrt{3}},$$

$$\cos 60^\circ = \frac{x}{r} = \frac{1}{2}, \qquad \sec 60^\circ = \frac{r}{x} = 2,$$

$$\tan 60^\circ = \frac{y}{x} = \sqrt{3}, \qquad \cot 60^\circ = \frac{x}{y} = \frac{1}{\sqrt{3}}.$$

Use of Reference Angles

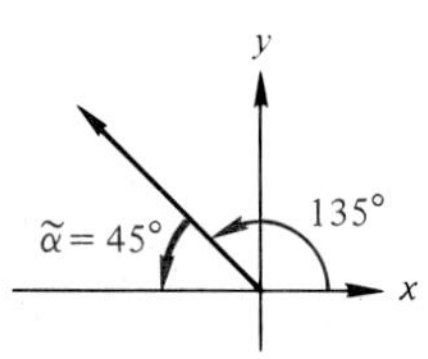

a.

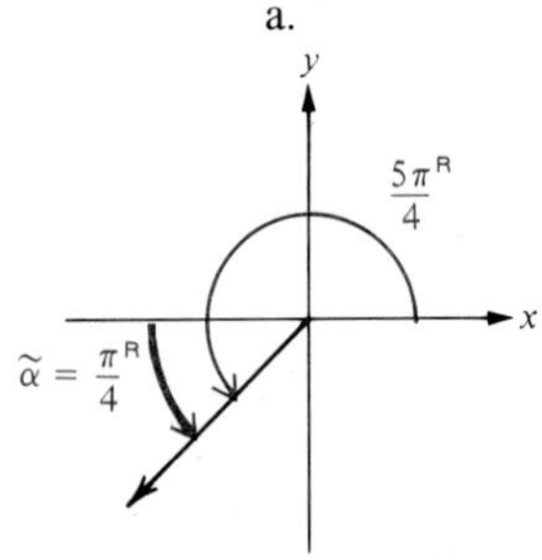

b.

A reference angle $\tilde{\alpha}$ can be used in the same way that a reference arc $\tilde{x}$ is used to find function values of angles with measure greater than 90° or less than 0°.

Examples a. $\cos 135^\circ = -\cos \tilde{\alpha} = -\cos 45^\circ$

From Table 3.1, we have that $\cos 45^\circ = 1/\sqrt{2}$. Hence,

$$\cos 135^\circ = -\cos 45^\circ = -\frac{1}{\sqrt{2}}.$$

b. $$\sin \frac{5\pi}{4}^{R} = -\sin \tilde{\alpha}$$

$$= -\sin \frac{\pi}{4}^{R} = -\frac{1}{\sqrt{2}}.$$

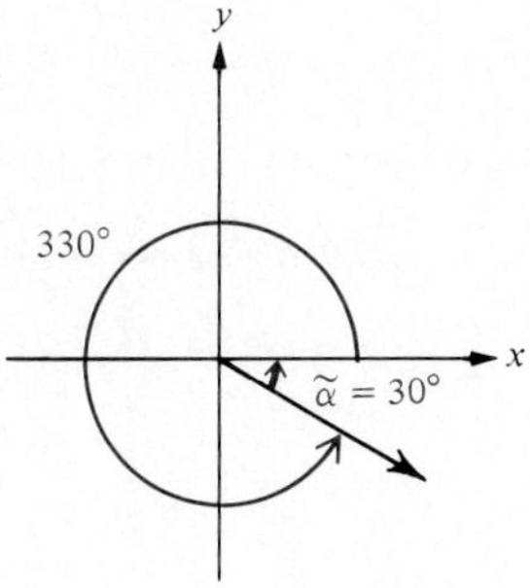

c.

c. $\tan 330° = -\tan \tilde{\alpha}$

$$= -\tan 30° = -\frac{1}{\sqrt{3}}$$

Using a Calculator

Approximations for trigonometric function values for angles, including angles with measures greater than 90° or less than 0°, can be obtained directly by using a calculator.

We will continue our practice of rounding off calculator readings of function values (circular or trigonometric functions) to four places. Measures of angles given in degrees and minutes should be changed to a decimal before using a calculator (See Appendix A.7).

Examples Using a calculator set for degrees, we have

a. $\cos 28° \, 12' = \cos 28.2° \approx 0.8813.$

b. $\sec 14° \, 18' = \dfrac{1}{\cos 14.3°} \approx 1.0320.$

Examples Using a calculator set for radians, we have

a. $\sin 1.32^{\mathrm{R}} \approx 0.9687$

b. $\cot 2.48^{\mathrm{R}} = \dfrac{1}{\tan 2.48^{\mathrm{R}}} \approx -1.2843.$

Periodic Properties

The tangent and cotangent trigonometric functions are periodic with period π^{R} or 180°. Hence, for these functions

$$\mathrm{Trig}(\alpha) = \mathrm{Trig}(\alpha + k\pi)^{\mathrm{R}} = \mathrm{Trig}(\alpha + k \cdot 180)°, \quad k \in J.$$

The trigonometric functions cosine, sine, secant, and cosecant are periodic with period $2\pi^{\mathrm{R}}$ or 360°. Hence, for these functions

$$\mathrm{Trig}(\alpha) = \mathrm{Trig}(\alpha + k \cdot 2\pi)^{\mathrm{R}} = \mathrm{Trig}(\alpha + k \cdot 360)°, \quad k \in J.$$

Thus, function values of angles with measures greater than 360° or less than −360° can also be obtained by using Table 3.1 or a calculator.

Examples A sketch is helpful.

a. $\cos \dfrac{7\pi^{R}}{3} = \cos\left(\dfrac{\pi}{3} + 2\pi\right)^{R}$

$= \cos \dfrac{\pi^{R}}{3} = \dfrac{1}{2}.$

b. $\tan 840^\circ = \tan(120 + 4 \cdot 180)^\circ$

$= \tan 120^\circ = -\tan 60^\circ$

$= -\sqrt{3}.$

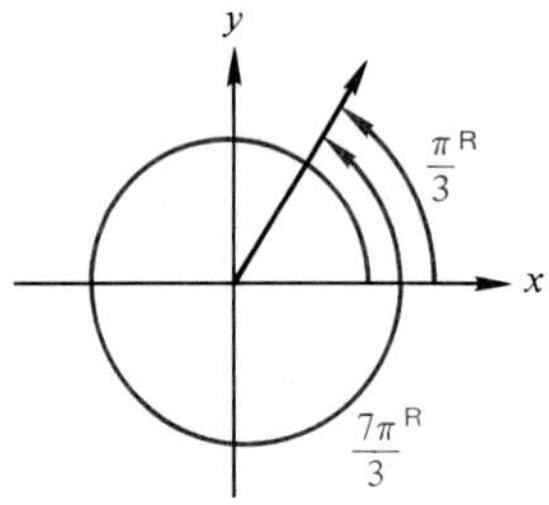

a.

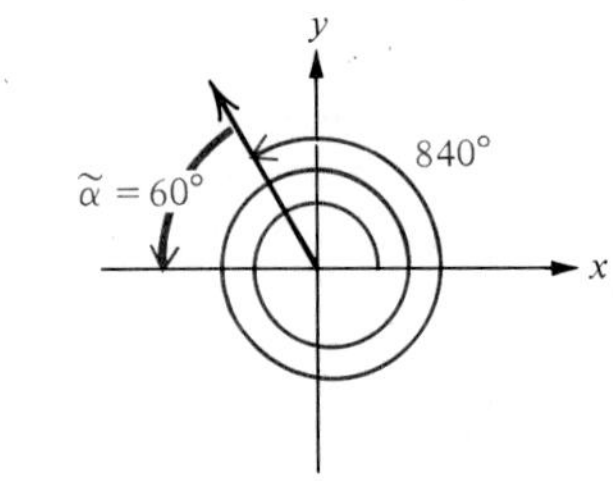

b.

Examples Using a calculator set for degrees, we have

a. $\tan 755^\circ \approx 0.7002$

b. $\cot(-385^\circ) = \dfrac{1}{\tan(-385^\circ)}$

$\approx -2.1445.$

Examples Using a calculator set for radians, we have

a. $\sin 7.49^{R} \approx 0.9345$

b. $\csc 11.38^{R} = \dfrac{1}{\sin 11.38^{R}}$

≈ -1.079

Inverse Trigonometric Functions

The inverse circular functions were defined in Section 2.6. The analogous inverse trigonometric functions are defined in a similar way. Thus an equation such as

$$\sin \alpha = \frac{1}{2}, \quad -90^\circ \le \alpha \le 90^\circ,$$

can be written using inverse notation as

$$\alpha = \text{Arcsin}\,\frac{1}{2} \quad \text{or} \quad \alpha = \text{Sin}^{-1}\,\frac{1}{2}.$$

You can interpret the right-hand member of each equation as another name (symbol) for the angle α.

Examples Find each value of α.

a. $\alpha = \text{Arccos}\,\dfrac{\sqrt{3}}{2}$ b. $\alpha = \text{Arctan}\,(-1.2002)$

Solutions

a. α is the angle such that $\cos\alpha = \sqrt{3}/2$ and $0° \le \alpha \le 180°$. From Table 3.1, $\alpha = 30°$.

b. α is the angle such that $\tan\alpha = -1.2002$ and $-90° < \alpha < 90°$. By using a calculator set for degrees,

$$\alpha \approx -50.2° = -50°\,12'.$$

EXERCISE SET 3.4

Find each of the following (exact value).

Examples a. $\sin 900°$ b. $\cot \dfrac{13\pi}{6}^{\mathsf{R}}$

Solutions Using the periodic property of the trigonometric functions and Table 3.1, we have

a. $\sin 900° = \sin(180 + 2 \cdot 360)° = \sin 180° = 0$

b. $\cot \dfrac{13\pi}{6}^{\mathsf{R}} = \cot\left(\dfrac{\pi}{6} + 2\pi\right)^{\mathsf{R}} = \cot \dfrac{\pi}{6}^{\mathsf{R}} = \dfrac{1}{\tan(\pi/6)^{\mathsf{R}}} = \sqrt{3}$

1. $\sin 390°$ **2.** $\cos 405°$ **3.** $\tan \dfrac{7\pi}{3}^{\mathsf{R}}$ **4.** $\sec \dfrac{5\pi}{2}^{\mathsf{R}}$

5. $\csc 780°$ **6.** $\tan 3\pi^{\mathsf{R}}$ **7.** $\cot 7\pi^{\mathsf{R}}$ **8.** $\cos \dfrac{15\pi}{2}^{\mathsf{R}}$

9. $\sec 900°$ **10.** $\tan 1110°$ **11.** $\csc 990°$ **12.** $\sin 1350°$

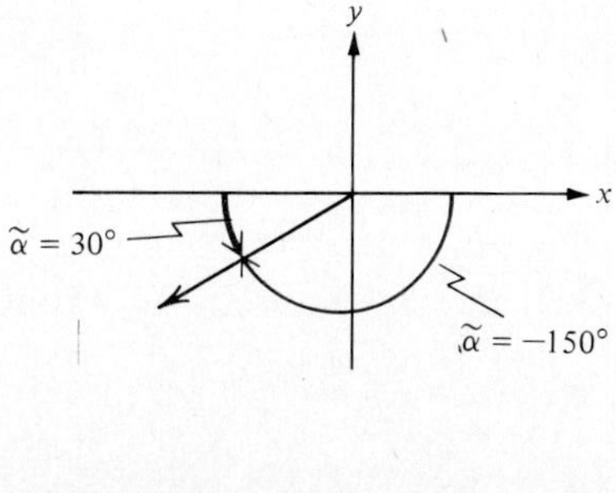

Example $\sin(-150°)$

Solution Using the reference angle $\tilde{\alpha}$ in the figure, we have that

$$\sin(-150°) = -\sin 30°.$$

(continued)

From Table 3.1,

$$\sin(-150^\circ) = -\sin 30^\circ = -\frac{1}{2}.$$

13. $\cos 120^\circ$ **14.** $\sin 135^\circ$ **15.** $\tan \frac{7\pi}{6}^{R}$

16. $\tan \frac{4\pi}{3}^{R}$ **17.** $\csc \frac{7\pi}{4}^{R}$ **18.** $\sec \frac{5\pi}{6}^{R}$

19. $\sin(-30^\circ)$ **20.** $\sec(-90^\circ)$ **21.** $\cos(-240^\circ)$

22. $\cos(-150^\circ)$ **23.** $\cos 480^\circ$ **24.** $\tan 855^\circ$

Find an approximation for each of the following.

Examples a. $\cos 2.35^{R}$ b. $\tan 624^\circ\ 36'$

Solutions

a. Using a calculator set for radians, we have

$$\cos 2.35^{R} \approx -0.7027.$$

b. Using a calculator set for degrees, we have

$$\tan 624^\circ\ 36' = \tan 624.6^\circ \approx 10.579.$$

25. $\sin 0.83^{R}$ **26.** $\tan(-1.24^{R})$ **27.** $\cos(-0.16^{R})$

28. $\csc 1.34^{R}$ **29.** $\cot 7.19^{R}$ **30.** $\sec 13.67^{R}$

31. $\cos 9.7^\circ$ **32.** $\sin 2.3^\circ$ **33.** $\tan 46.6^\circ$

34. $\sin 76.2^\circ$ **35.** $\tan 422.8^\circ$ **36.** $\cos 738.5^\circ$

37. $\tan 435^\circ\ 12'$ **38.** $\cos 339^\circ\ 54'$ **39.** $\cot 549^\circ\ 24'$

40. $\sec 986^\circ\ 6'$ **41.** $\cos(-83^\circ\ 30')$ **42.** $\sin(-23^\circ\ 12')$

Find the measure of α in degrees (exact value).

Examples a. $\alpha = \text{Arcsin} \frac{\sqrt{3}}{2}$ b. $\alpha = \text{Arcsin}\left(-\frac{\sqrt{3}}{2}\right)$

Solutions

a. α is the angle such that $\sin \alpha = \sqrt{3}/2$ and $-90 \le \alpha \le 90^\circ$. From Table 3.1, $\alpha = 60^\circ$.

b. α is the angle such that $\sin \alpha = -\sqrt{3}/2$ and $-90° \le \alpha \le 90°$. Thus, $\alpha = -60°$.

43. $\alpha = \text{Arctan } 1$ **44.** $\alpha = \text{Arccos } \frac{1}{2}$ **45.** $\alpha = \text{Arccsc } 2$

46. $\alpha = \text{Arccot } \sqrt{3}$ **47.** $\alpha = \text{Arcsin}\left(-\frac{1}{2}\right)$ **48.** $\alpha = \text{Arctan}(-1)$

Find an approximation in degrees for each α.

Examples a. $\alpha = \text{Arcsin } 0.4633$ b. $\alpha = \text{Arccos}(-0.4337)$

Solutions Using a calculator set for degrees, we have

a. $\alpha \approx 27.6°$ b. $\alpha \approx 115.7°$

49. $\alpha = \text{Arccos } 0.2773$ **50.** $\alpha = \text{Arcsin } 0.9367$

51. $\alpha = \text{Arctan } 0.5938$ **52.** $\alpha = \text{Arccos } 0.1305$

53. $\alpha = \text{Arcsin}(-0.7934)$ **54.** $\alpha = \text{Arctan}(-2.4383)$

B

Write each of the following composite function values as a numerical, or an algebraic, expression without using inverse notation. Assume values for x for which each function value exists.

Example $\sin\left(\text{Arccos } \frac{3}{5}\right)$

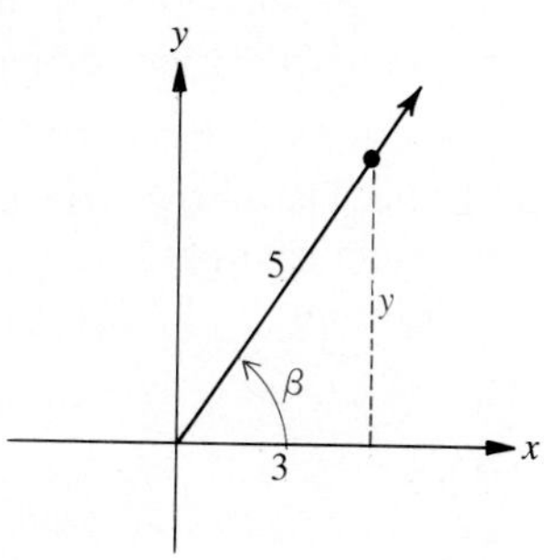

Solution Because the value of Arccos 3/5 is not evident by inspection, we let $\beta = \text{Arccos } 3/5$, from which $\cos \beta = 3/5$. The sketch shows a right triangle in which $\cos \beta = 3/5$. From the Pythagorean theorem, we have

$$y^2 + 3^2 = 5^2$$

$$y = 4.$$

Hence,

$$\sin\left(\text{Arccos } \frac{3}{5}\right) = \sin \beta = \frac{y}{5} = \frac{4}{5}.$$

55. $\cos\left(\text{Arcsin } \frac{3}{7}\right)$ **56.** $\sin\left(\text{Arccos } \frac{5}{13}\right)$ **57.** $\sin\left(\text{Arcsin } \frac{2}{5}\right)$

58. $\cos\left(\operatorname{Arccos}\frac{1}{3}\right)$ **59.** $\tan\left(\operatorname{Arccos}\frac{4}{5}\right)$ **60.** $\tan\left(\operatorname{Arcsin}\frac{5}{13}\right)$

61. $\cos\left(\operatorname{Arcsin}\frac{x}{3}\right)$ **62.** $\tan\left(\operatorname{Arccos}\frac{x}{4}\right)$ **63.** $\sin(\operatorname{Arccos} x)$

64. $\sin(\operatorname{Arctan} x)$

3.5 Right Triangles

When we use given information about the lengths of one, two, or three sides of a triangle and the measure of one or two angles to find the remaining, or unknown, lengths and measures, we are said to be *solving a triangle.* You may want to review some pertinent topics of geometry in Appendix A (Secs. A.7–A.9) before you begin this section.

Right triangles can be solved by using the trigonometric ratios of the acute angles of the triangle. Figure 3.15 shows that, in the first quadrant, any point (x, y) on the terminal side of an angle α in standard position determines a right triangle with sides of length x and y and with hypotenuse of length $\sqrt{x^2 + y^2}$. Therefore, in the special case where α is an acute angle in a right triangle,

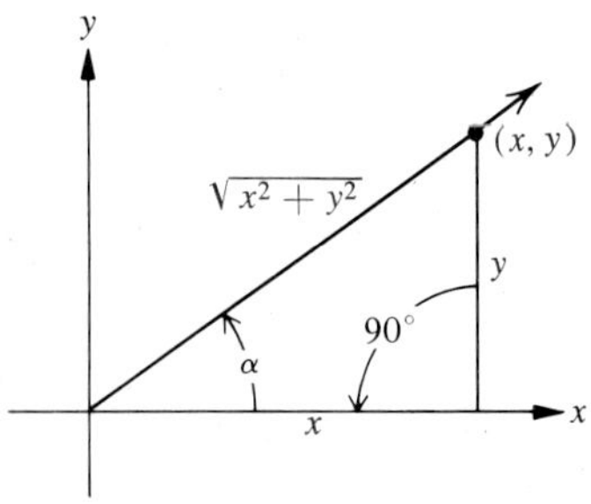

Figure 3.15

$$\sin\alpha = \frac{\text{length of side opposite }\alpha}{\text{length of hypotenuse}}, \qquad \csc\alpha = \frac{\text{length of hypotenuse}}{\text{length of side opposite }\alpha},$$

$$\cos\alpha = \frac{\text{length of side adjacent to }\alpha}{\text{length of hypotenuse}}, \qquad \sec\alpha = \frac{\text{length of hypotenuse}}{\text{length of side adjacent to }\alpha},$$

$$\tan\alpha = \frac{\text{length of side opposite }\alpha}{\text{length of side adjacent to }\alpha}, \qquad \cot\alpha = \frac{\text{length of side adjacent to }\alpha}{\text{length of side opposite }\alpha}.$$

It is not necessary that α be in standard position. The above relations involving the ratios of the lengths of the sides of a right triangle are equally applicable to a right triangle in any position. For example, in Figure 3.16a, the six trigonometric ratios of the angle α are

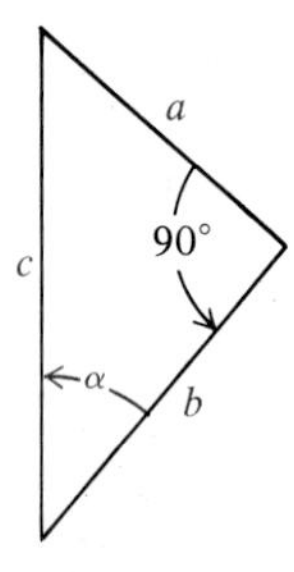

Figure 3.16a

$$\sin\alpha = \frac{a}{c}, \qquad \csc\alpha = \frac{c}{a},$$

$$\cos\alpha = \frac{b}{c}, \qquad \sec\alpha = \frac{c}{b},$$

$$\tan\alpha = \frac{a}{b}, \qquad \cot\alpha = \frac{b}{a}.$$

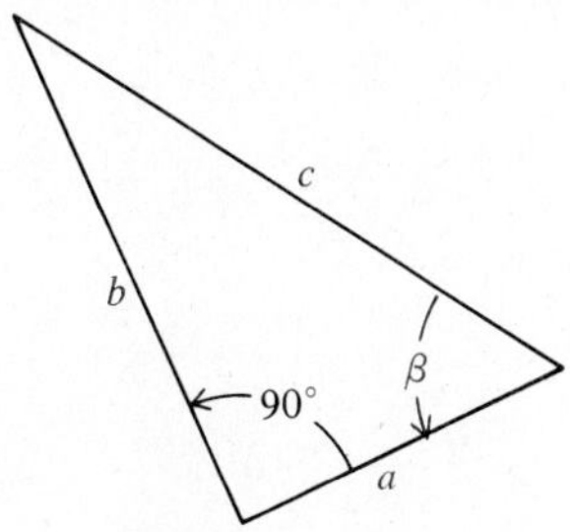

Figure 3.16b

In Figure 3.16b, the trigonometric ratios of the angle β are

$$\sin \beta = \frac{b}{c}, \qquad \csc \beta = \frac{c}{b},$$

$$\cos \beta = \frac{a}{c}, \qquad \sec \beta = \frac{c}{a},$$

$$\tan \beta = \frac{b}{a}, \qquad \cot \beta = \frac{a}{b}.$$

Generally, the sides of a triangle are labeled with their lengths, a, b, and c, and the angles opposite a, b, and c are labeled α, β, and γ, or A, B, and C, respectively, where γ or C names the right angle in a right triangle. The measures of angles in triangles are usually given in degree units.

Cofunctions

Two functions f and g are called **cofunctions** if

$$f(x) = g(y) \qquad \text{and} \qquad x + y = 90^\circ$$

In the trigonometric ratios above, observe that $\sin \alpha = \cos \beta$ and $\cos \alpha = \sin \beta$, where $\alpha + \beta = 90^\circ$. Hence, the sine and cosine are cofunctions. Furthermore, for the same reason, the secant and cosecant are cofunctions, as are the tangent and cotangent. For example,

$$\tan 32^\circ = 0.6249 \qquad \text{and} \qquad \cot(90^\circ - 32^\circ) = \cot 58^\circ = 0.6249.$$

Solution of Triangles

Right triangles can be solved by using the trigonometric ratios on page 122 and the fact that *the sum of the measures of the angles of any triangle equals* 180°.

Example One angle of a right triangle measures 34° and the side opposite this angle has a length of 12 inches. Solve the triangle.

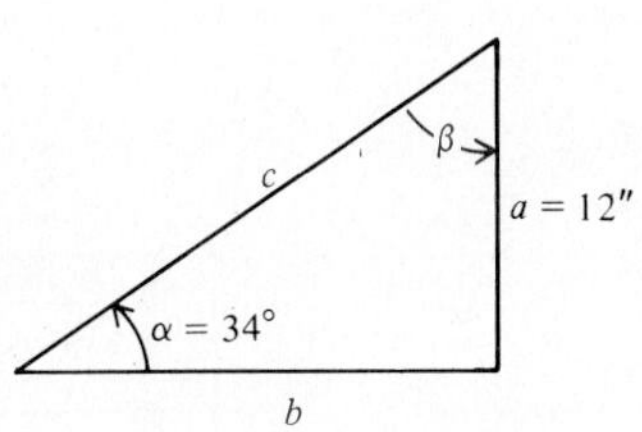

Solution Sketch the right triangle, showing the given information. First, from the fact that $\alpha + \beta + \gamma = 180^\circ$ and $\gamma = 90^\circ$,

$$\beta = 90^\circ - \alpha = 90^\circ - 34^\circ = 56^\circ.$$

By definition,

$$\tan 34^\circ = \frac{a}{b} = \frac{12}{b} \qquad \text{or} \qquad b = \frac{12}{\tan 34^\circ}$$

(continued)

and

$$\sin 34^\circ = \frac{a}{c} = \frac{12}{c} \quad \text{or} \quad c = \frac{12}{\sin 34^\circ}.$$

Thus,

$$b \approx \frac{12}{0.6745} \approx 17.8 \quad \text{and} \quad c \approx \frac{12}{0.5592} \approx 21.5.$$

Hence, $\beta = 56^\circ$, $b \approx 17.8$ in., and $c \approx 21.5$ in.

In the above example we have shown values for tan 34° and sin 34° to four decimal places as intermediate steps in the computation. The final results which have been rounded off to three significant digits* will be the same if a single key-stroke sequence is used to obtain values for b and c. We will continue to show function values in intermediate steps even though as soon as you develop calculator proficiency, you will be able to make a single key-stroke sequence to obtain most results.

Example The hypotenuse of a right triangle is 14.7 centimeters long and the length of a side is 6.8 centimeters. Solve the triangle.

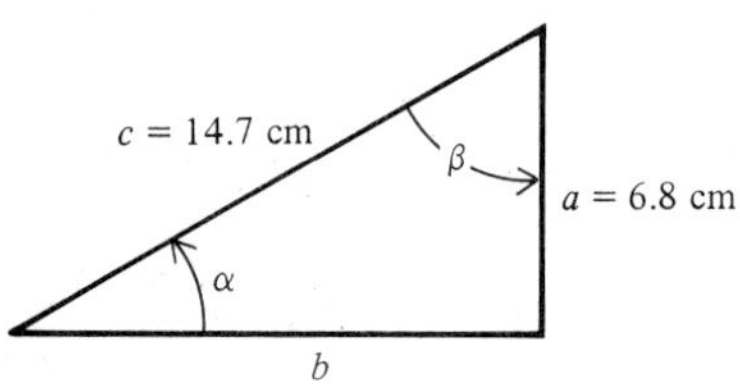

Solution Sketch the right triangle showing the given information. From the two given dimensions we could first obtain the measure of α using the ratios

$$\sin \alpha = \frac{6.8}{14.7} \quad \text{or} \quad \csc \alpha = \frac{14.7}{6.8}.$$

Alternatively we could first obtain the measure of β using the ratios

$$\cos \beta = \frac{6.8}{14.7} \quad \text{or} \quad \sec \beta = \frac{14.7}{6.8}.$$

We will use

$$\sin \alpha = \frac{6.8}{14.7} \approx 0.4626,$$

from which

$$\alpha \approx \text{Sin}^{-1}\, 0.4626 \approx 27.6^\circ.$$

* See Appendix B for a discussion of rounding-off procedures and significant digits.

Then we have that

$$\beta \approx 90^\circ - 27.6^\circ = 62.4^\circ.$$

We can find an approximation for b by using the Pythagorean theorem.

$$b^2 = \sqrt{(14.7)^2 - (6.8)^2};$$
$$b = \sqrt{216.09 - 46.24} \approx 13.0.$$

Alternatively, we can find b by using the results we obtained for the measure of α or β. If we use $\alpha \approx 27.6^\circ$, we can set up an equation using a variety of trigonometric ratios. One ratio is

$$\cos \alpha = \frac{b}{14.7},$$

from which

$$b = 14.7 \cos \alpha \approx 14.7 \cos 27.6^\circ$$
$$\approx 14.7(0.8862) \approx 13.0.$$

Hence $b \approx 13.0$ cm, $\alpha \approx 27.6^\circ$, $\beta \approx 62.4^\circ$, and $\gamma = 90^\circ$.

Accuracy of Results

In the foregoing example the lengths b and c were expressed by an approximation using three significant digits. Unless otherwise stated, we shall continue to specify lengths of line segments to this degree of accuracy when solutions are given in decimal notation, and we shall specify measures of angles to the *nearest tenth of a degree or nearest six minutes*. Furthermore, we shall assume that all data have similar accuracy. Thus, $a = 12$ in the above example implies that $a = 12.0$ and $\alpha = 34^\circ$ implies that $\alpha = 34.0^\circ = 34^\circ\ 00'$.

Special Names for Angles

Certain angles are given special names in the fields of surveying and navigation. For example, the angle formed by a horizontal ray and an observer's "line of sight" to any object above the horizontal is called an **angle of elevation** (Figure 3.17a on page 126). The angle formed by a horizontal ray and an observer's "line of sight" to any object below the horizontal is called an **angle of depression** (Figure 3.17b).

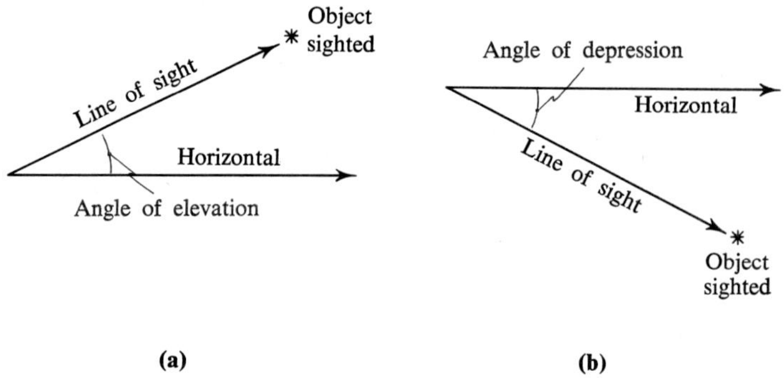

Figure 3.17

Example On level ground at point A the measure of the angle of elevation of the top of a tower is $57° 24'$. The distance from A to the base of the tower is 50 meters. How high is the tower?

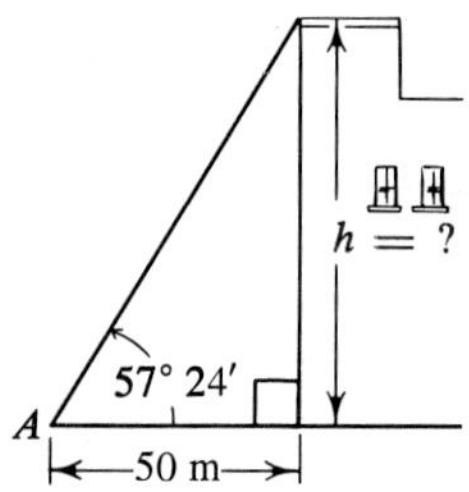

Solution By definition,

$$\tan 57° 24' = \frac{h}{50},$$

where h represents the height of the tower. Equivalently,

$$h = 50 \tan 57° 24' = 50 \tan 57.4°$$
$$\approx 50(1.5637) \approx 78.2.$$

Thus, the height of the tower is approximately 78.2 meters.

EXERCISE SET 3.5

A

Rewrite the given function of α as the corresponding cofunction of β.

Examples a. $\sin 83.6°$ b. $\sec 63° 24'$

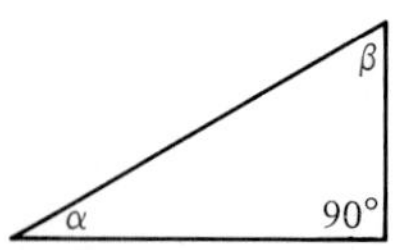

Solutions

a. $\sin 83.6° = \cos(90° - 83.6°) = \cos 6.4°$

b. $\sec 63° 24' = \csc(90° - 63° 24') = \csc 26° 36'$

1. $\cos 62°$ **2.** $\sec 51°$ **3.** $\tan 79.3°$

4. $\sin 59.6°$ **5.** $\csc 88° 12'$ **6.** $\cot 48° 48'$

The following measurements are lengths of sides and/or measures of angles of a right triangle, where in each case, $\gamma = 90°$.

Examples

a. $\alpha = 23.4°$; find β. b. $a = 6.3$ and $\beta = 47°$; find b.

c. $a = 5.31$ and $c = 7.27$; find α. d. $a = 6.2$ and $b = 4.3$; find c.

Solutions A sketch for each case is helpful.

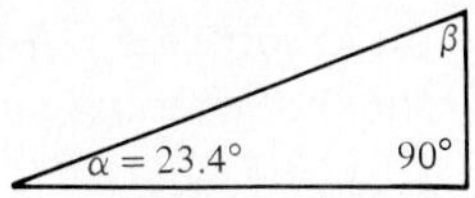

a.

a. Since α and β are complementary angles, $\beta = 90° - 23.4° = 66.6°$.

b. b is related to a and β by the tangent or cotangent functions. Using the tangent, we have

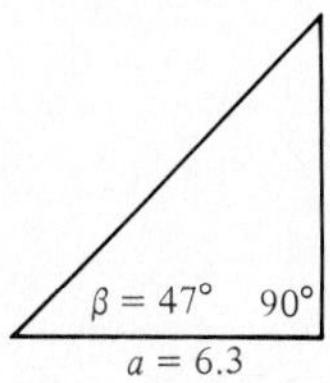

b.

$$\tan 47° = \frac{b}{6.3},$$

from which

$$b = 6.3 \tan 47° \approx 6.76.$$

c. α is related to a and c by the sine or cosecant functions. Using the sine, we have

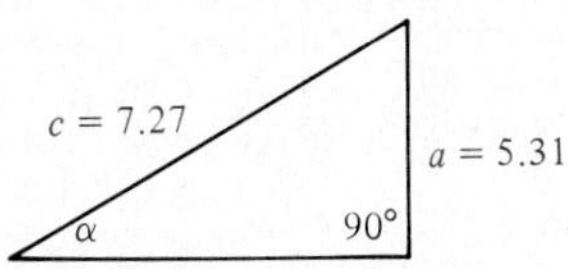

c.

$$\sin \alpha = \frac{a}{c} = \frac{5.31}{7.27}$$

from which

$$\alpha = \text{Arcsin}\,\frac{5.31}{7.27} \approx 46.9°.$$

d. From the Pythagorean theorem,

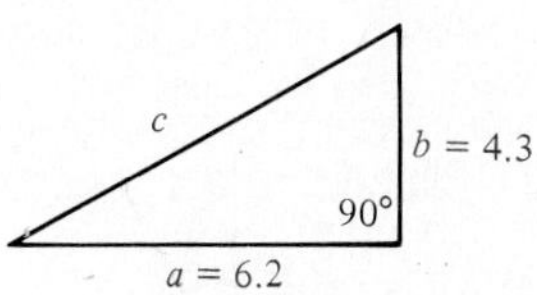

d.

$$c^2 = (6.2)^2 + (4.3)^2$$

Hence,

$$c = \sqrt{(6.2)^2 + (4.3)^2} \approx 7.55.$$

7. $\beta = 42°$; find α. **8.** $\alpha = 16.1°$; find β.

9. $b = 4.2$ and $\alpha = 23°$; find a. **10.** $a = 21.3$ and $\alpha = 24°$; find c.

11. $b = 2.9$ and $c = 3.5$; find β. **12.** $a = 4.9$ and $c = 7.3$; find α.

13. $a = 7$ and $\beta = 9.1°$; find c. **14.** $b = 6$ and $\alpha = 16.7°$; find a.

15. $a = 4.2$ and $b = 6.3$; find c.

16. $a = 7.3$ and $b = 2.1$; find c.

17. $a = 0.6$ and $c = 2.4$; find b.

18. $b = 0.4$ and $c = 1.3$; find a.

Solve each right triangle. In each case, $\gamma = 90°$.

Example $a = 4.1$ and $c = 7.9$

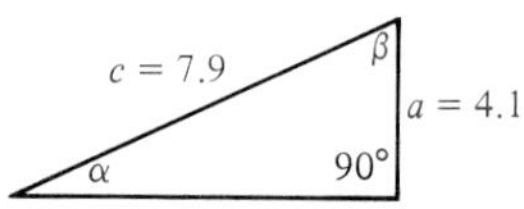

Solution A sketch is helpful. In this case, we want to solve for α, β, and b. Sine or cosecant functions relate the two known quantities a and c with α. Using the sine, we have

$$\sin \alpha = \frac{4.1}{7.9},$$

from which

$$\alpha = \text{Arcsin}\,\frac{4.1}{7.9} \approx 31.3°.$$

Then,

$$\beta = 90° - 31.3° = 58.7°.$$

The length b can now be obtained by using any one of several different trigonometric ratios (or the Pythagorean theorem). Using $\cos \alpha$, we have

$$\cos 31.3° = \frac{b}{7.9},$$

from which

$$b = 7.9 \cos 31.3° \approx 6.75.$$

19. $a = 2$ and $b = 3$

20. $c = 41$ and $b = 9$

21. $c = 17$ and $a = 15$

22. $c = 14$ and $\alpha = 17°$

23. $c = 23.5$ and $\beta = 51°$

24. $a = 6.1$ and $\alpha = 73°$

25. $a = 11$ and $\beta = 22.2°$

26. $b = 5$ and $\alpha = 64.7°$

27. $b = 15$ and $\alpha = 50°\ 18'$

28. $c = 20$ and $\beta = 27°\ 30'$

29. $b = 30$ and $\beta = 41°\ 36'$

30. $a = 32$ and $\beta = 75°\ 6'$

Solve.

31. The measure of an angle of elevation from a surveyor's position (ignoring the surveyor's height) to the top of a hill is 37° 24′. The top of the hill is 275 meters above a level line through the surveyor's position. How far is the surveyor from a point directly below the top of the hill?

32. From the top of a vertical cliff, 80 meters above the surface of the ocean, the measure of the angle of depression to a marker on the surface of the ocean is 18° 12′. How far is the marker from the foot of the cliff?

33. A 40-foot ladder is used to reach the top of a 24-foot wall. If the ladder extends 10 feet past the top of the wall, find the measure of the angle that the ladder forms with the horizontal ground.

34. The lengths of the sides of an isosceles triangle are 17, 17, and 30 centimeters. Find the measures of the angles in the triangle. *Hint:* Use the altitude of the triangle.

35. A ladder 30 decimeters long is leaning against the side of a building. If the ladder makes an angle of 25° 18′ with the side of the building, how far up from the ground does the ladder touch the building?

36. A tower 120 feet high casts a shadow 60 feet long. Find the angle of elevation of the sun.

37. A railroad track makes an angle of 4° 48′ with the horizontal. How far must a train go up the track to gain 20 meters in altitude?

38. A rectangle is 20 centimeters long and 14 centimeters wide. Find the angles that a diagonal makes with the sides.

39. A parachutist is descending vertically. How far does the parachutist fall as the angle of elevation from A changes for 50° to 30°?

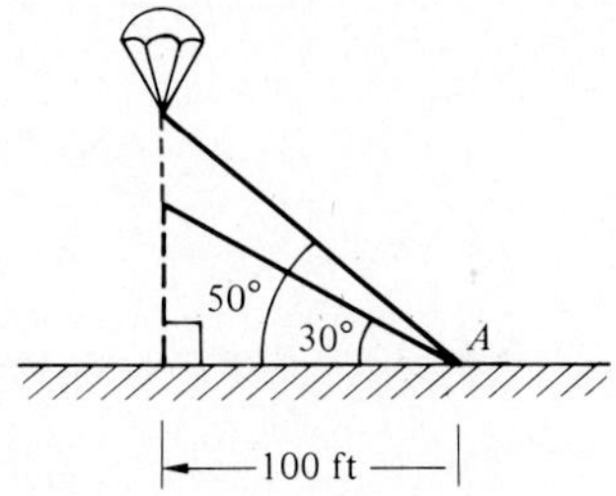

Ex. 39

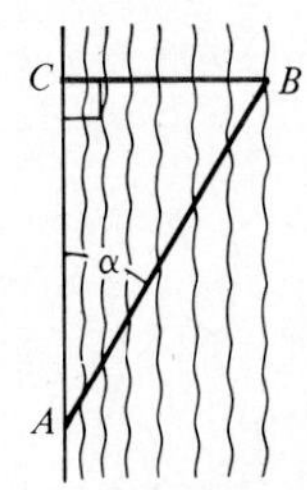

Ex. 40

40. A surveyor measures AC as 35.3 meters and α as 57.2°. How wide is the river?

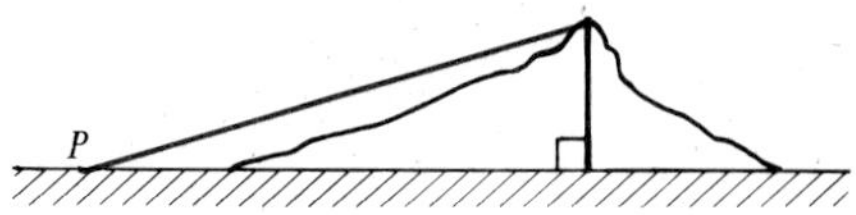

Ex. 41

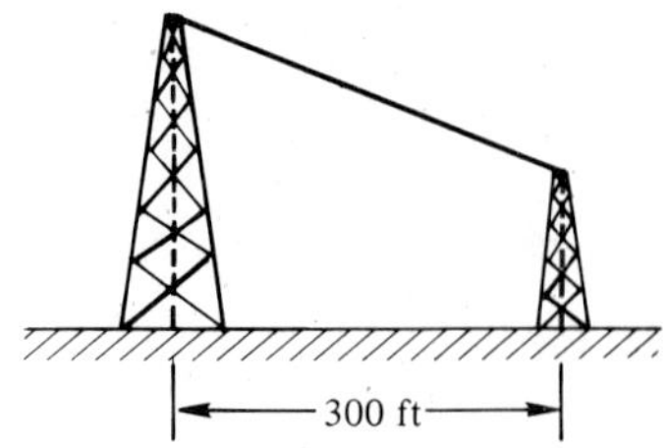

Ex. 42

41. An observer at P notes that the angle of elevation to the top of a mountain is 16.5°. If the mountain is 3500 feet high, how far is P from the top of the mountain?

42. In the figure, the angle of depression from the top of one tower to the top of the other is 20°. If the height of the taller tower is 150 feet, how high is the shorter?

B

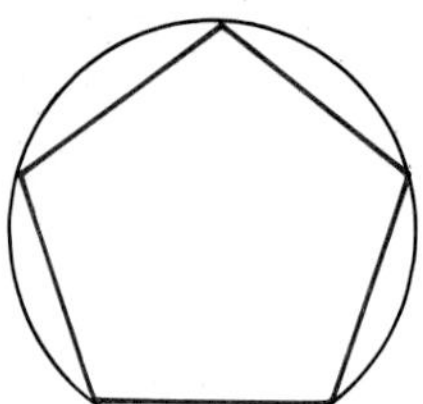

Ex. 43

43. A regular pentagon is inscribed in a circle of radius 5 centimeters. Find the length of a side of the pentagon.

44. The angle of elevation of the top of the tower is 62.3° from A and 37.1° from B. If AB is 60 feet, find the height of the tower.

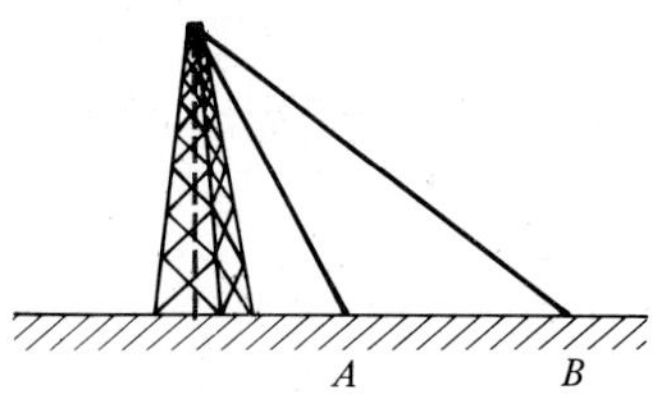

Ex. 44

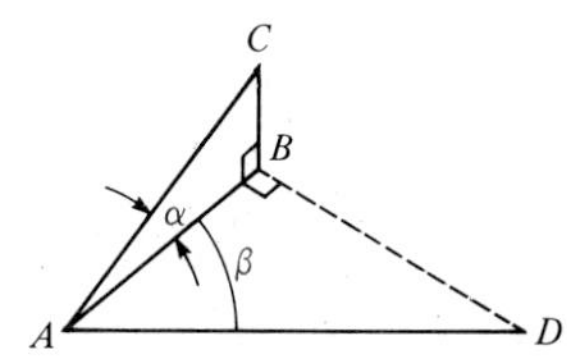

Ex. 45

45. If $\alpha = 30°$, $\beta = 60°$, and $BC = 2$, find AD.

46. Show that (see drawing for Exercise 45)

$$\cos \beta = \frac{BC \cot \alpha}{AD}.$$

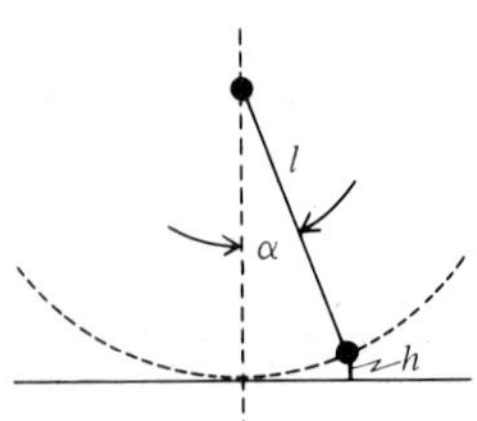

Ex. 47

47. Prove that when a simple pendulum of length l is inclined at an angle α with the vertical, the bob is at a height h above its lowest position, where $h = l(1 - \cos \alpha)$.

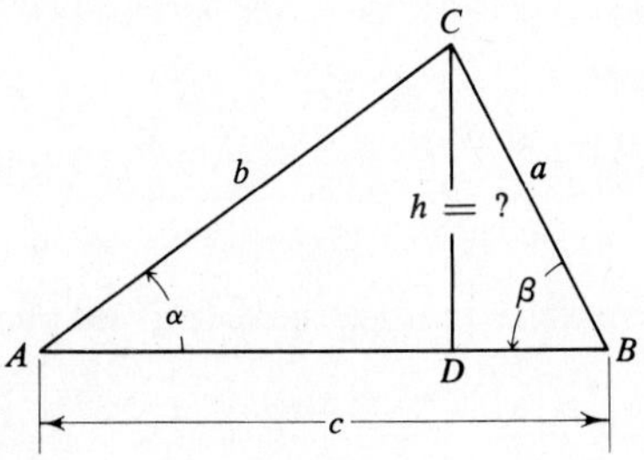

Ex. 48

48. Show that the length h of the perpendicular from C to the side c of any triangle ABC can be found by

$$h = \frac{c}{\cot \alpha + \cot \beta}.$$

Hint: $AD = h \cot \alpha$.

3.6 The Law of Sines

In Section 3.5 we solved right triangles using trigonometric ratios expressed in terms of the lengths of the sides of the triangle. In this section and in the next, special formulas will be derived to enable us to solve any triangle, right, acute, or obtuse.

Consider $\triangle ABC$ in Figure 3.18 (β may be either acute or obtuse), and introduce the altitude $\overline{CD}$. Thus, $\triangle ADC$ and $\triangle BDC$ are right triangles. In either figure, in $\triangle ADC$,

$$\sin \alpha = \frac{h}{b}, \qquad \text{from which} \qquad h = b \sin \alpha.$$

In $\triangle BDC$ in Figure 3.18a,

$$\sin \beta = \frac{h}{a}, \qquad \text{from which} \qquad h = a \sin \beta.$$

In $\triangle BDC$ in Figure 3.18b, $\angle CBD$ is equal to the reference angle for β; therefore,

$$\sin \beta = \sin \angle CBD = \frac{h}{a},$$

$$h = a \sin \beta.$$

Hence, we have in each case

$$h = b \sin \alpha = a \sin \beta.$$

Dividing each member of $b \sin \alpha = a \sin \beta$ by ab, we have

$$\frac{\sin \alpha}{a} = \frac{\sin \beta}{b}. \tag{1}$$

If in Figure 3.18 we use the altitude from vertex A and a similar argument,

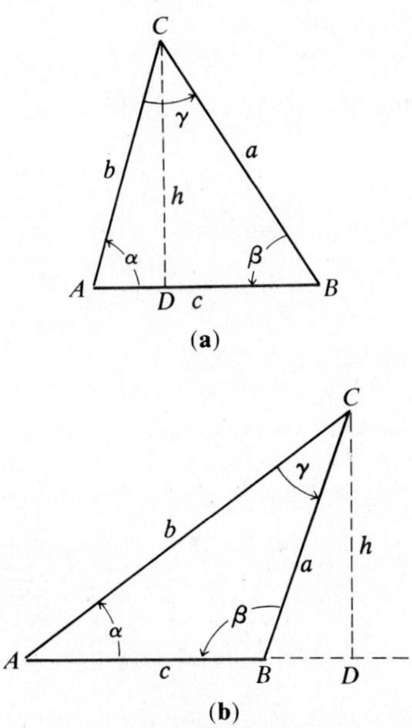

Figure 3.18

we will have

$$\frac{\sin \beta}{b} = \frac{\sin \gamma}{c}. \tag{2}$$

Comparing Equations (1) and (2), we observe that we have proved the following theorem.

Theorem 3.5 *If α, β, and γ are the angles of any triangle, and if a, b, and c are the lengths of the sides opposite α, β, and γ, respectively, then*

$$\frac{\sin \alpha}{a} = \frac{\sin \beta}{b} = \frac{\sin \gamma}{c}.$$

In many applications this relationship, called the **Law of Sines,** is conveniently used in the form

$$\frac{a}{\sin \alpha} = \frac{b}{\sin \beta} = \frac{c}{\sin \gamma}.$$

Two Angles and a Side

The Law of Sines can be used to solve triangles when several kinds of data are given. We consider first the case in which the measures of two angles and the length of one side of a triangle are given.

Example Solve the triangle for which $\alpha = 30°$, $\beta = 45°$, and $a = 20$.

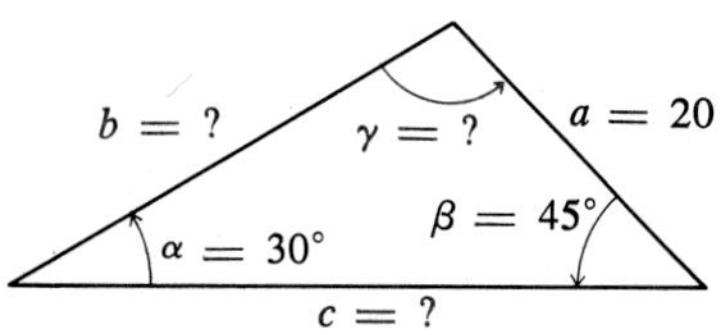

Solution A sketch of the triangle showing the given information is helpful. First, from the fact that $\alpha + \beta + \gamma = 180°$, it follows that

$$\begin{aligned} \gamma &= 180° - \alpha - \beta \\ &= 180° - 30° - 45° = 105° \end{aligned}$$

From the Law of Sines,

$$\frac{20}{\sin 30°} = \frac{b}{\sin 45°},$$

or, equivalently,

$$b = \frac{20 \sin 45°}{\sin 30°} \approx \frac{20(0.7071)}{0.5000} \approx 28.3.$$

Then, again from the Law of Sines,

$$\frac{20}{\sin 30^\circ} = \frac{c}{\sin 105^\circ},$$

or, equivalently,

$$c = \frac{20 \sin 105^\circ}{\sin 30^\circ} = \frac{20 \sin 75^\circ}{\sin 30^\circ} \approx \frac{20(0.9659)}{0.5000}$$
$$\approx 38.6.$$

Hence,

$$b \approx 28.3, \quad c \approx 38.6, \quad \text{and} \quad \gamma = 105^\circ.$$

Two Sides and an Opposite Angle

A second case in which the Law of Sines is used is one in which the lengths of two sides of a triangle, say a and b, and the measure of an angle opposite one of them, say, α, are given. Ambiguity may arise here, depending on a in relation to b and the measure of α. Figure 3.19, in which α is an *acute angle* and h is the length of a perpendicular segment, illustrates the following possibilities that may occur.

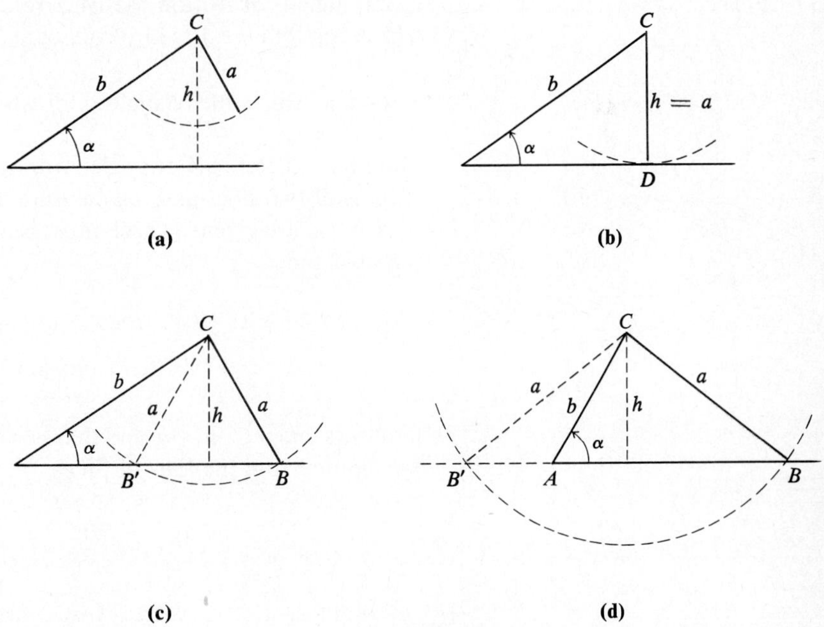

Figure 3.19

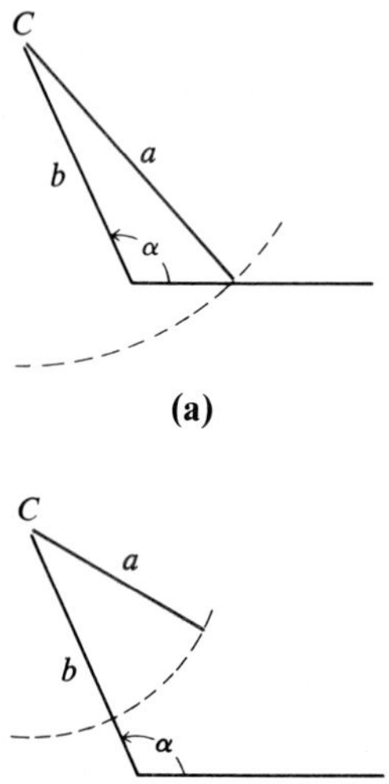

(a)

(b)

Figure 3.20

1. If $a < h$, the side with length a cannot intersect the side opposite C and no triangle is possible, as shown in Figure 3.19a. In this case, since $h = b \sin \alpha$, we have $a < b \sin \alpha$.
2. If $a = h = b \sin \alpha$, the side with length a intersects the side opposite C at D. In this case exactly one right triangle is possible, as shown in Figure 3.19b.
3. If $a > h$ (or $b \sin \alpha$) and $a < b$, or $b \sin \alpha < a < b$, two triangles are possible, as shown in Figure 3.19c.
4. If $a \geq b$, the triangle $AB'C$, shown in Figure 3.19d, does not contain the given angle α and therefore does not lead to a solution. Hence, in this case, only one triangle is possible. In particular, if $a = b$, the triangle is isosceles.

Figure 3.20, in which α is an *obtuse angle*, illustrates the following two possibilities.

1. If $a > b$, then one triangle is possible, as shown in Figure 3.20a.
2. If $a \leq b$, then no triangle is possible as shown in Figure 3.20b.

While we have concerned ourselves only with the variables a, b, and α in the above discussion, similar results are valid when other variables are involved in the given information. A sketch of the triangle using the given information will help to indicate the possibilities that exist. In particular, sketches of triangles like those shown in Figures 3.19 and 3.20, in which the given angle is drawn on the left, are usually the most helpful in deciding which possibility exists for given data.

Example Solve the triangle(s) for which $a = 7$, $b = 5$, and $\beta = 30°$.

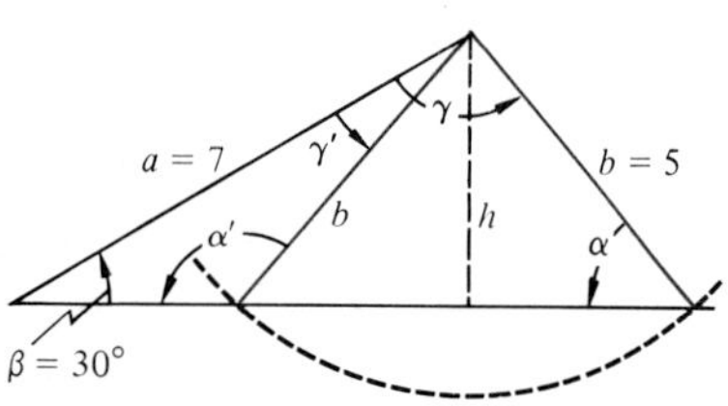

Solution Sketch the triangle(s) using the given information. Since the lengths of two sides and the measure of an angle β, opposite the side with measure b, are given, a check must be made to see how many triangles are possible. Since

$$h = a \sin \beta = 7 \sin 30°$$
$$= 7(0.5000) = 3.5,$$

it follows that $h < b < a$ and that two triangles are possible, as indicated in the figure. From the Law of Sines,

$$\frac{\sin \alpha}{7} = \frac{\sin 30°}{5}$$

$$\sin \alpha = \frac{7 \sin 30°}{5} = \frac{7(0.5000)}{5} = 0.7000.$$

Hence,

$$\alpha = \text{Sin}^{-1}\, 0.7000 \approx 44.4^\circ.$$

Since α is the base angle of an isosceles triangle and α' is the supplement of the other base angle,

$$\alpha' \approx 180^\circ - 44.4^\circ = 135.6^\circ.$$

For $\alpha \approx 44.4^\circ$, the solution of one of the two possible triangles is completed as follows. Since the sum of the measures of the angles of a triangle equals 180°,

$$\begin{aligned} \gamma &= 180^\circ - \alpha - \beta \\ &= 180^\circ - 44.4^\circ - 30^\circ = 105.6^\circ. \end{aligned}$$

From the Law of Sines

$$\begin{aligned} \frac{5}{\sin 30^\circ} &= \frac{c}{\sin 105.6^\circ}; \\ c &= \frac{5 \sin 105.6^\circ}{\sin 30^\circ} \\ &\approx \frac{5(0.9632)}{0.5000} \approx 9.63. \end{aligned}$$

For $\alpha' \approx 135.6^\circ$, the solution of the second possible triangle can be completed. Since

$$\gamma' \approx 180^\circ - 135.6^\circ - 30^\circ = 14.4^\circ,$$

then by the Law of Sines

$$\begin{aligned} \frac{5}{\sin 30^\circ} &= \frac{c'}{\sin 14.4^\circ}; \\ c' &\approx \frac{5(0.2487)}{0.5000} \approx 2.49. \end{aligned}$$

Thus, the two solutions are

$$\alpha \approx 44.4^\circ, \quad \gamma \approx 105.6^\circ, \quad \text{and} \quad c \approx 9.63;$$

$$\alpha' \approx 135.6^\circ, \quad \gamma' \approx 14.4^\circ, \quad \text{and} \quad c' \approx 2.49.$$

EXERCISE SET 3.6

A

Solve each triangle.

1. $a = 5$, $\alpha = 20°$, and $\beta = 75°$

2. $a = 32$, $\alpha = 26.3°$, and $\gamma = 81.8°$

3. $a = 58.4$, $\beta = 37.2°$, and $\gamma = 100°$

4. $b = 1.9$, $\alpha = 111.7°$, and $\beta = 5.1°$

5. $b = 0.42$, $\alpha = 35°\ 36'$, and $\gamma = 91°\ 30'$

6. $b = 0.88$, $\beta = 63°\ 54'$, and $\gamma = 34°\ 12'$

7. $c = 13.6$, $\alpha = 30°\ 24'$, and $\beta = 72°\ 6'$

8. $c = 1.4$, $\alpha = 135°\ 12'$, and $\gamma = 34°\ 48'$

Determine the number of triangles that satisfy each set of conditions and solve each triangle.

9. $a = 7$, $b = 5$, and $\alpha = 30°$

10. $a = 10$, $b = 9$, and $\beta = 60°$

11. $a = 4.2$, $c = 6.1$, and $\alpha = 32.2°$

12. $a = 38$, $c = 45$, and $\gamma = 35.6°$

13. $b = 16$, $c = 32$, and $\beta = 30°$

14. $b = 0.3$, $c = 0.4$, and $\gamma = 62°$

15. $b = 3.9$, $a = 5$, and $\beta = 42°\ 42'$

16. $b = 37$, $a = 51$, and $\alpha = 135°\ 30'$

Find the altitude to the given side and then find the area of the triangle.

17. $a = 3.4$, $\beta = 13°$, and $\alpha = 50°$

18. $c = 3.8$, $\beta = 15°\ 6'$, and $\gamma = 50°$

19. To find the distance from an observer's point A on one side of a river (see figure) to a point P on the other side, a line AP' measuring 2540 feet was laid out on one side and the angles $P'AP$ and $AP'P$ were measured and found to be $47°\ 18'$ and $78°\ 6'$, respectively. Find the distance from A to P to the nearest foot.

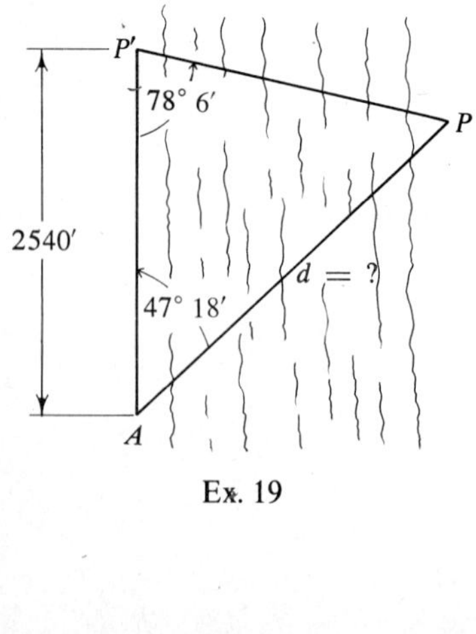

Ex. 19

20. Two persons, 400 feet apart, observe a balloon between them that is in a vertical plane with them. The respective angles of elevation of the balloon are observed by the men to be 75° 24′ and 49° 30′. Find the height of the balloon above the ground.

21. One diagonal of a parallelogram is 20 centimeters long and at one end forms angles to 20° and 40° with the sides of the parallelogram. Find the lengths of the sides.

22. A ship at sea is sighted from two points, *A* and *B*, on the shore. *A* and *B* are 5 kilometers apart, and the line of sight from *A* to the ship forms an angle of 30° 18′ with $\overline{AB}$, while the line of sight from *B* to the ship forms an angle of 41° 12′ with $\overline{AB}$. How far is the ship from point *A*?

23. A tower is built vertically on a hillside that slants upward 10° from the horizontal. From a point 200 meters uphill from the foot of the tower, the angle of elevation of the top of the tower is 12° 30′ (see figure). How high is the tower?

24. From a window 75 meters above the ground, the measure of the angle of elevation to the top of a nearby building is 29° 42′ (see figure). From a point on the ground directly below the window, the measure of the angle of elevation of the top of the same building is 52° 24′. Find the height of the building.

25. The angle of elevation of the building is 48° from *A*, and 61° from *B*. If *AB* is 20 feet, find the height of the building.

26. The pole is tilted toward the sun at an angle of 8° from the vertical. When the angle of elevation of the sun is 54°, the shadow of the pole is 12 feet long. How long is the pole?

27. A hill makes an angle of 10° with the horizontal. From a point 50 meters downhill from a vertical tower, the angle of elevation of the tower is 56°. Find the height of the tower.

28. From two points on shore, the angles α and β to a light at *C* are 16.4° and 58.6°, respectively. If *AB* is 1500 feet, find the distance, *d*, of the light from the shore.

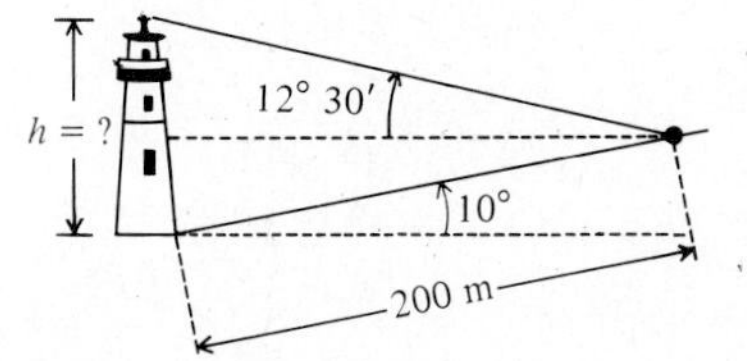

Ex. 23

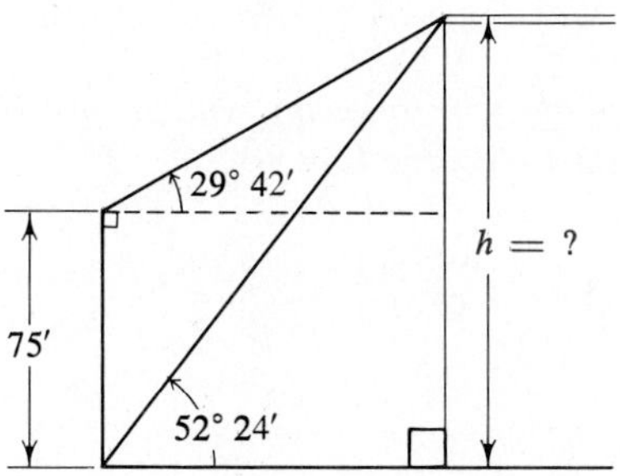

Ex. 24

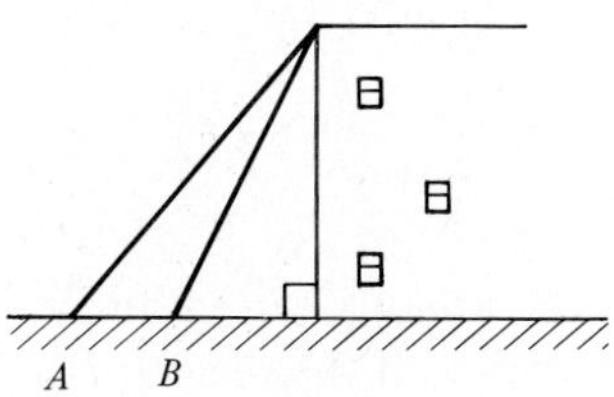

Ex. 25

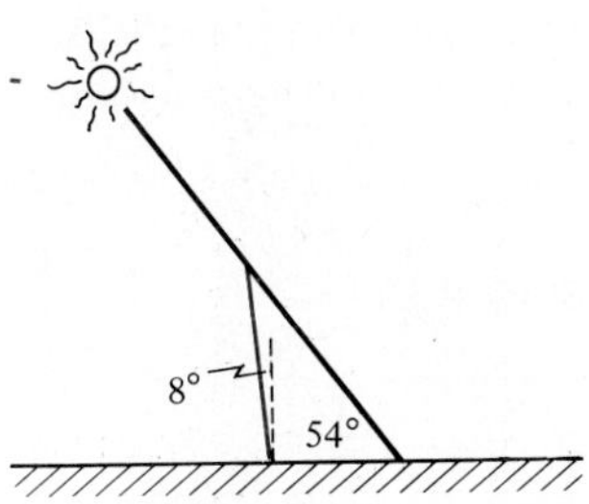

Ex. 26

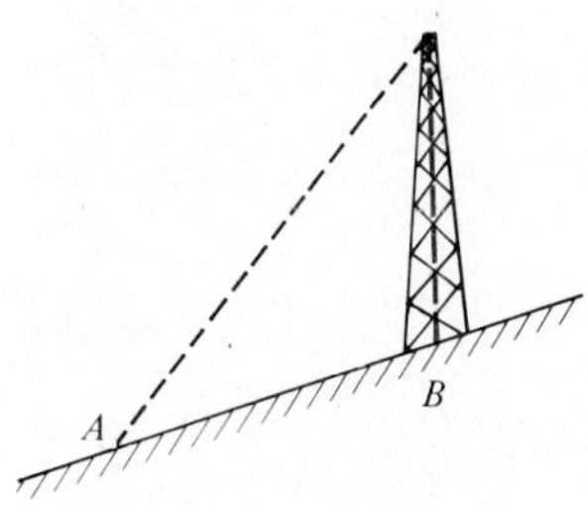

Ex. 27

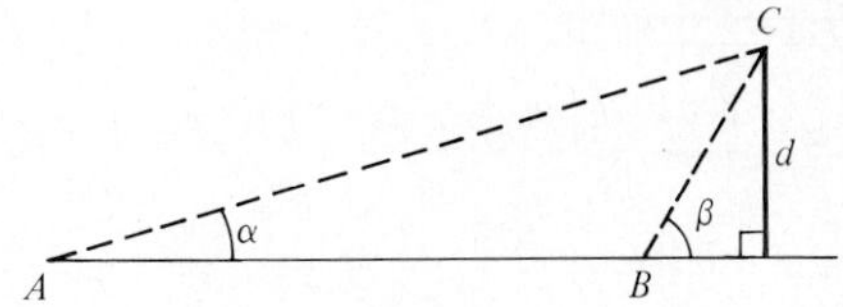

Ex. 28

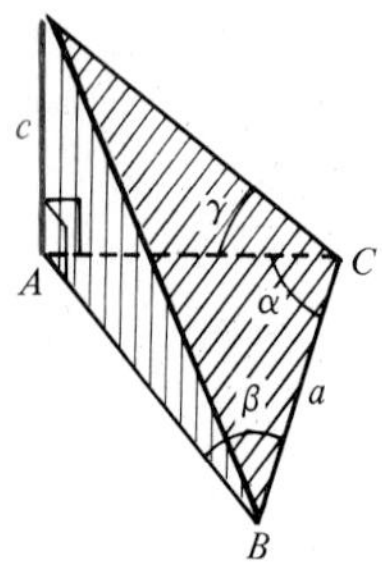

Ex. 29

B

29. Using the figure, show that

$$c = \frac{a \sin \beta \tan \gamma}{\sin(\alpha + \beta)}$$

30. Using the figure for Exercise 29, if $\alpha = 40°$, $\beta = 60°$, $\gamma = 30°$, and $a = 100$ feet, find c.

Show that, in any triangle ABC, each of the following statements is true. *Hint:* In Exercises 31 and 32 start with the Law of Sines.

31. $\dfrac{a+b}{b} = \dfrac{\sin \alpha + \sin \beta}{\sin \beta}$

32. $\dfrac{a-b}{b} = \dfrac{\sin \alpha - \sin \beta}{\sin \beta}$

33. $\dfrac{a-b}{a+b} = \dfrac{\sin \alpha - \sin \beta}{\sin \alpha + \sin \beta}$

34. $\dfrac{a+b+c}{b} = \dfrac{\sin \alpha + \sin \beta + \sin \gamma}{\sin \beta}$

3.7 The Law of Cosines

The Law of Sines considered in Section 3.6 cannot be used directly to solve triangles in which the lengths of two sides and the measure of the included angle are specified or in which only the lengths of the three sides are specified.

Consider a triangle ABC in which the lengths a and b and the measure of angle γ are known. Orienting triangle ABC with γ at the origin, as shown in Figure 3.21, we find that

$$\cos \gamma = \frac{x}{b} \qquad \text{and} \qquad \sin \gamma = \frac{y}{b},$$

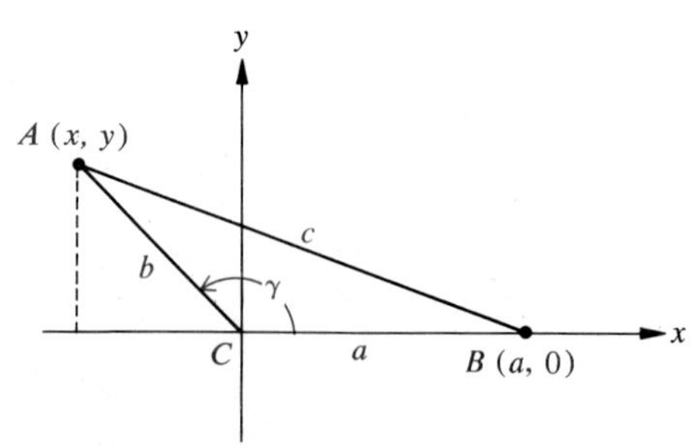

Figure 3.21

from which the coordinates of A are

$$x = b\cos\gamma \quad \text{and} \quad y = b\sin\gamma.$$

From the distance formula (see Appendix A.4),

$$\begin{aligned} c^2 &= (y-0)^2 + (x-a)^2 \\ &= (b\sin\gamma - 0)^2 + (b\cos\gamma - a)^2 \\ &= b^2\sin^2\gamma + b^2\cos^2\gamma - 2ab\cos\gamma + a^2 \\ &= b^2(\sin^2\gamma + \cos^2\gamma) - 2ab\cos\gamma + a^2. \end{aligned}$$

Since $\sin^2\gamma + \cos^2\gamma = 1$, we have

$$c^2 = a^2 + b^2 - 2ab\cos\gamma.$$

The discussion above is a proof of one of the parts of the following theorem, known as the Law of Cosines. By placing vertex A or vertex B at the origin, the proofs of the other parts can be completed in a similar way.

Theorem 3.6 *If α, β, and γ are the angles of any triangle, and a, b, and c are the lengths of the sides opposite α, β, and γ, respectively, then*

$$\begin{aligned} \mathbf{c^2} &= \mathbf{a^2 + b^2 - 2ab\cos\gamma}, \\ \mathbf{b^2} &= \mathbf{a^2 + c^2 - 2ac\cos\beta}, \\ \mathbf{a^2} &= \mathbf{b^2 + c^2 - 2bc\cos\alpha}. \end{aligned}$$

Notice that if $\gamma = 90°$, then $\cos\gamma = \cos 90° = 0$ and the Law of Cosines reduces to

$$\begin{aligned} c^2 &= a^2 + b^2 - 2ab(0), \\ c^2 &= a^2 + b^2, \end{aligned}$$

the familiar formula applicable to right triangles (the Pythagorean theorem).

Two Sides and the Included Angle

Now consider how the Law of Cosines can be used to solve a triangle when the measure of an angle and the lengths of the adjacent sides are known.

Example Given the measures $\beta = 75°$, $a = 7$, and $c = 5$ of an oblique triangle, find the length b.

(continued)

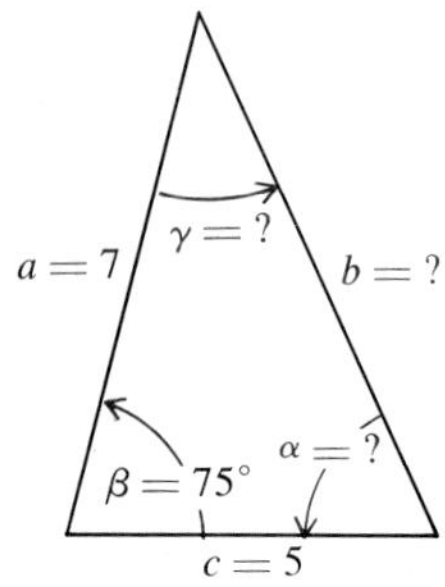

Solution Sketch a triangle showing the given information. The length b can be obtained by the use of the Law of Cosines expressed in the form $b^2 = a^2 + c^2 - 2ac \cos \beta$. Substituting the known values gives

$$b^2 = 7^2 + 5^2 - 2(7)(5) \cos 75^\circ$$
$$\approx 49 + 25 - 70(0.2588) \approx 55.9;$$
$$b \approx \sqrt{55.9} \approx 7.48.$$

Three Sides

The Law of Sines could be used to complete the solution of the triangle in the above example since the length b, opposite the given angle β, is known. However, we shall use the Law of Cosines in the following example in order to show how it can be used to determine the measure of an angle if the lengths of three sides of a triangle are known.

Example To find the measure of α in the foregoing example,

$$a^2 = b^2 + c^2 - 2bc \cos \alpha$$

is first written in the form

$$\cos \alpha = \frac{b^2 + c^2 - a^2}{2bc}.$$

From the previous calculations, $b^2 \approx 55.9$ and $b \approx 7.48$. Thus,

$$\cos \alpha \approx \frac{55.9 + 25 - 49}{2(7.48)(5)} \approx 0.4265.$$

Because α, an angle in a triangle, must have a measure less than 180°, and $\cos \alpha > 0$, then

$$\alpha \approx 64.8^\circ.$$

Because $\alpha + \beta + \gamma = 180^\circ$,

$$\gamma \approx 180^\circ - 64.8^\circ - 75^\circ = 40.2^\circ.$$

Thus, the solution to the triangle is

$$b \approx 7.48, \quad \alpha \approx 64.8^\circ, \quad \text{and} \quad \gamma \approx 40.2^\circ.$$

If, after first finding the length of the third side, $b \approx 7.48$ in the above example, the Law of Sines is used to find the measure of a second angle, the angle selected should be the smaller of the two remaining angles (the one opposite the shorter of the two given sides) and hence *an acute angle*. Thus, if the Law of Sines had been used to find the second angle in the above example, we would have found $\sin \gamma$ rather than $\sin \alpha$ because, of the two given sides, $c < a$ and γ is opposite c. This is necessary because, if we were not certain that the angle we seek using the Law of Sines is acute, we could not be sure whether the positive number we obtain is the measure of the angle we seek or of its supplement because the sines of an acute angle and an obtuse angle are both positive.

Given three sides of a triangle, any of the angles can be obtained by using one of the Equations (1), (2), or (3). However, it is best to use the formula that will give the cosine of the largest angle (the angle opposite the longest side) because if the triangle happens to be obtuse, this will be the obtuse angle and *the other two angles will be acute.*

EXERCISE SET 3.7

A

Solve each triangle.

1. $b = 4$, $c = 3.5$, and $\alpha = 71°$

2. $c = 0.3$, $a = 0.1$, and $\beta = 70°$

3. $a = 3.2$, $b = 2.2$, and $\gamma = 75.3°$

4. $b = 6.0$, $a = 5.1$, and $\gamma = 83.5°$

5. $a = 0.7$, $c = 0.8$ and $\beta = 141°\ 30'$

6. $b = 3.4$, $a = 2.1$, and $\gamma = 122°\ 12'$

7. $b = 1.6$, $c = 3.2$, and $\alpha = 100°\ 24'$

8. $c = 2.1$, $b = 4.3$, and $\alpha = 130°\ 36'$

Solve the triangles for which the lengths of the three sides are given.

Example $a = 3.4$, $b = 2.7$, and $c = 1.3$

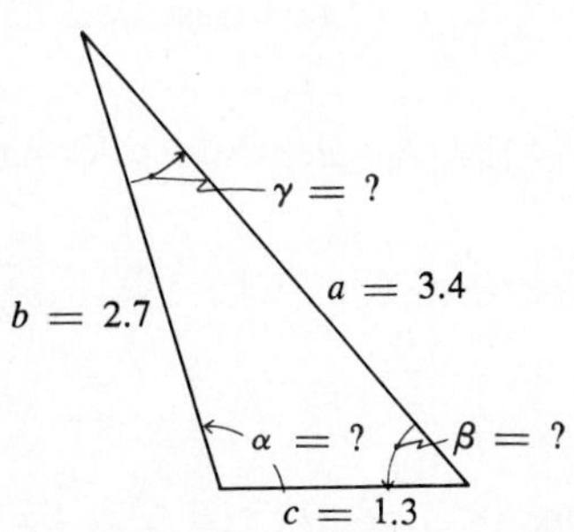

Solution Draw a sketch showing the given information. From the Law of Cosines,

$$\cos \alpha = \frac{b^2 + c^2 - a^2}{2bc} = \frac{(2.7)^2 + (1.3)^2 - (3.4)^2}{2(2.7)(1.3)}$$

$$\approx -0.3675.$$

(continued)

Hence, $\tilde{\alpha} \approx 68.4°$. Then, because $\cos \alpha < 0$, α is an angle such that $90° < \alpha < 180°$. Thus,

$$\alpha \approx 180° - 68.4° = 111.6°.$$

From the Law of Sines,

$$\frac{\sin \alpha}{a} = \frac{\sin \beta}{b},$$

$$\frac{\sin 111.6°}{3.4} = \frac{\sin \beta}{2.7},$$

from which

$$\sin \beta = \frac{2.7(\sin 111.6°)}{3.4} \approx \frac{2.7(0.9298)}{3.4} \approx 0.7384.$$

Thus,

$$\beta \approx 47.6°.$$

From the fact that $\gamma = 180° - \alpha - \beta$,

$$\gamma \approx 180° - 111.6° - 47.6°$$
$$\approx 20.8°.$$

9. $a = 9$, $b = 7$, and $c = 5$

10. $a = 2.7$, $b = 5.1$, and $c = 4.4$

11. $a = 1.2$, $b = 9$, and $c = 10$

12. $a = 4.5$, $b = 7.5$, and $c = 5.8$

13. $a = 6$, $b = 8$, and $c = 12$

14. $a = 6$, $b = 12$, and $c = 13$

15. $a = 4$, $b = 5$, and $c = 6$

16. $a = 5$, $b = 7$, and $c = 8$

17. Find the measure of the smallest angle of the triangle whose sides have lengths 4.3, 5.1, and 6.3.

18. Find the measure of the smallest angle of the triangle whose sides have lengths 3.0, 4.2, and 3.8.

19. Find the measure of the largest angle of the triangle whose sides have lengths 2.9, 3.3, and 4.1.

20. Find the measure of the largest angle of the triangle whose sides have lengths 6.0, 8.2, and 9.4.

21. Find the area of the triangle in Exercise 17.

22. Find the area of the triangle in Exercise 18.

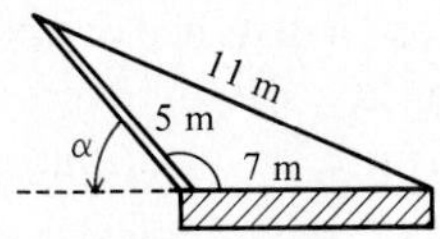

Ex. 27

23. A ship is supposed to travel directly from port A to port B, a distance of 10 kilometers. After traveling a distance of 5 kilometers, the captain discovers he has been traveling 15° off course. At this point, how far is the ship from port B?

24. Two ships leave the same port on courses that form an angle of 37° 12′. When one ship has traveled 12 kilometers, the other has traveled 18 kilometers. How far apart are the ships at that time?

25. Two sides of a parallelogram that is 6 centimeters by 8 centimeters include an angle of 67°. Find the length of the longer diagonal.

26. A tower 40 meters high is on top of a hill. A cable 50 meters long runs from the top of the tower to a point 20 meters down the hill. Find the angle the hillside forms with a horizontal line.

27. A crane at the edge of a dock is supported by a cable 11 meters long, attached to the dock 7 meters from the base of the crane (see figure). If the crane arm is 5 meters long, find the acute angle that the crane arm forms with the horizontal.

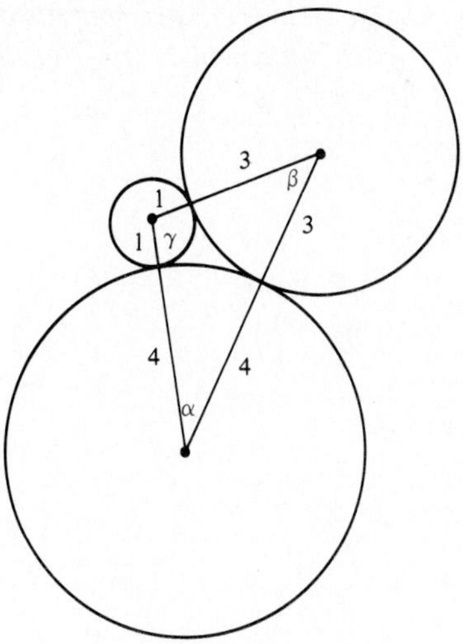

Ex. 28

28. Three circles with radii 1, 3, and 4 centimeters are tangent to each other (see figure). To the nearest 6′, find the three angles formed by the lines joining their centers.

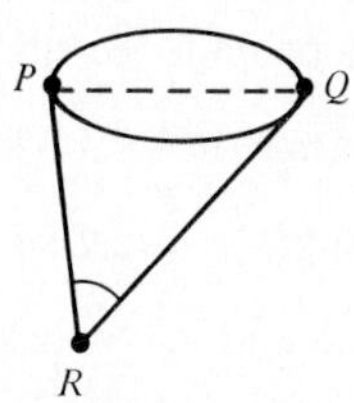

Ex. 29

29. Points P and Q are on opposite ends of a lake. If point R is 1200 feet from P and 1600 feet from Q, and the angle at R is 50°, how long is the lake?

30. A vertical tower 30 feet high is on a hill that forms an angle of 15° with the horizontal. How long is a guy wire that runs from the top of the tower to a point 20 feet downhill from the base of the tower?

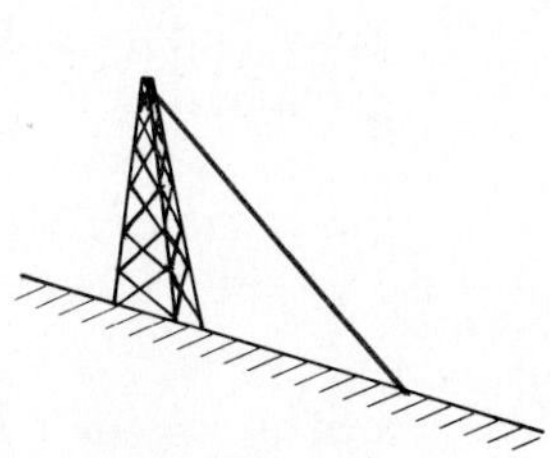

Ex. 30

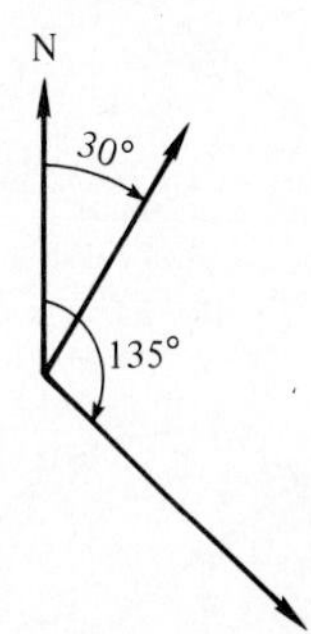

Ex. 31

31. Two airplanes leave a field at the same time. One flies 30° east of north at 250 kilometers per hour; the other 45° east of south at 300 kilometers per hour. How far apart are they at the end of 2 hours?

32. A pilot is flying from Salt Lake City to Denver, a distance of about 500 miles. When she is 300 miles from Salt Lake City, she discovers she has been flying 16° off course. How far is she from Denver at that time?

B

The formulas in Exercises 33–36 where $s = \frac{1}{2}(a + b + c)$, *are sometimes used in the solution of triangles. Show that each statement is true for all* $\alpha \in A$. *Hint:* Use the fact that $\cos \alpha = \dfrac{b^2 + c^2 - a^2}{2bc}$.

33. $1 + \cos \alpha = \dfrac{(b + c + a)(b + c - a)}{2bc}$

34. $1 - \cos \alpha = \dfrac{(a - b + c)(a + b - c)}{2bc}$

35. $\dfrac{1 + \cos \alpha}{2} = \dfrac{s(s - a)}{bc}$.

36. $\dfrac{1 - \cos \alpha}{2} = \dfrac{(s - b)(s - c)}{bc}$.

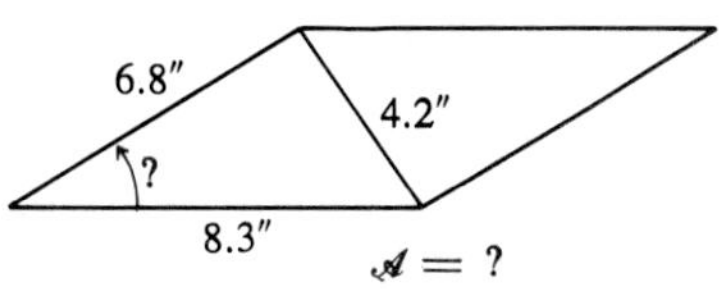

Ex. 40

37. Use the results of Exercises 35 and 36 to prove Hero's (or Heron's) formula, $\mathscr{A} = \sqrt{s(s - a)(s - b)(s - c)}$, which can be used to find the area of any triangle directly from the lengths of its three sides.

38. Use Hero's formula to find the area of the triangle in Exercise 9 above.

39. Use Hero's formula to find the area of the triangle in Exercise 11 above.

40. Two sides of a parallelogram are of lengths 6.8 and 8.3 inches, and one of the diagonals is of length 4.2 inches (see figure). Find the area of the parallelogram.

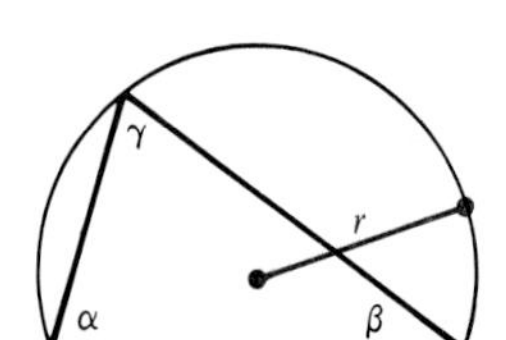

Ex. 41

41. Show that the radius r of an inscribed circle of a triangle (see figure) is given by

$$r = \sqrt{\frac{(s - a)(s - b)(s - c)}{s}}$$

where $s = \frac{1}{2}(a + b + c)$.

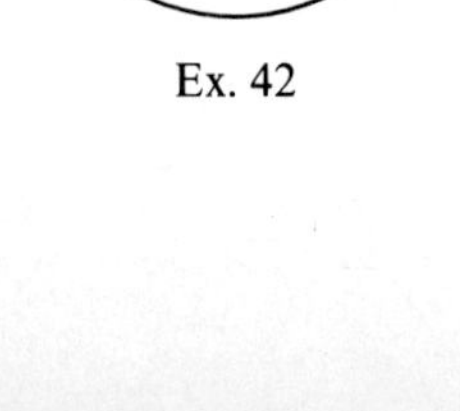

Ex. 42

42. Show that the radius r of a circumscribed circle of a triangle (see figure) is given by

$$r = \frac{a}{2 \sin \alpha} = \frac{b}{2 \sin \beta} = \frac{c}{2 \sin \gamma}.$$

Chapter Summary

3.1 An angle that has its initial side along the positive x-axis and its vertex at the origin is said to be in **standard position.** Such an angle is said to be in that quadrant in which its terminal side lies.

Two kinds of angle measure most commonly used are **degree measure** and **radian measure.**

The measure of an angle in radians is related to the measure of the angle in degrees by the equation

$$\frac{m^{\circ}(\alpha)}{360} = \frac{m^{R}(\alpha)}{2\pi}$$

or by the equations

$$m^{\circ}(\alpha) = \frac{180}{\pi} m^{R}(\alpha) \qquad \text{and} \qquad m^{R}(\alpha) = \frac{\pi}{180} m^{\circ}(\alpha).$$

Angles having the same initial side and the same terminal side are called **coterminal angles.**

3.2 An angle α whose vertex is at the center of a circle of radius r intercepts an arc of length

$$s = r \cdot m^{R}(\alpha).$$

The **linear velocity** v of a point moving along a circle at a constant speed is given by

$$v = \frac{s}{t} = \frac{r \cdot m^{R}(\alpha)}{t},$$

where t is the time required to move a distance $s = r \cdot m^{R}(\alpha)$. The **angular velocity** ω of the moving point is given by

$$\omega = \frac{m^{R}(\alpha)}{t},$$

where ω is given in radians per unit time.

The linear velocity v and the angular velocity ω of a point moving along a circle of radius with length r are related by the equation

$$v = r \cdot \omega,$$

where ω is given in radians per unit time.

3.3 If α is an angle in standard position, x and y (not both 0) are the coordinates of any point on the terminal side of α, and $r = \sqrt{x^2 + y^2}$, then

$$\text{cosine } \alpha = \frac{x}{r}, \qquad \text{secant } \alpha = \frac{r}{x} \quad (x \neq 0),$$

$$\text{sine } \alpha = \frac{y}{r}, \qquad \text{cosecant } \alpha = \frac{r}{y} \quad (y \neq 0),$$

$$\text{tangent } \alpha = \frac{y}{x} \quad (x \neq 0), \qquad \text{cotangent } \alpha = \frac{x}{y} \quad (y \neq 0).$$

These six ratios, called **trigonometric ratios,** are determined only by the position of the terminal side of an angle in standard position and not by the point selected on the terminal side. Each of these ratios determines a **trigonometric function** of the same name.

3.4 The values of the trigonometric functions are equal to the corresponding values of the analogous circular functions. That is,

$$\text{Circ}(s) = \text{Trig}(\alpha^\circ) = \text{Trig}(\alpha^{\text{R}}),$$

where the central angle α intercepts an arc of length s on a unit circle.

Many properties of circular functions are, with appropriate rewording, also valid for trigonometric functions.

Inverse trigonometric functions are defined in a manner similar to the definitions for the inverse circular functions.

3.5 The six trigonometric function values of an acute angle of a right triangle can be expressed as ratios of the sides of the triangle.

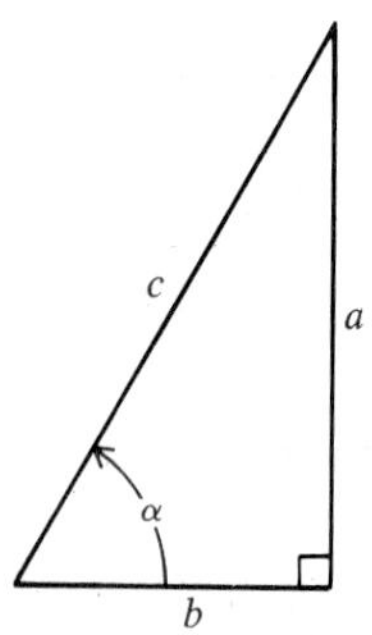

$$\sin \alpha = \frac{a}{c}, \qquad \csc \alpha = \frac{c}{a},$$

$$\cos \alpha = \frac{b}{c}, \qquad \sec \alpha = \frac{c}{b},$$

$$\tan \alpha = \frac{a}{b}, \qquad \cot \alpha = \frac{b}{a}.$$

3.6 The **Law of Sines:**

If α, β, and γ are the angles of any triangle, and if a, b, and c are the lengths of the sides opposite α, β, and γ, respectively, then

$$\frac{\sin \alpha}{a} = \frac{\sin \beta}{b} = \frac{\sin \gamma}{c}.$$

This law is used if the measures of two angles and the length of any side, or if the lengths of two sides and the measure of an angle opposite one of these sides, are given. In the latter case, certain ambiguities may arise. Two, one, or no triangles may exist.

3.7 The **Law of Cosines:**
If α, β, and γ are the angles of any triangle, and a, b, and c are the lengths of the sides opposite α, β, and γ, respectively, then

$$c^2 = a^2 + b^2 - 2ab \cos \gamma,$$

$$b^2 = a^2 + c^2 - 2ac \cos \beta,$$

$$a^2 = b^2 + c^2 - 2bc \cos \alpha.$$

This law is used if the lengths of two sides and the measure of the included angle, or if the lengths of three sides, are given.

A list of symbols introduced in this chapter is shown on the inside front covers and important properties are shown on the inside back covers.

Review Exercises

3.1 *Find the degree measure of α to the nearest tenth of a degree.*

1. $\alpha = \dfrac{2\pi}{5}^{\text{R}}$ **2.** $\alpha = 2.41^{\text{R}}$

Find the radian measure of α to the nearest hundredth of a radian.

3. $\alpha = 75^\circ$ **4.** $\alpha = 257^\circ$

Find the least positive angle coterminal with each angle α.

5. $\alpha = 938^\circ$ **6.** $\alpha = 7.23^{\text{R}}$

Write a general expression for all angles coterminal with α.

7. $\alpha = 131^\circ$ **8.** $\alpha = \dfrac{7\pi}{4}^{\text{R}}$

3.2 **9.** Find the length of the arc intercepted on a circle by a central angle α with measure 2.2^{R} if the radius of the circle is 1.3 meters in length.

10. Find the measure of the central angle that intercepts an arc with length 60.2 centimeters on a circle whose radius is 3.1 meters in length.

11. Find the linear velocity v of a point on a circle moving with constant speed if $r = 2$ decimeters and $\omega = 6$ radians per minute.

12. Find the angular velocity ω of a point on a circle moving with constant speed if $r = 0.5$ meter and $v = 19.7$ centimeters per second.

3.3 *Find the value of each of the six trigonometric functions of α, if the terminal side of α (in standard position) contains the point corresponding to the given ordered pair.*

13. $(4, 3)$ **14.** $(5, 12)$

15. $(-2\sqrt{3}, 2)$ **16.** $(-2, 5)$

Determine the quadrant in which each angle α lies.

17. $\sin \alpha < 0$ and $\cos \alpha > 0$ **18.** $\cos \alpha > 0$ and $\tan \alpha < 0$

19. $\tan \alpha > 0$ and $\sec \alpha < 0$ **20.** $\csc \alpha < 0$ and $\cot \alpha < 0$

3.4 *Find each of the following (exact value).*

21. $\sin 405^\circ$ **22.** $\cot 1110^\circ$ **23.** $\tan 5\pi^R$

24. $\sin \frac{15\pi^R}{2}$ **25.** $\cos(-120^\circ)$ **26.** $\csc\left(-\frac{7\pi^R}{4}\right)$

Find an approximation for each of the following.

27. $\cos 0.61$ **28.** $\tan(-1.07)$

29. $\sin 4.9^\circ$ **30.** $\sec 183.3^\circ$

31. $\cot 480.9^\circ$ **32.** $\csc(-74^\circ\ 36')$

Find each measure of α in degrees.

33. $\text{Arcsin}\ \frac{1}{2}$ **34.** $\text{Arccos}\ \frac{1}{\sqrt{2}}$

35. Arctan 2.3332 **36.** Arccos(−0.5344)

3.5 *Solve each right triangle. In each case, $\gamma = 90^\circ$.*

37. $c = 14,\ \beta = 51^\circ$ **38.** $c = 6.7,\ a = 5.4$

39. $a = 12.9,\ \alpha = 32^\circ\ 36'$ **40.** $b = 9.72,\ \alpha = 71^\circ\ 18'$

3.6 *Solve each triangle.*

41. $c = 13.6,\ \alpha = 30.3°,\ \beta = 72.2°$

42. $c = 1.4,\ \alpha = 135°\ 6',\ \gamma = 34°\ 54'$

Determine the number of triangles that satisfy the conditions in each exercise and solve each triangle.

43. $b = 3.9,\ a = 5,\ \beta = 42.7°$

44. $b = 37,\ a = 51,\ \alpha = 135°\ 30'$

3.7 *Solve each triangle.*

45. $a = 1.9,\ b = 2.3,\ \gamma = 58°\ 12'$

46. $b = 2.3,\ c = 5.7,\ \alpha = 23.1°$

47. $a = 3.6,\ b = 4.2,\ c = 6.1$

48. $a = 2.1,\ b = 3.4,\ c = 4.1$

4 IDENTITIES AND CONDITIONAL EQUATIONS

Equations such as

$$a \cdot 1 = a, \quad a + b = b + a, \quad \text{and} \quad a(b + c) = ab + ac,$$

which are true for all real numbers a, b, and c, are called **identities;** *the left-hand member and the right-hand member of each equation are equivalent expressions.** In algebra, we have used such identities to simplify algebraic expressions whenever we substituted one expression for an equivalent expression.

In this chapter we will review some basic identities that involve trigonometric (or circular) function values that have been introduced in previous chapters and we will introduce additional ones that will be useful in our work. We will then use these identities to rewrite expressions that involve trigonometric function values as equivalent expressions.

In our work with identities we shall not explicitly specify the replacement set of the variable in each case. The variables α, β, γ, and θ will represent elements of the set of all angles, and the variables x and y will be considered as representing elements either of the set of all angles or of the set of real numbers. Furthermore, it should be understood:

> *An equation that is specified to be an identity without noting the restrictions is in fact an identity only for all replacements of the variables for which both members of the equation are defined.*

* Also see Appendix A.3.

4.1 Basic Trigonometric Identities

The trigonometric identities

$$\csc \alpha = \frac{1}{\sin \alpha}, \quad \sec \alpha = \frac{1}{\cos \alpha}, \quad \cot \alpha = \frac{1}{\tan \alpha},$$

$$\tan \alpha = \frac{\sin \alpha}{\cos \alpha}, \quad \cot \alpha = \frac{\cos \alpha}{\sin \alpha}, \quad \text{and} \quad \sin^2 \alpha + \cos^2 \alpha = 1,$$

stated in Chapter 1, have been shown to be consequences of the definition of the trigonometric ratios.

Two other identities follow from the equation $\sin^2 \alpha + \cos^2 \alpha = 1$. For any α for which $\alpha \neq 0$, we multiply each member by $1/\cos^2 \alpha$ to obtain the equivalent equations

$$\frac{\sin^2 \alpha}{\cos^2 \alpha} + \frac{\cos^2 \alpha}{\cos^2 \alpha} = \frac{1}{\cos^2 \alpha},$$

$$\tan^2 \alpha + 1 = \sec^2 \alpha.$$

Multiplying each member of $\sin^2 \alpha + \cos^2 \alpha = 1$ by $1/\sin^2 \alpha$ for $\sin \alpha \neq 0$, we obtain

$$\frac{\sin^2 \alpha}{\sin^2 \alpha} + \frac{\cos^2 \alpha}{\sin^2 \alpha} = \frac{1}{\sin^2 \alpha},$$

$$1 + \cot^2 \alpha = \csc^2 \alpha.$$

The basic identities that we have stated above should be recognized in equivalent forms. For example, using the symbol $\leftrightarrow$ to mean "is equivalent to," we have

$$\sec \alpha = \frac{1}{\cos \alpha} \quad \leftrightarrow \quad (\cos \alpha)(\sec \alpha) = 1 \quad \leftrightarrow \quad \cos \alpha = \frac{1}{\sec \alpha};$$

$$\csc \alpha = \frac{1}{\sin \alpha} \quad \leftrightarrow \quad (\sin \alpha)(\csc \alpha) = 1 \quad \leftrightarrow \quad \sin \alpha = \frac{1}{\csc \alpha};$$

$$\cot \alpha = \frac{1}{\tan \alpha} \quad \leftrightarrow \quad (\tan \alpha)(\cot \alpha) = 1 \quad \leftrightarrow \quad \tan \alpha = \frac{1}{\cot \alpha};$$

$$\sin^2 \alpha + \cos^2 \alpha = 1 \quad \leftrightarrow \quad \sin^2 \alpha = 1 - \cos^2 \alpha \quad \leftrightarrow \quad \cos^2 \alpha = 1 - \sin^2 \alpha;$$

$$\tan^2 \alpha + 1 = \sec^2 \alpha \quad \leftrightarrow \quad \tan^2 \alpha = \sec^2 \alpha - 1;$$

$$\cot^2 \alpha + 1 = \csc^2 \alpha \quad \leftrightarrow \quad \cot^2 \alpha = \csc^2 \alpha - 1.$$

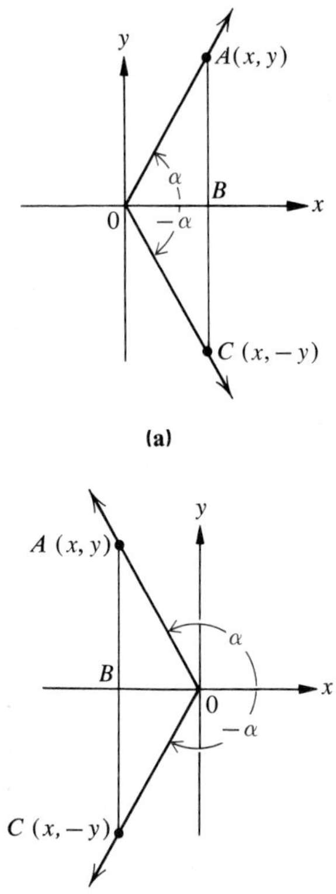

Figure 4.1

Identities for $\sin(-\alpha)$, $\cos(-\alpha)$, and $\tan(-\alpha)$

Several other useful trigonometric relationships can be obtained from geometric considerations. In Figure 4.1a,

$$\angle AOB = \angle COB, \quad OB = OB, \quad \text{and} \quad \angle OBA = \angle OBC.$$

Hence (see Appendix A.9),

$$\triangle AOB \cong \triangle COB,$$

from which we have that each pair of corresponding sides are congruent and the absolute values of the respective components of the ordered pairs corresponding to points A and C are equal. Because

$$\sin(-\alpha) = \frac{-y}{r} \quad \text{and} \quad \sin \alpha = \frac{y}{r},$$

$$\mathbf{\sin(-\alpha) = -\sin \alpha.} \tag{1}$$

Also, $\cos(-\alpha) = x/r$ and $\cos \alpha = x/r$, from which

$$\mathbf{\cos(-\alpha) = \cos \alpha,} \tag{2}$$

and $\tan(-\alpha) = -y/x$ and $\tan \alpha = y/x$, from which

$$\mathbf{\tan(-\alpha) = -\tan \alpha.} \tag{3}$$

A similar argument for angles α and $-\alpha$ that are in Quadrants II and III, as shown in Figure 4.1b, leads to the results stated by Equations (1), (2), and (3).

Because identities we have introduced are simply equations in which the left member and right member are equivalent expressions and the function values in these expressions are real numbers, we can sometimes simplify expressions that involve trigonometric function values by using these identities and applying the familiar algebraic properties for real numbers.

Examples Write each expression in terms of sine and cosine and simplify.

a. $(\tan \alpha)(\cos^2 \alpha)$ b. $\dfrac{\cos \beta}{\cot^2 \beta}$ c. $\tan(-\gamma) + \sec(-\gamma)$

Solutions

a. Substitute $\sin \alpha/\cos \alpha$ for $\tan \alpha$ and simplify.

$$(\tan \alpha)(\cos^2 \alpha) = \frac{\sin \alpha}{\cos \alpha} \cdot \cos^2 \alpha = \sin \alpha \cdot \cos \alpha$$

b. Substitute $\cos^2 \beta/\sin^2 \beta$ for $\cot^2 \beta$ and simplify.

$$\frac{\cos \beta}{\cot^2 \beta} = \frac{\cos \beta}{\dfrac{\cos^2 \beta}{\sin^2 \beta}} = \cos \beta \cdot \frac{\sin^2 \beta}{\cos^2 \beta} = \frac{\sin^2 \beta}{\cos \beta}$$

c. Substitute $\sin(-\gamma)/\cos(-\gamma)$ for $\tan(-\gamma)$ and $1/\cos(-\gamma)$ for $\sec(-\gamma)$ and simplify.

$$\tan(-\gamma) + \sec(-\gamma) = \frac{\sin(-\gamma)}{\cos(-\gamma)} + \frac{1}{\cos(-\gamma)}$$

$$= \frac{-\sin \gamma}{\cos \gamma} + \frac{1}{\cos \gamma} = \frac{1 - \sin \gamma}{\cos \gamma}$$

EXERCISE SET 4.1

A

Use the identities of this section to express each of the following in terms of a single function value.

Examples a. $1 - \sin^2 \alpha$ b. $\dfrac{\cos^2 \alpha}{\sin^2 \alpha}$

Solutions

a. Since $\sin^2 \alpha + \cos^2 \alpha = 1$,

$$1 - \sin^2 \alpha = \cos^2 \alpha.$$

b. Since $\dfrac{\cos \alpha}{\sin \alpha} = \cot \alpha$,

$$\frac{\cos^2 \alpha}{\sin^2 \alpha} = \cot^2 \alpha.$$

1. $1 + \tan^2 \alpha$ **2.** $1 + \cot^2 \alpha$ **3.** $1 - \cos^2 \alpha$

4. $\sec^2 \alpha - 1$ **5.** $\cos^2 \alpha - 1$ **6.** $1 - \sec^2 \alpha$

7. $\dfrac{\sin^3 \alpha}{\cos^3 \alpha}$ **8.** $\dfrac{\cos^4 \alpha}{\sin^4 \alpha}$

Use the identities of this section to write each expression in terms of sine and cosine and simplify.

Examples

a. $(\sin \alpha)(\cot^2 \alpha)(\sec^2 \alpha)$ b. $\dfrac{\cot^2 x}{\cos x}$

(continued)

Solutions

a. Substituting $\cos\alpha/\sin\alpha$ for $\cot\alpha$ and $1/\cos\alpha$ for $\sec\alpha$ yields

$$(\sin\alpha)(\cot^2\alpha)(\sec^2\alpha) = (\sin\alpha)\left(\frac{\cos^2\alpha}{\sin^2\alpha}\right)\left(\frac{1}{\cos^2\alpha}\right)$$

$$= \frac{1}{\sin\alpha}.$$

b. Since $\cot x = \cos x/\sin x$,

$$\frac{\cot^2 x}{\cos x} = \frac{\cos^2 x/\sin^2 x}{\cos x}$$

$$= \frac{\cos^2 x}{\sin^2 x}\cdot\frac{1}{\cos x} = \frac{\cos x}{\sin^2 x}.$$

9. $(\sec\alpha)(\cot\alpha)$

10. $(\csc\alpha)(\tan\alpha)$

11. $(\cot\alpha)(\csc\alpha)$

12. $(\tan\alpha)(\sec\alpha)$

13. $1 - \tan^2\alpha$

14. $1 - \cot^2\alpha$

15. $1 - \sec^2\alpha$

16. $1 - \csc^2\alpha$

17. $(\csc^2\beta)(\tan\beta)$

18. $(\sec^2\beta)(\cot\beta)$

19. $(\csc^2\beta)(\cot\beta)$

20. $(\sec^2\beta)(\tan\beta)$

21. $\dfrac{\sec\gamma}{\tan\gamma}$

22. $\dfrac{\csc\gamma}{\cot\gamma}$

23. $\dfrac{\cot\gamma}{\sec\gamma}$

24. $\dfrac{\tan\gamma}{\csc\gamma}$

25. $\dfrac{\cot^2 x}{\csc^2 x}$

26. $\dfrac{\tan^2 x}{\sec^2 x}$

27. $\dfrac{\sec^2 x}{\cot^2 x}$

28. $\dfrac{\csc^2 x}{\tan^2 x}$

29. $\tan y + \cot y$

30. $\sec y + \csc y$

31. $\tan^2\theta + \cot^2\theta$

32. $\sec^2\theta + \csc^2\theta$

33. $\sec\theta + \dfrac{\csc^2\theta}{\sec\theta}$

34. $\csc\theta + \dfrac{\sec^2\theta}{\csc\theta}$

35. $(\sin\alpha)(\cot^2\alpha)(\csc^2\alpha)$

36. $(\cos\alpha)(\tan^2\alpha)(\sec^2\alpha)$

37. $\tan\alpha - \dfrac{\cot^2\alpha}{\tan\alpha}$

38. $\sec\alpha - \dfrac{\csc^2\alpha}{\sec\alpha}$

Examples a. $\cos(-\alpha)\cdot\tan(-\alpha)$ b. $\cot(-\alpha)\cdot\sin\alpha$

Solutions

a. Since

$$\cos(-\alpha) = \cos\alpha \quad \text{and} \quad \tan(-\alpha) = -\tan\alpha,$$

then

$$\begin{aligned}\cos(-\alpha)\cdot\tan(-\alpha) &= \cos\alpha\cdot(-\tan\alpha)\\ &= \cos\alpha\cdot\left(\frac{-\sin\alpha}{\cos\alpha}\right)\\ &= -\sin\alpha.\end{aligned}$$

b. Since

$$\cot(-\alpha) = \frac{\cos(-\alpha)}{\sin(-\alpha)},$$

$$\cos(-\alpha) = \cos\alpha, \quad \text{and} \quad \sin(-\alpha) = -\sin\alpha,$$

then

$$\begin{aligned}\cot(-\alpha)\cdot\sin\alpha &= \frac{\cos\alpha}{-\sin\alpha}\cdot\sin\alpha\\ &= -\cos\alpha.\end{aligned}$$

39. $\csc(-\alpha)\cdot\tan\alpha$

40. $\sec(-\alpha)\cdot\cot(-\alpha)$

41. $\sec^2(-\beta)\cdot\cot(-\beta)$

42. $\csc^2\beta\cdot\tan(-\beta)$

43. $\dfrac{\csc x}{\cot(-x)}$

44. $\dfrac{\sec(-x)}{\tan(-x)}$

45. $\cot(-y) + \tan(-y)$

46. $\csc(-y) + \sec y$

B

Simplify each expression.

47. $\dfrac{(\cos\alpha + \sin\alpha)^2}{\cos\alpha - \sin\alpha} - \dfrac{1}{\dfrac{\cos\alpha - \sin\alpha}{2\cos\alpha\sin\alpha}}$

48. $\dfrac{\sin\alpha}{\dfrac{\sin\alpha}{\sin\alpha + \cos\alpha}} - \dfrac{2\sin\alpha\cos\alpha}{\sin\alpha + \cos\alpha}$

49. $\dfrac{\dfrac{1}{\sin\alpha - \cos\alpha}}{\dfrac{\sin\alpha}{\sin^2\alpha - \cos^2\alpha}}$

50. $\dfrac{\dfrac{1}{\sin^2\alpha} + \dfrac{1}{\cos^2\alpha}}{\dfrac{1}{\sin\alpha\cos\alpha}} - \dfrac{\cos\alpha}{\sin\alpha}$

A function satisfying $f(-x) = f(x)$ for all x in the domain is called an **even function** and a function satisfying $f(-x) = -f(x)$ is called an **odd function.**

Which of the following functions are even and which are odd?

51. $f(x) = \sin x$

52. $f(x) = \cos x$

53. $f(x) = \tan x$

54. $f(x) = \cot x$

55. $f(x) = \sec x$

56. $f(x) = \csc x$

4.2 Proving Identities

In Section 4.1 we introduced some basic identities that we used to rewrite and simplify expressions that involved trigonometric function values. In Section 4.1 we always expressed our results in terms of sine and cosine function values. Sometimes we may also want to rewrite trigonometric expressions in terms of other function values or constants.

Example Show that $\csc \alpha \cos \alpha$ is equivalent to $\cot \alpha$.

Solution Since $\csc \alpha = 1/\sin \alpha$,

$$\csc \alpha \cos \alpha = \frac{1}{\sin \alpha} \cdot \cos \alpha$$

$$= \frac{\cos \alpha}{\sin \alpha}.$$

Since $\cos \alpha/\sin \alpha = \cot \alpha$, we have that

$$\csc \alpha \cos \alpha = \cot \alpha.$$

It is customary to express the statement in the above example as follows.

"Show that $\csc \alpha \cos \alpha = \cot \alpha$ is an identity."

Of course for this identity the proof is exactly as given in the above solution.

Let us consider another example.

Example Show that

$$\frac{\cos \alpha - \sin \alpha}{\cos \alpha} = 1 - \tan \alpha \tag{1}$$

is an identity.

Solution We shall rewrite the right-hand member in terms of sines and cosines. Substituting $\sin \alpha/\cos \alpha$ for $\tan \alpha$ in the right-hand member,

$$\frac{\cos \alpha - \sin \alpha}{\cos \alpha} = 1 - \frac{\sin \alpha}{\cos \alpha}.$$

Writing the right-hand member as a single term yields

$$\frac{\cos \alpha - \sin \alpha}{\cos \alpha} = \frac{\cos \alpha - \sin \alpha}{\cos \alpha},$$

which is clearly an identity.

Example Show that

$$1 + \tan^2 \alpha = \frac{1}{\cos^2 \alpha} \tag{2}$$

is an identity.

Solution We shall rewrite the left-hand member in terms of sines and cosines. Substituting $\sin^2 \alpha/\cos^2 \alpha$ for $\tan^2 \alpha$ in the left-hand member,

$$1 + \frac{\sin^2 \alpha}{\cos^2 \alpha} = \frac{1}{\cos^2 \alpha}.$$

Writing the left-hand member as a single term yields

$$\frac{\cos^2 \alpha + \sin^2 \alpha}{\cos^2 \alpha} = \frac{1}{\cos^2 \alpha}.$$

Since $\cos^2 \alpha + \sin^2 \alpha$ equals 1,

$$\frac{1}{\cos^2 \alpha} = \frac{1}{\cos^2 \alpha}$$

which is clearly an identity.

Sometimes it is helpful to express *both* members of an identity in terms of sines and cosines.

Example Show that

$$\cot \alpha + \tan \alpha = \csc \alpha \sec \alpha \tag{3}$$

is an identity.

(continued)

Solution Substituting $\cos\alpha/\sin\alpha$ for $\cot\alpha$ and $\sin\alpha/\cos\alpha$ for $\tan\alpha$ in the left-hand member,

$$\frac{\cos\alpha}{\sin\alpha} + \frac{\sin\alpha}{\cos\alpha} = \csc\alpha\sec\alpha,$$

$$\frac{\cos^2\alpha + \sin^2\alpha}{\sin\alpha\cos\alpha} = \csc\alpha\sec\alpha,$$

$$\frac{1}{\sin\alpha\cos\alpha} = \csc\alpha\sec\alpha,$$

Substituting $1/\sin\alpha$ for $\csc\alpha$ and $1/\cos\alpha$ for $\sec\alpha$ in the right-hand member,

$$\frac{1}{\sin\alpha\cos\alpha} = \frac{1}{\sin\alpha}\cdot\frac{1}{\cos\alpha},$$

$$\frac{1}{\sin\alpha\cos\alpha} = \frac{1}{\sin\alpha\cos\alpha}.$$

The two members are identical, and the original equation is an identity.

In the above examples members of the equations were written in terms of sine and cosine function values to establish that the equations were indeed identities. Although there is no general method to prove an identity, as shown in the above examples there are three possible ways to proceed using algebraic properties and the basic identities that have been established.

1. Rewrite the right-hand member so that it is identical to the left-hand member. [See Equation (1) above.]
2. Rewrite the left-hand member so that it is identical to the right-hand member. [See Equation (2) above.]
3. Rewrite both members *independently* so that they are identical. [See Equation (3) above.]

Proving identities will provide you with the opportunity to use the basic identities and hence become familiar with these important relationships, and also to use the trigonometric function values (which simply represent real numbers) in conjunction with properties of algebra.

Following is a list of suggestions that may help in determining the best method to use to prove an identity.

1. If one member of an equation appears to be the more complicated, it should be transformed so that it reduces to the simpler member of the equation.
2. In each step of a proof, look for any possible application of algebraic procedures such as combining fractions, writing a single fraction as

two or more fractions, using the distributive law, factoring, or multiplying the numerator and denominator of a fraction by the conjugate of one or the other.

3. If an elegant approach in which an identity can be established in one or two steps is not obvious, as a last resort the identity can usually be proven by changing all the terms to expressions involving only sines and/or cosines, and then simplifying both members.

In the following examples, we illustrate several of these approaches.

Example Show that $\dfrac{\sin^2 x}{1 + \cos x} = \dfrac{\tan x - \sin x}{\tan x}$ is an identity.

Solution In this case we will simplify both members by appropriate substitutions. Expressing the right-hand member in terms of sines and cosines,

$$\begin{aligned}\frac{\sin^2 x}{1 + \cos x} &= \frac{\tan x - \sin x}{\tan x}\\ &= \frac{\tan x}{\tan x} - \frac{\sin x}{\tan x}\\ &= 1 - \frac{\sin x}{\dfrac{\sin x}{\cos x}}\\ &= 1 - \sin x \cdot \frac{\cos x}{\sin x}\\ &= 1 - \cos x.\end{aligned}$$

Rewriting the left-hand member by substituting $1 - \cos^2 x$ for $\sin^2 x$,

$$\frac{1 - \cos^2 x}{1 + \cos x} = 1 - \cos x.$$

Factoring $1 - \cos^2 x$ as $(1 - \cos x)(1 + \cos x)$ yields

$$\frac{(1 - \cos x)(1 + \cos x)}{1 + \cos x} = 1 - \cos x,$$

from which

$$1 - \cos x = 1 - \cos x.$$

Hence, the equation is an identity.

Example Show that $\dfrac{\cos x}{1 + \sin x} = \dfrac{1 - \sin x}{\cos x}$ is an identity.

Solution We first note that the numerator of the right-hand member is $1 - \sin x$. Hence, we try to obtain this expression in the numerator of the left-hand member by using an algebraic property. In this case, we can multiply the left-hand member by $(1 - \sin x)/(1 - \sin x)$ to obtain

$$\frac{\cos x(1 - \sin x)}{(1 + \sin x)(1 - \sin x)} = \frac{1 - \sin x}{\cos x},$$

from which

$$\frac{\cos x(1 - \sin x)}{1 - \sin^2 x} = \frac{1 - \sin x}{\cos x}.$$

Substituting $\cos^2 x$ for $1 - \sin^2 x$ in the left-hand member yields

$$\frac{\cos x(1 - \sin x)}{\cos^2 x} = \frac{1 - \sin x}{\cos x}.$$

Then, dividing the numerator and denominator of the left-hand member by $\cos x$,

$$\frac{1 - \sin x}{\cos x} = \frac{1 - \sin x}{\cos x},$$

which is clearly an identity.

Example Show that $\dfrac{\sec^2 \theta - \tan^2 \theta}{\sec \theta + \tan \theta} = \dfrac{1 - \sin \theta}{\cos \theta}$ is an identity.

Solution If we note that $\sec^2 \theta - \tan^2 \theta$ (of the form $a^2 - b^2$) can be factored as $(\sec \theta - \tan \theta)(\sec \theta + \tan \theta)$, we can make a substitution to obtain

$$\frac{(\sec \theta - \tan \theta)(\sec \theta + \tan \theta)}{\sec \theta + \tan \theta} = \frac{1 - \sin \theta}{\cos \theta}$$

from which

$$\sec \theta - \tan \theta = \frac{1 - \sin \theta}{\cos \theta}.$$

Now substituting $1/\cos\theta$ for $\sec\theta$ and $\sin\theta/\cos\theta$ for $\tan\theta$ and writing the left-hand member as a single fraction, we have that

$$\frac{1}{\cos\theta} - \frac{\sin\theta}{\cos\theta} = \frac{1-\sin\theta}{\cos\theta}$$

$$\frac{1-\sin\theta}{\cos\theta} = \frac{1-\sin\theta}{\cos\theta}.$$

Hence, the equation is an identity.

Counterexamples

In general, it is a relatively simple matter to show that an equation is not an identity, if such is the case, by finding a *counterexample.* For example, substituting 0 for α in the equation

$$\sin\alpha = \cos\alpha$$

yields

$$\sin 0 = \cos 0 \qquad \text{and} \qquad 0 = 1,$$

which is a false statement. Thus, the equation $\sin\alpha = \cos\alpha$ is not an identity.

Of course, in demonstrating a counterexample, we should select a replacement for the variable so that each member of the equation is defined.

EXERCISE SET 4.2

A

Simplify each expression on the left to show that it is equivalent to the expression on the right for all values of the variable for which both expressions are defined.

Example $(1-\cos^2\beta)(1+\cot^2\beta);\ 1$

Solution Substituting $\sin^2\beta$ for $1-\cos^2\beta$ and $\csc^2\beta$ for $1+\cot^2\beta$ yields

$$(1-\cos^2\beta)(1+\cot^2\beta) = (\sin^2\beta)(\csc^2\beta).$$

Since $\csc^2\beta = 1/\sin^2\beta$,

$$(1-\cos^2\beta)(1+\cot^2\beta) = (\sin^2\beta)\left(\frac{1}{\sin^2\beta}\right) = 1.$$

1. $\sec\alpha \sin\alpha$; $\tan\alpha$

2. $\tan\alpha \cos\alpha$; $\sin\alpha$

3. $\sin\beta \cot\beta$; $\cos\beta$

4. $\csc\beta \tan\beta$; $\sec\beta$

5. $\dfrac{\sin\gamma}{\tan\gamma}$; $\cos(-\gamma)$

6. $\dfrac{\sin(-\gamma)\sec\gamma}{\tan(-\gamma)}$; 1

7. $\cot^2\gamma \sin^2\gamma$; $\cos^2\gamma$

8. $\tan^2\alpha \cos^2\alpha$; $\sin^2\alpha$

9. $\cos^2\alpha(1 + \tan^2\alpha)$; 1

10. $(\csc^2\beta - 1)(\sin^2\beta)$; $\cos^2\beta$

11. $\dfrac{(1 + \sin\beta)(1 - \sin\beta)}{\cos\beta}$; $\cos\beta$

12. $\dfrac{(1 + \cos\gamma)(1 - \cos\gamma)}{\sin\gamma}$; $\sin\gamma$

13. $\dfrac{\cos^2\gamma}{\cot^2\gamma}$; $\sin^2\gamma$

14. $\dfrac{1 + \tan^2\gamma}{\csc^2\gamma}$; $\tan^2\gamma$

15. $\dfrac{\sin^2\alpha}{\tan^2\alpha - \sin^2\alpha}$; $\cot^2\alpha$

16. $\dfrac{\cos^2\alpha}{\cot^2\alpha - \cos^2\alpha}$; $\tan^2\alpha$

17. $\dfrac{1}{\tan\alpha + \cot\alpha}$; $\dfrac{\cos\alpha}{\csc\alpha}$

18. $\dfrac{1}{\csc\alpha - \sin\alpha}$; $\dfrac{\tan\alpha}{\cos\alpha}$

Verify that each equation is an identity.

19. $\cos x \tan x \csc x = 1$

20. $\cot x \sec x = \csc x$

21. $\sec y - \cos y = \tan y \sin y$

22. $\sec y - \tan y \sin y = \cos y$

23. $\cos^2 x - \sin^2 x = 2\cos^2 x - 1$

24. $(\sec^2 y - 1)(\csc^2 y - 1) = 1$

25. $\tan^2 y + \sec^2 y = \dfrac{2}{\cos^2 y} - 1$

26. $(\tan x + \cot x)\dfrac{\sin x}{\sec x} = 1$

27. $\dfrac{\cos x + \sin x}{\sin x} = 1 + \cot x$

28. $\dfrac{1 - \tan^2 y}{\tan y} = \cot y - \tan y$

29. $\tan^2 y - \sin^2 y = \tan^2 y \sin^2 y$

30. $\dfrac{1}{1 + \sin x} + \dfrac{1}{1 - \sin x} = \dfrac{2}{\cos^2 y}$

31. $\tan y + \cot y = \sec y \csc y$

32. $\dfrac{1 + \tan^2 y}{\tan^2 y} = \csc^2 y$

33. $\cos^4 x - \sin^4 x = \cos^2 x - \sin^2 x$ *Hint:* Factor left-hand member.

34. $\dfrac{1 + \sec x}{\sec x} = \dfrac{\sin^2 x}{1 - \cos x}$ *Hint:* $\sin^2 x = (1 - \cos x)(1 + \cos x)$.

35. $\dfrac{1 - \cos x}{\sin x} = \dfrac{\sin x}{1 + \cos x}$ *Hint:* $\dfrac{\sin x}{1 + \cos x} = \dfrac{\sin x(1 - \cos x)}{(1 + \cos x)(-\cos x)}$.

36. $\dfrac{1+\sin x}{\cos x} = \dfrac{\cos x}{1-\sin x}$ *Hint:* $\dfrac{1+\sin x}{\cos x} = \dfrac{(1+\sin x)(1-\sin x)}{\cos x(1-\sin x)}$

37. $\csc y = \dfrac{\sec y + \csc y}{1+\tan y}$

38. $\dfrac{1}{\sec y - \tan y} = \sec y + \tan y$

39. $\dfrac{\sec y - 1}{\sec y + 1} = \dfrac{1-\cos y}{1+\cos y}$

40. $\dfrac{\cos y}{\sec y - \tan y} = \dfrac{\cos^2 y}{1-\sin y}$

41. $\sec^2 y + \csc^2 y = \sec^2 y \csc^2 y$

42. $\sec^4 y - \sec^2 y = \tan^4 y + \tan^2 y$

43. $\cos(-\alpha)\tan(-\alpha)\csc\alpha = -1$

44. $\cot(-\alpha)\sec(-\alpha) = \csc(-\alpha)$

45. $\dfrac{\cos(-\alpha)+\sin\alpha}{\sin(-\alpha)} = -(1+\cot\alpha)$

46. $\dfrac{\cos(-\alpha)-\sin(-\alpha)}{\cos\alpha} = 1+\tan\alpha$

47. $\sec(-\alpha)\sin(-\alpha) = -\tan\alpha$

48. $\dfrac{\sin(-\alpha)}{\tan(-\alpha)} = \cos\alpha$

49. $\dfrac{\sin(-\alpha)\sec\alpha}{\tan(-\alpha)} = 1$

50. $\dfrac{1+\tan^2(-\alpha)}{\csc^2(-\alpha)} = \tan^2\alpha$

By using counterexamples show that each equation is not an identity.

51. $\sin x + \cos x \tan x = 2$

52. $\sin^2 x + 2\cos x - \cos^2 x = 1$

53. $2\sin y\cos y + \sin y = 0$

54. $\tan^2 y - 2\tan y = 0$

B

Verify that each equation is an identity.

55. $\sec\alpha = \dfrac{\cot\alpha + \tan\alpha}{\csc\alpha}$

56. $\dfrac{\cot\alpha + \tan\alpha}{\sec\alpha} = \csc\alpha$

57. $\sin^2\alpha = \dfrac{\cos^2\alpha + \tan^2\alpha - 1}{\tan^2\alpha}$

58. $\dfrac{\sin^2\alpha + \cot^2\alpha - 1}{\cot^2\alpha} = \cos^2\alpha$

59. $\sec\beta - 1 = \dfrac{\sec\beta + 1}{(\csc\beta + \cot\beta)^2}$

60. $\tan\beta + \sec\beta = \dfrac{1}{\sec\beta - \tan\beta}$

61. $\csc\gamma - \cot\gamma = \dfrac{1}{\cot\gamma + \csc\gamma}$

62. $\sec^2\gamma - \csc^2\gamma = \dfrac{\tan\gamma}{\cot\gamma} - \dfrac{\cot\gamma}{\tan\gamma}$

4.3 Sum and Difference Formulas for Cosine

In this section we shall establish several trigonometric identities involving the sum or difference of two angles or two real numbers. We first derive a formula for $\cos(\alpha + \beta)$. In Figure 4.2, angles α, $\alpha + \beta$, and $-\beta$ are in standard position, with their sides intersecting the circle of radius 1 at points P_1, P_2, P_3, and P_4. If P_1 has coordinates (x_1, y_1), then, since the radius r of the circle equals 1,

$$\cos\alpha = \frac{x_1}{1} \quad \text{and} \quad \sin\alpha = \frac{y_1}{1}.$$

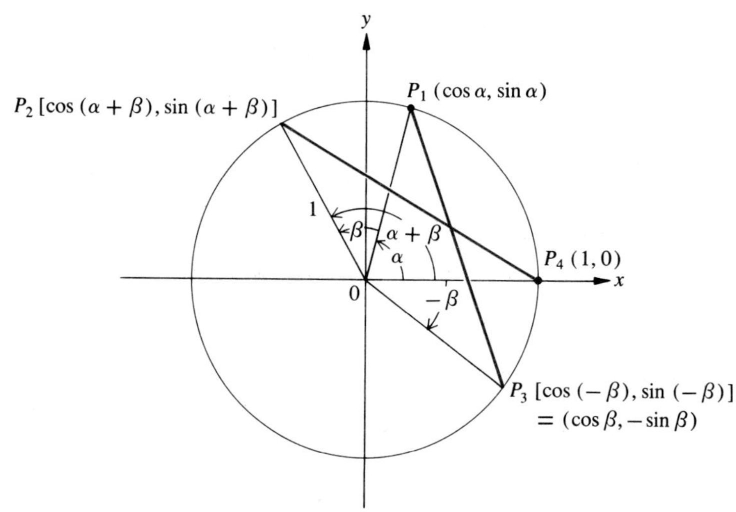

Figure 4.2

Hence, the coordinates at P_1 are $x_1 = \cos\alpha$ and $y_1 = \sin\alpha$. We obtain the coordinates at P_2 and P_3 in a similar way. Note that from Equations (1) and (2) in Section 4.1, we have $\sin(-\beta) = -\sin\beta$ and $\cos(-\beta) = \cos\beta$. Since all radii are equal and the central angles $\angle P_4OP_2$ and $\angle P_3OP_1$ are equal (see Appendix A.9),

$$\triangle P_4OP_2 \cong \triangle P_3OP_1$$

and

$$P_4P_2 = P_3P_1.$$

From the distance formula (see Appendix A.4), we have

$$\begin{aligned}(P_3P_1)^2 &= [\cos\alpha - \cos\beta]^2 + [\sin\alpha - (-\sin\beta)]^2 \\ &= (\cos\alpha - \cos\beta)^2 + (\sin\alpha + \sin\beta)^2 \\ &= \cos^2\alpha - 2\cos\alpha\cos\beta + \cos^2\beta \\ &\qquad + \sin^2\alpha + 2\sin\alpha\sin\beta + \sin^2\beta.\end{aligned}$$

Since $\cos^2\alpha + \sin^2\alpha = 1$ and $\cos^2\beta + \sin^2\beta = 1$,

$$(P_3P_1)^2 = 2 - 2\cos\alpha\cos\beta + 2\sin\alpha\sin\beta.$$

Again using the distance formula, we have

$$\begin{aligned}(P_4P_2)^2 &= [\cos(\alpha+\beta) - 1]^2 + [\sin(\alpha+\beta) - 0]^2 \\ &= \cos^2(\alpha+\beta) - 2\cos(\alpha+\beta) + 1 + \sin^2(\alpha+\beta).\end{aligned}$$

Since $\cos^2(\alpha+\beta) + \sin^2(\alpha+\beta) = 1$,

$$(P_4P_2)^2 = 2 - 2\cos(\alpha+\beta).$$

Since $P_4P_2 = P_3P_1$, we have

$$2 - 2\cos(\alpha+\beta) = 2 - 2\cos\alpha\cos\beta + 2\sin\alpha\sin\beta,$$

which gives

$$\boldsymbol{\cos(\alpha+\beta) = \cos\alpha\cos\beta - \sin\alpha\sin\beta.} \tag{1}$$

Substituting $-\beta$ for β in Equation (1), we have

$$\cos[\alpha + (-\beta)] = \cos\alpha\cos(-\beta) - \sin\alpha\sin(-\beta).$$

Since

$$\cos(-\beta) = \cos\beta \qquad \text{and} \qquad \sin(-\beta) = -\sin\beta,$$

we obtain

$$\boldsymbol{\cos(\alpha-\beta) = \cos\alpha\cos\beta + \sin\alpha\sin\beta.} \tag{2}$$

These very important relationships for $\cos(\alpha+\beta)$ and $\cos(\alpha-\beta)$ are called the **sum formula** and the **difference formula,** respectively, for the cosine function.

We can, as illustrated in the following examples, find some function values using the sum and difference formulas. Although using a calculator or tables is a more efficient way of finding these function values, such examples give us practice in applying the formulas.

Example Compute $\cos 15°$.

Solution Because we know the exact function values of 45° and 30°, we replace 15° with $45° - 30°$ in Equation (2) and obtain

$$\begin{aligned}\cos 15° &= \cos(45° - 30°)\\ &= \cos 45° \cos 30° + \sin 45° \sin 30°\\ &= \frac{1}{\sqrt{2}} \cdot \frac{\sqrt{3}}{2} + \frac{1}{\sqrt{2}} \cdot \frac{1}{2} = \frac{\sqrt{3}+1}{2\sqrt{2}}.\end{aligned}$$

Example Compute $\cos \dfrac{7\pi}{12}$.

Solution Since $\dfrac{1}{3} + \dfrac{1}{4} = \dfrac{7}{12}$, we have $\dfrac{7\pi}{12} = \dfrac{\pi}{3} + \dfrac{\pi}{4}$. By Equation (1),

$$\cos \frac{7\pi}{12} = \cos\left(\frac{\pi}{3} + \frac{\pi}{4}\right) = \cos \frac{\pi}{3} \cos \frac{\pi}{4} - \sin \frac{\pi}{3} \sin \frac{\pi}{4}.$$

The appropriate substitutions for the function values yields

$$\cos \frac{7\pi}{12} = \frac{1}{2} \cdot \frac{1}{\sqrt{2}} - \frac{\sqrt{3}}{2} \cdot \frac{1}{\sqrt{2}} = \frac{1-\sqrt{3}}{2\sqrt{2}}.$$

In general we shall leave such results in radical form. However, you can, if you wish, obtain a rational number approximation. For example, since $\sqrt{2} \approx 1.414$ and $\sqrt{3} \approx 1.732$,

$$\cos \frac{7\pi}{12} \approx \frac{1 - 1.732}{2(1.414)} \approx -0.2588.$$

The sum and difference formulas for the cosine function can be used to establish other identities. Three identities that we will be using are

$$\mathbf{\cos(180° - \alpha) = -\cos \alpha,} \tag{3}$$

$$\mathbf{\cos(180° + \alpha) = -\cos \alpha,} \tag{4}$$

$$\mathbf{\cos(360° - \alpha) = \cos \alpha.} \tag{5}$$

Identities such as these are known as **reduction formulas.** The proof that Equation (3) is an identity is shown here; the proofs of (4) and (5) are left as exercises.

Using the difference formula [Equation (2) on page 165] and substituting 180° for α and α for β, we obtain

$$\cos(180^\circ - \alpha) = (\cos 180^\circ)(\cos \alpha) + (\sin 180^\circ)(\sin \alpha).$$

Since $\cos 180^\circ = -1$ and $\sin 180^\circ = 0$, we have

$$\cos(180^\circ - \alpha) = (-1)\cos \alpha + (0)\sin \alpha = -\cos \alpha.$$

Reduction formulas can be used instead of relying on sketches of reference angles to rewrite a function value of an angle greater than 90° or less than 0° as a function value of an acute angle.

Examples Write each expression as the cosine of an acute angle.

a. $\cos 150^\circ$ b. $\cos 225^\circ$

Solutions

a. Since $150^\circ = 180^\circ - 30^\circ$, we use Equation (3) to obtain

$$\cos 150^\circ = \cos(180^\circ - 30^\circ) = -\cos 30^\circ.$$

b. Since $225^\circ = 180^\circ + 45^\circ$, we use Equation (4) to obtain

$$\cos 225^\circ = \cos(180^\circ + 45^\circ) = -\cos 45^\circ.$$

EXERCISE SET 4.3

A

Find the exact value of each expression using Equation (1) or Equation (2) of this section.

1. $\cos 75^\circ$ (Use: $75^\circ = 120^\circ - 45^\circ$)

2. $\cos 105^\circ$ (Use: $105^\circ = 60^\circ + 45^\circ$)

3. $\cos 165^\circ$ (Use: $165^\circ = 120^\circ + 45^\circ$)

4. $\cos 195^\circ$ (Use: $195^\circ = 135^\circ + 60^\circ$)

5. $\cos \frac{\pi}{12}$ $\left(\text{Use: } \frac{\pi}{12} = \frac{\pi}{3} - \frac{\pi}{4}\right)$

6. $\cos \frac{5\pi}{12}$ $\left(\text{Use: } \frac{5\pi}{12} = \frac{\pi}{6} + \frac{\pi}{4}\right)$

Verify that each equation is correct.

Example $\cos 110° \cos 70° - \sin 110° \sin 70° = -1$

Solution From Equation (1), if we replace α with $110°$ and β with $70°$, we obtain

$$\begin{aligned}\cos 110° \cos 70° - \sin 110° \sin 70° &= \cos(110° + 70°)\\ &= \cos 180° = -1.\end{aligned}$$

7. $\cos 40° \cos 140° - \sin 40° \sin 140° = -1$

8. $\cos 20° \cos 160° - \sin 20° \sin 160° = -1$

9. $\cos 35° \cos 55° - \sin 35° \sin 55° = 0$

10. $\cos 50° \cos 40° - \sin 50° \sin 40° = 0$

11. $\cos 100° \cos 55° + \sin 100° \sin 55° = \dfrac{1}{\sqrt{2}}$

12. $\cos 130° \cos 70° + \sin 130° \sin 70° = \dfrac{1}{2}$

Write each expression as a single function of kx, k an integer.

Example $\cos 2x \cos 3x - \sin 2x \sin 3x$

Solution From Equation (1),

$$\begin{aligned}\cos 2x \cos 3x - \sin 2x \sin 3x &= \cos(2x + 3x)\\ &= \cos 5x.\end{aligned}$$

13. $\cos 4x \cos 5x - \sin 4x \sin 5x$

14. $\cos x \cos 2x - \sin x \sin 2x$

15. $\cos 8x \cos 6x + \sin 8x \sin 6x$

16. $\cos 5x \cos x + \sin 5x \sin x$

17. $\cos \frac{2}{3} x \cos \frac{1}{3} x - \sin \frac{2}{3} x \sin \frac{1}{3} x$

18. $\cos \frac{3}{4} x \cos \frac{5}{4} x - \sin \frac{3}{4} x \sin \frac{5}{4} x$

19. $\cos(x + 70°) \cos(x - 70°) - \sin(x + 70°) \sin(x - 70°)$

20. $\cos(65° + x) \cos(65° - x) + \sin(65° + x) \sin(65° - x)$

If angles α and β have the given function values and are in the indicated quadrant, use Equation (1) to find $\cos(\alpha + \beta)$.

Example $\cos \alpha = -1/\sqrt{5}$ and $\sin \beta = 2/\sqrt{13}$; α in II and β in I

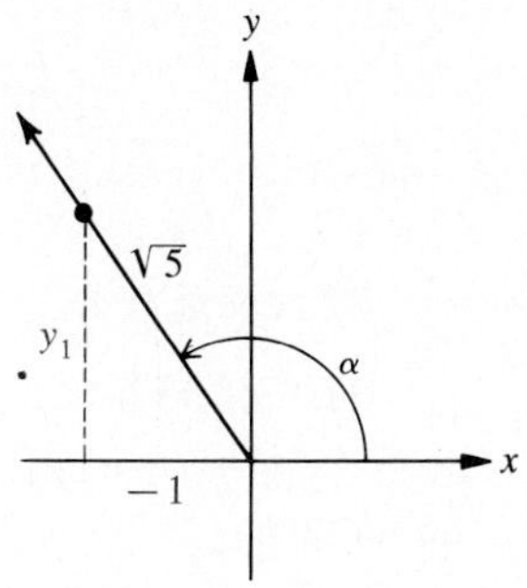

Solution Since α is in Quadrant II, as shown in the sketch, using the Pythagorean theorem.

$$y_1^2 + (-1)^2 = (\sqrt{5})^2,$$
$$y_1^2 + 1 = 5,$$
$$y_1^2 = 4,$$
$$y_1 = 2 \quad (-2 \text{ does not apply}).$$

Hence, $\sin \alpha = \dfrac{y_1}{\sqrt{5}} = \dfrac{2}{\sqrt{5}}.$

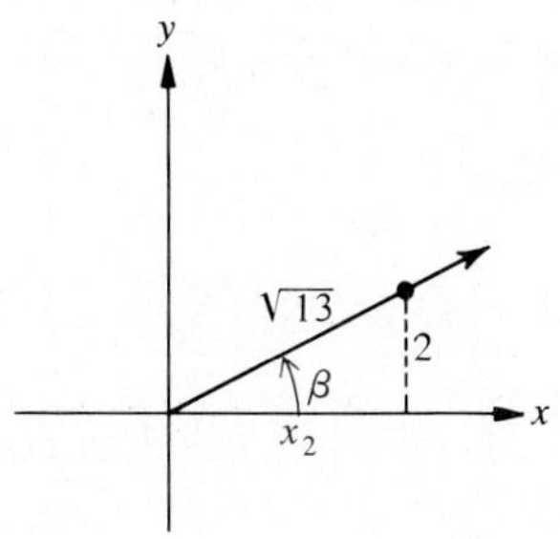

Since β is in Quadrant I, from the sketch,

$$x_2^2 + 2^2 = (\sqrt{13})^2,$$
$$x_2^2 + 4 = 13,$$
$$x_2^2 = 9,$$
$$x_2 = 3 \quad (-3 \text{ does not apply}).$$

Hence, $\cos \beta = \dfrac{x_2}{\sqrt{13}} = \dfrac{3}{\sqrt{13}}.$ Therefore,

$$\begin{aligned}\cos(\alpha + \beta) &= \cos \alpha \cos \beta - \sin \alpha \sin \beta \\ &= -\frac{1}{\sqrt{5}} \cdot \frac{3}{\sqrt{13}} - \frac{2}{\sqrt{5}} \cdot \frac{2}{\sqrt{13}} \\ &= -\frac{3}{\sqrt{65}} - \frac{4}{\sqrt{65}} = -\frac{7}{\sqrt{65}}.\end{aligned}$$

21. $\cos \alpha = 3/5$ and $\sin \beta = 5/13$; α and β in Quadrant I

22. $\cos \alpha = -4/5$ and $\sin \beta = -7/25$; α in Quadrant III and β in IV

23. $\sin \alpha = 1/\sqrt{10}$ and $\cos \beta = -4/\sqrt{17}$; α in Quadrant II and β in III

24. $\sin \alpha = -5/\sqrt{29}$ and $\cos \beta = 1/\sqrt{26}$; α in Quadrant IV and β in I

Use one of the reduction formulas [Equation (3), (4), or (5)] of this section to write each expression as a function value of an acute angle.

Example $\cos 135°$

Solution Using Equation (3),

$$\cos 135° = \cos(180° - 45°) = -\cos 45°.$$

25. $\cos 120°$ **26.** $\cos 240°$ **27.** $\cos 210°$

28. $\cos 300°$ **29.** $\cos 315°$ **30.** $\cos 330°$

Verify that each equation in Exercises 31–35 is an identity.

31. $\cos(180° + \alpha) = -\cos \alpha$ **32.** $\cos(360° - \alpha) = \cos \alpha$

33. $\cos(90° - \alpha) = \sin \alpha$ **34.** $\cos(90° + \alpha) = -\sin \alpha$

35. $\cos 2x = \cos^2 x - \sin^2 x.$ *Hint:* $\cos 2x = \cos(x + x).$

36. Use the results of Exercise 35 to show that

$$\cos 2x = 2 \cos^2 x - 1 = 1 - 2 \sin^2 x.$$

37. Show by counterexample that $\cos(\alpha + \beta) = \cos \alpha + \cos \beta$ is not an identity.

38. Show by counterexample that $\cos(\alpha - \beta) = \cos \alpha - \cos \beta$ is not an identity.

39. Show that $\cos\left(\frac{\pi}{2} - \alpha\right) = \sin \alpha.$

40. Use the results of Exercise 39 to show that $\sin\left(\frac{\pi}{2} - \alpha\right) = \cos \alpha.$

Verify that each equation in Exercises 41–48 is an identity. Equations whose graphs are related to these identities were given as Exercises 25–32 in Section 2.3.

41. $\sin\left(x + \frac{\pi}{2}\right) = \cos x$ **42.** $\cos\left(x - \frac{\pi}{2}\right) = \sin x$

43. $\cos\left(x + \frac{\pi}{2}\right) = -\sin x$ **44.** $\sin\left(x - \frac{\pi}{2}\right) = -\cos x$

45. $\sqrt{2}\sin\left(x+\frac{\pi}{4}\right)=\sin x+\cos x$

46. $\sqrt{2}\sin\left(x-\frac{\pi}{4}\right)=\sin x-\cos x$

47. $\sqrt{2}\cos\left(x+\frac{\pi}{4}\right)=\cos x-\sin x$

48. $\sqrt{2}\cos\left(x-\frac{\pi}{4}\right)=\sin x+\cos x$

B

Evaluate each of the following.

Example $\cos\left(\text{Arccot}\,\frac{1}{\sqrt{3}}+\text{Arcsin}\,\frac{1}{\sqrt{2}}\right)$

Solution Let

$$\alpha=\text{Arccot}\,\frac{1}{\sqrt{3}}\qquad\text{and}\qquad\beta=\text{Arcsin}\,\frac{1}{\sqrt{2}}.\tag{6}$$

Then, using symbols α and β and the sum formula on page 165, we have

$$\cos(\alpha+\beta)=\cos\alpha\cos\beta-\sin\alpha\sin\beta.$$

By inspection of Equations (6), $\alpha=(\pi/3)^{\mathsf{R}}$. Hence,

$$\cos\alpha=\frac{1}{2}\qquad\text{and}\qquad\sin\alpha=\frac{\sqrt{3}}{2}.$$

Also by inspection of (1), $\beta=(\pi/4)^{\mathsf{R}}$. Thus, we have that

$$\sin\beta=\frac{1}{\sqrt{2}}\qquad\text{and}\qquad\cos\beta=\frac{1}{\sqrt{2}}.$$

Therefore,

$$\begin{aligned}\cos\left(\text{Arccot}\,\frac{1}{\sqrt{3}}+\text{Arcsin}\,\frac{1}{\sqrt{2}}\right)&=\frac{1}{2}\cdot\frac{1}{\sqrt{2}}-\frac{\sqrt{3}}{2}\cdot\frac{1}{\sqrt{2}}\\&=\frac{1-\sqrt{3}}{2\sqrt{2}}.\end{aligned}$$

49. $\cos\left(\text{Arcsin}\,\frac{1}{2} + \text{Arcsec}\,2\right)$ **50.** $\cos\left(\text{Arccos}\,\frac{1}{2} + \text{Arccsc}\,2\right)$

51. $\cos\left(\text{Arctan}\,1 - \text{Arccos}\,\frac{\sqrt{3}}{2}\right)$ **52.** $\cos\left(\text{Arcsin}\,\frac{1}{\sqrt{2}} - \text{Arcsec}\,1\right)$

53. $\cos[\text{Csc}^{-1}(-2) - \text{Sec}^{-1}\sqrt{2}]$

54. $\cos\left[\text{Cos}^{-1}\left(-\frac{1}{\sqrt{2}}\right) - \text{Sin}^{-1}(-1)\right]$

4.4 Sum and Difference Formulas for Sine

Sum and difference formulas for the sine function can be obtained from the sum and difference formulas for the cosine function. First, however, we consider two relationships between function values for the cosine and sine functions. From the difference formula for the cosine function we have

$$\cos\left(\frac{\pi}{2} - \theta\right) = \cos\frac{\pi}{2}\cos\theta + \sin\frac{\pi}{2}\sin\theta.$$

Replacing $\cos(\pi/2)$ with 0 and $\sin(\pi/2)$ with 1, we have

$$\cos\left(\frac{\pi}{2} - \theta\right) = 0 \cdot \cos\theta + 1 \cdot \sin\theta,$$

$$\boldsymbol{\cos\left(\frac{\pi}{2} - \theta\right) = \sin\theta.} \tag{1}$$

Replacing θ with $\pi/2 - \theta$ in Equation (1), we obtain

$$\cos\left[\frac{\pi}{2} - \left(\frac{\pi}{2} - \theta\right)\right] = \sin\left(\frac{\pi}{2} - \theta\right),$$

$$\cos\left[\left(\frac{\pi}{2} - \frac{\pi}{2}\right) + \theta\right] = \sin\left(\frac{\pi}{2} - \theta\right),$$

$$\cos\theta = \sin\left(\frac{\pi}{2} - \theta\right),$$

$$\boldsymbol{\sin\left(\frac{\pi}{2} - \theta\right) = \cos\theta.} \tag{2}$$

Examples

a. $\sin\dfrac{\pi}{3} = \cos\left(\dfrac{\pi}{2} - \dfrac{\pi}{3}\right)$

$= \cos\dfrac{\pi}{6}$

b. $\cos 46^\circ = \sin(90^\circ - 46^\circ)$

$= \sin 44^\circ$

In Section 3.5 we noted that the sine and cosine are cofunctions. Equations (1) and (2) also establish this fact because in each case $(\pi/2 - \theta) + \theta = \pi/2$.

Now we consider the sum and difference formulas for the sine function. If we replace θ with $\alpha + \beta$ in Equation (1), we have

$$\begin{aligned}\sin(\alpha + \beta) &= \cos\left[\frac{\pi}{2} - (\alpha + \beta)\right]\\ &= \cos\left[\left(\frac{\pi}{2} - \alpha\right) - \beta\right],\end{aligned}$$

which by the difference formula for the cosine can be written as

$$\sin(\alpha + \beta) = \cos\left(\frac{\pi}{2} - \alpha\right)\cos\beta + \sin\left(\frac{\pi}{2} - \alpha\right)\sin\beta.$$

By Equations (1) and (2),

$$\cos\left(\frac{\pi}{2} - \alpha\right) = \sin\alpha \quad \text{and} \quad \sin\left(\frac{\pi}{2} - \alpha\right) = \cos\alpha.$$

Hence, we have

$$\boldsymbol{\sin(\alpha + \beta) = \sin\alpha\cos\beta + \cos\alpha\sin\beta.} \tag{3}$$

Replacing β with $-\beta$ in Equation (3) yields

$$\sin[\alpha + (-\beta)] = \sin\alpha\cos(-\beta) + \cos\alpha\sin(-\beta).$$

Substituting $\cos\beta$ for $\cos(-\beta)$ and $(-\sin\beta)$ for $\sin(-\beta)$, we have

$$\sin(\alpha - \beta) = \sin\alpha\cos\beta + \cos\alpha(-\sin\beta),$$

from which

$$\boldsymbol{\sin(\alpha - \beta) = \sin\alpha\cos\beta - \cos\alpha\sin\beta.} \tag{4}$$

Example Compute $\sin 15°$.

Solution Since $15° = 45° - 30°$, by Equation (4),

$$\sin 15° = \sin(45° - 30°) = \sin 45° \cos 30° - \cos 45° \sin 30°$$
$$= \frac{1}{\sqrt{2}} \cdot \frac{\sqrt{3}}{2} - \frac{1}{\sqrt{2}} \cdot \frac{1}{2} = \frac{\sqrt{3} - 1}{2\sqrt{2}}.$$

Example Compute $\sin \frac{\pi}{12}$.

Solution Observe that $\frac{\pi}{12} = \frac{\pi}{3} - \frac{\pi}{4}$. By using Equation (4),

$$\sin \frac{\pi}{12} = \sin\left(\frac{\pi}{3} - \frac{\pi}{4}\right) = \sin \frac{\pi}{3} \cos \frac{\pi}{4} - \cos \frac{\pi}{3} \sin \frac{\pi}{4}.$$

Making the appropriate substitutions for the function values yields

$$\sin \frac{\pi}{12} = \frac{\sqrt{3}}{2} \cdot \frac{1}{\sqrt{2}} - \frac{1}{2} \cdot \frac{1}{\sqrt{2}} = \frac{\sqrt{3} - 1}{2\sqrt{2}}.$$

The same results are obtained if $\pi/12$ is written as $(\pi/4) - (\pi/6)$.

Again we note that in practical applications, approximations for function values such as $\sin(\pi/12)$ can be found readily by using a calculator.

The sum and difference formulas for the sine function can be used to establish the following reduction formulas.

$$\mathbf{\sin(180° - \alpha) = \sin \alpha,} \tag{5}$$

$$\mathbf{\sin(180° + \alpha) = -\sin \alpha,} \tag{6}$$

$$\mathbf{\sin(360° - \alpha) = -\sin \alpha.} \tag{7}$$

The proof that Equation (5) is an identity is presented here; the proofs of (6) and (7) are left as exercises.

Using the difference formula, Equation (4), and substituting 180° for α and α for β, we obtain

$$\sin(180° - \alpha) = \sin 180° \cos \alpha - \cos 180° \sin \alpha.$$

Since

$$\sin 180° = 0 \quad \text{and} \quad \cos 180° = -1,$$

we have

$$\sin(180° - \alpha) = 0 \cdot \cos \alpha - (-1)\sin \alpha = \sin \alpha.$$

EXERCISE SET 4.4

A

Find the value of each expression using Equation (3) or (4).

1. $\sin 105°$ (Use: $105° = 60° + 45°$)

2. $\sin 75°$ (Use: $75° = 120° - 45°$)

3. $\sin 165°$ (Use: $165° = 210° - 45°$)

4. $\sin 195°$ (Use: $195° = 135° + 60°$)

5. $\sin \frac{5\pi}{12}$ $\left(\text{Use: } \frac{5\pi}{12} = \frac{\pi}{6} + \frac{\pi}{4}\right)$

6. $\sin \frac{\pi}{12}$ $\left(\text{Use: } \frac{\pi}{12} = \frac{\pi}{3} - \frac{\pi}{4}\right)$

Verify that each equation is correct.

Example $\sin 100° \cos 80° + \cos 100° \sin 80° = 0$

Solution From Equation (3) of this section, if we replace α with $100°$ and β with $80°$,

$$\sin 100° \cos 80° + \cos 100° \sin 80° = \sin(100° + 80°) = \sin 180° = 0.$$

7. $\sin 20° \cos 160° + \cos 20° \sin 160° = 0$

8. $\sin 40° \cos 140° + \cos 40° \sin 140° = 0$

9. $\sin 50° \cos 40° + \cos 50° \sin 40° = 1$

10. $\sin 35° \cos 55° + \cos 35° \sin 55° = 1$

11. $\sin 110° \cos 80° - \cos 110° \sin 80° = \frac{1}{2}$

12. $\sin 95° \cos 50° - \cos 95° \sin 50° = \frac{1}{\sqrt{2}}$

Write each expression as a single function of kx, where k is an integer.

Example $\sin 2x \cos 3x + \cos 2x \sin 3x$ *(continued)*

Solution Using Equation (3), we have

$$\sin 2x \cos 3x + \cos 2x \sin 3x = \sin(2x + 3x) = \sin 5x.$$

13. $\sin x \cos 2x + \cos x \sin 2x$

14. $\sin 4x \cos 5x + \cos 4x \sin 5x$

15. $\sin 5x \cos x - \cos 5x \sin x$

16. $\sin 8x \cos 6x - \cos 8x \sin 6x$

17. $\sin \frac{3}{4}x \cos \frac{5}{4}x + \cos \frac{3}{4}x \sin \frac{5}{4}x$

18. $\sin \frac{4}{3}x \cos \frac{1}{3}x - \cos \frac{4}{3}x \sin \frac{1}{3}x$

19. $\sin(50^\circ + x) \cos(50^\circ - x) - \cos(50^\circ + x) \sin(50^\circ - x)$

20. $\sin(x + 60^\circ) \cos(x - 60^\circ) + \cos(x + 60^\circ) \sin(x - 60^\circ)$

If angles α and β have the given function values and are in the indicated quadrants, use Equation (3) to find $\sin(\alpha + \beta)$.

Example $\cos \alpha = -1/\sqrt{5}$, $\sin \beta = 2/\sqrt{13}$, α in Quadrant II, β in I

Solution From the example on page 169, we have $\sin \alpha = 2/\sqrt{5}$ and $\cos \beta = 3/\sqrt{13}$. Therefore, from Equation (3),

$$\begin{aligned}\sin(\alpha + \beta) &= \sin \alpha \cos \beta + \cos \alpha \sin \beta \\ &= \frac{2}{\sqrt{5}} \cdot \frac{3}{\sqrt{13}} + \left(-\frac{1}{\sqrt{5}}\right) \cdot \frac{2}{\sqrt{13}} \\ &= \frac{6}{\sqrt{65}} - \frac{2}{\sqrt{65}} = \frac{4}{\sqrt{65}}.\end{aligned}$$

21. $\cos \alpha = -4/5$, $\sin \beta = -7/25$, α in Quadrant III, β in IV

22. $\cos \alpha = 3/5$, $\sin \beta = 5/13$, α and β in Quadrant I

23. $\sin \alpha = -5/\sqrt{29}$, $\cos \beta = 1/\sqrt{26}$, α in Quadrant IV, β in I

24. $\sin \alpha = 1/\sqrt{10}$, $\cos \beta = -4/\sqrt{17}$, α in Quadrant II, β in III

Use one of the reduction formulas [Equation (5), (6), or (7)] of this section to write each expression as a function of an acute angle.

Example $\sin 135^\circ$

Solution By Equation (5),

$$\sin 135^\circ = \sin(180^\circ - 45^\circ) = \sin 45^\circ.$$

25. $\sin 240°$ **26.** $\sin 120°$ **27.** $\sin 150°$

28. $\sin 210°$ **29.** $\sin 330°$ **30.** $\sin 315°$

Verify that each equation is an identity.

31. $\sin(180° + \alpha) = -\sin \alpha$

32. $\sin(360° - \alpha) = -\sin \alpha$

33. $\sin(270° - \alpha) = -\cos \alpha$

34. $\sin(270° + \alpha) = -\cos \alpha$

35. $\sin 2x = 2 \sin x \cos x$. *Hint:* $\sin 2x = \sin(x + x)$.

36. $\sin(\alpha - 30°) = \dfrac{\sqrt{3} \sin \alpha - \cos \alpha}{2}$

37. Show by a counterexample that $\sin(\alpha - \beta) = \sin \alpha - \sin \beta$ is not an identity.

38. Show by a counterexample that $\sin(\alpha + \beta) = \sin \alpha + \sin \beta$ is not an identity.

B

Evaluate each of the following.

Example $\sin\left(\text{Arcsin} \dfrac{1}{\sqrt{2}} + \text{Arccos} \dfrac{1}{2}\right)$

Solution Let

$$\alpha = \text{Arcsin} \frac{1}{\sqrt{2}} \quad \text{and} \quad \beta = \text{Arccos} \frac{1}{2}. \tag{8}$$

Then, by using the symbols α and β, and by the sum formula,

$$\sin\left(\text{Arcsin} \frac{1}{\sqrt{2}} + \text{Arccos} \frac{1}{2}\right) = \sin(\alpha + \beta)$$

$$= \sin \alpha \cos \beta + \cos \alpha \sin \beta.$$

By inspection of Equations (1), $\alpha = (\pi/4)^{\text{R}}$, and hence

$$\sin \alpha = \frac{1}{\sqrt{2}} \quad \text{and} \quad \cos \alpha = \frac{1}{\sqrt{2}}.$$

(continued)

Also, by inspection of (8), $\beta = (\pi/3)^{\mathrm{R}}$, and hence $\cos \beta = 1/2$ and $\sin \beta = \sqrt{3}/2$. Therefore,

$$\sin\left(\operatorname{Arcsin}\frac{1}{\sqrt{2}} + \operatorname{Arccos}\frac{1}{2}\right) = \frac{1}{\sqrt{2}} \cdot \frac{1}{2} + \frac{1}{\sqrt{2}} \cdot \frac{\sqrt{3}}{2}$$

$$= \frac{1}{2\sqrt{2}} + \frac{\sqrt{3}}{2\sqrt{2}} = \frac{1 + \sqrt{3}}{2\sqrt{2}}.$$

39. $\sin(\operatorname{Arcsin} 0 + \operatorname{Arccos} 1)$

40. $\sin(\operatorname{Arccos} 0 + \operatorname{Arcsin} 1)$

41. $\sin\left(\operatorname{Arcsin} 0 - \operatorname{Arcsec}\frac{2}{\sqrt{3}}\right)$

42. $\sin\left(\operatorname{Arccos} 1 - \operatorname{Arcsin}\frac{1}{\sqrt{2}}\right)$

43. $\sin\left(\operatorname{Arctan} 1 - \operatorname{Arccos}\frac{\sqrt{3}}{2}\right)$

44. $\sin\left(\operatorname{Arcsin}\frac{1}{2} - \operatorname{Arcsec} 2\right)$

4.5 Sum and Difference Formulas for Tangent

Sum and difference formulas for the tangent function follow from the sum and difference formulas for the sine and cosine functions. Since

$$\tan(\alpha + \beta) = \frac{\sin(\alpha + \beta)}{\cos(\alpha + \beta)},$$

by substituting appropriately, we have

$$\tan(\alpha + \beta) = \frac{\sin \alpha \cos \beta + \cos \alpha \sin \beta}{\cos \alpha \cos \beta - \sin \alpha \sin \beta}.$$

Multiplying each term of the numerator and denominator of the right-hand member by $1/\cos \alpha \cos \beta$ yields

$$\tan(\alpha + \beta) = \frac{\dfrac{\sin \alpha \cos \beta}{\cos \alpha \cos \beta} + \dfrac{\cos \alpha \sin \beta}{\cos \alpha \cos \beta}}{\dfrac{\cos \alpha \cos \beta}{\cos \alpha \cos \beta} - \dfrac{\sin \alpha \sin \beta}{\cos \alpha \cos \beta}},$$

from which, by simplifying the right-hand member, we obtain

$$\boldsymbol{\tan(\alpha + \beta) = \frac{\tan \alpha + \tan \beta}{1 - \tan \alpha \tan \beta}}. \tag{1}$$

By substituting $-\beta$ for β in Equation (1), we have

$$\tan[\alpha + (-\beta)] = \frac{\tan \alpha + \tan(-\beta)}{1 - \tan \alpha \tan(-\beta)}.$$

Substituting $-\tan \beta$ for $\tan(-\beta)$ yields

$$\tan[\alpha + (-\beta)] = \frac{\tan \alpha + (-\tan \beta)}{1 - \tan \alpha(-\tan \beta)},$$

$$\boldsymbol{\tan(\alpha - \beta) = \frac{\tan \alpha - \tan \beta}{1 + \tan \alpha \tan \beta}}. \tag{2}$$

Example Compute tan 75°.

Solution Observe that $75° = 45° + 30°$. Then, using Equation (1) and replacing α with 45° and β with 30°,

$$\tan 75° = \tan(45° + 30°) = \frac{\tan 45° + \tan 30°}{1 - \tan 45° \tan 30°}$$

$$= \frac{1 + \frac{1}{\sqrt{3}}}{1 - 1\left(\frac{1}{\sqrt{3}}\right)} = \frac{\left(1 + \frac{1}{\sqrt{3}}\right)\cdot\sqrt{3}}{\left(1 - \frac{1}{\sqrt{3}}\right)\cdot\sqrt{3}} = \frac{\sqrt{3} + 1}{\sqrt{3} - 1}.$$

The sum and difference formulas established in this section can be used to establish the following reduction formulas.

$$\boldsymbol{\tan(180° - \alpha) = -\tan \alpha,} \tag{3}$$

$$\boldsymbol{\tan(180° + \alpha) = \tan \alpha,} \tag{4}$$

$$\boldsymbol{\tan(360° - \alpha) = -\tan \alpha.} \tag{5}$$

The proof that Equation (3) is an identity is presented here; the proofs of (4) and (5) are left as exercises.

Using the difference formula [Equation (2)] and substituting 180° for α and α for β, we obtain

$$\tan(180° - \alpha) = \frac{\tan 180° - \tan \alpha}{1 + \tan 180° \tan \alpha}$$

$$= \frac{0 - \tan \alpha}{1 + (0)\tan \alpha}$$

$$= -\tan \alpha.$$

EXERCISE SET 4.5

A

Use Equation (1) or (2) to find the value of each expression.

1. $\tan 105°$ (Use: $105° = 45° + 60°$)

2. $\tan 15°$ (Use: $15° = 45° - 30°$)

3. $\tan 195°$ (Use: $195° = 240° - 45°$)

4. $\tan 165°$ (Use: $165° = 120° + 45°$)

5. $\tan \frac{\pi}{12} \left(\text{Use: } \frac{\pi}{12} = \frac{\pi}{3} - \frac{\pi}{4} \right)$

6. $\tan \frac{3\pi}{4} \left(\text{Use: } \frac{3\pi}{4} = \pi - \frac{\pi}{4} \right)$

Verify that each equation is correct.

Example $\dfrac{\tan 20° + \tan 25°}{1 - \tan 20° \tan 25°} = 1$

Solution From Equation (1), if we replace α with $20°$ and β with $25°$,

$$\frac{\tan 20° + \tan 25°}{1 - \tan 20° \tan 25°} = \tan(20° + 25°)$$

$$= \tan 45° = 1.$$

7. $\dfrac{\tan 50° + \tan 130°}{1 - \tan 50° \tan 130°} = 0$

8. $\dfrac{\tan 100° + \tan 50°}{1 - \tan 100° \tan 50°} = -\dfrac{1}{\sqrt{3}}$

9. $\dfrac{\tan 140° + \tan 100°}{1 - \tan 140° \tan 100°} = \sqrt{3}$

10. $\dfrac{\tan 95° + \tan 40°}{1 - \tan 95° \tan 40°} = -1$

11. $\dfrac{\tan 110° - \tan 50°}{1 + \tan 110° \tan 50°} = \sqrt{3}$

12. $\dfrac{\tan 115° - \tan 70°}{1 + \tan 115° \tan 70°} = 1$

Write each expression as a single function of kx where k is an integer.

Example $\dfrac{\tan 2x + \tan 3x}{1 - \tan 2x \tan 3x}$

Solution Using Equation (1), we obtain

$$\frac{\tan 2x + \tan 3x}{1 - \tan 2x \tan 3x} = \tan(2x + 3x)$$

$$= \tan 5x.$$

13. $\dfrac{\tan 4x + \tan 5x}{1 - \tan 4x \tan 5x}$

14. $\dfrac{\tan 2x + \tan x}{1 - \tan 2x \tan x}$

15. $\dfrac{\tan 8x - \tan 6x}{1 + \tan 8x \tan 6x}$

16. $\dfrac{\tan 5x - \tan x}{1 + \tan 5x \tan x}$

17. $\dfrac{\tan \frac{4}{3}x - \tan \frac{1}{3}x}{1 + \tan \frac{4}{3}x \tan \frac{1}{3}x}$

18. $\dfrac{\tan \frac{5}{4}x + \tan \frac{3}{4}x}{1 - \tan \frac{5}{4}x \tan \frac{3}{4}x}$

Use one of the reduction formulas [Equation (3), (4), or (5)] of this section to write each expression as a function of an acute angle.

Example $\tan 135°$

Solution Using Equation (3), we obtain

$$\tan 135° = \tan(180° - 45°) = -\tan 45°.$$

19. $\tan 120°$ **20.** $\tan 240°$ **21.** $\tan 210°$

22. $\tan 150°$ **23.** $\tan 315°$ **24.** $\tan 330°$

Verify that each equation is an identity.

25. $\tan(180° + \alpha) = \tan \alpha$

26. $\tan(360° - \alpha) = -\tan \alpha$

27. $\tan(90° - \alpha) = \cot \alpha$

28. $\cot(90° - \alpha) = \tan \alpha$

29. $\tan 2x = \dfrac{2 \tan x}{1 - \tan^2 x}$. *Hint:* $\tan 2x = \tan(x + x)$.

30. $\tan\left(x + \dfrac{\pi}{4}\right) = \dfrac{1 + \tan x}{1 - \tan x}$

B

31. $\dfrac{\tan(45° - \alpha)}{\tan(45° + \alpha)} = \dfrac{(1 - \tan \alpha)^2}{(1 + \tan \alpha)^2}$

32. $\tan \beta = \dfrac{\tan(45° + \beta) - 1}{1 + \tan(45° + \beta)}$

33. $\cot(\alpha + \beta) = \dfrac{\cot \alpha \cot \beta - 1}{\cot \beta + \cot \alpha}$

34. $\cot(\alpha - \beta) = \dfrac{\cot \alpha \cot \beta + 1}{\cot \beta - \cot \alpha}$

4.6 Double-Angle Formulas

Additional identities follow from the sum and difference formulas developed in Sections 4.3 and 4.4. For example, if α is substituted for β in

$$\cos(\alpha + \beta) = \cos \alpha \cos \beta - \sin \alpha \sin \beta,$$

we obtain

$$\cos(\alpha + \alpha) = \cos\alpha\cos\alpha - \sin\alpha\sin\alpha,$$

$$\mathbf{\cos 2\alpha = \cos^2\alpha - \sin^2\alpha.} \tag{1}$$

The right-hand member can be written in a variety of convenient forms. Thus, if $1 - \sin^2\alpha$ is substituted for $\cos^2\alpha$, we have

$$\cos 2\alpha = (1 - \sin^2\alpha) - \sin^2\alpha,$$

$$\mathbf{\cos 2\alpha = 1 - 2\sin^2\alpha.} \tag{2}$$

If $1 - \cos^2\alpha$ is substituted for $\sin^2\alpha$ in Equation (1), we have

$$\cos 2\alpha = \cos^2\alpha - (1 - \cos^2\alpha),$$

$$\mathbf{\cos 2\alpha = 2\cos^2\alpha - 1.} \tag{3}$$

Identities (1), (2), and (3), and similar identities that follow for the sine and tangent functions, called **double-angle formulas,** are often helpful for rewriting expressions in more useful forms. You will prove additional identities in the exercises to become familiar with these important relationships.

Example Show that $\dfrac{1 - \cos 2x}{2\sin x\cos x} = \tan x$ is an identity.

Solution From Equation (2), we can substitute $1 - 2\sin^2 x$ for $\cos 2x$ in the left-hand member to obtain

$$\frac{1 - (1 - 2\sin^2 x)}{2\sin x\cos x} = \tan x$$

$$\frac{2\sin^2 x}{2\sin x\cos x} = \tan x$$

$$\frac{\sin x}{\cos x} = \tan x$$

$$\tan x = \tan x.$$

Hence, the original equation is an identity.

Example Show that $2\sec^2\beta = \dfrac{5 + \cos 2\beta}{1 + \cos 2\beta} - 1$ is an identity

Solution From Equation (3), we substitute $2\cos^2\beta - 1$ for $\cos 2\beta$ in the right-hand member to obtain

$$\begin{aligned}2\sec^2\beta &= \frac{5 + (2\cos^2\beta - 1)}{1 + (2\cos^2\beta - 1)} - 1\\ &= \frac{4 + 2\cos^2\beta}{2\cos^2\beta} - 1\\ &= \frac{4 + 2\cos^2\beta - 2\cos^2\beta}{2\cos^2\beta}\\ &= \frac{4}{2\cos^2\beta}\\ 2\sec^2\beta &= 2\sec^2\beta.\end{aligned}$$

Hence, the original equation is an identity.

If α is substituted for β in

$$\sin(\alpha + \beta) = \sin\alpha\cos\beta + \cos\alpha\sin\beta,$$

we obtain

$$\sin(\alpha + \alpha) = \sin\alpha\cos\alpha + \cos\alpha\sin\alpha,$$

$$\boldsymbol{\sin 2\alpha = 2\sin\alpha\cos\alpha.} \tag{4}$$

Example Show that $\dfrac{2}{\sin 2\theta} = \tan\theta + \cot\theta$ is an identity.

Solution Substituting $2\sin\theta\cos\theta$ for $\sin 2\theta$ in the left-hand member, and $\sin\theta/\cos\theta$ for $\tan\theta$ and $\cos\theta/\sin\theta$ for $\cot\theta$ in the right-hand member yields

$$\begin{aligned}\frac{2}{2\sin\theta\cos\theta} &= \frac{\sin\theta}{\cos\theta} + \frac{\cos\theta}{\sin\theta}\\ &= \frac{\sin^2\theta + \cos^2\theta}{\sin\theta\cos\theta}\\ &= \frac{1}{\sin\theta\cos\theta}\\ &= \frac{2}{2\sin\theta\cos\theta}.\end{aligned}$$

Hence, the original equation is an identity.

If α is substituted for β in

$$\tan(\alpha + \beta) = \frac{\tan \alpha + \tan \beta}{1 - \tan \alpha \tan \beta},$$

we obtain

$$\tan(\alpha + \alpha) = \frac{\tan \alpha + \tan \alpha}{1 - \tan \alpha \tan \alpha},$$

$$\boldsymbol{\tan 2\alpha = \frac{2 \tan \alpha}{1 - \tan^2 \alpha}}. \tag{5}$$

Example Show that $\tan 2x = \dfrac{2 \tan x}{2 - \sec^2 x}$ is an identity.

Solution Substituting $\tan^2 x + 1$ for $\sec^2 x$ yields

$$\tan 2x = \frac{2 \tan x}{2 - (\tan^2 x + 1)}$$

$$= \frac{2 \tan x}{1 - \tan^2 x},$$

which is an identity by Equation (5).

Relationships similar to those developed in this section for sine, cosine, and tangent can be derived for cotangent, secant, and cosecant. However, they are of less importance and will not be considered.

EXERCISE SET 4.6

A

Express each of the following as a single function of kx, or $k\alpha$, where k is a positive integer. Assume that x or α takes on no value for which the expression is undefined.

Examples a. $1 - 2 \sin^2 x$ b. $\dfrac{\tan 3x}{1 - \tan^2 3x}$

Solutions

a. From Equation (2),

$$1 - 2 \sin^2 x = \cos 2x.$$

b. From Equation (5),

$$\frac{\tan 3x}{1 - \tan^2 3x} = \frac{1}{2} \cdot \frac{2 \tan 3x}{1 - \tan^2 3x}$$

$$= \frac{1}{2} \tan 2(3x) = \frac{1}{2} \tan 6x.$$

1. $2 \sin x \cos x$

2. $2 \cos^2 x - 1$

3. $\cos^2 x - \sin^2 x$

4. $\dfrac{2 \tan x}{1 - \tan^2 x}$

5. $2 \sin 2x \cos 2x$

6. $\dfrac{4 \tan 2x}{1 - \tan^2 2x}$

7. $\sin \dfrac{x}{2} \cos \dfrac{x}{2}$

8. $4 \sin 3x \cos 3x$

9. $\cos^2 5\alpha - \sin^2 5\alpha$

10. $1 - 2 \sin^2 4\alpha$

11. $4 \cos^2 6\alpha - 2$

12. $2 \sin 5\alpha \cos 5\alpha$

13. $\sin 2\alpha \cos \alpha - \cos 2\alpha \sin \alpha$

14. $\sin 3\alpha \cos 2\alpha + \cos 3\alpha \sin 2\alpha$

15. $\cos 3x \cos 5x + \sin 3x \sin 5x$

16. $\cos 4x \cos 2x - \sin 4x \sin 2x$

Verify that each equation is an identity.

17. $\cos^2 \alpha = \dfrac{1 + \cos 2\alpha}{2}$

18. $\sin^2 \alpha = \dfrac{1 - \cos 2\alpha}{2}$

19. $\cos 2x + 2 \sin^2 x = 1$

20. $\sin 2x \csc x = 2 \cos x$

21. $\dfrac{\sin 2x}{1 + \cos 2x} = \tan x$

22. $\sin 2x = \dfrac{2 \tan x}{1 + \tan^2 x}$

23. $\sec^2 x = \dfrac{2}{1 + \cos 2x}$

24. $\cot x = \dfrac{1 + \cos 2x}{\sin 2x}$

25. $\cos 2\beta = \dfrac{1 - \tan^2 \beta}{1 + \tan^2 \beta}$

26. $\cot^2 \beta = \dfrac{1 + \cos 2\beta}{1 - \cos 2\beta}$

27. $\cot 4x = \dfrac{1 - \tan^2 2x}{2 \tan 2x}$

28. $\cos^4 x - \sin^4 x = \cos 2x$

29. $\sin 3x = 3 \sin x - 4 \sin^3 x$
Hint: First rewrite $\sin 3x$ as $\sin(2x + x)$.

30. $\cos 3x = 4 \cos^3 x - 3 \cos x$

31. $\sin 4\alpha = (\cos \alpha)(4 \sin \alpha - 8 \sin^3 \alpha)$

32. $\cos 4\alpha = 1 - 8 \sin^2 \alpha \cos^2 \alpha$

Example If $\sin x = 3/5$ and $\cos x < 0$, find $\sin 2x$.

Solution To evaluate

$$\sin 2x = 2 \sin x \cos x, \tag{1}$$

the values for $\cos x$ for those values of x for which $\sin x = 3/5$ and $\cos x < 0$ are needed. Substituting 3/5 for $\sin x$ in

$$\sin^2 x + \cos^2 x = 1$$

gives

$$\left(\frac{3}{5}\right)^2 + \cos^2 x = 1,$$

$$\cos^2 x = 1 - \frac{9}{25} = \frac{16}{25}.$$

Because $\cos x < 0$, $\cos x = -4/5$. Substituting 3/5 for $\sin x$ and $-4/5$ for $\cos x$ in $\sin 2x = 2 \sin x \cos x$ yields

$$\sin 2x = 2\left(\frac{3}{5}\right)\left(-\frac{4}{5}\right) = -\frac{24}{25}.$$

33. If $\sin x = 4/5$ and $\cot x > 0$, find $\sin 2x$.

34. If $\sin x = 12/13$ and $\sec x < 0$, find $\cos 2x$.

35. If $\cos 2\alpha = -8/17$ and $180° \leq 2\alpha \leq 270°$, find $\sin \alpha$.

36. If $\cos 2\alpha = 5/13$ and $270° \leq 2\alpha \leq 360°$, find $\cos \alpha$.

37. If $\tan \beta = -3/4$, find $\tan 2\beta$.

38. If $\tan \beta = 5/12$, find $\tan 2\beta$.

4.7 Additional Identities

The identities in this section are less important than some of the identities introduced in the preceding sections. However, they are sometimes useful in rewriting expressions involving trigonometric function values.

In Section 4.6 we observed that

$$\cos 2\alpha = 2 \cos^2 \alpha - 1.$$

Solving this for $\cos^2 \alpha$, we have

$$1 + \cos 2\alpha = 2 \cos^2 \alpha,$$

from which

$$\boldsymbol{\cos^2 \alpha = \frac{1 + \cos 2\alpha}{2}.} \tag{1}$$

We also noted in Section 4.5 that

$$\cos 2\alpha = 1 - 2 \sin^2 \alpha.$$

Solving this for $\sin^2 \alpha$, we have

$$2 \sin^2 \alpha = 1 - \cos 2\alpha,$$

from which

$$\boldsymbol{\sin^2 \alpha = \frac{1 - \cos 2\alpha}{2}.} \tag{2}$$

These identities enable us to write the second-degree terms, $\cos^2 \alpha$ and $\sin^2 \alpha$, in terms of first-degree terms, They also enable us to write additional useful identities.

Half-Angle Formulas

Substituting $\alpha/2$ for α in Equation (1) yields

$$\cos^2 \frac{\alpha}{2} = \frac{1 + \cos 2(\alpha/2)}{2},$$

from which

$$\boldsymbol{\cos^2 \frac{\alpha}{2} = \frac{1 + \cos \alpha}{2}.} \tag{3}$$

Substituting $\alpha/2$ for α in Equation (2) yields

$$\sin^2 \frac{\alpha}{2} = \frac{1 - \cos 2(\alpha/2)}{2},$$

from which

$$\boldsymbol{\sin^2 \frac{\alpha}{2} = \frac{1 - \cos \alpha}{2}.} \tag{4}$$

Values for $\cos(\alpha/2)$ and $\sin(\alpha/2)$ are positive or negative, depending on the value of $\alpha/2$. Table 1.1 (page 5) can be used to help you select the appropriate sign.

Example Use Equation (4) to find $\sin \frac{\pi}{8}$.

Solution With $\alpha/2 = \pi/8$ and $\alpha = 2(\pi/8) = \pi/4$, we obtain

$$\sin^2 \frac{\pi}{8} = \frac{1 - \cos \frac{\pi}{4}}{2}.$$

Because $0 < \pi/8 < \pi/2$, we use the positive square root for $\sin(\pi/8)$. Then, since $\cos(\pi/4) = 1/\sqrt{2}$,

$$\sin \frac{\pi}{8} = \sqrt{\frac{1 - \frac{1}{\sqrt{2}}}{2}} = \sqrt{\frac{\sqrt{2} - 1}{2\sqrt{2}}} = \frac{1}{2}\sqrt{2 - \sqrt{2}}.$$

Formulas for $\tan(\alpha/2)$ also follow directly from identities we have developed. Substituting $\alpha/2$ for α in $\tan \alpha = (\sin \alpha)/(\cos \alpha)$, we have

$$\tan \frac{\alpha}{2} = \frac{\sin \frac{\alpha}{2}}{\cos \frac{\alpha}{2}}.$$

Multiplying both the numerator and the denominator of the right-hand member by $2 \sin(\alpha/2)$ yields

$$\tan \frac{\alpha}{2} = \frac{2 \sin^2 \frac{\alpha}{2}}{2 \sin \frac{\alpha}{2} \cos \frac{\alpha}{2}}.$$

Substituting $(1 - \cos \alpha)/2$ for $\sin^2(\alpha/2)$ and substituting $\sin 2(\alpha/2)$ for $2 \sin(\alpha/2) \cos(\alpha/2)$, we obtain

$$\tan \frac{\alpha}{2} = \frac{2\left(\frac{1 - \cos \alpha}{2}\right)}{\sin 2\left(\frac{\alpha}{2}\right)},$$

$$\boldsymbol{\tan \frac{\alpha}{2} = \frac{1 - \cos \alpha}{\sin \alpha}}. \qquad (5)$$

By multiplying the numerator and the denominator of the right-hand member of Equation (5) by $1 + \cos\alpha$, we obtain

$$\tan\frac{\alpha}{2} = \frac{(1-\cos\alpha)(1+\cos\alpha)}{\sin\alpha(1+\cos\alpha)}$$

$$= \frac{\sin^2\alpha}{\sin\alpha(1+\cos\alpha)},$$

from which

$$\mathbf{\tan\frac{\alpha}{2} = \frac{\sin\alpha}{1+\cos\alpha}}. \tag{6}$$

Example Use Equation (5) to find $\tan\dfrac{\pi}{8}$.

Solution If $\pi/8 = \alpha/2$, then $\alpha = \pi/4$ and

$$\tan\frac{\pi}{8} = \frac{1-\cos\dfrac{\pi}{4}}{\sin\dfrac{\pi}{4}} = \frac{1-\dfrac{1}{\sqrt{2}}}{\dfrac{1}{\sqrt{2}}}$$

$$= \sqrt{2} - 1.$$

Just as identities such as $\sin 2\alpha = 2\sin\alpha\cos\alpha$ are sometimes called double-angle formulas, the identities developed above are sometimes called **half-angle formulas.**

Relationships similar to those developed above for sine, cosine, and tangent can be derived for cotangent, secant, and cosecant. They are of less important and will not be considered.

Product, Sum, and Difference Formulas

The following identities are derived from the sum and difference formulas for the cosine and sine that were developed in Sections 4.2 and 4.3. The proofs are left as exercises.

$$\cos\alpha\cos\beta = \tfrac{1}{2}[\cos(\alpha+\beta) + \cos(\alpha-\beta)], \tag{7}$$

$$\sin\alpha\sin\beta = -\tfrac{1}{2}[\cos(\alpha+\beta) - \cos(\alpha-\beta)], \tag{8}$$

$$\sin\alpha\cos\beta = \tfrac{1}{2}[\sin(\alpha+\beta) + \sin(\alpha-\beta)], \tag{9}$$

$$\cos\alpha\sin\beta = \tfrac{1}{2}[\sin(\alpha+\beta) - \sin(\alpha-\beta)]. \tag{10}$$

Example Express $\cos 5x \sin 2x$ as a sum or difference.

Solution From Equation (10),

$$\begin{aligned}\cos 5x \sin 2x &= \tfrac{1}{2}[\sin(5x + 2x) - \sin(5x - 2x)] \\ &= \tfrac{1}{2}(\sin 7x - \sin 3x).\end{aligned}$$

The following four identities enable us to express sums and differences of function values in terms of products.

$$\cos \alpha + \cos \beta = 2 \cos\left(\frac{\alpha + \beta}{2}\right) \cos\left(\frac{\alpha - \beta}{2}\right), \tag{11}$$

$$\cos \alpha - \cos \beta = -2 \sin\left(\frac{\alpha + \beta}{2}\right) \sin\left(\frac{\alpha - \beta}{2}\right), \tag{12}$$

$$\sin \alpha + \sin \beta = 2 \sin\left(\frac{\alpha + \beta}{2}\right) \cos\left(\frac{\alpha - \beta}{2}\right), \tag{13}$$

$$\sin \alpha - \sin \beta = 2 \cos\left(\frac{\alpha + \beta}{2}\right) \sin\left(\frac{\alpha - \beta}{2}\right). \tag{14}$$

Example Express $\sin 4\alpha + \sin 2\alpha$ as a product.

Solution From Equation (13),

$$\begin{aligned}\sin 4\alpha + \sin 2\alpha &= 2 \sin\left(\frac{4\alpha + 2\alpha}{2}\right) \cos\left(\frac{4\alpha - 2\alpha}{2}\right) \\ &= 2 \sin 3\alpha \cos \alpha.\end{aligned}$$

EXERCISE SET 4.7

A

Use the half-angle formulas to find the exact value for each of the following.

Example $\sin 105°$

Solution From Equation (4), with $\alpha/2 = 105°$ and $\alpha = 210°$,

$$\sin 105° = \sqrt{\frac{1 - \cos 210°}{2}},$$

where the positive square root is chosen because 105° is in Quadrant II and

$\sin 105°$ is therefore positive. Since $\cos 210° = -\sqrt{3}/2$,

$$\sin 105° = \sqrt{\frac{1-(-\sqrt{3}/2)}{2}} = \frac{\sqrt{2+\sqrt{3}}}{2}.$$

1. $\sin 15°$ **2.** $\cos 15°$ **3.** $\tan 15°$ **4.** $\sin 75°$

5. $\cos 75°$ **6.** $\tan 75°$ **7.** $\cos \dfrac{\pi}{8}$ **8.** $\sin \dfrac{\pi}{12}$

9. $\tan \dfrac{5\pi}{12}$ **10.** $\cos \dfrac{\pi}{12}$ **11.** $\cos \dfrac{7\pi}{12}$ **12.** $\sin \dfrac{7\pi}{12}$

Verify each identity.

13. $\tan \dfrac{x}{2} = \csc x - \cot x$

14. $\sin \dfrac{\theta}{2} \cos \dfrac{\theta}{2} = \dfrac{\sin \theta}{2}$

15. $\left(\cos \dfrac{\alpha}{2} - \sin \dfrac{\alpha}{2}\right)^2 = 1 - \sin \alpha$

16. $\sin \beta - \cos \beta \tan \dfrac{\beta}{2} = \tan \dfrac{\beta}{2}$

17. $\sin^2 \dfrac{\gamma}{2} = \dfrac{\sec \gamma - 1}{2 \sec \gamma}$

18. $\tan \dfrac{x}{2} \sin x = \dfrac{\tan x - \sin x}{\sin x \sec x}$

In Exercises 19–24, assume $0° \leq x < 360°$.

Example If $\tan x = -\dfrac{3}{4}$ and $\cos x$ is negative, find $\sin \dfrac{x}{2}$.

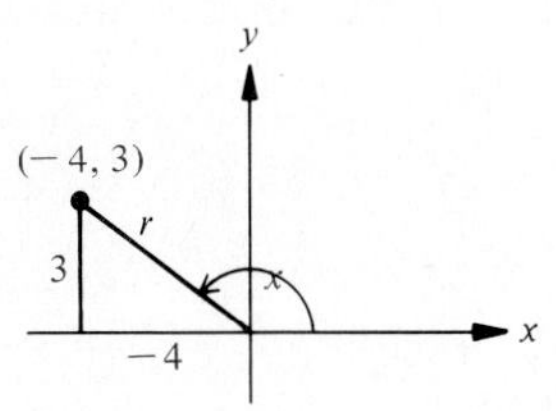

Solution Since $\cos x$ and $\tan x$ are both negative, x is an angle in Quadrant II, and because $\tan x = -3/4$, a point on the terminal ray has the x-coordinate -4 and the y-coordinate 3. From the Pythagorean formula (see Appendix A.8),

$$\begin{aligned} r &= \sqrt{x^2 + y^2} \\ &= \sqrt{(-4)^2 + 3^2} = 5. \end{aligned}$$

From Equation (4) with $\cos x = -4/5$, we obtain

$$\sin \frac{x}{2} = \sqrt{\frac{1-(-4/5)}{2}} = \frac{3}{\sqrt{10}},$$

where the positive square root is taken because x is in Quadrant II and therefore $x/2$ must be in Quadrant I.

19. If $\sin x = \dfrac{\sqrt{3}}{2}$ and $\cos x > 0$, find $\tan \dfrac{x}{2}$.

20. If $\sin \theta = \dfrac{1}{2}$ and $\sec \theta > 0$, find $\sin \dfrac{\theta}{2}$.

21. If $\cos \alpha = -\dfrac{1}{3}$ and $\tan \alpha > 0$, find $\sin \dfrac{\alpha}{2}$.

22. If $\tan \beta = 3$ and $\sin \beta > 0$, find $\cos \dfrac{\beta}{2}$.

23. If $\tan x = -1$ and $\sin x < 0$, find $\cos \dfrac{x}{2}$.

24. If $\sin \theta = \dfrac{1}{\sqrt{2}}$ and $\cos \theta < 0$, find $\tan \dfrac{\theta}{2}$.

Write each expression as a sum or difference of function values.

Example $\sin 3x \sin x$

Solution From Equation (8),

$$\begin{aligned} \sin 3x \sin x &= -\tfrac{1}{2}[\cos(3x + x) - \cos(3x - x)] \\ &= -\tfrac{1}{2}(\cos 4x - \cos 2x). \end{aligned}$$

25. $\sin 5^\circ \cos 12^\circ$ **26.** $\cos 7^\circ \cos 20^\circ$ **27.** $\cos 2x \sin x$

28. $\sin 2x \sin 5x$ **29.** $\cos 2x \cos 5x$ **30.** $\sin 6x \cos x$

Write each expression as a product.

Example $\cos 3\alpha - \cos \alpha$

Solution From Equation (12),

$$\begin{aligned} \cos 3\alpha - \cos \alpha &= -2 \sin\left(\frac{3\alpha + \alpha}{2}\right) \sin\left(\frac{3\alpha - \alpha}{2}\right) \\ &= -2 \sin 2\alpha \sin \alpha. \end{aligned}$$

31. $\cos 20^\circ + \cos 4^\circ$ **32.** $\sin 18^\circ - \sin 6^\circ$ **33.** $\sin 3\alpha + \sin 5\alpha$

34. $\cos \alpha + \cos 5\alpha$ **35.** $\sin 5\alpha - \sin \alpha$ **36.** $\cos 6\alpha - \cos 2\alpha$

B

Verify each of the following identities that are listed in this section.

37. Equation (7) **38.** Equation (8) **39.** Equation (9)

40. Equation (10) **41.** Equation (11) **42.** Equation (12)

43. Equation (13) **44.** Equation (14)

4.8 Conditional Equations

In Sections 4.1–4.7 we considered identities—equations that are satisfied by all angle or all real number replacements for which both of the members are defined. We now consider equations that are conditional—that is, they are satisfied only for particular replacements for the variable.

Finding Values of Angles α, $0° \le \alpha < 360°$

We have already solved many simple conditional equations, such as

$$\sin \alpha = \frac{1}{2}, \tag{1}$$

in Chapter 1 over the interval $0° \le \alpha \le 90°$. In this case

$$\alpha = \text{Arcsin}\,\frac{1}{2} = 30°.$$

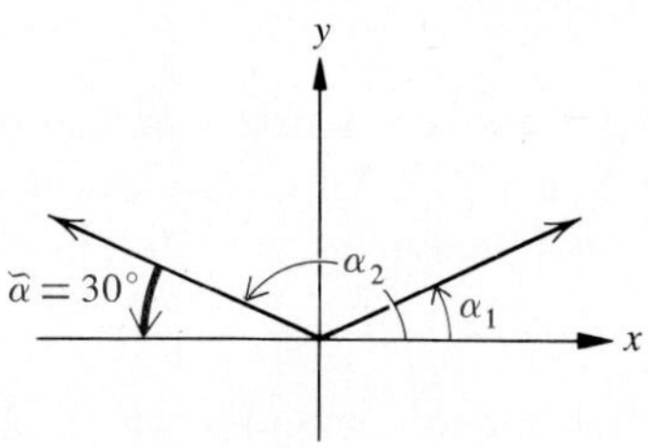

Figure 4.3

However, over the interval 0° to 360°, there are two elements in the domain of the sine function corresponding to 1/2 in the range; either

$$\alpha_1 = 30° \qquad \text{or} \qquad \alpha_2 = 150°$$

satisfies Equation (1), as shown in Figure 4.3, where $\tilde{\alpha} = 30°$.

In general, for most elements in the range of a trigonometric function, there are two corresponding elements in its domain over the interval 0° to 360° or 0^R to $2\pi^R$. A reference angle can be used to find all elements in the domain of a trigonometric function over those intervals that satisfy a given condition. For any positive function value, the angle obtained from a table or a calculator is an angle in the first quadrant and is the reference angle that can be used to find a possible second value that satisfies the specified condition.

Example Given that $\tan \alpha = 1$, find the following for which the statement is true.

a. The least nonnegative angle α in degrees;

b. $\{\alpha \mid 0° \leq \alpha < 360°]$.

Solution The reference angle $\tilde{\alpha}$, for which

$$\tan \tilde{\alpha} = \tan \alpha = 1,$$

is

$$\tilde{\alpha} = \text{Tan}^{-1}\ 1 = 45°$$

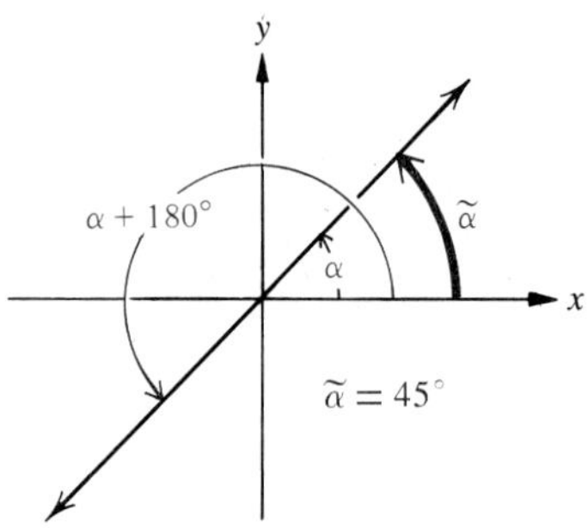

Since $\tan \alpha$ (equal to 1) is positive, α is in Quadrant I or Quadrant III. A sketch showing a reference angle is helpful.

a. The least nonnegative angle α_1 is

$$\alpha = \tilde{\alpha} = 45°.$$

b. Because the period of the tangent function is 180°,

$$\alpha + 180° = 45° + 180° = 225°.$$

Hence,

$$\{\alpha \mid \tan \alpha = 1, \quad 0° \leq \alpha < 360°\} = \{45°, 225°\}.$$

A procedure similar to the one used in the above example can also be used in cases in which the function value is negative. In these cases it is helpful to work with the absolute value of the function value to obtain the reference angle.

Example Given that $\sin \alpha = -0.6494$, find the following for which the statement is true.

a. The least nonnegative angle α in degrees;

b. $\{\alpha \mid 0° \leq \alpha < 360°\}$.

Solution The reference angle $\tilde{\alpha}$, for which

$$\sin \tilde{\alpha} = |\sin \alpha| = |-0.6494| = 0.6494,$$

is

$$\tilde{\alpha} = \text{Sin}^{-1}\ 0.6494 \approx 40.5°.$$

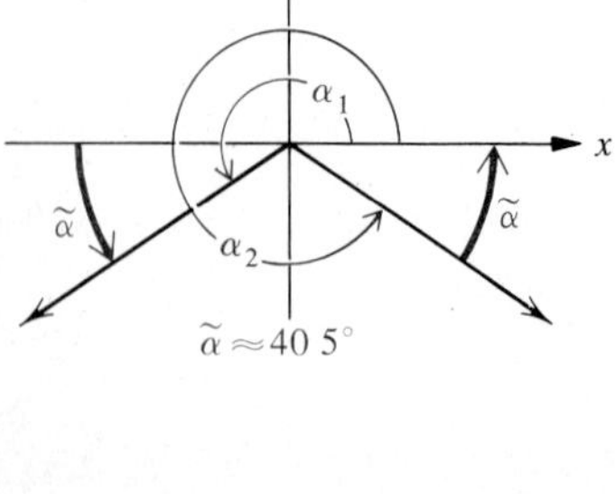

Since $\sin \alpha$ is negative, α is in Quadrant III or IV. A sketch showing the reference angle in each of these quadrants is helpful.

a. The least nonnegative angle is

$$\alpha_1 = 180^\circ + \tilde{\alpha} \approx 180^\circ + 40.5^\circ = 220.5^\circ.$$

b. In Quadrant IV,

$$\alpha_2 = 360^\circ - \tilde{\alpha} \approx 360^\circ - 40.5^\circ = 319.5^\circ.$$

Hence,

$$\{\alpha \mid \sin \alpha = -0.6494, \quad 0^\circ \le \alpha < 360^\circ\} = \{220.5^\circ, 319.5^\circ\}.$$

Note that in the above example the function value was *negative*. Had we used a calculator directly rather than first finding a reference angle, we would have obtained

$$\alpha = \text{Sin}^{-1}(-0.6494) \approx -40.5^\circ,$$

from which we could then have obtained the reference angle $\tilde{\alpha} = 40.5^\circ$, and then the desired solutions.

Most likely, in such cases where the function values are negative, you will prefer to find the reference angle first as shown in the solution to the example.

Example Given that $\cos \alpha = -0.7518$, find the following for which the statement is true.

a. The least nonnegative angle α in radians;

b. $\{\alpha \mid 0^{\mathsf{R}} \le \alpha < 2\pi^{\mathsf{R}}\}$.

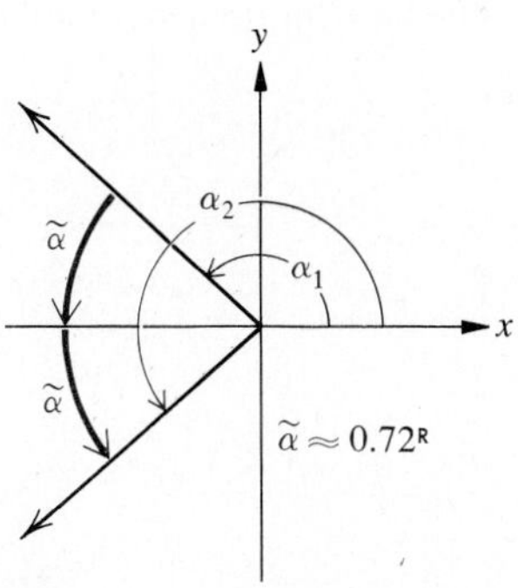

Solution The reference angle $\tilde{\alpha}$, for which

$$\cos \tilde{\alpha} = |\cos \alpha| = |-0.7518| = 0.7518,$$

is

$$\tilde{\alpha} = \text{Cos}^{-1}\, 0.7518 \approx 0.72^{\mathsf{R}}.$$

Since $\cos \alpha$ is negative, α is in Quadrant II or Quadrant III. A sketch showing the reference angle in each of these quadrants is helpful.

a. The least nonnegative angle is

$$\alpha_1 = \pi^{\mathsf{R}} - \tilde{\alpha}$$
$$\approx (3.14 - 0.72)^{\mathsf{R}} = 2.42^{\mathsf{R}}.$$

(continued)

b. In Quadrant III.

$$\begin{aligned}\alpha_2 &= \pi^{R} + \tilde{\alpha} \\ &\approx (3.14 + 0.72)^{R} = 3.86^{R}.\end{aligned}$$

Hence,

$$\{\alpha \mid \cos \alpha = -0.7518, \quad 0^{R} \leq \alpha < 2\pi^{R}\} = \{2.42^{R}, 3.86^{R}\}.$$

Finding All Values of Angles

The trigonometric functions are periodic. Hence, we can obtain the infinite set of values for angles α for a given trigonometric function value by first finding the value of α in the interval $0° \leq \alpha < 360°$ (or $0^{R} \leq \alpha < 2\pi^{R}$) by the methods used in the above examples.

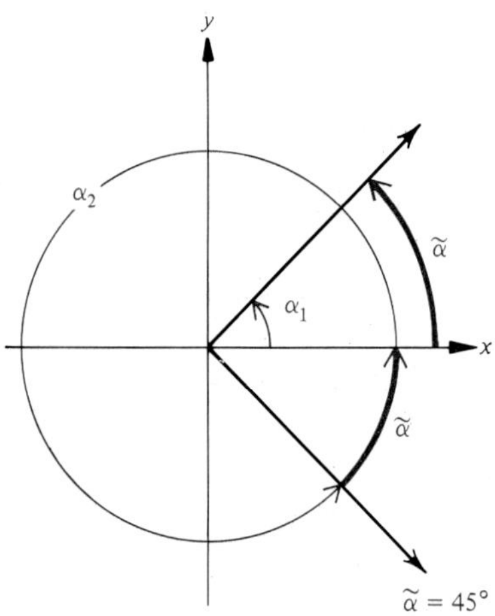

Example Find the set of all angles α (in degree measure) for which $\cos \alpha = 1/\sqrt{2}$.

Solution The reference angle $\tilde{\alpha}$, for which

$$\cos \tilde{\alpha} = |\cos \alpha| = \frac{1}{\sqrt{2}}$$

is

$$\tilde{\alpha} = \mathrm{Cos}^{-1} \frac{1}{\sqrt{2}} = 45°.$$

Since $\cos \alpha$ is positive, α is in Quadrants I or IV. The figure shows the reference angles in each of these quadrants. The least nonnegative angle is

$$\alpha_1 = 45°.$$

In Quadrant IV,

$$\begin{aligned}\alpha_2 &= 360° - \tilde{\alpha} \\ &= 360° - 45° = 315°.\end{aligned}$$

Hence,

$$\left\{\alpha \mid \cos \alpha = \frac{1}{\sqrt{2}}\right\} = \{\alpha \mid \alpha = 45° + k \cdot 360° \quad \text{or} \quad \alpha = 315° + k \cdot 360°\}.$$

Finding Values of x, $x \in R$

In the above discussion and examples we viewed the elements in the domain associated with a given element in the range as angles with degree or radian measures. Similar procedures are used if the elements in the domain are viewed as real numbers.

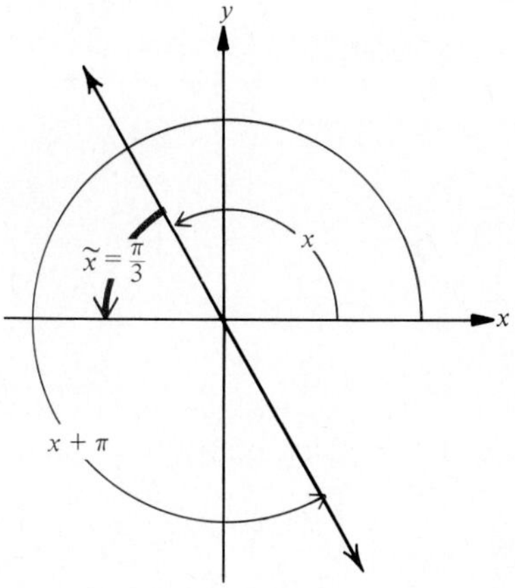

Example Find the set of all real numbers for which $\tan x = -\sqrt{3}$.

Solution The reference arc $\tilde{x}$, for which

$$\tan \tilde{x} = |\tan x| = \sqrt{3}$$

is

$$\tilde{x} = \text{Tan}^{-1} \sqrt{3} = \frac{\pi}{3}.$$

Since $\tan x$ is negative, x is a real number that can be associated with an angle in Quadrants II or IV. The figure shows $\tilde{x}$. The least nonnegative x is

$$x = \pi - \frac{\pi}{3} = \frac{2\pi}{3}.$$

Since the tangent function has period π,

$$\{x \mid \tan x = -\sqrt{3}\} = \left\{x \mid x = \frac{2\pi}{3} + k \cdot \pi\right\}.$$

EXERCISE SET 4.8

A

For each function value, find **(a)** *the exact value of the least nonnegative angle α in degrees and* **(b)** $\{\alpha \mid 0° \leq \alpha < 360°\}$ *for which the statement is true.*

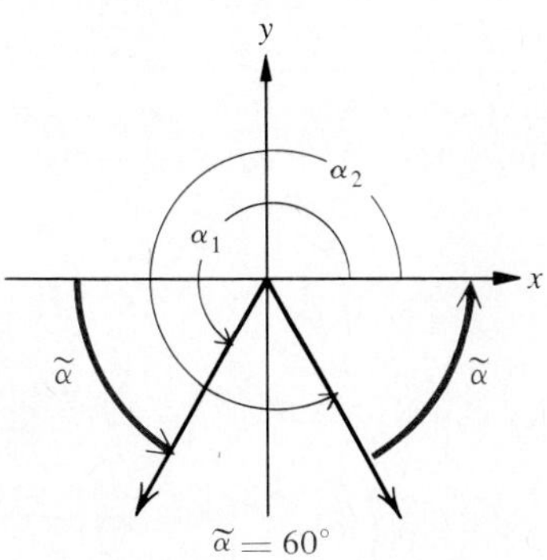

Example $\sin \alpha = -\dfrac{\sqrt{3}}{2}$

Solution The reference angle $\tilde{\alpha}$, for which

$$\sin \tilde{\alpha} = |\sin \alpha| = \left|-\frac{\sqrt{3}}{2}\right| = \frac{\sqrt{3}}{2},$$

is

$$\tilde{\alpha} = \text{Sin}^{-1} \frac{\sqrt{3}}{2} = 60°.$$

Since $\sin \alpha$ (equal to $-\sqrt{3}/2$) is negative, α is in Quadrant III or Quadrant IV. A sketch showing the reference angle in each of these quadrants is helpful.

(continued)

a. The least nonnegative angle in Quadrant III is

$$\alpha_1 = 180° + \tilde{\alpha} = 180° + 60° = 240°.$$

b. The angle in Quadrant IV is

$$\alpha_2 = 360° - 60° = 300°.$$

Hence,

$$\left\{\alpha \,|\, \sin\alpha = -\frac{\sqrt{3}}{2},\quad 0° \le \alpha < 360°\right\} = \{240°, 300°\}.$$

1. $\cos\alpha = \frac{\sqrt{3}}{2}$ **2.** $\sin\alpha = \frac{1}{\sqrt{2}}$ **3.** $\tan\alpha = -\frac{1}{\sqrt{3}}$

4. $\cos\alpha = \frac{1}{2}$ **5.** $\cot\alpha = \sqrt{3}$ **6.** $\tan\alpha = 1$

7. $\sec\alpha = \sqrt{2}$ **8.** $\cot\alpha = -1$ **9.** $\csc\alpha = -2$

10. $\sec\alpha = 2$ **11.** $\sin\alpha = -\frac{1}{2}$ **12.** $\csc\alpha = \frac{2}{\sqrt{3}}$

13. $\cos\alpha = -\frac{1}{\sqrt{2}}$ **14.** $\sin\alpha = -\frac{\sqrt{3}}{2}$ **15.** $\csc\alpha = \sqrt{2}$

16. $\cos\alpha = -1$ **17.** $\sin\alpha = 0$ **18.** $\tan\alpha = -\sqrt{3}$

19. $\sec\alpha = -\frac{2}{\sqrt{3}}$ **20.** $\cot\alpha = 0$ **21.** $\csc\alpha$ undefined

22. $\sec\alpha$ undefined **23.** $\cot\alpha$ undefined **24.** $\tan\alpha$ undefined

For each function value, find an approximation in degrees for (**a**) *the least nonnegative angle* α *and* (**a**) $\{\alpha \,|\, 0° \le \alpha < 360°\}$ *for which each statement is true.*

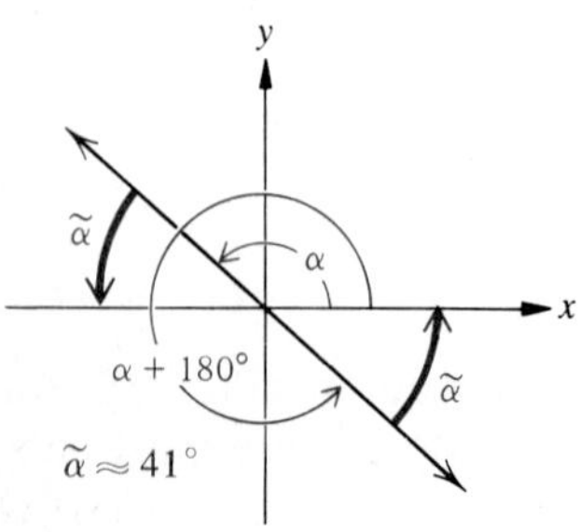

Example $\tan\alpha = -0.8693$

Solution The reference angle $\tilde{\alpha}$, for which

$$\tan\tilde{\alpha} = |\tan\alpha| = |-0.8693| = 0.8693,$$

is

$$\tilde{\alpha} = \text{Tan}^{-1}\, 0.8693 \approx 41°.$$

Since $\tan\alpha$ (equal to -0.8693) is negative, α is in Quadrant II or Quadrant IV. A sketch showing a reference angle is helpful.

a. The least nonnegative angle in Quadrant II is

$$\alpha = 180^\circ - \tilde{\alpha}$$
$$\approx 180^\circ - 41^\circ = 139^\circ.$$

b. Because the period of the tangent function is 180°,

$$\alpha + 180^\circ \approx 139^\circ + 180^\circ = 319^\circ.$$

Hence,

$$\{\alpha \mid \tan\alpha = -0.8693, \quad 0^\circ \le \alpha \le 360^\circ\} = \{139^\circ, 319^\circ\}.$$

25. $\sin\alpha = 0.2419$ **26.** $\cos\alpha = 0.5150$ **27.** $\tan\alpha = -2.6051$

28. $\cot\alpha = -2.1445$ **29.** $\sec\alpha = 1.0320$ **30.** $\csc\alpha = 1.1570$

31. $\cos\alpha = -0.9085$ **32.** $\sin\alpha = -0.3535$ **33.** $\cot\alpha = 2.1060$

34. $\tan\alpha = 0.3134$ **35.** $\csc\alpha = -1.2796$ **36.** $\sec\alpha = -3.8140$

For each function value, find an approximation in radians for (**a**) *the least nonnegative angle* α *and* (**b**) $\{\alpha \mid 0^R \le \alpha < 2\pi^R\}$ *for which each statement is true.*

Example $\cos\alpha = -0.9131$

Solution The reference angle $\tilde{\alpha}$, for which

$$\cos\tilde{\alpha} = |\cos\alpha| = |-0.9131| = 0.9131,$$

is

$$\tilde{\alpha} = \text{Cos}^{-1}\, 0.9131 \approx 0.42^R.$$

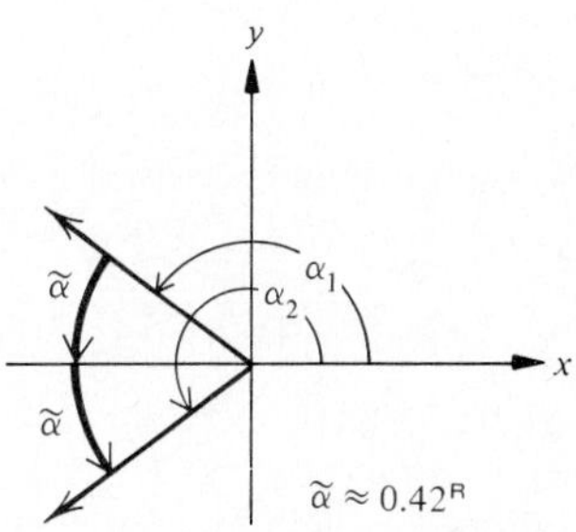

Since $\cos\alpha$ (equal to -0.9131) is negative, α is in Quadrant II or Quadrant III. A sketch showing the reference angle in each of these quadrants is helpful.

a. The least nonnegative angle in Quadrant II is

$$\alpha_1 = \pi^R - \tilde{\alpha} \approx \pi^R - 0.42^R \approx 2.72^R.$$

b. The angle in Quadrant III is

$$\alpha_2 \approx \pi^R + 0.42^R \approx 3.56^R.$$

(continued)

Hence,

$$\{\alpha \mid \cos \alpha = -0.9131, \quad 0^R \leq \alpha \leq 2\pi^R\} = \{2.72^R, 3.56^R\}.$$

37. $\sin \alpha = 0.6442$ **38.** $\cos \alpha = 0.5898$ **39.** $\tan \alpha = 0.2341$

40. $\sin \alpha = 0.2280$ **41.** $\sec \alpha = 1.795$ **42.** $\csc \alpha = 1.467$

43. $\cos \alpha = -0.9689$ **44.** $\sin \alpha = -0.5480$ **45.** $\sin \alpha = -0.1197$

46. $\tan \alpha = -1.050$ **47.** $\csc \alpha = -2.125$ **48.** $\sec \alpha = -5.273$

For each function value, find an approximation to the nearest tenth of a degree for (**a**) *the least nonnegative angle* α *and* (**b**) $\{\alpha \mid \alpha \in A\}$ *for which each statement is true.*

Example $\tan \alpha = -0.2162$

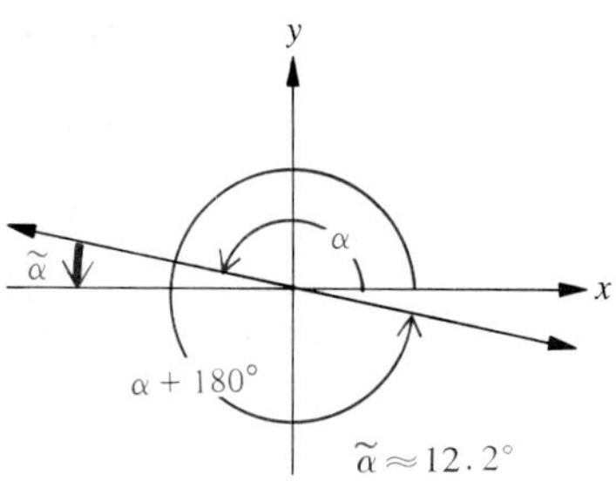

Solution The reference angle $\tilde{\alpha}$, for which

$$\tan \tilde{\alpha} = |\tan \alpha| = |-0.2162| = 0.2162,$$

is

$$\tilde{\alpha} = \text{Tan}^{-1}\, 0.2162 \approx 12.2°.$$

Since $\tan \alpha$ (equal to -0.2162) is negative, α is in Quadrant II or IV. The figure shows a reference angle. The least nonnegative angle is

$$\alpha = 180° - \tilde{\alpha}$$
$$\approx 180° - 12.2° = 167.8°.$$

Because the period of the tangent function is 180°,

$$\{\alpha \mid \tan \alpha = -0.2162\} = \{\alpha \mid \alpha \approx 167.8 + k \cdot 180°\}, \ k \in J.$$

49. $\sin \alpha = 0.2790$ **50.** $\cos \alpha = 0.9311$

51. $\cot \alpha = -1.2799$ **52.** $\tan \alpha = -1.4882$

53. $\csc \alpha = -1.2813$ **54.** $\sec \alpha = -1.1164$

55. $\tan \alpha = 0.3019$ **56.** $\sin \alpha = -0.9426$

For each function value, find $\{x \mid x \in R\}$ *for which each statement is true. Find exact values if possible; otherwise obtain approximations.*

Example $\cos x = -0.7324$

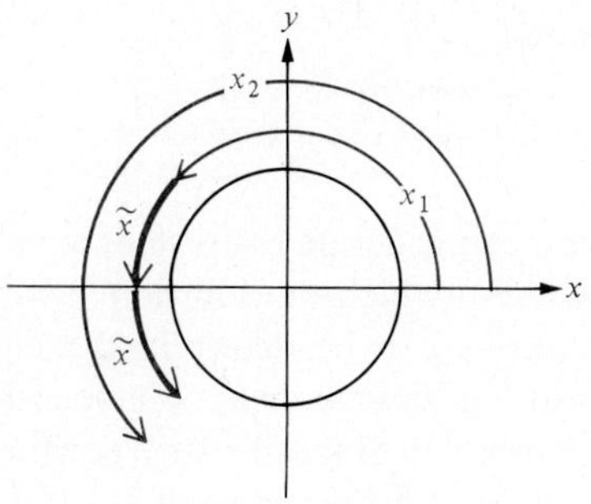

Solution The notion of a reference arc (or reference angle) is helpful. The reference arc $\tilde{x}$ for which

$$\cos \tilde{x} = |\cos x| = |-0.7324| = 0.7324,$$

is

$$\tilde{x} = \text{Cos}^{-1}\, 0.7324 \approx 0.75.$$

(The radian setting on a calculator is used.) Since $\cos x$ (equal to -0.7324) is negative, x is in Quadrant II or III. Hence, using $\pi \approx 3.14$,

$$x_1 \approx \pi - 0.75 \approx 3.14 - 0.75 = 2.39$$

and

$$x_2 \approx \pi + 0.75 \approx 3.14 + 0.75 = 3.89.$$

Because the period of the cosine function is 2π,

$$\{x \mid x \in R\} = \{x \mid x \approx 2.39 + k \cdot 2\pi\} \cup \{x \mid x \approx 3.89 + k \cdot 2\pi\}, \quad k \in J.$$

57. $\cos x = \dfrac{1}{\sqrt{2}}$

58. $\sin x = \dfrac{1}{\sqrt{2}}$

59. $\sin x = \dfrac{-\sqrt{3}}{2}$

60. $\cos x = \dfrac{-\sqrt{3}}{2}$

61. $\tan x = -1$

62. $\tan x = \sqrt{3}$

63. $\sin x = 0.3427$

64. $\sin x = -0.2372$

65. $\cos x = -0.4682$

66. $\cos x = 0.7301$

67. $\tan x = 1.642$

68. $\tan x = -1.874$

69. $\tan x = -0.7692$

70. $\tan x = -0.4931$

4.9 Additional Conditional Equations

In Section 4.8 we obtained the solution set over various domains of simple conditional equations such as

$$\sin x = \frac{1}{2}, \quad \cos x = -0.3824, \quad \text{and} \quad \tan x = 1.752$$

in which the left-hand member contained only one function value and that function value was of first degree. In the case $\sin x = 1/2$, we obtained solutions by inspection.

Perhaps you can determine the solution of

$$\sin x = \cos x, \quad 0 < x < \pi/2, \tag{1}$$

by inspection—perhaps not. However, you can generate equivalent equations and hopefully you will generate one from which the solution(s) will be evident by inspection. Generally, solutions can be determined most easily from equations that involve *values of one function only*. The identities developed in the preceding sections are useful in obtaining such equivalent equations from equations involving values of more than one function. For example, in (1) above, for $0 < x < \pi/2$, $\cos x \neq 0$. Thus, we can multiply each member by $1/\cos x$ to obtain

$$\frac{\sin x}{\cos x} = 1.$$

Substituting $\tan x$ for $\sin x/\cos x$, we obtain

$$\tan x = 1.$$

Since, in this case, values of x are restricted to $0 < x < \pi/2$,

$$x = \text{Tan}^{-1}\, 1 = \frac{\pi}{4}.$$

Example Solve $2 \sin x \cos x - \sin x = 0$ over the interval $0 \leq x \leq \pi/2$.

Solution Factoring the left-hand member we obtain the equivalent equation

$$\sin x(2 \cos x - 1) = 0.$$

The left-hand member equals zero if

$$\sin x = 0 \quad \text{or} \quad \cos x = \frac{1}{2},$$

from which

$$x = \text{Sin}^{-1}\, 0 = 0 \quad \text{or} \quad x = \text{Cos}^{-1}\, \frac{1}{2} = \frac{\pi}{3},$$

and the solution set over the interval $0 \leq x \leq \pi/2$ is $\left\{0, \frac{\pi}{3}\right\}$.

Equations that are quadratic in a given trigonometric function value, such as

$$\cos^2 x + \cos x - 2 = 0, \qquad x \in A, \tag{2}$$

can be solved by first solving the quadratic equation for the function value involved. To emphasize that we are using a standard algebraic method for solving a quadratic equation, we also show each step of the process, where we let $\cos x = u$ and write Equation (2) as

$$u^2 + u - 2 = 0. \tag{3}$$

Since the left-hand members of Equations (2) and (3) are factorable, the equations can be written equivalently as

$$(\cos x + 2)(\cos x - 1) = 0 \qquad \text{or} \qquad (u + 2)(u - 1) = 0.$$

Since $\cos x + 2$ cannot equal 0 (there is no angle x such that $\cos x = -2$), we seek values of x for which

$$\cos x - 1 = 0 \qquad \text{or} \qquad u - 1 = 0.$$

From either equation, we have

$$\cos x = 1.$$

The least nonnegative value of x is given by

$$x = \text{Cos}^{-1}\, 1 = 0.$$

Hence the solution set of Equation (2) is

$$\{x \mid x = k \cdot 360°,\ k \in J\} \qquad \text{or} \qquad \{x \mid x = k \cdot 2\pi^{\text{R}},\ k \in J\}.$$

EXERCISE SET 4.9

A

Solve each equation. Find all solutions in the interval $0 \le x \le \pi/2$.

Examples

a. $(2\cos x - \sqrt{3})(\sin x - 1) = 0$ b. $\sqrt{2}\sin x \cos x - \cos x = 0$

Solutions

a. The left-hand member equals 0 if

$$2\cos x - \sqrt{3} = 0 \qquad \text{or} \qquad \sin x - 1 = 0.$$

(continued)

From these equations

$$\cos x = \frac{\sqrt{3}}{2} \qquad \text{or} \qquad \sin x = 1,$$

and the solution set is $\{\pi/6, \pi/2\}$.

b. Factoring the left-hand member

$$\cos x(\sqrt{2} \sin x - 1) = 0.$$

The left-hand member equals 0 if

$$\cos x = 0 \qquad \text{or} \qquad \sin x = \frac{1}{\sqrt{2}},$$

and the solution set is $\{\pi/2, \pi/4\}$.

1. $2 \sin x - 1 = 0$

2. $2 \cos x - 1 = 0$

3. $2 \cos^2 x = 1$

4. $4 \sin^2 x = 3$

5. $3 \csc^2 x - 3 = 0$

6. $\sec^2 x = 4$

7. $(2 \cos^2 x - 1)(\sin x - 1) = 0$

8. $(\sin^2 x - 1)(\sec x - 2) = 0$

9. $\cos x \sin x = 0$

10. $3 \tan x \sec x = 0$

11. $2 \cos^3 x - \cos x = 0$

12. $2 \sin^3 x - \sin x = 0$

13. $2 \sin x \cos x + \sin x = 0$

14. $2 \sin x \cos x - \cos x = 0$

Solve each equation. Find all solutions in the interval $0 \leq x < 2\pi$.

Example $2 \cos^2 x + \cos x - 1 = 0$

Solution Factoring the left-hand member,

$$(2 \cos x - 1)(\cos x + 1) = 0,$$

$$\cos x = \frac{1}{2} \qquad \text{or} \qquad \cos x = -1,$$

and the solution set is $\{\pi/3, 5\pi/3, \pi\}$.

15. $2 \cos^2 x - \cos x - 1 = 0$

16. $2 \cos^2 x + 3 \cos x - 2 = 0$

17. $9 \cos^2 x - 21 \cos x - 8 = 0$

18. $25 \sin^2 x - 20 \sin x + 4 = 0$

Solve each equation (**a**) *in the interval* $0° \le x < 360°$ *and* (**b**) *for all* $x \in A$.

Example $\sec x - 2 \cos x = 1$

Solution

a. First write each item using the same function value. Substitute $1/\cos x$ for $\sec x$ to obtain

$$\frac{1}{\cos x} - 2 \cos x = 1.$$

Multiplying each member by $\cos x$ ($\cos x \neq 0$),

$$1 - 2 \cos^2 x = \cos x,$$

from which

$$2 \cos^2 x + \cos x - 1 = 0.$$

This equation is the same as in the preceding example. Thus, the solution set over the interval $0° \le x < 360°$ is $\{60°, 300°, 180°\}$.

b. For all $x \in A$, the solution set is

$$\{x \mid x = 60° + k \cdot 360°\} \cup \{x \mid x = 180° + k \cdot 360°\} \cup \{x \mid x = 300° + k \cdot 360°\}, \quad k \in J,$$

19. $\csc x + \sin x + 2 = 0$

20. $2 \cos x + \sec x - 3 = 0$

21. $2 \sin^2 x - \cos x - 1 = 0$

22. $2 \cos^2 x - 3 \sin x - 3 = 0$

23. $2 \sin x + \csc x - 3 = 0$

24. $\tan x + \cot x + 2 = 0$

Solve each equation for all $x \in R$.

25. $2 \cos^2 x + \sin x = 2$

26. $\tan^2 x + \sec x = 1$

27. $\sec^2 x = 1 + \sqrt{3} \tan x$

28. $\cos^2 x - \sin x = \sin^2 x$

29. $2 \sin x \cos x - 1 = \cos x - 2 \sin x$

30. $4 - \cos^2 x = 4 \sin x$

B

Solve each equation (**a**) *in the interval* $0° \le x < 360°$ *and* (**b**) *for all* $x \in A$. *Approximate answers to the nearest tenth of a degree.* Hint: *Use the quadratic formula.*

31. $\sin^2 x - \sin x - 1 = 0$

32. $\cos^2 x + 5 \cos x + 2 = 0$

33. $3 \cos x - 1 = \sec x$

34. $4 \sin x \cos x - 2 \sin x = \tan x$

Solve each equation in the interval $0 \le x \le 2\pi$. Hint: *Square each member.*

35. $1 + \cos x = \sqrt{3} \sin x$

36. $\cos x + \sin x = 1$

37. $\cot x = 1 + \csc x$

38. $1 + \tan x = -\sqrt{2} \sec x$

Solve each equation for all $x \in R$.

39. $2 \cot x \sec x + \cot x + 2 \sec x = -1$

40. $\cos x \tan x - \cos x + \sin x \tan x - \sin x = 0$

4.10 Conditional Equations for Multiples

Conditional equations such as

$$2 \sin x \cos x = \frac{\sqrt{3}}{2}, \tag{1}$$

$$\tan 3x = 1, \tag{2}$$

and

$$\cos 2x = \cos^2 x - 1 \tag{3}$$

need further discussion.

The identities involving multiples of the variable are often useful to help solve certain kinds of equations. For example, the solution set of

$$2 \sin x \cos x = \frac{\sqrt{3}}{2}, \tag{1}$$

over the interval $0 \le x \le \pi/2$, can readily be found if $\sin 2x$ is substituted for the left-hand member, $2 \sin x \cos x$, to obtain

$$\sin 2x = \frac{\sqrt{3}}{2}.$$

By inspection we have that

$$2x = \frac{\pi}{3} + k \cdot 2\pi \quad \text{or} \quad 2x = \frac{2\pi}{3} + k \cdot 2\pi,$$

$$x = \frac{\pi}{6} + k \cdot \pi \quad \text{or} \quad x = \frac{\pi}{3} + k \cdot \pi.$$

If $k = 0$, $x = \pi/6$ or $x = \pi/3$; if $k \geq 1$, then $x > \pi$. Hence, the solution set of Equation (1) over the interval from 0 to $\pi/2$ is $\{\pi/6, \pi/3\}$.

Now consider the solution set of

$$\tan 3x = 1 \tag{2}$$

over the interval $0 \leq x < 2\pi$. By inspection we have that

$$3x = \frac{\pi}{4} + k \cdot \pi,$$

and, dividing each term by 3,

$$x = \frac{\pi}{12} + k \cdot \frac{\pi}{3}, \quad k \in J.$$

To find the solution set over the interval $0 \leq x < 2\pi$, we take 0, 1, 2, 3, 4, and 5 as replacements for k and obtain

$$\left\{\frac{\pi}{12}, \frac{5\pi}{12}, \frac{3\pi}{4}, \frac{13\pi}{12}, \frac{17\pi}{12}, \frac{7\pi}{4}\right\}.$$

Note that if $k > 5$, then $x > 2\pi$.

Now consider the equation

$$\cos 2x = \cos^2 x - 1. \tag{3}$$

One method of solving this equation for $x \in R$ consists of substituting $2 \cos^2 x - 1$ for $\cos 2x$ to obtain the equivalent equations

$$2 \cos^2 x - 1 = \cos^2 x - 1,$$

$$\cos^2 x = 0.$$

Then, we have as the solution set

$$\left\{x \mid x = \frac{\pi}{2} + k\pi, \ k \in J\right\}.$$

Alternatively, we could have substituted $(1 + \cos 2x)/2$ for $\cos^2 x$ in the right-hand member of Equation (3) to obtain the equivalent equations

$$\cos 2x = \frac{1 + \cos 2x}{2} - 1,$$

$$2 \cos 2x = 1 + \cos 2x - 2,$$

$$\cos 2x = -1.$$

By inspection, we have that

$$2x = \pi + k \cdot 2\pi, \quad k \in J,$$

from which the solution set is given by

$$\left\{x \mid x = \frac{\pi}{2} + k\pi, \quad k \in J\right\},$$

as we found above using the other method.

EXERCISE SET 4.10

A

Solve each equation for all x, such that $0 \le x \le \pi/2$.

Example $\cos 3x = \dfrac{1}{2}$

Solution Using Table 1.2 and the fact that the cosine of an angle in Quadrant I or IV is positive, we first consider all possible solutions and obtain

$$3x = \frac{\pi}{3} + k \cdot 2\pi \qquad \text{or} \qquad 3x = \frac{5\pi}{3} + k \cdot 2\pi.$$

From which, dividng each term by 3,

$$x = \frac{\pi}{9} + k \cdot \frac{2\pi}{3} \qquad \text{or} \qquad x = \frac{5\pi}{9} + k \cdot \frac{2\pi}{3}.$$

Since the problem restricts x to the interval $0 \le x \le \pi/2$, we need only the value 0 for k. If $k = 0$, $x = \pi/9$ or $x = 5\pi/9$. Furthermore, since $5\pi/9$ is not in the required interval, the solution set is $\{\pi/9\}$.

1. $\sin 3x = 0$

2. $\cos 3x = 1$

3. $\tan 4x = -\sqrt{3}$

4. $\cot 4x = -1$

5. $\sec 5x = 2$

6. $\csc 5x = \sqrt{2}$

7. $\cos 2x - \sin x = 0$

8. $\cos 2x + \cos x = 0$

9. $\sin 2x + \sin x = 0$

10. $\sin 2x - \cos x = 0$

11. $\sin 2x = \cos 2x$

12. $\sin \dfrac{x}{2} = \tan \dfrac{x}{2}$

13. $\tan \frac{x}{2} - \cos x = -1$

14. $\cos x = \sin \frac{x}{2}$

Solve each equation in the interval $0° \leq x < 360°$.

Example $\sin 4x = \frac{\sqrt{3}}{2}$

Solution By inspection we have $4x = 60°$. However, since $\sin \theta$ is also positive for angles in Quadrant II, we know that $4x = 120°$ will also satisfy equation. Thus for $k \in J$,

$$4x = 60° + k \cdot 360° \qquad \text{or} \qquad 4x = 120° + k \cdot 360°.$$

Dividing each member of both equations by 4, we obtain

$$x = 15° + k \cdot 90° \qquad \text{or} \qquad x = 30° + k \cdot 90°.$$

Since the problem restricts x to the interval $0° \leq x < 360°$, we need only the values of 0, 1, 2, and 3 for k.

If $k = 0$, $x = 15°$ or $x = 30°$,
if $k = 1$, $x = 105°$ or $x = 120°$,
if $k = 2$, $x = 195°$ or $x = 210°$,
if $k = 3$, $x = 285°$ or $x = 300°$,

and the solution set is $\{15°, 30°, 105°, 120°, 195°, 210°, 285°, 300°\}$.

15. $\sin 3x = \frac{1}{2}$

16. $\cos 4x = -1$

17. $\sin 3x = \cos 3x$

18. $\tan 2x = \cot 2x$

19. $\sin^2 2x = 1$

20. $\cos^2 2x = 1$

21. $\tan^2 3x = 1$

22. $\cot^2 3x = 3$

23. $\left(1 + \cos \frac{x}{2}\right) \cos \frac{x}{2} = 0$

24. $\sin^2 \frac{x}{4} - \frac{1}{2} \sin \frac{x}{4} = 0$

25. $5 \sec 3x = 2 \cos 3x - 3$

26. $\tan x \tan 2x = 1$

B

Solve each equation for all x, such that $x \in R$.

27. $\sin 2x = \cos x$

28. $2 \cos 2x - 1 = -3 \sin^2 x$

29. $\sin x \cos 3x = 0$

30. $(\sin 2x - 1)(\cos 3x + 1) = 0$

31. $\tan 2x - 3 \sec^2 2x = -5$

32. $\cos^2 2x + 3 \sin 2x - 3 = 0$

33. $\sin 2x \cos x + \cos 2x \sin x = \dfrac{1}{\sqrt{2}}$

34. $\sin 3x \cos x - \cos 3x \sin x = \dfrac{\sqrt{3}}{2}$

35. Use graphical methods to approximate a positive solution for

$$\cos 2x = x, \quad x \in R.$$

Hint: Graph $y = \cos 2x$ and $y = x$ on the same axes.

36. Use graphical methods to approximate a positive solution for

$$x^2 - 3 \sin \frac{x}{2} = 0, \quad x \in R.$$

Chapter Summary

4.1–4.7 Equations are **identities** if they are satisfied for all real number replacements of the variable for which both members of the equation are defined.

For convenience, the basic identities that have been established are listed inside the back cover.

An equation can be shown *not* to be an identity by replacing the variable with a real number for which the resulting statement is false.

4.8–4.10 **Conditional equations** involving circular function values can be solved by inspection or by generating equivalent equations from which the solutions are evident by inspection. The identities introduced in this chapter are useful in obtaining such equivalent equations.

A list of symbols introduced in this chapter is shown inside the front cover, and important properties are shown inside the back cover.

Review Exercises

A

4.1–4.5 *Express each of the following in terms of* $\sin \alpha$ *and/or* $\cos \alpha$ *and simplify the expression.*

1. $\csc \alpha \tan \alpha$

2. $\cot^2 \alpha \sec^2 \alpha$

3. $\sin^2 \alpha(\csc^2 \alpha - 1)$

4. $\cot \alpha \cos \alpha \sec^2 \alpha$

5. $\sin(-\alpha)\cot\alpha$

6. $\sec^2(-\alpha)\cos(-\alpha)$

Verify that each equation is an identity.

7. $\sin^2\alpha(1+\cot^2\alpha) = 1$

8. $\dfrac{\cos\alpha\csc\alpha}{\cot\alpha} = 1$

9. $\tan\alpha = \dfrac{1}{\cos\alpha\csc\alpha}$

10. $1+\tan^2\alpha = \csc^2\alpha\tan^2\alpha$

11. $\sin(-\alpha) = -\dfrac{\tan\alpha}{\sec\alpha}$

12. $\tan(-\alpha) = \dfrac{\cos(-\alpha)\sin(-\alpha)}{1-\sin^2\alpha}$

13. $\cos(270^\circ + x) = \sin x$

14. $\cos(270^\circ - x) = -\sin x.$

15. $\sin(90^\circ + x) = \cos x$

16. $\sin(\alpha - 45^\circ) = \dfrac{\sin\alpha - \cos\alpha}{\sqrt{2}}$

17. $\cos(\alpha - 45^\circ) = \dfrac{\cos\alpha + \sin\alpha}{\sqrt{2}}$

18. $\tan(\alpha - 45^\circ) = \dfrac{\tan\alpha - 1}{1+\tan\alpha}$

4.6 **19.** $\cos 2x + 1 = 2\cos^2 x$

20. $\sin 2x = 2\sin^2 x\cos x\csc x$

21. $\csc x = 2\cos x\csc 2x$

22. $\cos 2x = \cos^4 x - \sin^4 x$

By counterexamples, show that each equation is not an identity.

23. $\sin\alpha\sec\alpha = 2\tan\alpha$

24. $\dfrac{1-\cos\alpha}{\sin\alpha} = \dfrac{1+\sin\alpha}{\cos\alpha}$

4.7 *Use half-angle formulas to find the exact value for each of the following.*

25. $\cos 105^\circ$

26. $\tan 165^\circ$

Verify the identities in Exercises 27 and 28.

27. $\cos^2\dfrac{x}{2} = \dfrac{\sin^2 x}{2(1-\cos x)}$

28. $\sin^2\dfrac{x}{2} = \dfrac{\csc x - \cot x}{2\csc x}$

29. Express $\cos x\cos 5x$ in terms of sums of function values.

30. Express $\sin 6x + \sin 4x$ in terms of products of function values.

4.8 *For each function value, find* **(a)** *the exact value of the least nonnegative angle* α *in degrees and* **(b)** $\{\alpha \mid 0^\circ \le \alpha < 360^\circ\}$ *for which the statement is true.*

31. $\sin\alpha = \dfrac{\sqrt{3}}{2}$

32. $\tan\alpha = -1$

For each function value, find an approximation in degrees for (**a**) *the least nonnegative angle* α *and* (**b**) $\{\alpha \mid 0° \le \alpha < 360°\}$ *for which the statement is true.*

33. $\cos \alpha = -0.4695$

34. $\csc \alpha = 1.6972$

For each function value, find an approximation in radians for (**a**) *the least nonnegative angle* α *and* (**b**) $\{\alpha \mid 0^R \le \alpha < 2\pi^R\}$ *for which the statement is true.*

35. $\sec \alpha = 1.0720$

36. $\cot \alpha = -0.2439$

37. $\sin \alpha = 0.3107$

38. $\cos \alpha = 0.8965$

For each function value, find an approximation in real numbers for (*a*) *the least nonnegative number x and* (*b*) $\{x \mid x \in R\}$ *for which the statement is true.*

39. $\cos x = 0.6704$

40. $\sin x = 0.2118$

41. $\tan x = -1.0476$

42. $\tan x = 2.5011$

4.9 *Solve each equation in the interval* $0 \le x < \pi/2$.

43. $2 \sin x - \sqrt{3} = 0$

44. $\csc^2 x = 4$

45. $\sec x - 2 \cos x = 0$

46. $\cot x \csc x = 0$

Solve each equation in the interval $0° \le x < 360°$.

47. $\cot^2 x - 2 \csc^2 x + 5 = 0$

48. $\sec^2 - 4 \cos^2 x = 0$

Solve each equation for $x \in R$.

49. $16 \sin^2 x - 9 \csc^2 x = 0$

50. $5 \sec^2 x - 6 \tan^2 x = 4$

4.10 *Solve each equation in the interval* $0° \le x < 360°$.

51. $\cos 3x = \dfrac{1}{\sqrt{2}}$

52. $\sin 4x = \dfrac{\sqrt{3}}{2}$

53. $\sin 2x - \tan x = 0$

54. $\tan 2x = \tan x$

Solve each equation for $x \in R$.

55. $\tan 4x = \dfrac{1}{\sqrt{3}}$

56. $\cos 2x + \cos x = 0$

5 VECTORS

5.1 Geometric Vectors

Some physical quantities, such as length and mass, are characterized by *magnitude alone*. If units of measure are specified, the quantities can be represented by real numbers and associated with points in a number line. In the description of other physical quantities, such as force, velocity, and acceleration, *direction as well as magnitude* is required. Such a quantity can be represented by a pair of real numbers and associated with a line segment in the form of an arrow with length representing the magnitude and pointing in an assigned direction.

Definition 5.1 *A two-dimensional* **geometric vector*** *designated by* $\vec{\mathbf{v}}$ *is a line segment with a specified direction.*

If the end points of a geometric vector are named, it is sometimes convenient to specify the vector by using these names. For example, in Figure 5.1 the geometric vector can also be designated by $\overrightarrow{\mathbf{PQ}}$, where the point named P is called the **initial point** and the point named Q is called the **terminal point.** For discussion purposes, the geometric vector in Figure 5.1 is shown in the framework of a rectangular coordinate system.

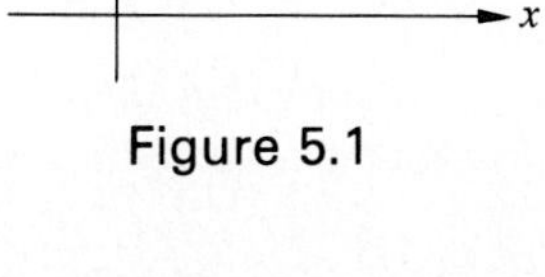

Figure 5.1

Sometimes a geometric vector is simply represented by **v** without using the arrow above the symbol. We are using the arrow to differentiate this symbol for a geometric vector from the symbol v as a variable for a real

* The term *geometric vector* is used to differentiate this notion of a directed line segment from a vector as an ordered pair of real numbers introduced in Section 5.3.

number. Also the arrow emphasizes the fact that a geometric vector has *magnitude* and *direction.*

Magnitude and Direction

Each geometric vector in the plane has a length that is a real number that depends on the unit of measurement.

Definition 5.2 *For all geometric vectors* $\vec{v}$, *the* **norm** (*or* **magnitude**) *of* $\vec{v}$ *is the length of* $\vec{v}$ *and is designated by* $\|\vec{v}\|$.

When a geometric vector is considered in relation to a rectangular coordinate system, the following concept is useful.

Definition 5.3 *The* **direction angle** *of a geometric vector* $\vec{v}$ *is an angle* α, $-180° < \alpha \leq 180°$, *such that the initial side of* α *is the ray from the initial point of* $\vec{v}$ *parallel to the x-axis and directed in the positive x direction; the terminal side of* α *is the ray from the initial point of* $\vec{v}$ *and containing* $\vec{v}$.

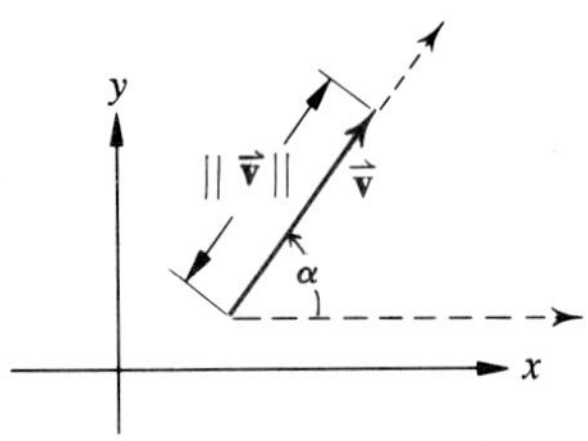

Figure 5.2

Figure 5.2 shows $\|\vec{v}\|$ and the direction angle α associated with a vector $\vec{v}$. From a vector viewpoint, the real number $\|\vec{v}\|$ is called a **scalar.**

Definition 5.4 *For all geometric vectors* $\vec{v}$, *the geometric vector* $-\vec{v}$ *has the same magnitude as* $\vec{v}$ *and has a direction angle which differs in measure from the measure of the direction angle of* $\vec{v}$ *by* 180°.

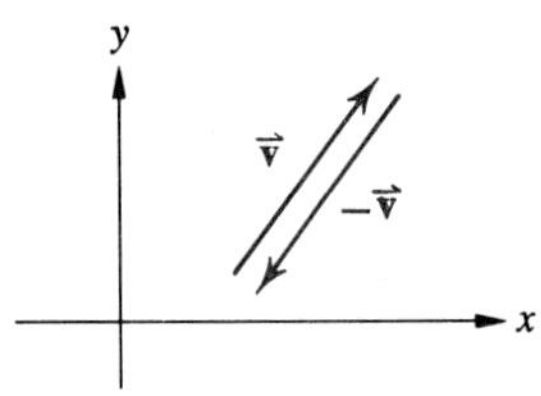

Figure 5.3

Figure 5.3 shows two vectors $\vec{v}$ and $-\vec{v}$.

Definition 5.5 *Two geometric vectors* $\vec{v}_1$ *and* $\vec{v}_2$ *are* **equivalent** *if they have the same magnitude and the same direction angle.*

Figure 5.4 illustrates three equivalent vectors. We use the symbol = to indicate that two vectors are equivalent.

If we can imagine "sliding" a geometric vector so that it remains parallel to its original position in the plane, then any geometric vector can "slide" onto any vector equivalent to it.

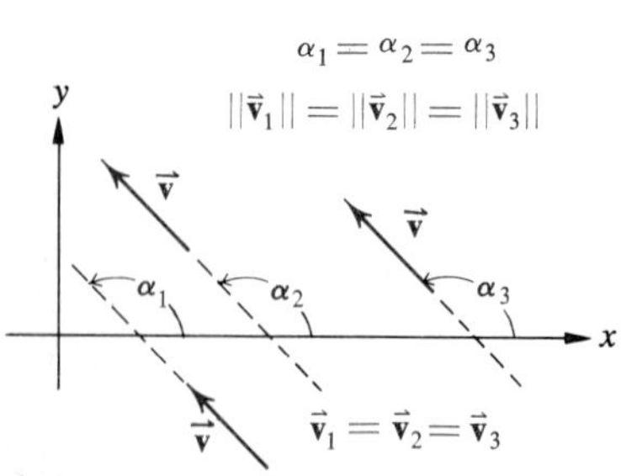

Figure 5.4

Operations

Geometric vectors are very useful in setting up mathematical models of practical problems. However, before this can be done, certain operations must be defined.

Definition 5.6 *For all geometric vectors* $\vec{v}_1$ *and* $\vec{v}_2$, *such that the terminal point of* $\vec{v}_1$ *is the initial point of* $\vec{v}_2$, $\vec{v}_1 + \vec{v}_2$ *is the geometric vector having the* *same initial point as* $\vec{v}_1$ *and the same terminal point as* $\vec{v}_2$. $\vec{v}_1 + \vec{v}_2$ *is called the* **sum** (*or* **resultant**) *of* $\vec{v}_1$ *and* $\vec{v}_2$.

As indicated in Figure 5.5a, if $\vec{v}_1$ and $\vec{v}_2$ are not in the relative locations required by Definition 5.6, we "slide" one of them, say $\vec{v}_2$, parallel to its original position into the correct location and then find the sum $\vec{v}_1 + \vec{v}_2$.

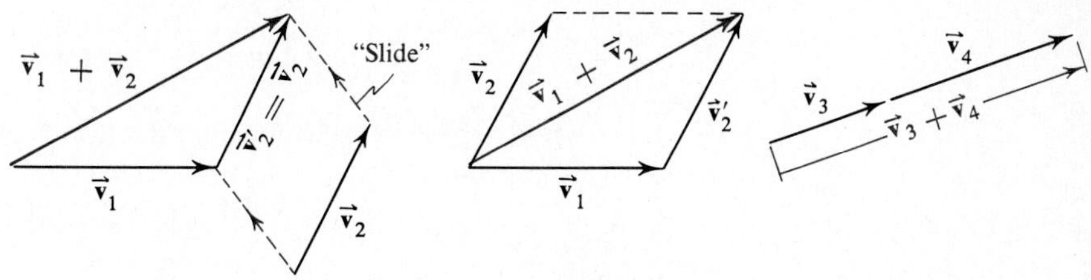

Figure 5.5

Because a triangle is formed by $\vec{v}_1, \vec{v}_2$, and $\vec{v}_1 + \vec{v}_2$, we sometimes say that geometric vectors are added according to "the triangle law." From the properties of a parallelogram, we have that, if $\vec{v}_1$ and $\vec{v}_2$ have the same initial point, as shown in Figure 5.5b, the sum $\vec{v}_1 + \vec{v}_2$ is a diagonal of the parallelogram with adjacent sides $\vec{v}_1$ and $\vec{v}_2$. Therefore, we sometimes also say that geometric vectors are added according to "the parallelogram law." Figure 5.5c shows the sum $\vec{v}_3 + \vec{v}_4$ if $\vec{v}_3$ and $\vec{v}_4$ are collinear (vectors lie on the same line).

Note that the symbol "+" in a vector sum is being used in a different sense from that of its ordinary usage to pair two real numbers in a sum, although the operation is commonly referred to as the "addition" of geometric vectors.

Example Construct the sum $\vec{v}_1 + \vec{v}_2$.

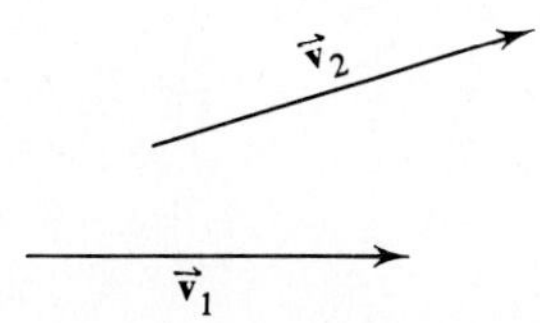

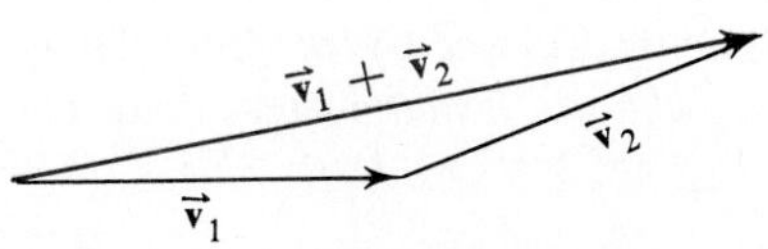

a.

Solution "Slide" the initial point of $\vec{v}_2$ onto the terminal point of $\vec{v}_1$. The sum $\vec{v}_1 + \vec{v}_2$ is the geometric vector from the initial point of $\vec{v}_1$ to the terminal point of $\vec{v}_2$ (figure a).

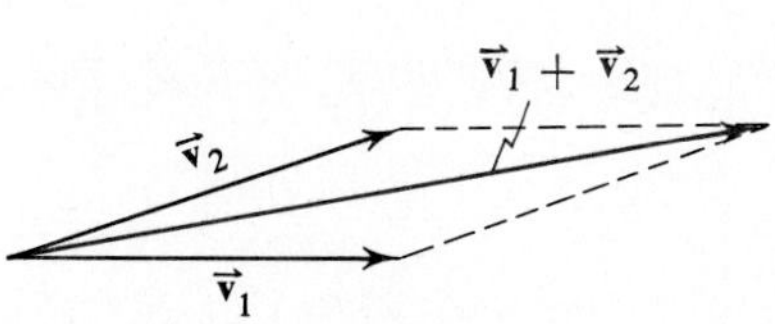

b.

Alternate Solution "Slide" the initial point of $\vec{v}_2$ onto the initial point of $\vec{v}_1$ and form a parallelogram. The sum $\vec{v}_1 + \vec{v}_2$ is the diagonal of the parallelogram (figure b).

Definition 5.7 *For all geometric vectors $\vec{v}$ and any real number c, $c\vec{v}$ is a geometric vector with magnitude $\|c\vec{v}\| = |c| \cdot \|\vec{v}\|$, having the same direction angle as $\vec{v}$ if $c > 0$ and the same direction angle as $-\vec{v}$ if $c < 0$. The geometric vector $c\vec{v}$ is called a* **scalar multiple** *of $\vec{v}$.*

Example As shown in the figure, $2\vec{v}$ and $\frac{1}{2}\vec{v}$ are geometric vectors in the same direction as $\vec{v}$ such that

$$\|2\vec{v}\| = 2\|\vec{v}\| \qquad \text{and} \qquad \|\tfrac{1}{2}\vec{v}\| = \tfrac{1}{2}\|\vec{v}\|,$$

while $-2\vec{v}$ is in the opposite direction such that

$$\|-2\vec{v}\| = |-2|\|\vec{v}\| = 2\|\vec{v}\|.$$

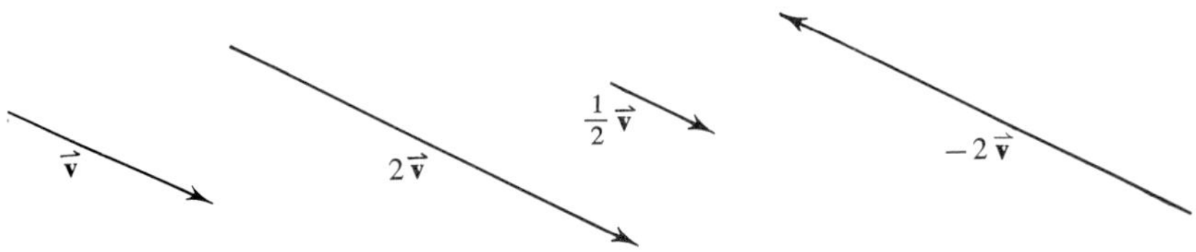

EXERCISE SET 5.1

A

In Exercises 1–40, use the geometric vectors below.

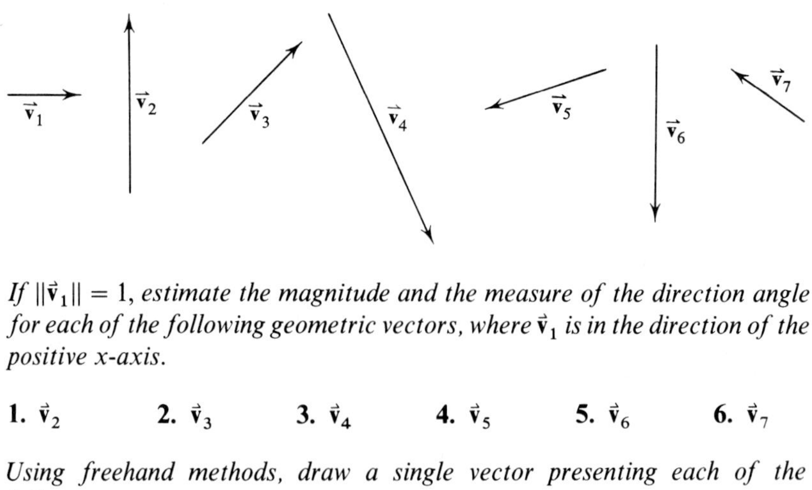

If $\|\vec{v}_1\| = 1$, estimate the magnitude and the measure of the direction angle for each of the following geometric vectors, where $\vec{v}_1$ is in the direction of the positive x-axis.

1. $\vec{v}_2$ **2.** $\vec{v}_3$ **3.** $\vec{v}_4$ **4.** $\vec{v}_5$ **5.** $\vec{v}_6$ **6.** $\vec{v}_7$

Using freehand methods, draw a single vector presenting each of the following sums.

Example $(\vec{v}_1 + \vec{v}_3) + \vec{v}_7$

Solution See figure.

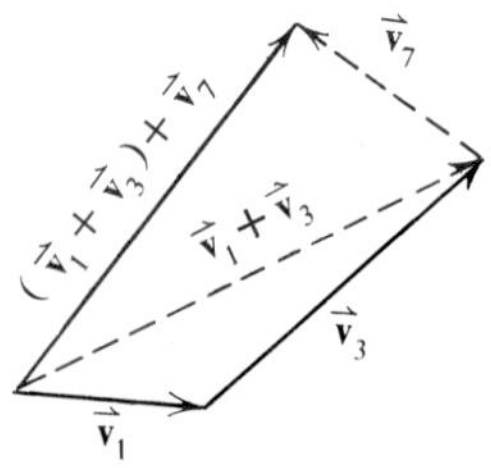

7. $\vec{v}_1 + \vec{v}_2$ **8.** $\vec{v}_1 + \vec{v}_3$ **9.** $\vec{v}_1 + \vec{v}_4$

10. $\vec{v}_1 + \vec{v}_5$ **11.** $\vec{v}_1 + \vec{v}_6$ **12.** $\vec{v}_1 + \vec{v}_7$

13. $\vec{v}_2 + \vec{v}_3$ **14.** $\vec{v}_2 + \vec{v}_4$ **15.** $(\vec{v}_2 + \vec{v}_5) + \vec{v}_6$

16. $(\vec{v}_2 + \vec{v}_6) + \vec{v}_7$ **17.** $(\vec{v}_2 + \vec{v}_3) + \vec{v}_5$ **18.** $\vec{v}_3 + (\vec{v}_4 + \vec{v}_7)$

Using freehand methods, draw each of the following.

19. $-\vec{v}_1$ **20.** $-\vec{v}_3$ **21.** $2\vec{v}_7$

22. $4\vec{v}_1$ **23.** $-2\vec{v}_5$ **24.** $-\vec{v}_6$

25. $\frac{1}{2}\vec{v}_3$ **26.** $-\frac{1}{3}\vec{v}_4$ **27.** $2\vec{v}_1 + 3\vec{v}_5$

28. $4\vec{v}_7 + 2\vec{v}_6$ **29.** $\frac{1}{2}\vec{v}_4 + \vec{v}_7$ **30.** $\vec{v}_2 + \frac{3}{4}\vec{v}_3$

31. $2\vec{v}_3 + (-3\vec{v}_1)$ **32.** $\vec{v}_7 + (-2\vec{v}_2)$

B

By appropriate freehand sketches show that each statement is true.

33. $\vec{v}_2 + \vec{v}_3 = \vec{v}_3 + \vec{v}_2$ **34.** $\vec{v}_5 + \vec{v}_4 = \vec{v}_4 + \vec{v}_5$

35. $(\vec{v}_6 + \vec{v}_7) + \vec{v}_2 = \vec{v}_6 + (\vec{v}_7 + \vec{v}_2)$ **36.** $\vec{v}_3 + (\vec{v}_7 + \vec{v}_5) = (\vec{v}_3 + \vec{v}_7) + \vec{v}_5$

37. $2(\vec{v}_3 + \vec{v}_5) = 2\vec{v}_3 + 2\vec{v}_5$ **38.** $(2 + 3)\vec{v}_7 = 2\vec{v}_7 + 3\vec{v}_7$

39. $\frac{1}{3}\vec{v}_1 + 2\vec{v}_1 = \frac{7}{3}\vec{v}_1$ **40.** $\vec{v}_2 + (-2\vec{v}_2) = -\vec{v}_2$

5.2 Applications of Geometric Vectors

The solution to many kinds of practical problems involving such quantities as velocities, accelerations, forces, and directed line segments can be found by means of geometric vectors.

Projections

First, consider the *projections* of a geometric vector on the x- and y-axes of a Cartesian coordinate system. Figure 5.6 depicts a geometric vector $\vec{v}$ and perpendicular lines drawn to the x- and y-axes from the endpoints P_1 and P_2 of $\vec{v}$. The geometric vector projections $\vec{v}_x$ and $\vec{v}_y$ of $\vec{v}$ on the x-axis and y-axis, respectively, are called the **rectangular components** of $\vec{v}$; $\vec{v}_x$ is called the **horizontal component** of $\vec{v}$, and $\vec{v}_y$ is called the **vertical component.** These components are determined by the magnitude of $\vec{v}$ and its direction. Furthermore,

$$\vec{v} = \vec{v}_x + \vec{v}_y.$$

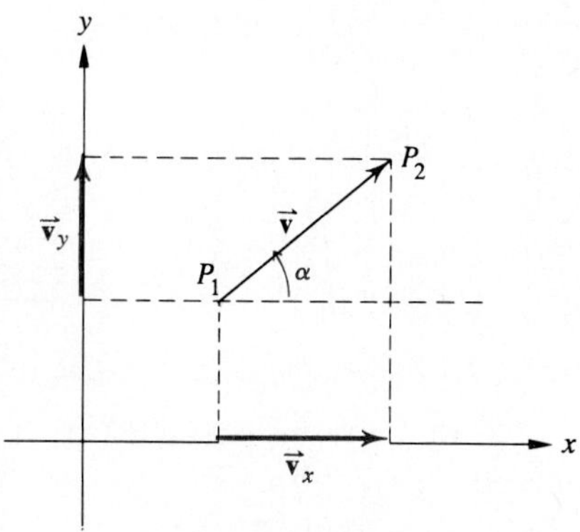

Figure 5.6

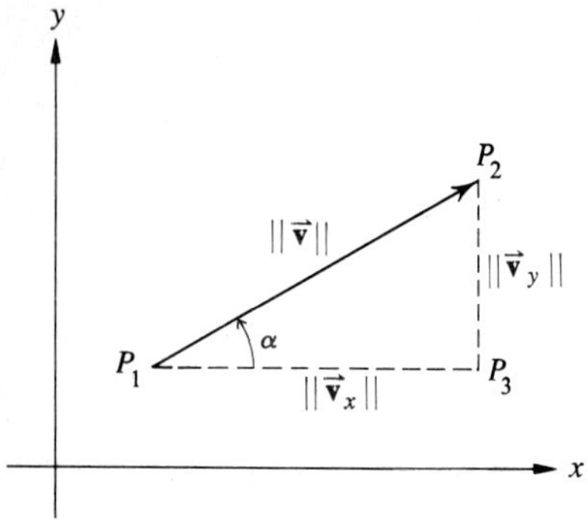

Figure 5.7

In Figure 5.7, $\overline{P_1P_3}$ is drawn parallel to the x-axis, and $\overline{P_2P_3}$ is drawn parallel to the y-axis. Since

$$|\sin \alpha| = \frac{||\vec{v}_y||}{||\vec{v}||},$$

$$||\vec{v}_y|| = ||\vec{v}|| \cdot |\sin \alpha|. \tag{1}$$

Also, since

$$|\cos \alpha| = \frac{||\vec{v}_x||}{||\vec{v}||},$$

$$||\vec{v}_x|| = ||\vec{v}|| \cdot |\cos \alpha|. \tag{2}$$

In Figure 5.7, α is shown to be a positive acute angle. From Definition 5.3, we have that $-180° < \alpha \leq 180°$ and hence $\cos \alpha$ or $\sin \alpha$ may be negative. Therefore because $||\vec{v}_x||$, $||\vec{v}_y||$, and $||\vec{v}||$ are all positive, we use $|\sin \alpha|$ and $|\cos \alpha|$ in Equations (1) and (2).

If the geometric vectors $\vec{v}_x$ and $\vec{v}_y$ are known, the magnitude $||\vec{v}||$ and direction angle α of $\vec{v}$ can be determined from the relationships in a right triangle, as shown in Figure 5.7.

$$||\vec{v}|| = \sqrt{||\vec{v}_x||^2 + ||\vec{v}_y||^2},$$

and

$$|\sin \alpha| = \frac{||\vec{v}_y||}{||\vec{v}||} \quad \text{or} \quad |\cos \alpha| = \frac{||\vec{v}_x||}{||\vec{v}||}.$$

Example The horizontal component, $\vec{v}_x$, has its initial point at (4, 0) with $||\vec{v}_x|| = 6$. The vertical component, $\vec{v}_y$, has its initial point at (0, 3) with $||\vec{v}_y|| = 8$. Find α and $||\vec{v}||$ and sketch $\vec{v}$. Assume $\vec{v}_x$ and $\vec{v}_y$ are directed in a positive direction.

Solution The magnitude

$$||\vec{v}|| = \sqrt{||\vec{v}_x||^2 + ||\vec{v}_y||^2}$$
$$= \sqrt{6^2 + 8^2} = \sqrt{100} = 10.$$

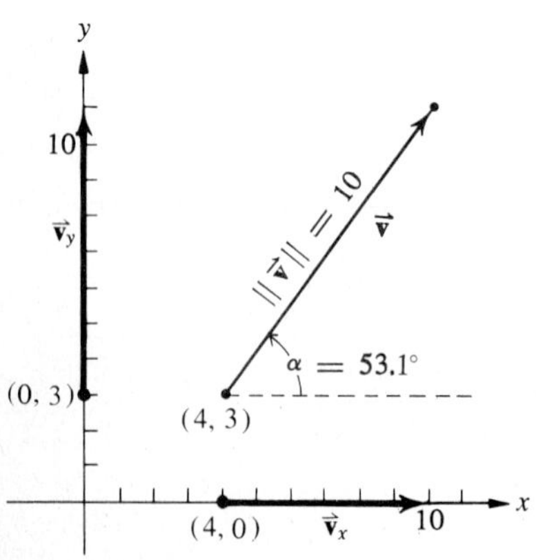

Because

$$\cos \alpha = \frac{6}{10} = 0.6000,$$

$$\alpha \approx 53.1°.$$

Thus, the geometric vector $\vec{v}$ has its initial point at (4, 3), a direction angle whose measure is approximately 53.1°, and a magnitude of 10.

Force

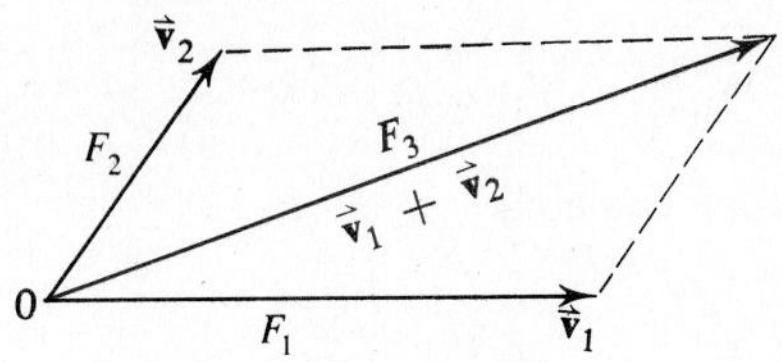

Figure 5.8

Consider two forces, F_1 and F_2, acting on a particle at the point O, as in Figure 5.8. Since a force has magnitude and direction, we can represent it with a geometric vector, with magnitude given in terms of the units of force and with direction angle given by the direction of the force. It is shown in physics that the two forces represented by $\vec{v}_1$ and $\vec{v}_2$ can be replaced by a single equivalent force, F_3, associated with the vector $\vec{v}_1 + \vec{v}_2$, the resultant of the forces represented by $\vec{v}_1$ and $\vec{v}_2$. The resultant $\vec{v}_1 + \vec{v}_2$ can be obtained, as discussed in Section 5.1, by constructing the parallelogram with adjacent sides $\vec{v}_1$ and $\vec{v}_2$ and drawing the diagonal; $\vec{v}_1 + \vec{v}_2$ is then associated with the resultant force F_3.

Example Two forces, the first one of 5 pounds and the second of 12 pounds, act on a body at right angles to each other. Find the magnitude of the resultant force and the measure of the angle that the resultant force makes with the 5-pound force.

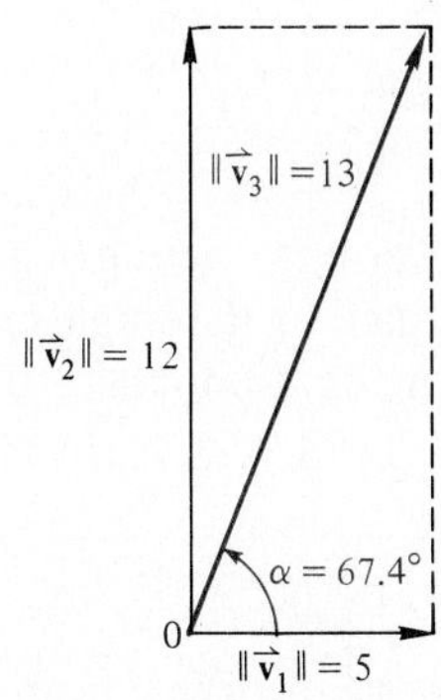

Solution Because $\|\vec{v}_1\| = 5$ and $\|\vec{v}_2\| = 12$,

$$\|\vec{v}\| = \sqrt{5^2 + 12^2} = \sqrt{169} = 13,$$

and

$$\cos \alpha = \frac{5}{13} \approx 0.3846.$$

Hence, $\alpha = 67.4°$. Thus, the magnitude of the resultant force is 13 pounds, and the angle it makes with the 5-pound force has a measure of approximately 67.4°.

In cases where the measure of an angle between two forces is not 90°, it is sometimes convenient to use the Law of Cosines and the Law of Sines in order to find the resultant force.

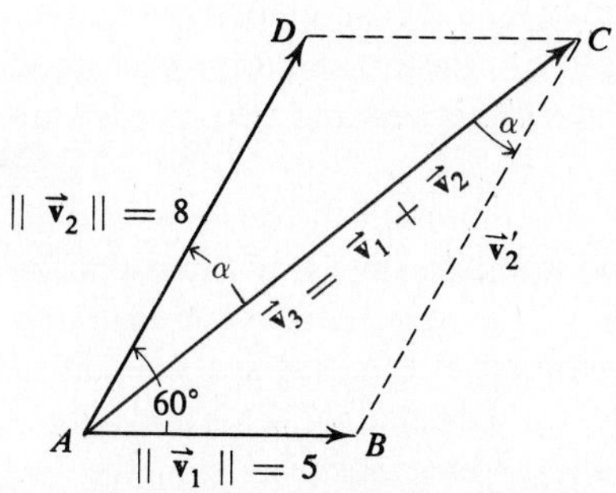

Example A force of 5 pounds and another force of 8 pounds act on an object at an angle of 60° with respect to each other. Find the magnitude of the resultant force and the angle it forms with respect to the 8-pound force.

Solution The figure shows the forces of 5 pounds and 8 pounds represented by the geometric vectors $\vec{v}_1$ and $\vec{v}_2$, respectively, and the resultant $\vec{v}_3 = \vec{v}_1 + \vec{v}_2$. Since $\vec{v}_2'$ is equivalent to $\vec{v}_2$, and since the consecutive

(continued)

angles of a parallelogram are supplementary,

$$\angle ABC = 180^\circ - 60^\circ = 120^\circ,$$

and it follows from the Law of Cosines that

$$\begin{aligned} \|\vec{v}_3\|^2 &= \|\vec{v}_1\|^2 + \|\vec{v}_2\|^2 - 2\|\vec{v}_1\| \cdot \|\vec{v}_2\| \cos 120^\circ \\ &= 5^2 + 8^2 - 2(5)(8)(-\cos 60^\circ) \\ &= 25 + 64 + 80(0.5000) = 129. \end{aligned}$$

Therefore, $\|\vec{v}_3\| = \sqrt{129} \approx 11.4$. From the law of Sines,

$$\frac{\sin \alpha}{\|\vec{v}_1\|} = \frac{\sin 120^\circ}{\|\vec{v}_3\|},$$

or, equivalently,

$$\begin{aligned} \sin \alpha &= \frac{\|\vec{v}_1\| \sin 120^\circ}{\|\vec{v}_3\|} \\ &\approx \frac{5(0.8660)}{11.4} \approx 0.3798. \end{aligned}$$

Hence, $\alpha \approx 22.3^\circ$. Thus, the forces of 5 and 8 pounds, acting at an angle of 60° with respect to each other, can be represented by the single resultant force of 11.4 pounds. The angle it makes with the 8-pound force has a measure of approximately 22.3°.

Velocity

Geometric vectors also can facilitate the solution of problems involving velocities and accelerations. As examples, we shall look at several air navigation problems. First, however, we consider some terminology that is commonly used. The **heading** of an airplane is the direction in which it is pointed, and the **course** is the direction in which it is actually moving over the ground. Its **air speed** is the speed relative to the air, and its **ground speed** is the speed relative to the ground. The two directions and two speeds may differ because of wind effect.

If we use one geometric vector, say $\vec{v}_1$, to represent the heading and air speed of an airplane (see Figure 5.9) and another, say $\vec{v}_2$, to represent the wind direction and speed, then $\vec{v}_3 = \vec{v}_1 + \vec{v}_2$ represents the course and ground speed. The angle α between the geometric vectors $\vec{v}_1$ and $\vec{v}_3$ is called the **drift angle.** The measures of the angles for the heading, wind, and course directions in air navigation are generally taken clockwise from the ray

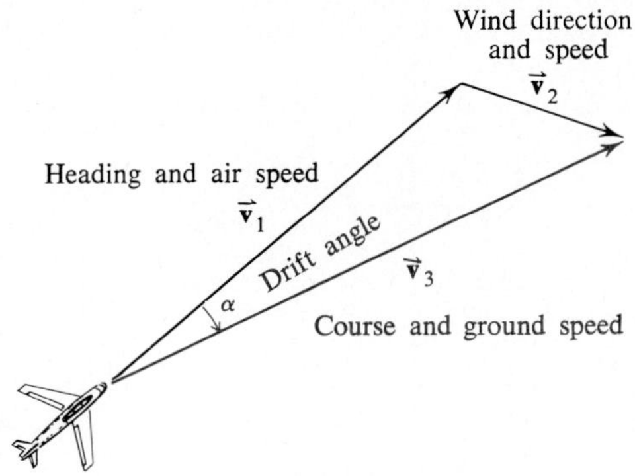

Figure 5.9

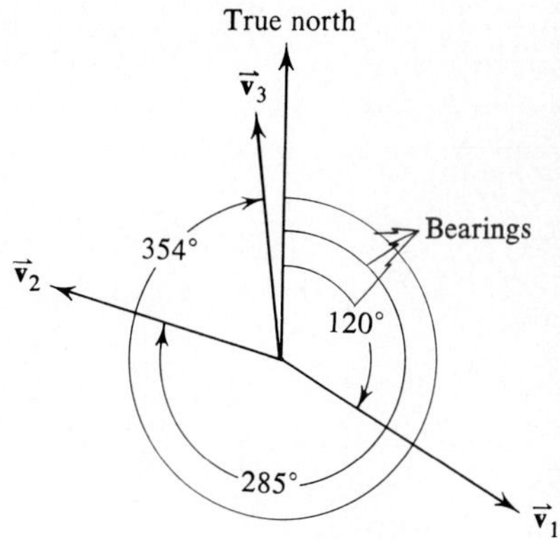

Figure 5.10

directed toward true north. The angle is called the **bearing** and has a positive measure. In graphical representations, we shall use the positive y-axis as the true north direction, and the positive x-axis as the true east direction. For example, in Figure 5.10, the bearing of the geometric vector $\vec{v}_1$ is 120°, that of $\vec{v}_2$ is 285°, and that of $\vec{v}_3$ is 354°.*

Example An airplane is headed northeast (bearing 45°) with an air speed of 500 miles per hour, with a wind blowing from the southeast (bearing 315°) at a speed of 75 miles per hour. Find the drift angle, the ground speed, and the course of the plane.

Solution The figure on page 222 shows three vectors: $\vec{v}_1$ representing the heading and air speed, $\vec{v}_2$ representing the wind direction and speed, and $\vec{v}_3$ representing the course and ground speed. From geometry we have that

$$\angle ABC = 315° - 45° - 180° = 90°.$$

Because $\angle ABC$ is a right angle,

$$\tan \alpha = \frac{75}{500} = 0.1500,$$

and the drift angle $\alpha \approx 8.5°$. The bearing, or direction, of the course is

$$45° - \alpha = 45° - 8.5° = 36.5°.$$

* Other types of bearings, each of which has its own reference, or initial, ray from which the angle is measured, are also defined. A bearing used in surveying is designated by an acute angle measured to the east or west of a ray pointing toward true north or true south, for example, N 35° E, N 38° W, S 22° E, and S 47° W.

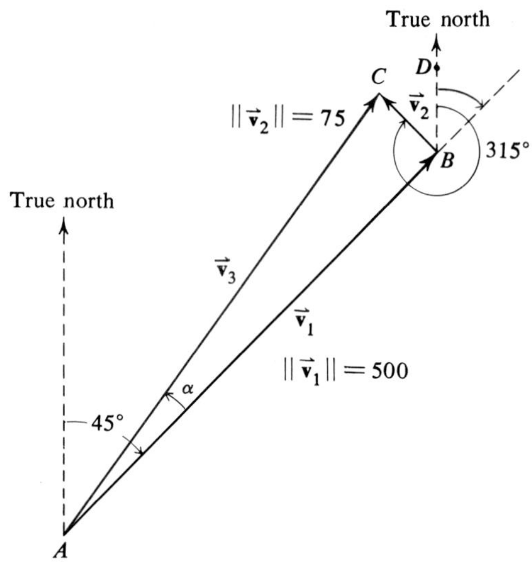

Now, because $\cos \alpha = 500/\|\vec{v}_3\|$,

$$\|\vec{v}_3\| = \frac{500}{\cos 8.5^\circ} \approx \frac{500}{0.9890} \approx 506.$$

Thus, the ground speed (magnitude of the course vector) is approximately 506 miles per hour.

EXERCISE SET 5.2

A

The direction angle α and $\|\vec{v}\|$ are given. Find $\|\vec{v}_x\|$ and $\|\vec{v}_y\|$.

Example $\alpha = 60^\circ$ and $\|\vec{v}\| = 8$

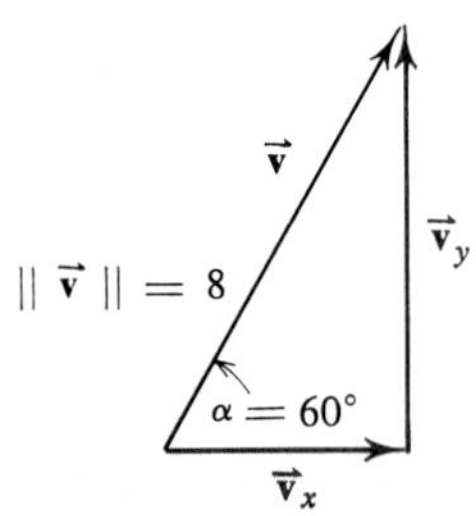

Solution The figure shows the geometric vector $\vec{v}$ and its rectangular components $\vec{v}_x$ and $\vec{v}_y$. From Equations (1) and (2) on page 218,

$$\|\vec{v}_x\| = \|\vec{v}\||\cos 60^\circ| = 8\left(\frac{1}{2}\right) = 4,$$

$$\|\vec{v}_y\| = \|\vec{v}\||\sin 60^\circ| = 8\left(\frac{\sqrt{3}}{2}\right) = 4\sqrt{3}.$$

1. $\alpha = 45^\circ$ and $\|\vec{v}\| = 5\sqrt{2}$

2. $\alpha = 30^\circ$ and $\|\vec{v}\| = 6$

3. $\alpha = 90^\circ$ and $\|\vec{v}\| = 15$

4. $\alpha = 120^\circ$ and $\|\vec{v}\| = 2$

5. $\alpha = -30°$ and $\|\vec{\mathbf{v}}\| = 12$

6. $\alpha = -45°$ and $\|\vec{\mathbf{v}}\| = 2\sqrt{2}$

7. $\alpha = -120°$ and $\|\vec{\mathbf{v}}\| = 8$

8. $\alpha = -150°$ and $\|\vec{\mathbf{v}}\| = 4$

In Exercises 9–14, the two given forces act on a point in a plane at an angle with the specified measure. Find the magnitude of the resultant force and the angles the resultant force makes with the given forces.

Example Forces of 10 and 20 pounds act on a point at an angle of 45° with respect to each other.

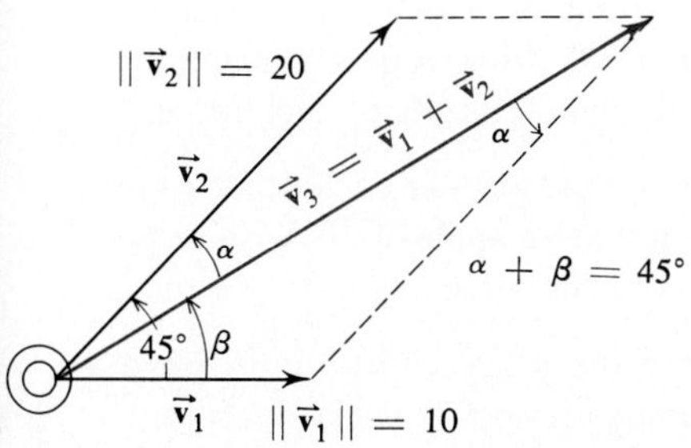

Solution The figure shows the given forces $\vec{\mathbf{v}}_1$ and $\vec{\mathbf{v}}_2$ and the resultant $\vec{\mathbf{v}}_3 = \vec{\mathbf{v}}_1 + \vec{\mathbf{v}}_2$. Since $\alpha + \beta = 45°$, the measure of the angle opposite $\vec{\mathbf{v}}_3$ is 135°. From the fact that $\cos 135° = -\cos 45°$ and from the Law of Cosines,

$$\|\vec{\mathbf{v}}_3\|^2 = 10^2 + 20^2 - 2(10)(20)(-\cos 45°)$$

$$\approx 100 + 400 + 283 \approx 783,$$

and by means of a calculator or by using interpolation from Table I, $\|\vec{\mathbf{v}}_3\| \approx 28.0$. From the Law of Sines,

$$\frac{\sin \alpha}{\|\vec{\mathbf{v}}_1\|} = \frac{\sin 135°}{\|\vec{\mathbf{v}}_3\|},$$

so

$$\sin \alpha = \frac{\|\vec{\mathbf{v}}_1\| \sin 135°}{\|\vec{\mathbf{v}}_3\|} \approx \frac{10(0.7071)}{28.0} \approx 0.2525.$$

Hence, $\alpha \approx 14.6°$. If $\alpha + \beta = 45°$, then

$$\beta \approx 45° - 14.6° = 30.4°.$$

Thus, the resultant force is approximately 28.0 pounds. The angles this force makes with the force of 10 pounds and with the force of 20 pounds have measures of approximately 30.4° and 14.6°, respectively.

9. Forces of 3 and 4 pounds, acting at an angle of 90° with respect to each other.

10. Forces of 5 and 7 pounds, acting at an angle of 30° with respect to each other.

11. Forces of 15 and 20 pounds, acting at an angle of 75° with respect to each other.

12. Forces of 8 and 11 pounds, acting at an angle of 132° 48′ with respect to each other.

13. Forces of 6 and 10 newtons acting at an angle of 120° with respect to each other.

14. Forces of 20 and 30 newtons, acting at an angle of 100° with respect to each other.

15. The resultant of two forces acting at an angle of 90° with respect to each other is 50 newtons. If one of the forces is 30 newtons, find the other force and the angle it makes with the resultant.

16. Two equal forces acting at an angle of 60° with respect to each other have a resultant of 80 newtons. Find the magnitude of each force.

17. Two forces acting at an angle of 60° with respect to each other have a resultant of 50 pounds. If one force acts at an angle of 20° with respect to the resultant, find the magnitude of each force.

18. Two forces, one of 100 pounds and the other 200 pounds, have a resultant of 250 pounds. Find the angle formed by the two forces.

Example An auto weighing 3000 pounds is on a slope that makes an angle of 9° 30′ with the horizontal. Find the force that pulls the auto down the hill.

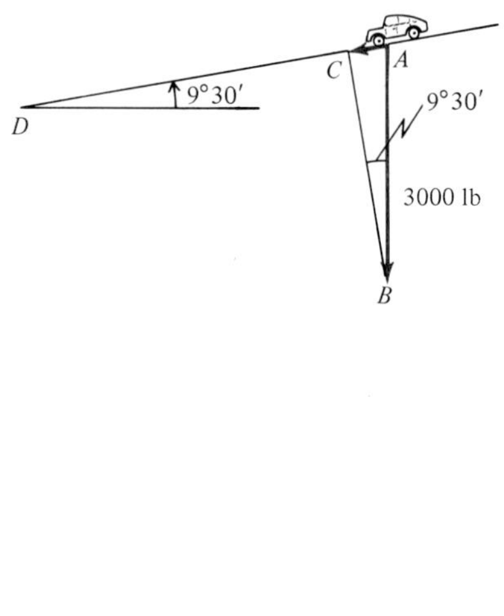

Solution The force due to gravity can be resolved into two forces, one force in the direction of the incline and one force perpendicular to the plane. See the figure. In this case, $\overrightarrow{AB}$, the force due to gravity, is resolved into $\overrightarrow{AC}$ along the incline and $\overrightarrow{CB}$, perpendicular to $\overrightarrow{AC}$. Hence, $\angle B = \angle D$ because the sides of $\angle B$ are perpendicular to the sides of $\angle D$. Thus,

$$\sin 9^\circ\ 30' = \frac{\|\overrightarrow{AC}\|}{3000},$$

from which

$$\|\overrightarrow{AC}\| = 3000 \sin 9^\circ\ 30' \approx 495.$$

Hence, a force of 495 pounds pulls the auto down the hill.

19. An auto, weighing 3500 pounds, is on a slope that makes an angle of 8° 24′ with the horizontal. Find the force that pulls the auto down the hill.

20. A steel ball, weighing 50 pounds, is on an inclined surface. What angle does the plane make with the horizontal if a force of 10 pounds is pulling the ball down the surface?

21. A force of 110 pounds is needed to keep a weight of 200 pounds from sliding down an inclined plane. What angle does the plane make with the horizontal? *Note:* The force that keeps it from sliding is opposite to the force pulling it down the plane, and has the same magnitude.

22. A force of 200 pounds is needed to keep a weight of 320 pounds from sliding down an inclined ramp. What angle does the plane make with the horizontal?

23. A bullet has an acceleration of 2 meters per second^2 in a direction making an angle of 35° with the horizontal. Find the horizontal and vertical components ($\|\vec{v}_x\|$ and $\|\vec{v}_y\|$) of its acceleration.

24. A space rocket has an acceleration of 4.2 meters per second^2 in a direction making an angle of 73° with the horizontal. Find the horizontal and vertical components of the rocket's acceleration.

25. A crew can row a boat at a speed of 9.2 kilometers per hour in still water. If the crew heads across a river at right angles to the current and finds its "drift angle" to be 6°, find the speed of the current.

26. A person can swim at a speed of 2 kilometers per hour in still water. If he heads across a river at right angles to a current of 5 kilometers per hour, find his speed in relation to the land and the direction in which he actually moves.

27. An airplane is headed southeast (bearing 135°) with an air speed of 600 miles per hour, with the wind blowing from the northeast (bearing 225°) at a speed of 120 miles per hour. Find the drift angle, the ground speed, and the course of the airplane.

28. Solve Exercise 27 if the wind is from the southwest (bearing 45°) at a speed of 60 miles per hour.

29. An airplane is headed on a bearing of 250° with an air speed of 425 miles per hour. The course has a bearing of 262°. The ground speed is 475 miles per hour. Find the drift angle, the wind direction, and the wind speed.

30. Solve Exercise 29 if the course has a bearing of 241°.

31. A pilot wishes to fly on a course bearing 90° and with a ground speed of 600 kilometers per hour. If a wind is blowing from the north (bearing 180°) with a speed of 50 kilometers an hour, what must be the heading and air speed of the aircraft?

32. Solve Exercise 31 if the wind is from the northwest (bearing 135°).

B

33. Two ships leave a harbor at the same time, the first traveling at 20 knots (nautical miles per hour) on a course bearing 40°, and the second at

25 knots on a course bearing 300°. After one hour, how far apart are they and what is the bearing of a course from the first ship to the second?

34. Three forces of 25, 35, and 45 pounds act in directions having bearings of 240°, 290°, and 340°, respectively. Find a force, acting due east, and a force, acting due south, that would counterbalance these three forces.

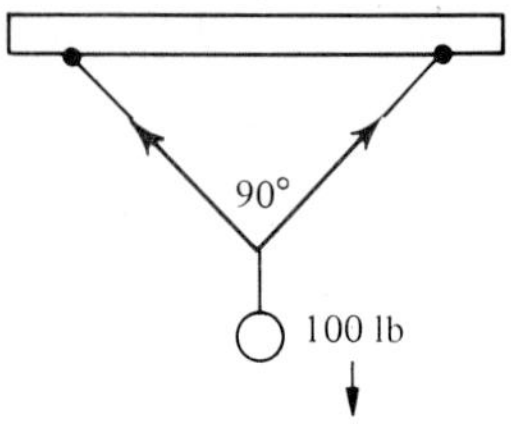

Ex. 35

35. A 100-pound weight hangs from a rope that makes an angle of 90° where it is attached to the weight (see figure). What is the tension (force) on each half of the rope?

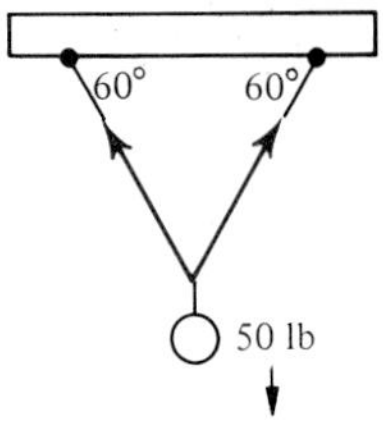

Ex. 36

36. A 50-pound weight hangs from a rope whose two ends are tied at 60° angles to a support (see figure). What is the tension on each half of the rope?

5.3 Vectors as Ordered Pairs

In Sections 5.1 and 5.2 we considered geometric vectors. In this section we will consider the more general notion of a two-dimensional vector as an ordered pair of real numbers, some of its properties, and the close relationship that exists between such a vector and a geometric vector in the plane.

Geometric vectors whose initial points are located at a fixed position in the plane are called **bound geometric vectors,** or **position vectors.** The terminal point of such a geometric vector with initial point at the origin has coordinates that are the components of exactly one ordered pair (a, b)* of real numbers. This relationship, shown in Figure 5.11, suggests the following definition.

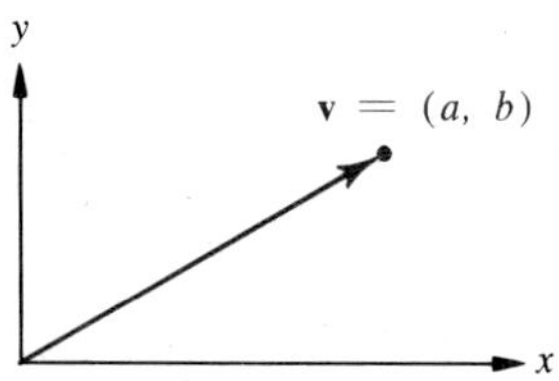

Figure 5.11

Definition 5.8 *A* **two-dimensional vector v** *is an ordered pair of real numbers*

$$\mathbf{v} = (a, b), \quad \textit{where } a, b \in R.$$

The set of all such vectors is designated by **V**.

For a vector $\mathbf{v} = (a, b)$, *the numbers a* and *b* are called the **scalar components** of **v**.

Notice that boldface type *without* an arrow above the symbol is being used to differentiate a vector **v**, an ordered pair of real numbers, from a geometric vector $\vec{\mathbf{v}}$.

* The notation $\langle a, b\rangle$ is often used for a two-dimensional vector.

Example The figure shows the geometric vectors associated with the vectors $\mathbf{v}_1 = (2, 3)$, $\mathbf{v}_2 = (-3, 5)$, $\mathbf{v}_3 = (-5, -1)$, and $\mathbf{v}_4 = (0, -4)$, respectively.

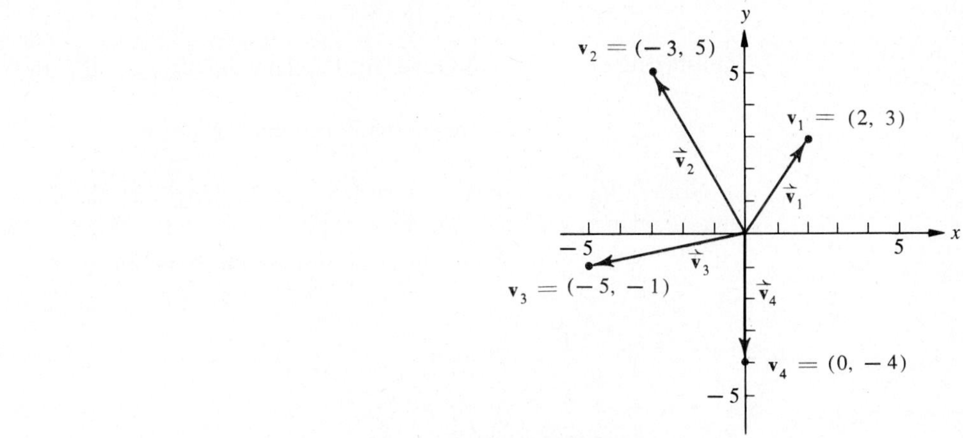

The following definitions are suggested by the close association of a vector with a geometric vector.

Definition 5.9 *For all* $\mathbf{v}_1, \mathbf{v}_2 \in \mathbf{V}$, *where* $\mathbf{v}_1 = (a_1, b_1)$ *and* $\mathbf{v}_2 = (a_2, b_2)$,

$$\mathbf{v}_1 = \mathbf{v}_2 \text{ if and only if } a_1 = a_2 \text{ and } b_1 = b_2.$$

Definition 5.10 *For all* $\mathbf{v} \in \mathbf{V}$, *where* $\mathbf{v} = (a, b)$, *the* **norm** *or* **magnitude** *of* $\mathbf{v}$ *is*

$$\|\mathbf{v}\| = \|(a, b)\| = \sqrt{a^2 + b^2}.$$

Definition 5.11 *For all* $\mathbf{v} \in \mathbf{V}$, *where* $\mathbf{v} = (a, b)$ *and* $\|\mathbf{v}\| \neq 0$, *the* **direction angle** *of* $\mathbf{v}$ *is the angle* α *such that*

$$\cos \alpha = \frac{a}{\|\mathbf{v}\|} \quad \text{and} \quad \sin \alpha = \frac{b}{\|\mathbf{v}\|},$$

where $-180° < \alpha \leq 180°$.

Although Definition 5.11 defines the direction angle of $\mathbf{v} = (a, b)$ in terms of both $\cos \alpha$ and $\sin \alpha$, in practice we generally determine only one function value in order to find α. This is possible because the quadrant in which $\mathbf{v}$ lies can readily be determined from the ordered pair (a, b).

Example By Definition 5.10, the norm of the vector $\mathbf{v} = (2, -5)$ is

$$\|\mathbf{v}\| = \sqrt{2^2 + (-5)^2} = \sqrt{29}.$$

(continued)

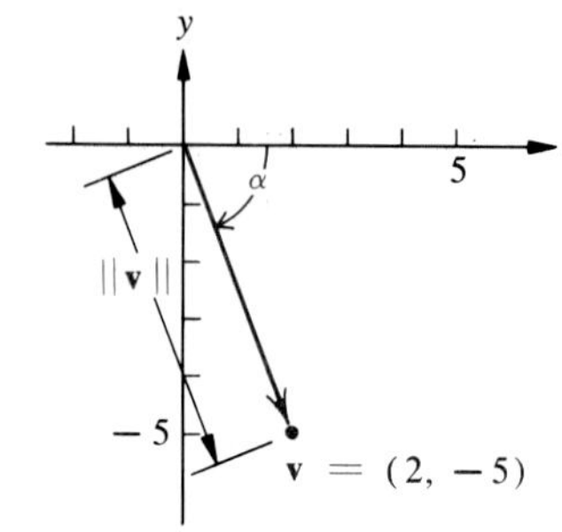

By Definition 5.11,

$$\sin \alpha = \frac{-5}{\sqrt{29}} \approx -0.9285,$$

and because $(2, -5)$ is in Quadrant IV, then $-90^\circ < \alpha < 0^\circ$, and

$$\alpha \approx -68.2^\circ = -68^\circ\ 12'.$$

Notice that the norm of a vector $\mathbf{v}$ is the length of the corresponding geometric vector $\vec{\mathbf{v}}$, and its direction angle is the angle α such that $-180^\circ < \alpha \leq 180^\circ$, associated with the geometric vector.

EXERCISE SET 5.3

A

Find real numbers x and y for which each statement is true.

Example $(x, -y) = (-2, -8)$

Solution By Definition 5.9, $x = -2$ and $y = 8$.

1. $(x, y) = (-3, 6)$
2. $(-x, y) = (5, 1)$
3. $(2x, 7) = (-7, -y)$
4. $(6, 3y) = (x, -6)$
5. $(x + y, x - y) = (-4, 0)$
6. $(2x - y, x + 2y) = (0, -5)$

For each ordered pair **(a)** *sketch each corresponding geometric vector with initial point at the origin and* **(b)** *find the norm and the direction angle (to the nearest six minutes) of each vector.*

Example $(-2, 3)$

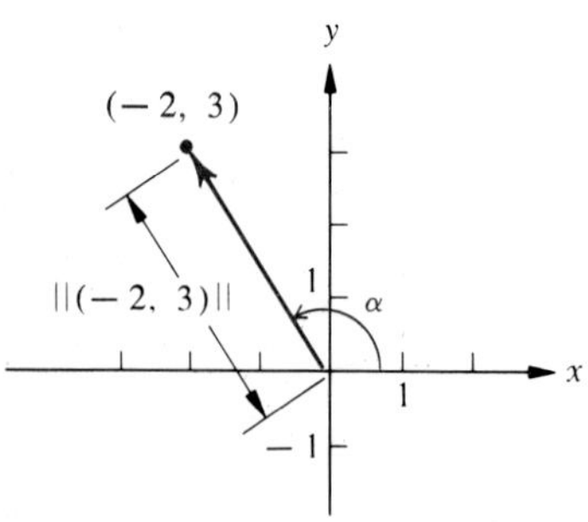

Solution

a. See the figure.

b. By Definition 5.10.

$$\|(-2, 3)\| = \sqrt{(-2)^2 + (3)^2} = \sqrt{13}.$$

By Definition 5.11,

$$\sin \alpha = \frac{3}{\sqrt{13}} \approx 0.8321,$$

and because $(-2, 3)$ is in Quadrant II, $90° < \alpha < 180°$, and

$$\alpha \approx 123.7° = 123°\ 42'.$$

7. $(3, 4)$ **8.** $(-3, 4)$ **9.** $(1, 1)$ **10.** $(-1, 1)$

11. $(-3, -3)$ **12.** $(0, -5)$ **13.** $(4, 0)$ **14.** $(\sqrt{3}, 1)$

Given the following norms and direction angles, find the vector **v**, *and sketch the corresponding geometric vector.*

Example $\|\mathbf{v}\| = 5$ and $\alpha = 30°$

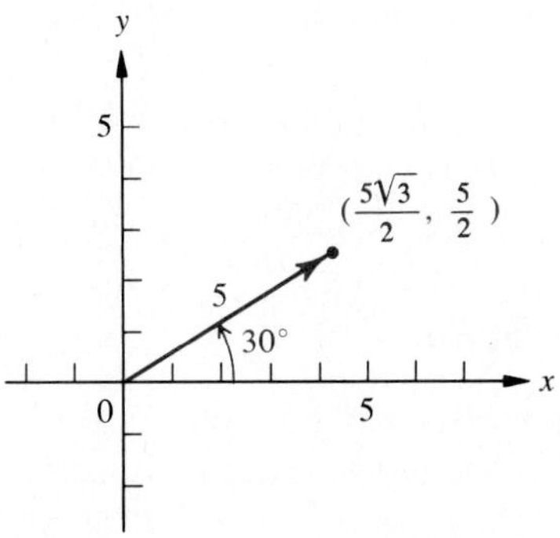

Solution By Definition 5.11, if $\mathbf{v} = (a, b)$, then

$$a = \|\mathbf{v}\| \cos \alpha,$$
$$b = \|\mathbf{v}\| \sin \alpha.$$

Thus,

$$\mathbf{v} = (5 \cos 30°, 5 \sin 30°) = \left(\frac{5\sqrt{3}}{2}, \frac{5}{2}\right).$$

15. $\|\mathbf{v}\| = 2\sqrt{2}$ and $\alpha = 45°$ **16.** $\|\mathbf{v}\| = 2$ and $\alpha = 60°$

17. $\|\mathbf{v}\| = 1$ and $\alpha = 90°$ **18.** $\|\mathbf{v}\| = 4$ and $\alpha = 150°$

19. $\|\mathbf{v}\| = 5$ and $\alpha = -150°$ **20.** $\|\mathbf{v}\| = 3\sqrt{2}$ and $\alpha = -30°$

B

21. Use Definition 5.11 to show that the direction angles of (a, b) and $(-b, a)$ differ by $90°$, given $a, b > 0$.

22. Show that if $a, b, c, d > 0$ and $ac - bd = 0$, then the direction angles of (a, b) and $(-c, d)$ differ by $90°$.

5.4 Fundamental Operations

Two basic operations on vectors are closely related to the operations on geometric vectors that have been defined in Section 5.1.

Vector Addition

As was the case for geometric vectors, the operation of forming the sum of two vectors is called **vector addition.**

Definition 5.12 *For all* $\mathbf{v}_1, \mathbf{v}_2 \in \mathbf{V}$, *where* $\mathbf{v}_1 = (a_1, b_1)$ *and* $\mathbf{v}_2 = (a_2, b_2)$ *the* **sum** $\mathbf{v}_1 + \mathbf{v}_2$ *is a vector such that*

$$\mathbf{v}_1 + \mathbf{v}_2 = (a_1, b_1) + (a_2, b_2) = (a_1 + a_2, b_1 + b_2).$$

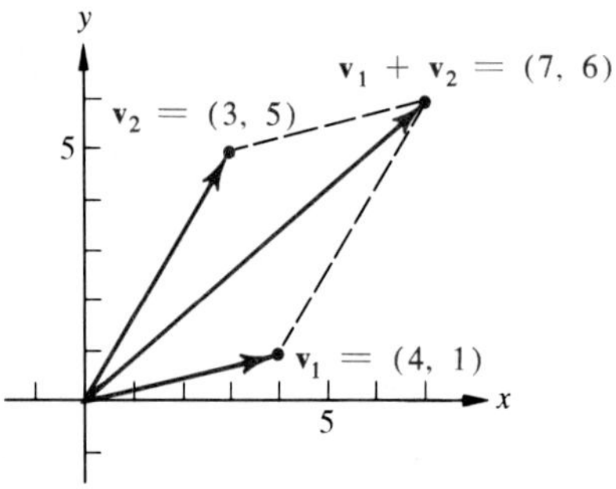

Example Given that $\mathbf{v}_1 = (4, 1)$ and $\mathbf{v}_2 = (3, 5)$,

$$\begin{aligned} \mathbf{v}_1 + \mathbf{v}_2 &= (4, 1) + (3, 5) \\ &= (4 + 3, 1 + 5) \\ &= (7, 6). \end{aligned}$$

The corresponding geometric vectors for $\mathbf{v}_1$, $\mathbf{v}_2$, and their sum $\mathbf{v}_1 + \mathbf{v}_2$ are shown in the figure.

Note that, in the above example, the resultant $\mathbf{v}_1 + \mathbf{v}_2$ is associated with the geometric vector that is one of the diagonals of a parallelogram.

Multiplication of a Vector by a Scalar

The second basic operation on vectors concerns multiplication of a scalar (real number) and a vector and is also related to the similar operation on geometric vectors.

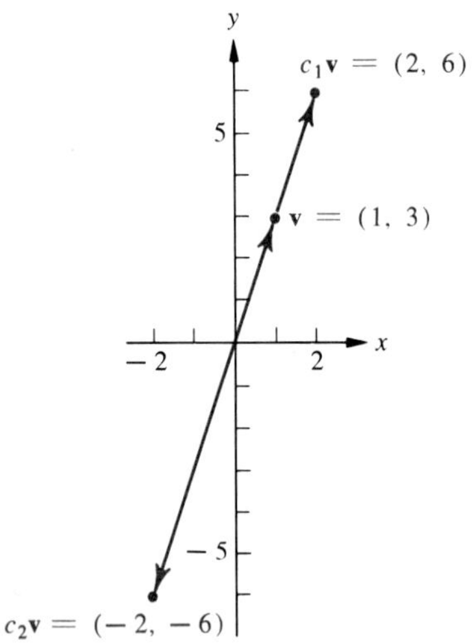

Definition 5.13 *For all* $\mathbf{v} \in \mathbf{V}$, *where* $\mathbf{v} = (a, b)$ *and* $c \in R$,

$$c\mathbf{v} = c(a, b) = (ca, cb).$$

Example If $\mathbf{v} = (1, 3)$, $c_1 = 2$, and $c_2 = -2$, then

$$c_1\mathbf{v} = 2(1, 3) = (2, 6)$$

and

$$c_2\mathbf{v} = -2(1, 3) = (-2, -6).$$

Notice from the above example the $c\mathbf{v}$ has a magnitude $|c| \cdot \|\mathbf{v}\|$ if c is a positive or negative real number. The direction angles of $\mathbf{v}$ and $c\mathbf{v}$ are the same if c is positive, and they differ by 180° if c is negative.

Properties of Vectors

The following two definitions establish the identity element and the additive inverse element in our system of vectors.

Definition 5.14 *The* **zero vector** *is*

$$\mathbf{0} = (0, 0).$$

Any direction may be assigned to the zero vector.

Since

$$\begin{aligned}\mathbf{v} + \mathbf{0} &= (a, b) + (0, 0)\\ &= (a + 0, b + 0)\\ &= (a, b) = \mathbf{v},\end{aligned}$$

the zero vector is an *additive identity element.*

Definition 5.15 *For all* $\mathbf{v} \in \mathbf{V}$, *where* $\mathbf{v} = (a, b)$,

$$-\mathbf{v} = -(a, b) = (-a, -b).$$

Since

$$\begin{aligned}\mathbf{v} + (-\mathbf{v}) &= (a, b) + (-a, -b)\\ &= [a + (-a), b + (-b)]\\ &= (0, 0) = \mathbf{0},\end{aligned}$$

$-\mathbf{v}$ is an *additive inverse element.*

Difference of Two Vectors

The difference of two vectors, $\mathbf{v}_2 - \mathbf{v}_1$, is defined similarly to the difference of two real numbers.

Definition 5.16 *For all vectors* $\mathbf{v}_1$ *and* $\mathbf{v}_2$, *the* **difference** $\mathbf{v}_2 - \mathbf{v}_1$ *is a vector such that*

$$\mathbf{v}_2 - \mathbf{v}_1 = \mathbf{v}_2 + (-\mathbf{v}_1).$$

Example Given that $\mathbf{v}_1 = (2, 5)$ and $\mathbf{v}_2 = (7, 1)$,

$$\begin{aligned}\mathbf{v}_2 - \mathbf{v}_1 &= (7, 1) - (2, 5)\\ &= (7, 1) + (-2, -5)\\ &= (7 - 2, 1 - 5)\\ &= (5, -4).\end{aligned}$$

Figure overleaf

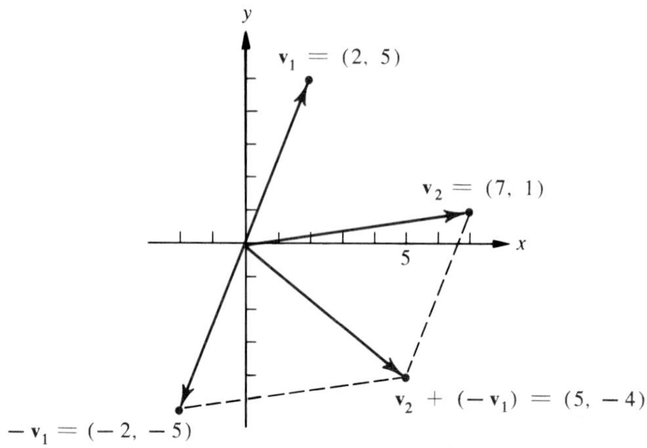

Quotient of a Vector and a Scalar

The quotient of a vector divided by a scalar is defined in terms of a product in much the same way as is the quotient of two real numbers.

Definition 5.17 *For all* $\mathbf{v} \in \mathbf{V}$ *and* $c \in R$ $(c \neq 0)$,

$$\frac{\mathbf{v}}{c} = \frac{1}{c}\mathbf{v}.$$

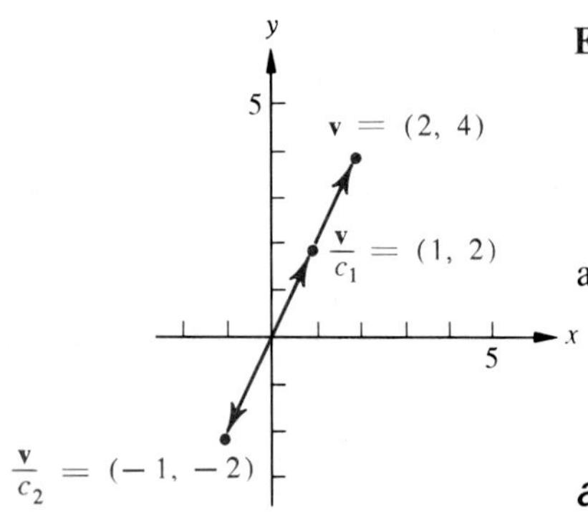

Example If $\mathbf{v} = (2, 4)$, $c_1 = 2$, and $c_2 = -2$, then

$$\frac{\mathbf{v}}{c_1} = \frac{(2, 4)}{2} = \frac{1}{2}(2, 4) = (1, 2),$$

and

$$\frac{\mathbf{v}}{c_2} = \frac{(2, 4)}{-2} = -\frac{1}{2}(2, 4) = (-1, -2).$$

$a\mathbf{i} + b\mathbf{j}$ Form

If vectors $\mathbf{v}_1 = (1, 0)$ and $\mathbf{v}_2 = (0, 1)$, called **unit vectors,** are designated $\mathbf{i}$ and $\mathbf{j}$, respectively, then each vector

$$\begin{aligned}\mathbf{v} &= (a, b)\\ &= (a, 0) + (0, b)\\ &= a(1, 0) + b(0, 1)\\ &= a\mathbf{i} + b\mathbf{j}.\end{aligned}$$

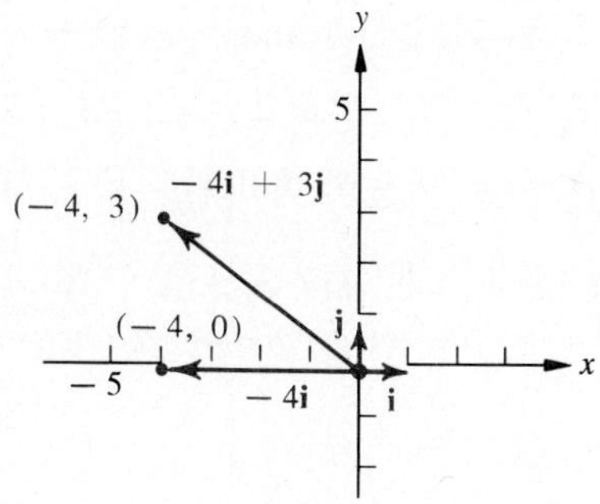

Here again the real numbers a and b are called the **scalar components** of the vector.

Examples Express each vector as the sum of scalar multiples of **i** and **j** and sketch the corresponding geometric vectors.

a. $(-4, 0)$ b. $(-4, 3)$

Solutions

a. $(-4, 0) = -4i + 0\mathbf{j}$
$= -4\mathbf{i}$

b. $(-4, 3) = -4\mathbf{i} + 3\mathbf{j}$

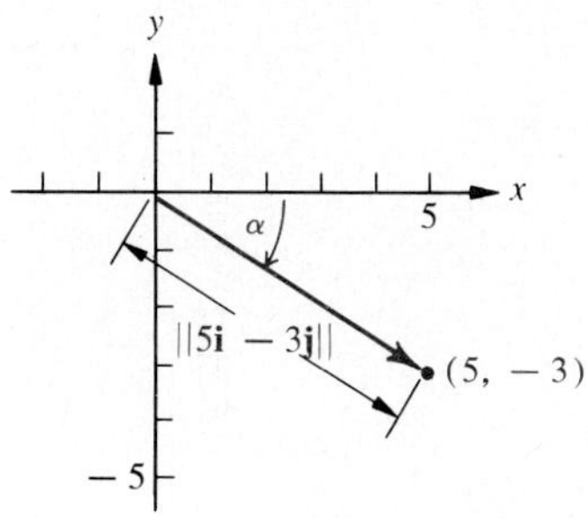

Example Find the norm and the direction angle of $5\mathbf{i} - 3\mathbf{j}$.

Solution $5\mathbf{i} - 3\mathbf{j} = (5, -3)$. Thus,

$$\|5\mathbf{i} - 3\mathbf{j}\| = \|(5, -3)\|,$$

from which, by Definition 5.10, we have

$$\|5\mathbf{i} - 3\mathbf{j}\| = \sqrt{5^2 + (-3)^2} = \sqrt{34}.$$

From Definition 5.11,

$$\sin\alpha = \frac{-3}{\sqrt{34}} \approx -0.5145.$$

Furthermore, $(5, -3)$ corresponds to a point in Quadrant IV. Hence

$$\alpha \approx -31.0°.$$

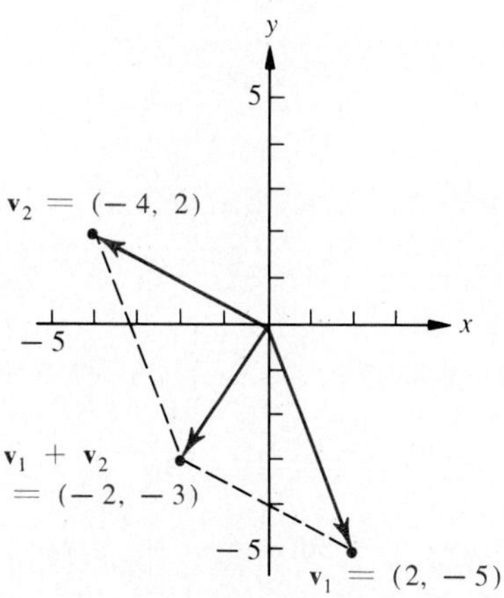

EXERCISE SET 5.4

A

Write the sum of each pair of vectors as an ordered pair and show the corresponding geometric vectors.

Example $\mathbf{v}_1 = (2, -5)$ and $\mathbf{v}_2 = (-4, 2)$

Solution
$$\mathbf{v}_1 + \mathbf{v}_2 = (2, -5) + (-4, 2)$$
$$= (2 + (-4), -5 + 2)$$
$$= (-2, -3)$$

1. $\mathbf{v}_1 = (-4, 0)$ and $\mathbf{v}_2 = (0, 6)$

2. $\mathbf{v}_1 = (-6, 0)$ and $\mathbf{v}_2 = (0, 1)$

3. $\mathbf{v}_1(1, -4)$ and $\mathbf{v}_2 = (2, 5)$

4. $\mathbf{v}_1 = (7, -8)$ and $\mathbf{v}_2 = (-3, -7)$

5. $\mathbf{v}_1 = (-5, 3)$ and $\mathbf{v}_2 = (1, -2)$

6. $\mathbf{v}_1 = (-6, 6)$ and $\mathbf{v}_2 = (3, -3)$

Let $\mathbf{v}_1 = (3, 6)$, $\mathbf{v}_2 = (5, -1)$, $\mathbf{v}_3 = (-4, 2)$, $c_1 = 2$, *and* $c_2 = -3$. *Write each expression as an ordered pair and show the corresponding geometric vectors.*

Example $c_1\mathbf{v}_2 + c_1\mathbf{v}_3$

Solution

$$\begin{aligned} c_1\mathbf{v}_2 + c_1\mathbf{v}_3 &= 2(5, -1) + 2(-4, 2) \\ &= (10, -2) + (-8, 4) \\ &= (10 + (-8), -2 + 4) \\ &= (2, 2) \end{aligned}$$

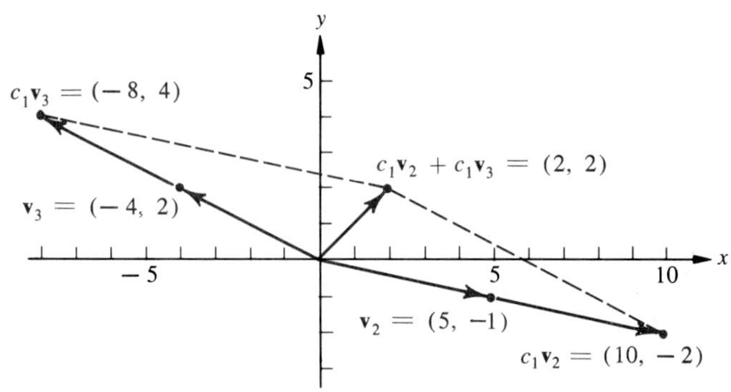

7. $-\mathbf{v}_1$

8. $-\mathbf{v}_2$

9. $\mathbf{v}_3 + \mathbf{0}$

10. $\mathbf{0} - \mathbf{v}_1$

11. $c_1\mathbf{v}_1$

12. $c_2\mathbf{v}_2$

13. $c_2\mathbf{v}_1 + \mathbf{v}_3$

14. $c_1\mathbf{v}_2 + \mathbf{v}_1$

15. $c_1\mathbf{v}_1 + c_2\mathbf{v}_2$

16. $c_1\mathbf{v}_2 + c_2\mathbf{v}_3$

17. $c_1(\mathbf{v}_1 + \mathbf{v}_2)$

18. $c_2(\mathbf{v}_2 + \mathbf{v}_3)$

19. $c_1\mathbf{v}_2 - \mathbf{v}_1$

20. $c_2\mathbf{v}_3 - \mathbf{v}_2$

21. $c_2\mathbf{v}_2 - c_1\mathbf{v}_3$

22. $c_2\mathbf{v}_3 - c_1\mathbf{v}_1$

23. $c_2(\mathbf{v}_2 - \mathbf{v}_3)$

24. $c_1(\mathbf{v}_3 - \mathbf{v}_1)$

25. $\dfrac{\mathbf{v}_2}{c_1}$

26. $\dfrac{\mathbf{v}_3}{c_2}$

27. $\dfrac{\mathbf{v}_1}{c_1} - \dfrac{\mathbf{v}_2}{c_2}$

28. $\dfrac{\mathbf{v}_2}{c_2} - \dfrac{\mathbf{v}_3}{c_1}$

29. $\dfrac{1}{4}c_1\mathbf{v}_1 - \dfrac{1}{3}c_2\mathbf{v}_3$

30. $\dfrac{1}{8}c_1\mathbf{v}_2 + \dfrac{1}{6}c_2\mathbf{v}_1$

Express each vector as a sum of scalar multiples of **i** *and* **j**.

Example (2, 9)

Solution $(2, 9) = (2, 0) + (0, 9)$

$$= 2(1, 0) + 9(0, 1)$$

$$= 2\mathbf{i} + 9\mathbf{j}$$

31. (3, 5) **32.** (−2, 1) **33.** (1, −4)
34. (−2, −3) **35.** (−3, 0) **36.** (0, 4)

Find the norm and the direction angle, to the nearest tenth, of each vector. Sketch the corresponding geometric vector.

Example $2\mathbf{i} - 3\mathbf{j}$

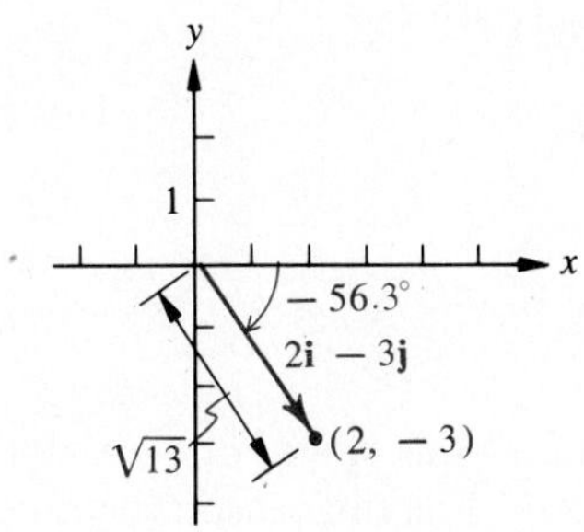

Solution $2\mathbf{i} - 3\mathbf{j} = (2, -3)$

Hence, by Definition 5.10,

$$\|2\mathbf{i} - 3\mathbf{j}\| = \|(2, -3)\|$$

$$= \sqrt{2^2 + (-3)^2}$$

$$= \sqrt{13}.$$

From Definition 5.11,

$$\sin \alpha = \frac{-3}{\sqrt{13}} \approx -0.8321.$$

Since $(2, -3)$ is in Quadrant IV, $-90° < \alpha < 0°$. Thus,

$$\alpha \approx -56.3°.$$

37. $4\mathbf{i} - 4\mathbf{j}$ **38.** $\sqrt{3}\mathbf{i} + \mathbf{j}$ **39.** $2\mathbf{i} + \mathbf{j}$ **40.** $-\mathbf{i} - 5\mathbf{j}$
41. $-\mathbf{i} - 3\mathbf{j}$ **42.** $-6\mathbf{i} + \mathbf{j}$ **43.** $-\mathbf{i} + \sqrt{3}\mathbf{j}$ **44.** $3\mathbf{i} - 2\mathbf{j}$

5.5 Inner Products

In Section 5.4 we considered two basic operations, vector addition and the multiplication of a vector by a scalar. The result of either operation is a vector and the set of vectors **V** is said to be closed for each of these operations. Another operation on two vectors is defined in such a way as to be useful in a variety of problems related to physical and geometric concepts.

Definition 5.18 *For all* $\mathbf{v}_1, \mathbf{v}_2 \in \mathbf{V}$, $\mathbf{v}_1 = (a_1, b_1)$ *and* $\mathbf{v}_2 = (a_2, b_2)$, *the* **inner product**

$$\mathbf{v}_1 \cdot \mathbf{v}_2 = (a_1, b_1) \cdot (a_2, b_2) = a_1 a_2 + b_1 b_2. \tag{1}$$

Since the symbol for the operation is a dot, the inner product is also called a **dot product.** Notice that the dot as used here has a different meaning than when it is used for the operation of multiplication in the set of real numbers.

Examples

a. If $\mathbf{v}_1 = (2, 3)$ and $\mathbf{v}_2 = (-5, 4)$, then

$$\begin{aligned} \mathbf{v}_1 \cdot \mathbf{v}_2 &= (2, 3) \cdot (-5, 4) \\ &= (2)(-5) + (3)(4) = 2. \end{aligned}$$

b. If $\mathbf{v}_1 = 3\mathbf{i} - 4\mathbf{j}$ and $\mathbf{v}_2 = \mathbf{i} + 2\mathbf{j}$, then

$$\begin{aligned} \mathbf{v}_1 \cdot \mathbf{v}_2 &= (3\mathbf{i} - 4\mathbf{j}) \cdot (\mathbf{i} + 2\mathbf{j}) \\ &= (3)(1) + (-4)2 = -5. \end{aligned}$$

Notice that the set of vectors $\mathbf{V}$ is not closed for this operation. The inner product of two vectors is a scalar, not a vector. For this reason the inner product is called the **scalar product.**

Another useful form of the inner product can be obtained from the Law of Cosines.

Theorem 5.2 *If* $\mathbf{v}_1 = (a_1, b_1)$ *and* $\mathbf{v}_2 = (a_2, b_2)$, *where* $\mathbf{v}_1$ *and* $\mathbf{v}_2$ *are nonzero vectors, and* γ *is the angle formed by the geometric vectors corresponding to* $\mathbf{v}_1$ *and* $\mathbf{v}_2$, then

$$\mathbf{v}_1 \cdot \mathbf{v}_2 = \|\mathbf{v}_1\| \, \|\mathbf{v}_2\| \cos \gamma. \tag{2}$$

Proof Consider the line segment with length c drawn between the points corresponding to (a_1, b_1) and (a_2, b_2), as shown in Figure 5.12. From the Law of Cosines,

$$c^2 = \|\mathbf{v}_1\|^2 + \|\mathbf{v}_2\|^2 - 2\|\mathbf{v}_1\| \, \|\mathbf{v}_2\| \cos \gamma,$$

from which, for nonzero vectors $\mathbf{v}_1$ and $\mathbf{v}_2$,

$$\cos \gamma = \frac{\|\mathbf{v}_1\|^2 + \|\mathbf{v}_2\|^2 - c^2}{2\|\mathbf{v}_1\| \, \|\mathbf{v}_2\|}.$$

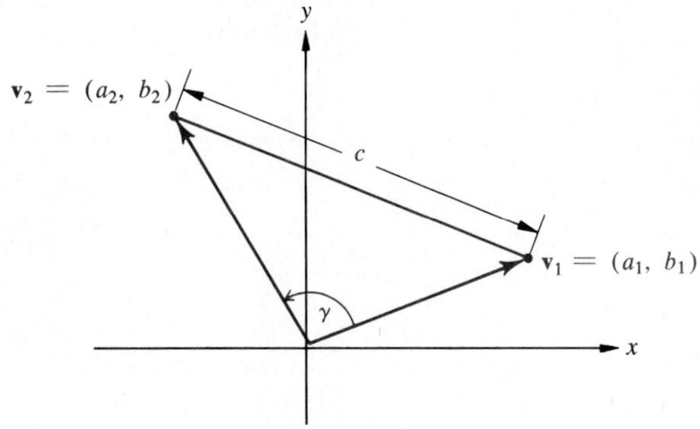

Figure 5.12

Substituting $\sqrt{a_1^2 + b_1^2}$ for $\|\mathbf{v}_1\|$, $\sqrt{a_2^2 + b_2^2}$ for $\|\mathbf{v}_2\|$, and

$$\sqrt{(a_2 - a_1)^2 + (b_2 - b_1)^2}$$

for c and simplifying, we obtain

$$\cos \gamma = \frac{a_1 a_2 + b_1 b_2}{\|\mathbf{v}_1\| \, \|\mathbf{v}_2\|}.$$

Substituting $\mathbf{v}_1 \cdot \mathbf{v}_2$ for $a_1 a_2 + b_1 b_2$ from Definition 5.18 yields

$$\mathbf{v}_1 \cdot \mathbf{v}_2 = \|\mathbf{v}_1\| \, \|\mathbf{v}_2\| \cos \gamma.$$

We can now express the inner product of two vectors $\mathbf{v}_1 = (a_1, b_1)$ and $\mathbf{v}_2 = (a_2, b_2)$ either, as shown in Equation (1), in terms of their scalar components or, as in Equation (2), in terms of the norms of the two vectors and the cosine of the angle γ formed by their corresponding geometric vectors.

Example Given that $\|\mathbf{v}_1\| = 3$, and $\mathbf{v}_1$ has a direction angle, $\alpha_1 = 20°$, $\|\mathbf{v}_2\| = 8$, and $\mathbf{v}_2$ has a direction angle, $\alpha_2 = 170°$, find $\mathbf{v}_1 \cdot \mathbf{v}_2$.

Solution From Theorem 5.2,

$$\mathbf{v}_1 \cdot \mathbf{v} = (3)(8)\cos(170° - 20°) = (3)(8)\cos 150°$$
$$= 24\left(-\frac{\sqrt{3}}{2}\right) = -12\sqrt{3}.$$

Scalar Projections

The property of inner products exhibited for two nonzero vectors in Equation (2) is particularly useful in several equivalent forms.

Theorem 5.3 *If* $\mathbf{v}_1$ *and* $\mathbf{v}_2$ *are nonzero vectors, and* γ *is the angle formed by the geometric vectors corresponding to* $\mathbf{v}_1$ *and* $\mathbf{v}_2$*, then*

$$\text{I. } \cos \gamma = \frac{\mathbf{v}_1 \cdot \mathbf{v}_2}{\|\mathbf{v}_1\| \, \|\mathbf{v}_2\|},$$

$$\text{II. } \|\mathbf{v}_1\| \cos \gamma = \frac{\mathbf{v}_1 \cdot \mathbf{v}_2}{\|\mathbf{v}_2\|},$$

$$\text{III. } \|\mathbf{v}_2\| \cos \gamma = \frac{\mathbf{v}_1 \cdot \mathbf{v}_2}{\|\mathbf{v}_1\|}.$$

Theorem 5.3-I enables us to find the angle γ between any two non-zero vectors.

Example The angle γ between the geometric vectors corresponding to (2, 3) and (0, 2) is given by

$$\cos \gamma = \frac{(2, 3) \cdot (0, 2)}{\|(2, 3)\| \, \|(0, 2)\|} = \frac{2 \cdot 0 + 3 \cdot 2}{\sqrt{4+9}\sqrt{0+4}}$$

$$= \frac{3}{\sqrt{13}} \approx 0.8321.$$

Thus, $\gamma \approx 33.7°$.

In Section 5.2 we obtained projections of a vector on the x- and y-axes of a Cartesian coordinate system. Parts II and III of Theorem 5.3 now give us a way to find the projection of a vector on any other vector.

The product $\|\mathbf{v}_1\| \cos \gamma$ is called the **scalar projection** of $\mathbf{v}_1$ on $\mathbf{v}_2$ and the product $\|\mathbf{v}_2\| \cos \gamma$ is called the scalar projection of $\mathbf{v}_2$ on $\mathbf{v}_1$. In a geometric sense, these can be interpreted as the lengths of the line segments shown in Figure 5.13 (a and b).

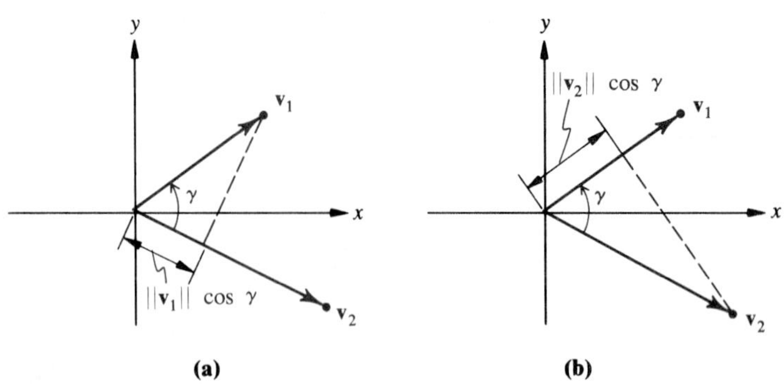

Figure 5.13

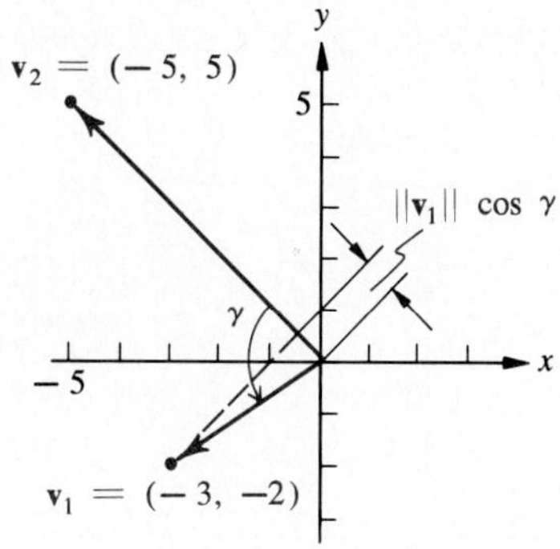

Example Given that $\mathbf{v}_1 = (-3, -2)$ and $\mathbf{v}_2 = (-5, 5)$, the scalar projection of $\mathbf{v}_1$ on $\mathbf{v}_2$ (Theorem 5.3-II) is

$$\|\mathbf{v}_1\| \cos \gamma = \frac{\mathbf{v}_1 \cdot \mathbf{v}_2}{\|\mathbf{v}_2\|} = \frac{(-3, -2) \cdot (-5, 5)}{\|(-5, 5)\|}$$

$$= \frac{15 - 10}{\sqrt{(-5)^2 + 5^2}} = \frac{5}{\sqrt{50}}$$

$$= \frac{5}{5\sqrt{2}} = \frac{1}{\sqrt{2}}.$$

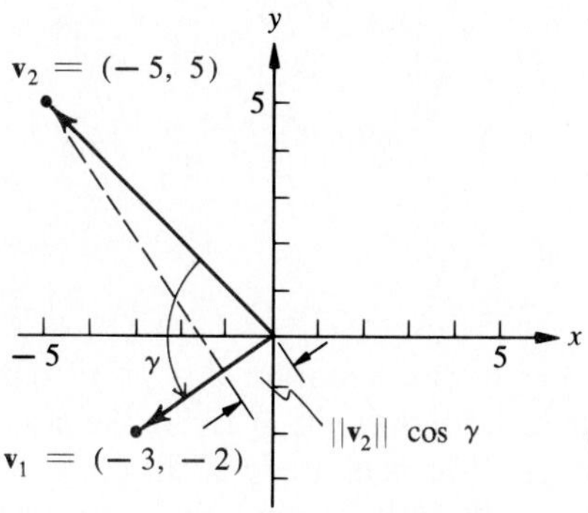

The scalar projection of $\mathbf{v}_2$ on $\mathbf{v}_1$ (Theorem 5.3-III) is

$$\|\mathbf{v}_2\| \cos \gamma = \frac{\mathbf{v}_1 \cdot \mathbf{v}_2}{\|\mathbf{v}_1\|} = \frac{(-3, -2) \cdot (-5, 5)}{\|(-3, -2)\|}$$

$$= \frac{15 - 10}{\sqrt{(-3)^2 + (-2)^2}} = \frac{5}{\sqrt{13}}.$$

Orthogonal Vectors

If $\mathbf{v}_1$ and $\mathbf{v}_2$ are nonzero vectors and are perpendicular, then $\|\mathbf{v}_1\| \neq 0$, $\|\mathbf{v}_2\| \neq 0$, $\cos \gamma = 0$, and

$$\mathbf{v}_1 \cdot \mathbf{v}_2 = \|\mathbf{v}_1\| \, \|\mathbf{v}_2\| \cos \gamma = 0.$$

We also note that for $0° \leq \gamma \leq 180°$, $\cos \gamma = 0$ only if $\gamma = 90°$. From Definition 5.18,

$$\mathbf{v}_1 \cdot \mathbf{v}_2 = a_1 a_2 + b_1 b_2.$$

Hence, we can test for the orthogonality (perpendicularity) of two vectors directly from their scalar components.

Theorem 5.4 *The vectors* $\mathbf{v}_1 = (a_1, b_1)$ *and* $\mathbf{v}_2 = (a_2, b_2)$ *are orthogonal if and only if*

$$\mathbf{v}_1 \cdot \mathbf{v}_2 = \boldsymbol{a_1 a_2 + b_1 b_2 = 0}.$$

Examples

a. $(2, -4)$ and $(6, 3)$ are orthogonal because

$$(2, -4) \cdot (6, 3) = (2)(6) + (-4)(3) = 0.$$

(continued)

b. $-\mathbf{i} - 2\mathbf{j}$ and $4\mathbf{i} - 2\mathbf{j}$ are orthogonal because

$$(-\mathbf{i} - 2\mathbf{j}) \cdot (4\mathbf{i} - 2\mathbf{j}) = (-1)(4) + (-2)(-2) = 0.$$

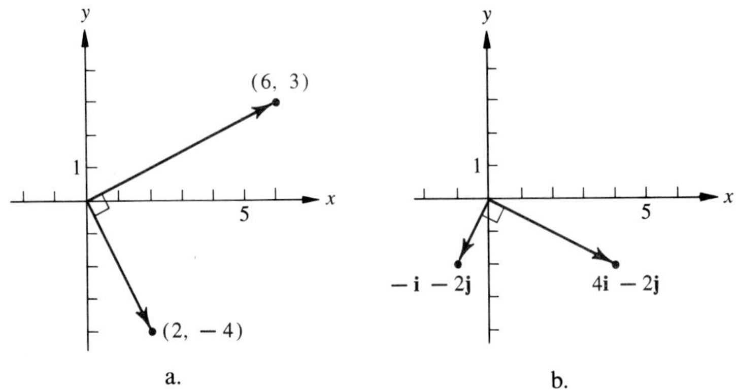

Applications

If a force is applied to an object so as to cause motion in a particular direction, and if the force is applied throughout the motion and in the same direction as the motion, then the product of the magnitude of the force (F) and the distance (d) the object moves is called **work.** In many situations, the direction of an applied force is not the same as the direction in which the object moves. In this case, the work is the product of the magnitude of the *component of the force in the direction of motion* and the distance the object moves. Since the inner product of two vectors can be used to find such a product whether or not the two vectors have the same direction, we provide a more general definition of work in Definition 5.19.

Definition 5.19 *If* $\mathbf{F}$ *is a force vector and* $\mathbf{d}$ *a displacement vector, then the work* W *is given by*

$$W = \mathbf{F} \cdot \mathbf{d}.$$

Example A force of 40 pounds at an angle of 30° to the horizontal is used to move a box 8 feet on a level floor. Find the work done.

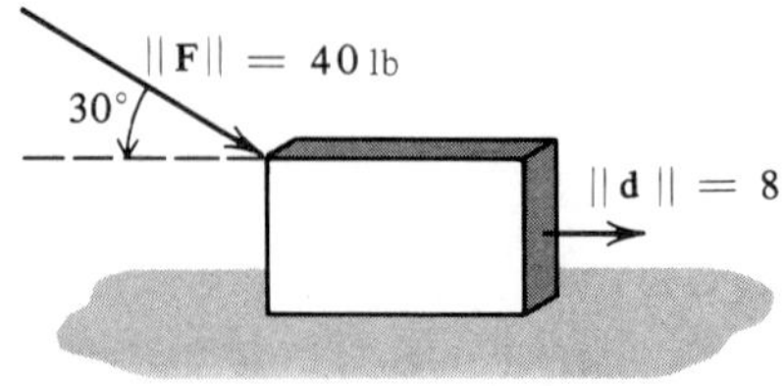

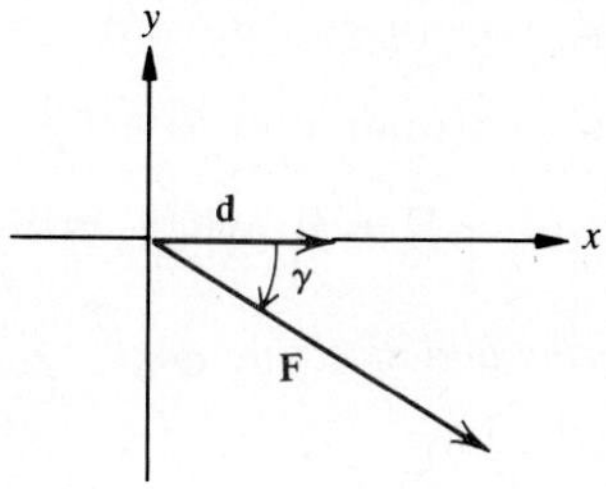

Solution The figure shows the vectors in a Cartesian system. Since $\|\mathbf{F}\| = 40$ pounds, $\|\mathbf{d}\| = 8$ feet, and $\gamma = 30°$,

$$
\begin{aligned}
W = \mathbf{F} \cdot \mathbf{d} &= \|\mathbf{F}\|\,\|\mathbf{d}\| \cos \gamma \\
&= (40)(8) \cos 30° \\
&= (40)(8)\left(\frac{\sqrt{3}}{2}\right) \\
&= 160\sqrt{3} \approx 277.
\end{aligned}
$$

If the force magnitude is measured in pounds and the displacement magnitude in feet, as in this case, the work is measured in foot-pounds. Hence the work done is approximately 277 foot-pounds.

Example How much work is done in lifting an 80-pound weight to a height of 3 feet?

Solution Since the force and displacement vectors are in the same direction, the angle $\gamma = 0°$ and $\cos \gamma = 1$. Hence,

$$W = \mathbf{F} \cdot \mathbf{d} = \|\mathbf{F}\|\,\|\mathbf{d}\| \cos \gamma = (80)(3)(1) = 240.$$

Hence, the work done is 240 foot-pounds. In this case the work is simply the product of the force times the distance.

EXERCISE SET 5.5

A

Find the inner product, $\mathbf{v}_1 \cdot \mathbf{v}_2$, of the given vectors.

Examples

a. $\mathbf{v}_1 = (7, -9)$ and $\mathbf{v}_2 = (-6, -5)$

b. $\mathbf{v}_1 = -9\mathbf{i} + 8\mathbf{j}$ and $\mathbf{v}_2 = 3\mathbf{i} - 2\mathbf{j}$

Solutions From Definition 5.18,

a. $\mathbf{v}_1 \cdot \mathbf{v}_2 = (7, -9) \cdot (-6, -5)$
$\quad = (7)(-6) + (-9)(-5) = 3$

b. $\mathbf{v}_1 \cdot \mathbf{v}_2 = (-9\mathbf{i} + 8\mathbf{j}) \cdot (3\mathbf{i} - 2\mathbf{j})$
$\quad = (-9)(3) + (8)(-2) = -43$

1. $\mathbf{v}_1 = (6, 0)$ and $\mathbf{v}_2 = (8, 10)$

2. $\mathbf{v}_1 = (9, 3)$ and $\mathbf{v}_2 = (8, -5)$

3. $\mathbf{v}_1 = (7, -8)$ and $\mathbf{v}_2 = (2, 2)$

4. $\mathbf{v}_1 = (3, 0)$ and $\mathbf{v}_2 = (0, -5)$

5. $\mathbf{v}_1 = 9\mathbf{i} - 4\mathbf{j}$ and $\mathbf{v}_2 = 5\mathbf{i}$

6. $\mathbf{v}_1 = 8\mathbf{j}$ and $\mathbf{v}_2 = 7\mathbf{i} + 7\mathbf{j}$

7. $\mathbf{v}_1 = \mathbf{i} - 2\mathbf{j}$ and $\mathbf{v}_2 = 3\mathbf{i} - 5\mathbf{j}$

8. $\mathbf{v}_1 = -4\mathbf{i} + 3\mathbf{j}$ and $\mathbf{v}_2 = 10\mathbf{j}$

Find the exact value of $\mathbf{v}_1 \cdot \mathbf{v}_2$ *given the norms and direction angles for* $\mathbf{v}_1$ *and* $\mathbf{v}_2$.

Example $\|\mathbf{v}_1\| = 10,\ \alpha_1 = 75°$ and $\|\mathbf{v}_2\| = 3,\ \alpha_2 = 45°$

Solution From Theorem 5.2,

$$\begin{aligned} \mathbf{v}_1 \cdot \mathbf{v}_2 &= (10)(3)\cos(75° - 45°) \\ &= 30 \cos 30° \\ &= 30\left(\frac{\sqrt{3}}{2}\right) = 15\sqrt{3}. \end{aligned}$$

9. $\|\mathbf{v}_1\| = 8,\ \alpha_1 = 60°$ and $\|\mathbf{v}_2\| = 9,\ \alpha_2 = 15°$

10. $\|\mathbf{v}_1\| = 7,\ \alpha_1 = 160°$ and $\|\mathbf{v}_2\| = 6,\ \alpha_2 = -170°$

11. $\|\mathbf{v}_1\| = 5,\ \alpha_1 = -155°$ and $\|\mathbf{v}_2\| = \frac{3}{10},\ \alpha_2 = 145°$

12. $\|\mathbf{v}_1\| = \frac{1}{2},\ \alpha_1 = -60°$ and $\|\mathbf{v}_2\| = 10,\ \alpha_2 = -150°$

Determine the angle γ, *where* $0° \le \gamma \le 180°$, *to the nearest tenth, formed by each pair of vectors.*

Example $\mathbf{v}_1 = (1, -7)$ and $\mathbf{v}_2 = (9, -1)$

Solution From Theorem 5.3-I,

$$\begin{aligned} \cos\gamma &= \frac{(1, -7)\cdot(9, -1)}{\sqrt{1^2 + (-7)^2}\sqrt{9^2 + (-1)^2}} \\ &= \frac{16}{\sqrt{50}\sqrt{82}} = \frac{16}{5\sqrt{2}\cdot\sqrt{2}\cdot\sqrt{41}} = \frac{8}{5\sqrt{41}} \approx 0.2499, \end{aligned}$$

from which $\gamma \approx 75.5°$.

13. $\mathbf{v}_1 = (3, 0)$ and $\mathbf{v}_2 = (-7, 3)$

14. $\mathbf{v}_1 = (1, -3)$ and $\mathbf{v}_2 = (8, -2)$

15. $\mathbf{v}_1 = (4, 0)$ and $\mathbf{v}_2 = (0, -2)$

16. $\mathbf{v}_1 = (-2, 0)$ and $\mathbf{v}_2 = (0, -3)$

17. $\mathbf{v}_1 = 9\mathbf{i} + 4\mathbf{j}$ and $\mathbf{v}_2 = 2\mathbf{i}$

18. $\mathbf{v}_1 = -4\mathbf{i} - 8\mathbf{j}$ and $\mathbf{v}_2 = \mathbf{i} - 5\mathbf{j}$

Find the scalar projection of $\mathbf{v}_1$ on $\mathbf{v}_2$ and of $\mathbf{v}_2$ on $\mathbf{v}_1$ for the given vectors. Sketch the corresponding geometric vectors and show the scalar projections.

Example $\mathbf{v}_1 = (1, 2)$ and $\mathbf{v}_2 = (-3, 4)$

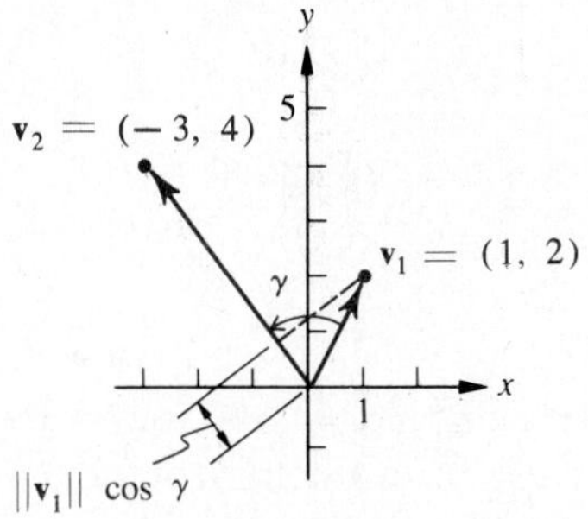

Solution From Theorem 5.3-II, the scalar projection of $\mathbf{v}_1$ on $\mathbf{v}_2$ is

$$\|\mathbf{v}_1\| \cos \gamma = \frac{\mathbf{v}_1 \cdot \mathbf{v}_2}{\|\mathbf{v}_2\|} = \frac{(1)(-3) + (2)(4)}{\sqrt{(-3)^2 + (4)^2}}$$

$$= \frac{-3 + 8}{\sqrt{25}} = \frac{5}{5} = 1.$$

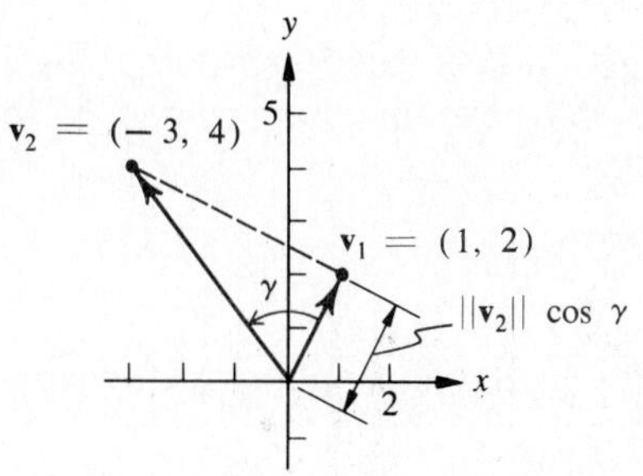

From Theorem 5.3-III, the scalar projection of $\mathbf{v}_2$ on $\mathbf{v}_1$ is

$$\|\mathbf{v}_2\| \cos \gamma = \frac{\mathbf{v}_1 \cdot \mathbf{v}_2}{\|\mathbf{v}_1\|} = \frac{5}{\sqrt{1^2 + 2^2}}$$

$$= \frac{5}{\sqrt{5}} = \sqrt{5}.$$

19. $\mathbf{v}_1 = -4\mathbf{i}$ and $\mathbf{v}_2 = -5\mathbf{i} - 2\mathbf{j}$

20. $\mathbf{v}_1 = \mathbf{i} - 6\mathbf{j}$ and $\mathbf{v}_2 = -5\mathbf{j}$

21. $\mathbf{v}_1 = (2, 3)$ and $\mathbf{v}_2 = (-4, 5)$

22. $\mathbf{v}_1 = (-9, 10)$ and $\mathbf{v}_2 = (2, 4)$

23. $\mathbf{v}_1 = (-3, 5)$ and $\mathbf{v}_2 = (3, -1)$

24. $\mathbf{v}_1 = (-7, 4)$ and $\mathbf{v}_2 = (3, -6)$

25. $\mathbf{v}_1 = 3\mathbf{i}$ and $\mathbf{v}_2 = 2\mathbf{j}$

26. $\mathbf{v}_1 = 2\mathbf{i}$ and $\mathbf{v}_2 = -2\mathbf{j}$

Verify that the following pairs of vectors are orthogonal.

Example $\mathbf{i} + 4\mathbf{j}$ and $8\mathbf{i} - 2\mathbf{j}$

Solution $(\mathbf{i} + 4\mathbf{j}) \cdot (8\mathbf{i} - 2\mathbf{j}) = (1)(8) + (4)(-2)$

$$= 8 - 8 = 0.$$

Hence, by Theorem 5.4, the vectors are orthogonal.

27. $(0, 5)$ and $(-3, 0)$

28. $(-2, 2)$ and $(5, 5)$

29. $(-4, 1)$ and $(1, 4)$

30. $(9, 1)$ and $(-1, 9)$

31. $6\mathbf{i} - 9\mathbf{j}$ and $3\mathbf{i} + 2\mathbf{j}$

32. $-3\mathbf{i} + \sqrt{3}\mathbf{j}$ and $\mathbf{i} + \sqrt{3}\mathbf{j}$

Find the work done, given each force vector and displacement vector.

Example $\mathbf{F} = (1, 5)$ and $\mathbf{d} = (-2, 3)$

Solution By Definition 5.19,

$$W = \mathbf{F} \cdot \mathbf{d} = (1, 5) \cdot (-2, 3)$$
$$= 1 \cdot (-2) + 5 \cdot 3 = 13$$

33. $\mathbf{F} = (3, 6)$ and $\mathbf{d} = (4, 2)$

34. $\mathbf{F} = (5, -2)$ and $\mathbf{d} = (-1, -7)$

35. $\mathbf{F} = 7\mathbf{i} - 2\mathbf{j}$ and $\mathbf{d} = 4\mathbf{i} - 4\mathbf{j}$

36. $\mathbf{F} = -12\mathbf{i} + 3\mathbf{j}$ and $\mathbf{d} = -3\mathbf{i} + 12\mathbf{j}$

37. A small wagon is pulled horizontally a distance of 100 meters by a short rope. If the rope is at an angle of 30° to the horizontal and a constant force of 20 newtons is used, find the work done.

38. How much work is done in moving a string of cars 500 meters along a track by a truck exerting a force of 2000 newtons at an angle of 20° to the direction of the track?

39. In Exercise 38 how much work is done if the force is applied in the direction of motion?

40. How much work is done by a 175-pound man climbing 12 feet up a vertical ladder?

B

41. Show that $\mathbf{v}_1 \cdot \mathbf{v}_1 = \|\mathbf{v}_1\|^2$.

42. Show that $\mathbf{v}_1 \cdot \mathbf{v}_2 = \mathbf{v}_2 \cdot \mathbf{v}_1$.

43. Show that $c(\mathbf{v}_1 \cdot \mathbf{v}_2) = (c\mathbf{v}_1) \cdot \mathbf{v}_2$.

44. Show that $\mathbf{v}_1 \cdot (\mathbf{v}_2 + \mathbf{v}_3) = \mathbf{v}_1 \cdot \mathbf{v}_2 + \mathbf{v}_1 \cdot \mathbf{v}_3$.

Chapter Summary

5.1 A **geometric vector** $\vec{v}$ is a line segment with a specified direction. Its **norm** or **magnitude** is designated by $\|\vec{v}\|$. The **direction angle** is the angle α, where $-180° < \alpha \leq 180°$, such that

> The initial side of α is the ray from the initial point of $\vec{v}$ parallel to the x-axis and directed in the positive x direction; the terminal side of α is the ray from the initial point of $\vec{v}$ and containing $\vec{v}$.

The **sum** $\vec{v}_1 + \vec{v}_2$ is defined as the geometric vector having as its initial point the initial point of $\vec{v}_1$ and as its terminal point the terminal point of $\vec{v}_2$, and where the terminal point of $\vec{v}_1$ is the initial point of $\vec{v}_2$. When viewed in this way, geometric vectors are said to be added according to "the triangle law." When viewed as the diagonal of a parallelogram with adjacent sides $\vec{v}_1$ and $\vec{v}_2$, geometric vectors are said to be added according to "the parallelogram law."

If $\vec{v}$ is any geometric vector and c is any real number (scalar), then $c\vec{v}$ is a geometric vector collinear with $\vec{v}$ with magnitude $|c| \cdot \|\vec{v}\|$.

5.2 The magnitudes of the **geometric vector projections,** $\vec{v}_x$ and $\vec{v}_y$, of $\vec{v}$ on the x-axis and y-axis, respectively, are given by

$$\|\vec{v}_x\| = \|\vec{v}\| \cdot |\cos \alpha| \qquad \text{and} \qquad \|\vec{v}_y\| = \|\vec{v}\| \cdot |\sin \alpha|.$$

Geometric vectors can help to visualize and thus facilitate the solution of problems involving such quantities as force, velocity, and acceleration, since these quantities have both magnitude and direction.

5.3 A **two-dimensional vector** $\mathbf{v}$ is an ordered pair of real numbers. That is, if $a, b \in R$, then $\mathbf{v} = (a, b)$. The set of all such vectors is designated by $\mathbf{V}$. Each vector in the set corresponds to a unique bound geometric vector with initial point at the origin.

For all $\mathbf{v}_1, \mathbf{v}_2 \in \mathbf{V}$, where $\mathbf{v}_1 = (a_1, b_1)$ and $\mathbf{v}_2 = (a_2, b_2)$,

$$\mathbf{v}_1 = \mathbf{v}_2 \text{ if and only if } a_1 = a_2 \text{ and } b_1 = b_2;$$

the **norm,** or **magnitude,** of $\mathbf{v} = (a, b)$ is the real number (scalar)

$$\|\mathbf{v}\| = \sqrt{a^2 + b^2};$$

for $\|\mathbf{v}\| \neq 0$, the **direction angle** of $\mathbf{v} = (a, b)$ is the angle α such that

$$\cos \alpha = \frac{a}{\|\mathbf{v}\|} \qquad \text{and} \qquad \sin \alpha = \frac{b}{\|\mathbf{v}\|} \qquad (-180^\circ < \alpha \leq 180^\circ).$$

The **norm** (or **magnitude**) of a vector, $\|\mathbf{v}\|$, is the length of the associated geometric vector $\vec{v}$ and the **direction angle** of the vector is the angle α, where $-180^\circ < \alpha \leq 180^\circ$, measured from the positive x-axis to the geometric vector corresponding to $\mathbf{v}$.

5.4 For all $\mathbf{v}_1, \mathbf{v}_2 \in \mathbf{V}$, where $\mathbf{v}_1 = (a_1, b_1)$, $\mathbf{v}_2 = (a_2, b_2)$, and $c \in R$,

$$\mathbf{v}_1 + \mathbf{v}_2 = (a_1, b_1) + (a_2, b_2) = (a_1 + a_2, b_1 + b_2);$$
$$c\mathbf{v} = c(a_1, b_1) = (ca_1, cb_1).$$

The **zero vector** $\mathbf{0} = (0, 0)$ is the additive identity for addition and the vector $-\mathbf{v} = (-a, -b)$ is the additive inverse of $\mathbf{v} = (a, b)$.

For all $\mathbf{v}_1, \mathbf{v}_2 \in \mathbf{V}$, where $\mathbf{v}_1 = (a_1, b_1)$, $\mathbf{v}_2 = (a_2, b_2)$, and $c \in R$,

$$\mathbf{v}_1 - \mathbf{v}_2 = \mathbf{v}_1 + (-\mathbf{v}_2);$$

$$\frac{\mathbf{v}_1}{c} = \frac{1}{c}\mathbf{v}_1 \quad (c \neq 0).$$

If $\mathbf{i} = (1, 0)$ and $\mathbf{j} = (0, 1)$, then each vector $\mathbf{v}$ can be expressed as

$$\mathbf{v} = (a, b) = a\mathbf{i} + b\mathbf{j}.$$

5.5 The **inner (dot, scalar) product** of vectors $\mathbf{v}_1$ and $\mathbf{v}_2$ is

$$\mathbf{v}_1 \cdot \mathbf{v}_2 = (a_1, b_1) \cdot (a_2, b_2) = a_1a_2 + b_1b_2.$$

If γ is the angle between the geometric vectors corresponding to $\mathbf{v}_1$ and $\mathbf{v}_2$, then

$$\mathbf{v}_1 \cdot \mathbf{v}_2 = \|\mathbf{v}_1\| \, \|\mathbf{v}_2\| \cos \gamma.$$

Equivalently,

$$\cos \gamma = \frac{\mathbf{v}_1 \cdot \mathbf{v}_2}{\|\mathbf{v}_1\| \, \|\mathbf{v}_2\|};$$

$$\|\mathbf{v}_1\| \cos \gamma = \frac{\mathbf{v}_1 \cdot \mathbf{v}_2}{\|\mathbf{v}_2\|}, \text{ the scalar projection of } \mathbf{v}_1 \text{ on } \mathbf{v}_2;$$

$$\|\mathbf{v}_2\| \cos \gamma = \frac{\mathbf{v}_1 \cdot \mathbf{v}_2}{\|\mathbf{v}_1\|}, \text{ the scalar projection of } \mathbf{v}_2 \text{ on } \mathbf{v}_1.$$

The vectors $\mathbf{v}_1 = (a_1, b_1)$ and $\mathbf{v}_2 = (a_2, b_2)$ are **orthogonal** if and only if the inner product

$$\mathbf{v}_1 \cdot \mathbf{v}_2 = a_1a_2 + b_1b_2 = 0.$$

A list of symbols introduced in this chapter is shown inside the front cover, and important properties are shown inside the back cover.

Review Exercises

A

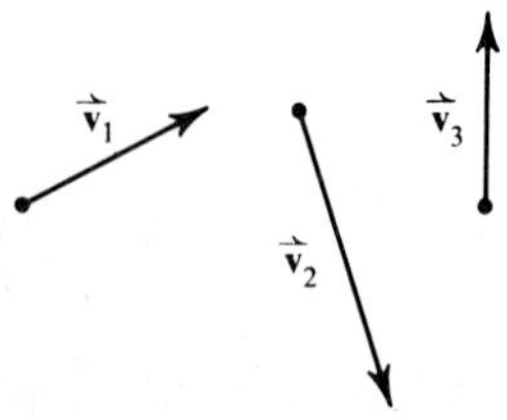

5.1 *Use free-hand methods and the geometric vectors shown at the left to draw a single vector to represent each of the following.*

1. $(\vec{v}_1 + \vec{v}_2) + \vec{v}_3$ **2.** $-3\vec{v}_1$ **3.** $2\vec{v}_2 + \vec{v}_3$ **4.** $2\vec{v}_1 - 3\vec{v}_3$

5.2 5. If the direction angle $\alpha = 60°$, and $\|\vec{v}\| = 12$, find $\|\vec{v}_x\|$ and $\|\vec{v}_y\|$.

6. If $\vec{v}_x$ and $\vec{v}_y$ are directed in a positive direction; $\|\vec{v}_x\| = 4$, with initial point at (1, 0); and $\|\vec{v}_y\| = 3$, with initial point at (0, 2), find $\|\vec{v}\|$ and the direction angle α.

7. The resultant of two forces acting at an angle of 90° with respect to each other is 130 newtons. If one of the forces is 50 newtons, find the other force and the angle it makes with the resultant.

8. A person can swim at a speed of 3 miles per hour in still water. If he heads across a river at right angles to a current of 6 miles per hour, find his speed in relation to the land and the direction in which he actually moves.

9. An airplane is headed on a bearing of 280° with an air speed of 800 kilometers per hour. The course has a bearing of 292°. The ground speed is 860 kilometers per hour. Find the wind direction and wind speed.

10. A force of 150 pounds is needed to keep a steel ball weighing 250 pounds from rolling down an inclined plane. What angle does the plane make with the horizontal?

5.3 *Find real numbers x and y for which each statement is true.*

11. $(x, y) = (-1, 3)$ **12.** $(x + y, x - y) = (5, 3)$

(a) *Sketch the geometric vector corresponding to each vector, and*
(b) *find the norm and the direction angle (to the nearest tenth) of each vector.*

13. $(-4, 3)$ **14.** $(5, -5)$

Given the following norms and direction angles, find the vector **v**.

15. $\|\mathbf{v}\| = 10$ and $\alpha = -135°$ **16.** $\|\mathbf{v}\| = 2$ and $\alpha = 150°$

5.4 *Write the sum of each pair of vectors as an ordered pair.*

17. $\mathbf{v}_1 = (4, -5)$ and $\mathbf{v}_2 = (8, 1)$ **18.** $\mathbf{v}_1 = (-3, 7)$ and $\mathbf{v}_2 = (-7, 3)$

Let $\mathbf{v}_1 = (3, -2)$, $\mathbf{v}_2 = (5, 4)$, $\mathbf{v}_3 = (1, -3)$, $c_1 = 2$, *and* $c_2 = -1$. *Express each of the following as an ordered pair.*

19. $\mathbf{v}_2 - c_2\mathbf{v}_1$ **20.** $c_1(\mathbf{v}_3 + \mathbf{v}_2)$ **21.** $\dfrac{\mathbf{v}_3}{c_2}$ **22.** $\dfrac{\mathbf{v}_2}{c_1\|\mathbf{v}_1\|}$

Express each vector as the sum of scalar multiples of **i** *and* **j**.

23. $(7, -9)$

24. $(-8, -1)$

Find the norm and direction angle (to the nearest tenth) of each vector.

25. $-10\mathbf{i} + 4\mathbf{j}$

26. $7\mathbf{i} - 9\mathbf{j}$

5.5 *Find the inner product* $\mathbf{v}_1 \cdot \mathbf{v}_2$ *for given vectors.*

27. $\mathbf{v}_1 = (6, 6)$ and $\mathbf{v}_2 = (-5, -4)$

28. $\mathbf{v}_1 = 8\mathbf{i} - \mathbf{j}$ and $\mathbf{v}_2 = 9\mathbf{i} - 3\mathbf{j}$

29. $\|\mathbf{v}_1\| = 8$ and $\alpha_1 = 60°$; $\|\mathbf{v}_2\| = 4$ and $\alpha_2 = 30°$

30. $\|\mathbf{v}_1\| = 10$ and $\alpha_1 = -45°$; $\|\mathbf{v}_2\| = 5$ and $\alpha_2 = -15°$

Determine the angle γ *to the nearest tenth, where* $0° \le \gamma \le 180°$, *formed by each pair of vectors.*

31. $\mathbf{v}_1 = (-7, 5)$ and $\mathbf{v}_2 = (1, -5)$

32. $\mathbf{v}_1 = -7\mathbf{i} - \mathbf{j}$ and $\mathbf{v}_2 = -8\mathbf{i} + 9\mathbf{j}$

Find the scalar projections of $\mathbf{v}_1$ *on* $\mathbf{v}_2$ *and of* $\mathbf{v}_2$ *on* $\mathbf{v}_1$.

33. $\mathbf{v}_1 = (1, 7)$ and $\mathbf{v}_2 = (10, -1)$

34. $\mathbf{v}_1 = 9\mathbf{i} + 2\mathbf{j}$ and $\mathbf{v}_2 = -8\mathbf{i} - 5\mathbf{j}$

Verify that each pair of vectors is orthogonal.

35. $(-4, -5)$ and $\left(4, -\frac{16}{5}\right)$

36. $7\mathbf{i} + 6\mathbf{j}$ and $5\mathbf{i} - \frac{35}{6}\mathbf{j}$

6 COMPLEX NUMBERS; POLAR COORDINATES

The set of real numbers does not contain an element whose square is a negative real number. Thus, any symbol of the form $\sqrt{-b}$, $b \in R$, $b > 0$, does not represent a real number, and equations such as

$$x^2 + 4 = 0 \qquad \text{equivalent to} \qquad x = \pm\sqrt{-4} \tag{1}$$

and

$$x^2 - 2x + 5 = 0 \qquad \text{equivalent to} \qquad x = \frac{2 \pm \sqrt{-16}}{2} \tag{2}$$

have no solutions in this set. In this chapter we consider a set of numbers, called the **set of complex numbers** and designated by C, that does provide solutions to such equations.

6.1 Sums and Products

The elements in the set C that we now consider are represented by various symbols. In this section we consider one form that is frequently used. In the following sections we consider other symbolic forms for these numbers.

Definition 6.1 *For all* $a, b \in R$, *a* **complex number** *is*

$$z = a + bi, \qquad \textit{where } i^2 = -1.$$

In this chapter we shall restrict the use of the variable z to represent elements in the set C. It is sometimes convenient to refer to a as the **real part** of the complex number and b as the **imaginary part,** with i the **imaginary unit.**

Properties of Equality

Equality in the set of complex numbers is established as follows.

Definition 6.2 *For every pair of complex numbers* $z_1 = a_1 + b_1 i$ *and* $z_2 = a_2 + b_2 i$,

$$z_1 = z_2 \quad \textit{if and only if} \quad a_1 = a_2 \textit{ and } b_1 = b_2.$$

Example $2x + y^2 i = 8 + 9i$ if and only if $2x = 8$ and $y^2 = 9$ or, equivalently, $x = 4$ and $y = \pm 3$.

Properties similar to the equality axioms in the set of real numbers R (see Appendix A.2) and the substitution law are valid for the set of complex numbers. These laws follow from the corresponding properties for real numbers. Thus, if $z_1, z_2, z_3 \in C$, then

a. $z_1 = z_1$ — Reflexive law
b. If $z_1 = z_2$, then $z_2 = z_1$ — Symmetric law
c. If $z_1 = z_2$ and $z_2 = z_3$, then $z_1 = z_3$ — Transitive law

Computing Sums and Products

If complex numbers are treated as if i were a variable and $a + bi$ were a binomial in the variable i, then the sum or product of two complex numbers can be found by applying the ordinary rules of the algebra of real numbers. Thus,

$$(2 + 3i) + (5 - 2i) = (2 + 5) + (3 - 2)i$$
$$= 7 + i;$$
$$(2 + 3i)(5 - 2i) = 10 + 11i - 6i^2.$$

However, replacing i^2 with -1, we have

$$(2 + 3i)(5 - 2i) = 10 + 11i - 6(-1)$$
$$= 16 + 11i.$$

More formally we have the following definition.

Definition 6.3 *If* $z_1 = a_1 + b_1 i$ *and* $z_2 = a_2 + b_2 i$, *then*

$$\mathbf{z_1 + z_2} = (a_1 + a_2) + (b_1 + b_2)i,$$

and

$$\mathbf{z_1 z_2} = (a_1 a_2 - b_1 b_2) + (a_1 b_2 + a_2 b_1)i.$$

Using such an approach for sums and products, we find that

1. C will contain elements $a + 0i$, or a, that can be identified with the real numbers for these operations, and
2. C will contain elements $0 + bi$, or bi, whose squares are negative real numbers.

For example,

$$\begin{aligned}(2) + (3) &= (2 + 0i) + (3 + 0i)\\ &= 5 + 0i = 5,\\ (2)(3) &= (2 + 0i)(3 + 0i)\\ &= 6 + 0i + 0i^2 = 6,\\ (3i)^2 &= 9i^2 = -9.\end{aligned}$$

Because the complex number $a + bi$, where $b = 0$, behaves like the real number a, we shall simply write a for $a + 0i$ as convenient and refer to a or $a + 0i$ as a real number.

If $b \neq 0$, the complex number $a + bi$ is called an **imaginary number** and for the special case where $a = 0$ and $b \neq 0$, the complex number is called a **pure imaginary number.** The relationship of the set C and other sets of numbers is shown in Figure 6.1. For example, 3 and -5 are elements of C that are real; $2 + 5i$ and $5i$ are elements of C that are imaginary numbers. Furthermore, $5i$ is a pure imaginary number.

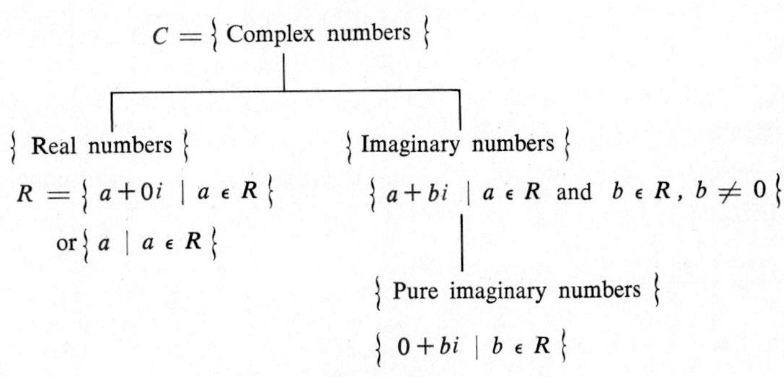

Figure 6.1

Periodic Property of i

Powers of the imaginary unit i have an interesting periodic property. Observe that

$$
\begin{aligned}
i^1 &= i, \\
i^2 &= -1 \\
i^3 &= i^2 \cdot i = -1 \cdot i = -i, \\
i^4 &= i^2 \cdot i^2 = (-1)(-1) = 1, \\
i^5 &= i^4 \cdot i = 1 \cdot i = i, \\
i^6 &= i^4 \cdot i^2 = 1 \cdot (-1) = -1, \\
i^7 &= i^4 \cdot i^3 = 1 \cdot (-i) = -i, \\
&\text{etc.}
\end{aligned}
$$

Thus, any integral power of i can be expressed as one of the numbers i, -1, $-i$, or 1.

Radical Notation

Another notation used to represent a complex number involves the square root symbol. In the set C, we make the following definition.

Definition 6.4 *For all* $b \in R$, $b > 0$,

$$\sqrt{-b} = \sqrt{-1}\sqrt{b} = i\sqrt{b}.$$

Observe that for the special case where $b = 1$,

$$\sqrt{-1} = \sqrt{-1}\sqrt{1} = i \cdot 1 = i.$$

Note that for a and b, positive real numbers,

$$\sqrt{-a}\sqrt{-b} = (i\sqrt{a})(i\sqrt{b}) = i^2\sqrt{a}\sqrt{b} = -\sqrt{ab},$$

a negative real number. Hence, the property of real numbers

$$\sqrt{a}\sqrt{b} = \sqrt{ab}, \quad a, b \geq 0,$$

is not applicable for imaginary numbers. For example,

$$\sqrt{-4}\sqrt{-9} \neq \sqrt{(-4)(-9)}.$$

In this case,

$$\sqrt{-4}\sqrt{-9} = 2i \cdot 3i = -6.$$

To avoid difficulty in rewriting products, expressions of the form $\sqrt{-b}$ $(b > 0)$ should first be expressed in the form $\sqrt{b}\,i$ or $i\sqrt{b}$.

Examples

a. $\sqrt{-3}(5 + \sqrt{-3})$
$= i\sqrt{3}(5 + i\sqrt{3})$
$= 5i\sqrt{3} + i^2(3)$
$= 5i\sqrt{3} + (-1)(3)$
$= -3 + 5i\sqrt{3}$

b. $(3 - \sqrt{-2})(3 + \sqrt{-2})$
$= (3 - i\sqrt{2})(3 + i\sqrt{2})$
$= 9 - i^2(2)$
$= 9 - (-1)(2)$
$= 11$

EXERCISE SET 6.1

A

Find real numbers x and y for which each statement is true.

Example By Definition 6.2, $x + 2yi = -5 + 8i$ if and only if

$$x = -5 \qquad \text{and} \qquad 2y = 8,$$

or, equivalently, if

$$x = -5 \qquad \text{and} \qquad y = 4.$$

1. $-5x + yi = -5 - 3i$

2. $x + 4i = 2yi$

3. $y + 4xi = 8 - 2i$

4. $x - 4i = 2y - x^2 i$

Write each expression in the form a + bi.

Example $(3 + 2i) + (1 - 3i) = (3 + 1) + [2 + (-3)]i$
$= 4 - i.$

5. $(4 + 8i) + (-3 + 2i)$

6. $(-6 + 6i) + (4 + 2i)$

7. $(-9 - 5i) + (-1 - 6i)$

8. $(-4 + 3i) + (5 - i)$

9. $(3 + 3i) + (-4 + 3i)$

10. $(-1 + 6i) + (2 - 2i)$

11. $(a + bi) + (-a - bi)$

12. $(a + bi) + (a - bi)$

Examples

a. $2(1 - 5i) = 2(1) + 2(-5)i$
$= 2 - 10i$

b. $-3(4 + i) = -3(4) - 3(1)i$
$= -12 - 3i$

13. $4(3 - 3i)$ **14.** $5(2 + 4i)$ **15.** $-6(6 + 2i)$

16. $-4(8 - i)$ **17.** $\frac{1}{2}(4 + 6i)$ **18.** $\frac{2}{3}(9 - 6i)$

Example $(-3 + 4i)(1 + 6i) = -3 + 4i - 18i + 24i^2$
$= -3 - 14i + 24(-1)$
$= -27 - 14i.$

19. $(-3 + 8i)(1 + 2i)$ **20.** $(4 + 2i)(-8 - 2i)$

21. $(-6 - i)(-9 + 8i)$ **22.** $(-3 - 9i)(3 + 4i)$

23. $(5 - 6i)(4 + i)$ **24.** $(5 + i)(-1 + i)$

25. $(a + bi)(a - bi)$ **26.** $(a + bi)(a + bi)$

Write each expression as one of the complex numbers, i, -1, $-i$, or 1.

27. i^5 **28.** i^8 **29.** i^{10} **30.** i^{11}

31. i^{14} **32.** i^{16} **33.** i^{19} **34.** i^{23}

Write each expression in the form bi or $a + bi$.

Examples

a. $\sqrt{-16} = \sqrt{16}\sqrt{-1}$
$= 4i$

b. $4 + \sqrt{-8} = 4 + \sqrt{4}\sqrt{2}\sqrt{-1}$
$= 4 + 2\sqrt{2}i$

35. $\sqrt{-25}$ **36.** $\sqrt{-49}$ **37.** $\sqrt{-12}$ **38.** $\sqrt{-27}$

39. $1 + \sqrt{-9}$ **40.** $2 - \sqrt{-64}$ **41.** $3 - \sqrt{-20}$ **42.** $5 + \sqrt{-48}$

Examples

a. $\sqrt{-9}(2 - 3\sqrt{-9}) = 3i(2 - 3 \cdot 3i)$
$= 3i(2 - 9i)$
$= 6i - 27i^2$
$= 27 + 6i$

b. $(1+\sqrt{-5})(1-2\sqrt{-5}) = (1+\sqrt{5}i)(1-2\sqrt{5}i)$

$$= 1+\sqrt{5}i - 2\sqrt{5}i - 2(\sqrt{5})^2 i^2$$
$$= 1-\sqrt{5}i - 2\cdot 5(-1)$$
$$= 11-\sqrt{5}i$$

43. $\sqrt{-4}(1-2\sqrt{-4})$

44. $\sqrt{-8}(3+4\sqrt{-2})$

45. $\sqrt{-16}(2+\sqrt{-4})$

46. $\sqrt{-9}(3-\sqrt{-25})$

47. $(\sqrt{-50}+2)\sqrt{-10}$

48. $(\sqrt{-48}-1)\sqrt{-21}$

49. $(5+2\sqrt{-4})(3-\sqrt{-4})$

50. $(1-4\sqrt{-9})(7+2\sqrt{-9})$

51. $(2+\sqrt{-3})(3-\sqrt{-3})$

52. $(3-\sqrt{-7})(3+\sqrt{-7})$

53. Show that $2i$ and $-2i$ satisfy $x^2+4=0$.

54. Show that $1+2i$ and $1-2i$ satisfy $x^2-2x+5=0$.

B

55. Show that $\left(-\frac{1}{2}+i\frac{\sqrt{3}}{2}\right)^3 = 1$ and $\left(-\frac{1}{2}-i\frac{\sqrt{3}}{2}\right)^3 = 1$.

56. Show that $\left(\frac{1}{2}+i\frac{\sqrt{3}}{2}\right)^3 = -1$ and $\left(\frac{1}{2}-i\frac{\sqrt{3}}{2}\right)^3 = -1$.

Find a and b for which each statement in Exercises 57 and 58 is true.

57. $(a+bi)^2 = -1$. *Hint:* First express the given equation as $a^2-b^2+2abi = -1$ and then use Definition 6.2.

58. $(a+bi)^2 = 1$.

The formula $e^{ix} = \cos x + i\sin x$, *where* $e = 2.718\ldots$, *is called* **Euler's formula** (*it is usually derived in a course in calculus*). *Use this formula to show that the following equations are identities.*

59. $e^{\pi i} = -1$

60. $e^{-ix} = \cos x - i\sin x$

61. $\sin x = \dfrac{e^{ix}-e^{-ix}}{2i}$

62. $\cos x = \dfrac{e^{ix}+e^{-ix}}{2}$

63. $\sin^2 x + \cos^2 x = 1$. *Hint:* Use Exercises 61 and 62.

6.2 Differences and Quotients

The operations of addition and multiplication of complex numbers have been defined so that many of the properties of real numbers are valid in the set of complex numbers.

Two other useful operations, subtraction and division, are defined in terms of addition and multiplication. These definitions are similar to the definitions for a difference and a quotient in the set of real numbers.

Definition 6.5 *For all* $z_1, z_2 \in C$, *the* **difference**

$$\mathbf{z_1 - z_2 = z_1 + (-z_2).}$$

Example $(3 + 4i) - (5 + 2i) = (3 + 4i) + (-5 - 2i)$

$$= (3 - 5) + (4 - 2)i = -2 + 2i.$$

Definition 6.6 *For all* $z_1, z_2 \in C$ *where* $z_2 \neq 0$, *the* **quotient**

$$\frac{z_1}{z_2} = z_1\left(\frac{1}{z_2}\right).$$

Since the axioms used for the real numbers together with equality axioms are also valid in C, for each property that follows from these axioms in the set R (see Appendix A.2), there is a corresponding property in the set C. One such property,

If $z_1, z_2, z_3 \in C$, *such that* $z_2, z_3 \neq 0$, *then*

$$\frac{z_1}{z_2} = \frac{z_1 z_3}{z_2 z_3}, \tag{1}$$

is particularly useful in rewriting quotients in the set C. We first make the following definition.

Definition 6.7 *For all* $z = a + bi$, *the* **conjugate** *of* z *is*

$$\bar{z} = a - bi.$$

Examples

a. The conjugate of $5 + 4i$ is $5 - 4i$.

b. The conjugate of $2 - 3i$ is $2 + 3i$.

A quotient of complex numbers, z_1/z_2, can be expressed in the form $a + bi$ by first multiplying both the numerator and the denominator by the conjugate of the denominator. If the denominator is a pure imaginary number, then it is necessary only to multiply the numerator and denominator by i. If the denominator is a real number and the numerator is of the form $a + bi$, then it is necessary only to write the quotient as two terms.

Examples

a. $$\frac{3}{1+2i} = \frac{3(1-2i)}{(1+2i)(1-2i)} = \frac{3-6i}{1-4i^2} = \frac{3-6i}{5} = \frac{3}{5} - \frac{6}{5}i$$

b. $$\frac{4-i}{2-3i} = \frac{(4-i)(2+3i)}{(2-3i)(2+3i)} = \frac{8+10i-3i^2}{4-9i^2} = \frac{11+10i}{13} = \frac{11}{13} + \frac{10}{13}i$$

c. $$\frac{3-i}{3i} = \frac{(3-i)i}{3i \cdot i} = \frac{3i-i^2}{3i^2} = \frac{3i+1}{-3} = -\frac{1}{3} - i$$

In Example a, note that the product of the complex number $1 + 2i$ and its conjugate $1 - 2i$ is the real number 5. In Example b, note that the product of the complex number $2 - 3i$ and its conjugate $2 + 3i$ is the real number 13. In fact, observe that for all $a + bi \in C$,

$$(a + bi)(a - bi) = a^2 - b^2i^2 = a^2 + b^2,$$

a real number.

As we noted in Section 6.1, to avoid difficulty in rewriting products, expressions of the form $\sqrt{-b}$ $(b > 0)$ should first be expressed in the from $\sqrt{b}i$ or $i\sqrt{b}$. Hence, quotients of this form should also be rewritten before using Equation (1).

Examples

a. $$\frac{5}{\sqrt{-4}} = \frac{5}{2i} = \frac{5 \cdot i}{2i \cdot i} = \frac{5i}{2(-1)} = -\frac{5}{2}i$$

b. $$\frac{4}{3-\sqrt{-2}} = \frac{4(3+i\sqrt{2})}{(3-i\sqrt{2})(3+i\sqrt{2})} = \frac{12+4i\sqrt{2}}{9-i^2(2)} = \frac{12+4i\sqrt{2}}{9-(-1)(2)} = \frac{12}{11} + \frac{4\sqrt{2}}{11}i$$

EXERCISE SET 6.2

A

Express each difference in the form a + bi.

Example $(1-4i)-(-1+10i) = (1-4i)+(1-10i)$
$= (1+1)+(-4-10)i$
$= 2-14i$

1. $(6+i)-(-7+5i)$

2. $(-7-3i)-(-5+5i)$

3. $(3-i)-(1-3i)$

4. $(3+2i)-(-5+8i)$

5. $(5+2i)-3i$

6. $4-(2-3i)$

Write the conjugate of each complex number in the form a + bi.

Examples a. $z=-3+4i$ b. $z=5-2i$

Solutions a. $\bar{z}=-3-4i$ b. $\bar{z}=5+2i$

7. $6+3i$

8. $5-5i$

9. $-3-3i$

10. $-4+i$

11. $3i$

12. 7

Express each quotient in the form a + bi.

Examples

a. $\dfrac{8-2i}{8i}=\dfrac{(8-2i)i}{8i\cdot i}$
$=\dfrac{8i-2i^2}{8i^2}$
$=\dfrac{8i+2}{-8}$
$=-\dfrac{1}{4}-i$

b. $\dfrac{-4}{-3+i}=\dfrac{(-4)(-3-i)}{(-3+i)(-3-i)}$
$=\dfrac{12+4i}{9-i^2}$
$=\dfrac{6}{5}+\dfrac{2}{5}i$

13. $\dfrac{-9-6i}{-4i}$

14. $\dfrac{-2+3i}{3i}$

15. $\dfrac{4}{2-9i}$

16. $\dfrac{2}{-8+5i}$

17. $\dfrac{2+i}{4+i}$

18. $\dfrac{2-5i}{-1+6i}$

19. $\dfrac{-5+9i}{6-2i}$

20. $\dfrac{3-2i}{3+2i}$

Examples

a. $\dfrac{3}{\sqrt{-2}} = \dfrac{3(i)}{i\sqrt{2}(i)}$

$= \dfrac{3i}{i^2\sqrt{2}}$

$= \dfrac{3i}{(-1)\sqrt{2}}$

$= -\dfrac{3}{\sqrt{2}}i$

b. $\dfrac{5}{2+\sqrt{-3}} = \dfrac{5(2-i\sqrt{3})}{(2+i\sqrt{3})(2-i\sqrt{3})}$

$= \dfrac{10-5i\sqrt{3}}{4-i^2(3)}$

$= \dfrac{10-5i\sqrt{3}}{4-(-1)(3)}$

$= \dfrac{10}{7} - \dfrac{5\sqrt{3}}{7}i$

21. $\dfrac{1}{\sqrt{-5}}$ **22.** $\dfrac{2}{\sqrt{-7}}$ **23.** $\dfrac{4}{\sqrt{-8}}$

24. $\dfrac{7}{\sqrt{-12}}$ **25.** $\dfrac{3}{4+\sqrt{-5}}$ **26.** $\dfrac{7}{5-\sqrt{-3}}$

27. $\dfrac{6}{\sqrt{-2}+3}$ **28.** $\dfrac{4}{\sqrt{-7}-6}$ **29.** $\dfrac{6}{\sqrt{2}+\sqrt{-5}}$

30. $\dfrac{10}{\sqrt{3}-\sqrt{-6}}$ **31.** $\dfrac{2}{\sqrt{5}+\sqrt{-8}}$ **32.** $\dfrac{3}{\sqrt{6}-\sqrt{-12}}$

6.3 Graphical Representation

The elements in $\{a + bi \mid a, b \in R\}$ can be placed in a one-to-one correspondence with the set of points in a plane by associating a with the abscissa of a point and b with the ordinate. In this way each point in a plane can be viewed as the graph of a complex number. A plane on which complex numbers are thus represented is called a **complex plane.** The elements in $\{a + 0i \mid a \in R\}$, or simply $\{a \mid a \in R\}$, are associated with the points on the x-axis, called the **real axis** and the elements in $\{0 + bi \mid b \in R\}$, or simply $\{bi \mid b \in R\}$, are associated with points on the y-axis, called the **imaginary axis.**

Examples Graph each number z, its negative, and its conjugate.

a. $z = 3 - 2i$ b. $z = -2 - 4i$

Solutions

a. The negative of $3 - 2i$ is $-3 + 2i$; the conjugate is $3 + 2i$.

b. The negative of $-2 - 4i$ is $2 + 4i$; the conjugate is $-2 + 4i$.

(continued)

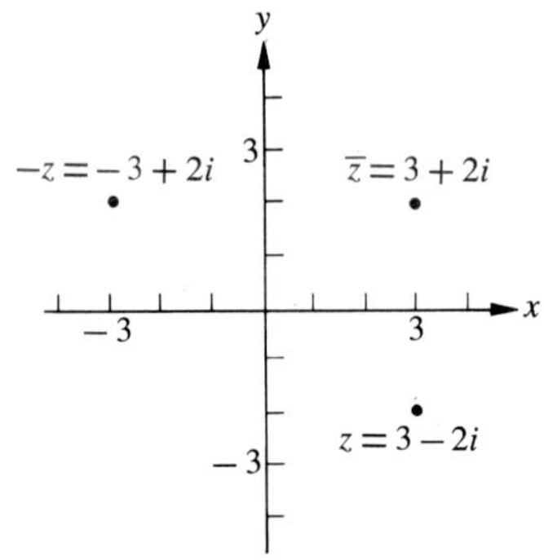

a.

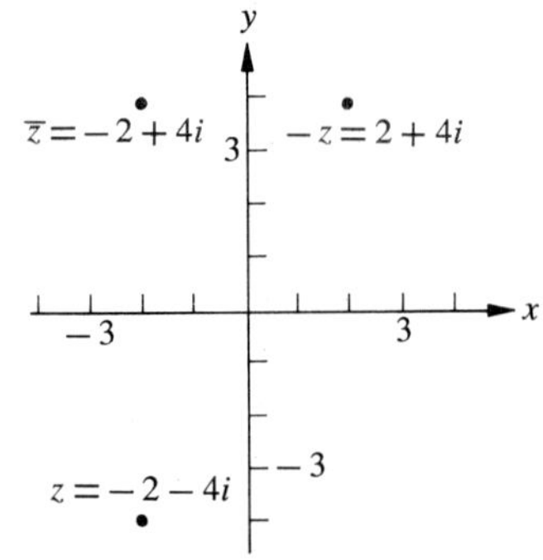

b.

Sometimes the graph of a complex number is shown as a geometric vector with its initial point located at the origin. With this representation, complex numbers can be added graphically by the parallelogram law applicable to geometric vectors.

Examples Represent each sum graphically using geometric vectors and check the results by analytic methods.

a. $(2 + 3i) + (4 - 5i)$ b. $(2 + 2i) + (-4 + 5i)$

Solutions

a.

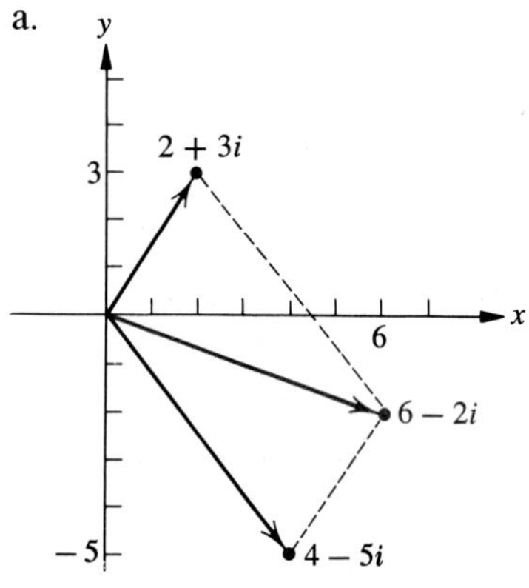

b.

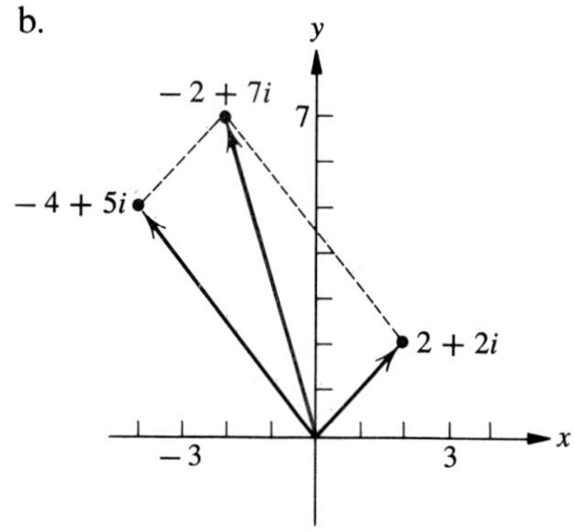

a. Check: $(2 + 3i) + (4 - 5i) = 6 - 2i$

b. Check: $(2 + 2i) + (-4 + 5i) = -2 + 7i$

Complex Numbers as Ordered Pairs

The elements in both the set $C = \{a + bi \mid a, b \in R\}$ and the elements in the set $\{(a, b) \mid a, b \in R\}$ have now been associated with the points in a plane.

Hence, elements in C can also be associated with, and represented by, ordered pairs of real numbers (a, b).

Examples

a. $2 + 3i = (2, 3)$

b. $-5 = -5 + 0i = (-5, 0)$

c. $6i = 0 + 6i = (0, 6)$

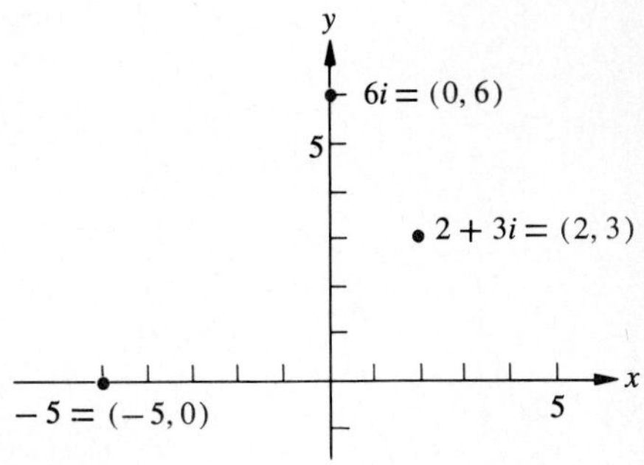

Absolute Value and Argument

Analogous to the norm $\|\vec{v}\|$ and the direction angle α for each geometric vector $\vec{v}$, there are a real number and a *set* of angles associated with each complex number.

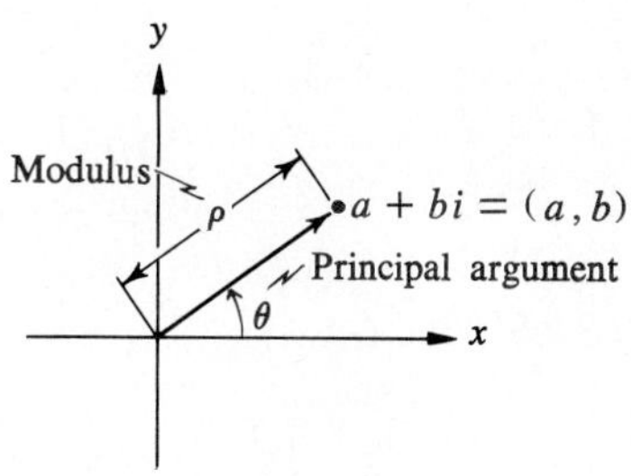

Figure 6.2

Definition 6.8 *For all* $z \in C$, *the* **absolute value** *or* **modulus** *of* z, *where* $z = a + bi = (a, b)$, *is*

$$\begin{aligned}\rho = |z| &= |a + bi| \\ &= |(a, b)| = \sqrt{a^2 + b^2}.\end{aligned}$$

The modulus ρ (the Greek letter "rho") is the distance from the origin to the graph of (a, b), as shown in Figure 6.2.

Definition 6.9 *For all* $z \in C$, *an* **argument** or **amplitude** *of* z, *where* $z = a + bi = (a, b)$, *denoted by* $\arg(a + bi)$ *or* $\arg(a, b)$, *is an angle* θ *such that*

$$\cos\theta = \frac{a}{\sqrt{a^2 + b^2}} = \frac{a}{\rho} \quad \text{and} \quad \sin\theta = \frac{b}{\sqrt{a^2 + b^2}} = \frac{b}{\rho} \quad (a^2 + b^2 \neq 0).$$

Thus, $\arg(a + bi)$ is an angle θ with initial side the positive x-axis and terminal side the ray from the origin through the graph of (a, b), as shown in Figure 6.2.

If θ is an argument of $a + bi$, then for $k \in J$, so is $(\theta + k \cdot 360)°$. The angle θ with the least nonnegative measure is sometimes called the **principal argument.** For $a + bi = 0 + 0i$, any angle θ can be used as an argument.

Example Find the absolute value and principle argument of $\sqrt{3} + i$.

Solution $\rho = |\sqrt{3} + i| = \sqrt{(\sqrt{3})^2 + 1^2} = \sqrt{3 + 1} = 2;$

$$\cos\theta = \frac{a}{\rho} = \frac{\sqrt{3}}{2} \quad \text{and} \quad \sin\theta = \frac{b}{\rho} = \frac{1}{2}.$$

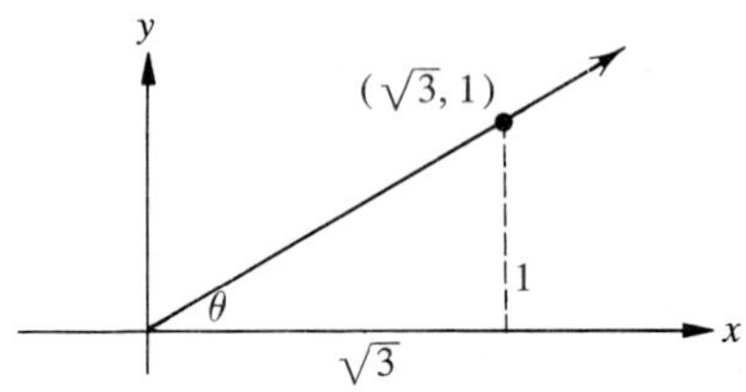

Because both $\cos\theta$ and $\sin\theta$ are positive, θ is in Quadrant I, and hence $\theta = 30°$.

There is another method involving the tangent function that is often used to determine θ. Consider the number $a + bi = \sqrt{3} + i$ from the example above. First plot the point $(a, b) = (\sqrt{3}, 1)$ and form angle θ. From the sketch, θ is in Quadrant I, and $\tan\theta = 1/\sqrt{3}$. Hence $\theta = 30°$.

EXERCISE SET 6.3

A

Express each complex number as an ordered pair.

1. $-1 + 6i$ **2.** $5 - 9i$ **3.** $8 + i$

4. $-2 + 7i$ **5.** $6i$ **6.** -5

Express each complex number in the form $a + bi$.

7. $(-8, 7)$ **8.** $(0, 5)$ **9.** $(-4, 0)$

10. $(6, -3)$ **11.** $(7, -\sqrt{2})$ **12.** $(-\sqrt{3}, 1)$

Graph each complex number z, its negative $-z$, and its conjugate $\bar{z}$.

Examples a. $-5 - 4i$ b. $(3, 2)$

Solutions

a. The negative is $5 + 4i$; the conjugate is $-5 + 4i$.

b. The negative is $(-3, -2)$; the conjugate is $(3, -2)$.

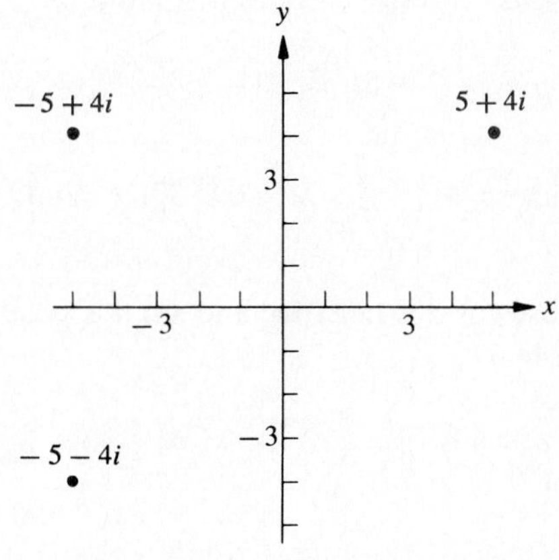

a.

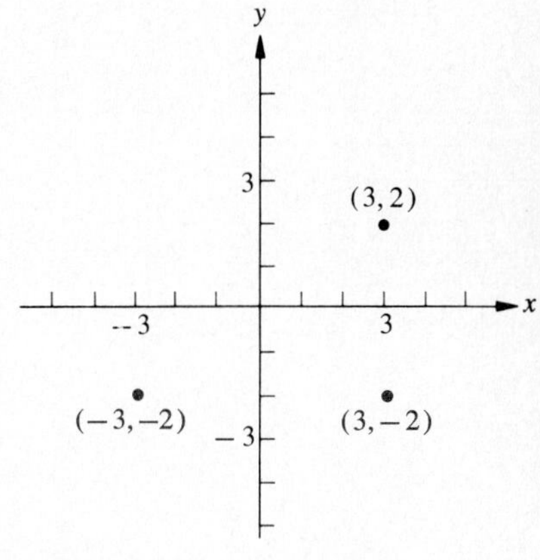

b.

13. $2 - 3i$ **14.** $-6 + 6i$ **15.** 5 **16.** $2i$

17. $(4, 6)$ **18.** $(-5, -3)$ **19.** $(-1, 0)$ **20.** $(0, 3)$

Represent each sum graphically. Check results by algebraic methods.

Example $(8 + 3i) + (1 - 5i)$

Solution

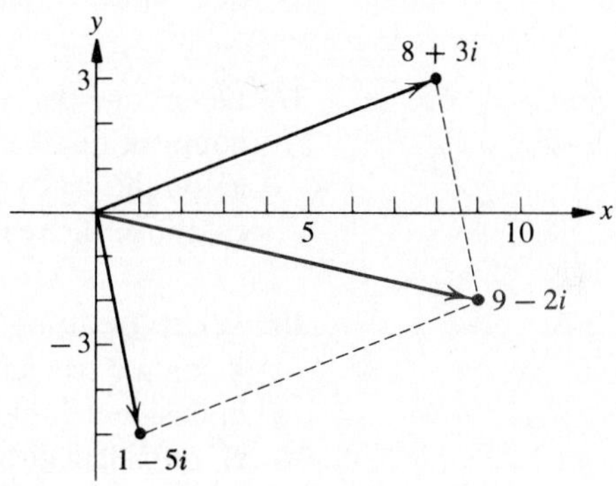

Check: $(8 + 3i) + (1 - 5i) = (8 + 1) + (3 - 5)i = 9 - 2i.$

21. $2 + (-3 + 6i)$ **22.** $(7 - i) + 2i$

23. $(4 - 7i) + (-8 + 2i)$ **24.** $(2 + 4i) + (1 - 5i)$

25. $(-6 + 7i) + (5 - 3i)$ **26.** $(-3 - 5i) + (3 - 2i)$

Find the absolute value and principal argument of each complex number.

Examples a. $-7 + 5i$ b. $(3, 9)$

(continued)

Solutions From Definitions 6.8 and 6.9,

a. $\rho = |-7 + 5i| = \sqrt{(-7)^2 + 5^2} = \sqrt{74};$

$$\cos\theta = -\frac{7}{\sqrt{74}} \approx -0.8137 \quad \text{and} \quad \sin\theta = \frac{5}{\sqrt{74}} \approx 0.5812.$$

Because $\cos\theta$ is negative and $\sin\theta$ is positive, θ is in Quadrant II; hence, $\theta \approx 144.5°$.

b. $\rho = |(3, 9)| = \sqrt{3^2 + 9^2} = \sqrt{90} = 3\sqrt{10};$

$$\cos\theta = \frac{3}{3\sqrt{10}} \approx 0.3162 \quad \text{and} \quad \sin\theta = \frac{9}{3\sqrt{10}} \approx 0.9487.$$

Because $\cos\theta$ and $\sin\theta$ are both positive, θ is in Quadrant I; hence, $\theta \approx 71.6°$.

27. $5 + 2i$ **28.** $6 + 8i$ **29.** $-2 - 2i$ **30.** $8 - 3i$

31. 7 **32.** -4 **33.** $-5i$ **34.** $6i$

35. $(3, -9)$ **36.** $(-6, -2)$ **37.** $(4, 0)$ **38.** $(0, -7)$

39. Determine the conditions that a and b must satisfy if the graph of the complex number $a + bi$ is
a. on the real axis. **b.** on the imaginary axis.
c. above the real axis. **d.** below the real axis.

40. What condition does $a^2 + b^2$ satisfy if the graph of $a + bi$ is
a. on a circle with radius of length 5?
b. inside a circle with radius of length 5?
c. outside a circle with radius of length 5?

B

Graphically, the multiplication or division of a complex number $a + bi$ by the complex number i produces a 90° rotation of the geometric vector corresponding to $a + bi$, counterclockwise for multiplication and clockwise for division. Graph

(a) *the given complex number $a + bi$,* **(b)** $i \cdot (a + bi)$, *and* **(c)** $\dfrac{a + bi}{i}$.

Example $2 + 3i$

Solutions

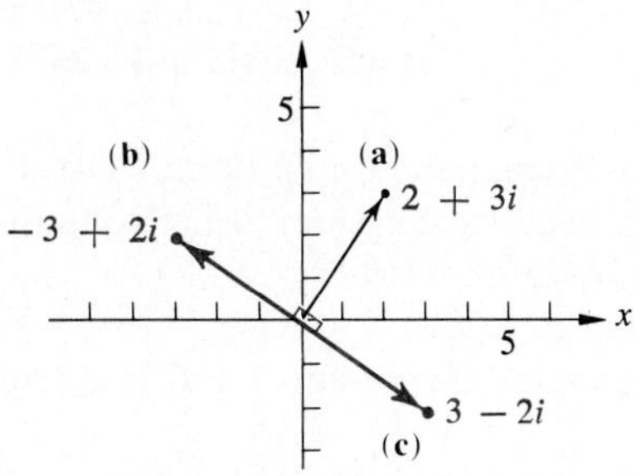

b. $i \cdot (2 + 3i) = 2i + 3i^2$

$= -3 + 2i$

c. $\dfrac{2 + 3i}{i} = \dfrac{(2 + 3i) \cdot i}{i \cdot i}$

$= \dfrac{2i + 3i^2}{i^2}$

$= \dfrac{2i - 3}{-1} = 3 - 2i$

41. $5 + 4i$ **42.** $1 - 2i$ **43.** $-6 + i$

44. $-3 - 4i$ **45.** 2 **46.** $-5i$

6.4 Trigonometric Form

In previous sections we have represented complex numbers in the form $a + bi$ and (a, b). It is sometimes convenient to represent a complex number in terms of its absolute value $\rho = \sqrt{a^2 + b^2}$ and its argument θ. From Definition 6.9,

$$\cos\theta = \frac{a}{\rho} \quad \text{and} \quad \sin\theta = \frac{b}{\rho}.$$

Therefore, as shown in Figure 6.3,

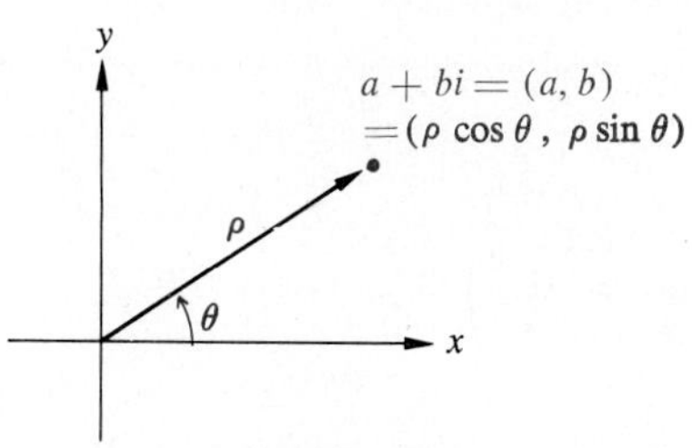

Figure 6.3

$$a = \rho\cos\theta \quad \text{and} \quad b = \rho\sin\theta,$$

from which we have

$$a + bi = \rho\cos\theta + (\rho\sin\theta)i,$$

$$\boldsymbol{a + bi = \rho(\cos\theta + i\sin\theta).}$$

(The expression $\cos\theta + i\sin\theta$ is sometimes abbreviated as cis θ.)

The sine and cosine functions are periodic with period of 360° or $2\pi^{R}$. Hence, it is also true that

$$\boldsymbol{a + bi = \rho[\cos(\theta + k \cdot 360°) + i\sin(\theta + k \cdot 360°)],\ k \in J,}$$

or

$$a + bi = \rho[\cos(\theta + k \cdot 2\pi^R) + i\sin(\theta + k \cdot 2\pi^R)], \quad k \in J.$$

When using these equations, we shall consider θ to be the angle of smallest nonnegative measure. Either equation is called the **trigonometric form** of a complex number.

Example Represent $-2 + 2i$ in trigonometric form.

Solution From Definition 6.8,

$$\rho = |-2 + 2i| = \sqrt{(-2)^2 + 2^2} = \sqrt{8} = 2\sqrt{2}.$$

From Definition 6.9,

$$\cos\theta = \frac{-2}{2\sqrt{2}} = -\frac{1}{\sqrt{2}} \quad \text{and} \quad \sin\theta = \frac{2}{2\sqrt{2}} = \frac{1}{\sqrt{2}}.$$

Thus, θ is in Quadrant II and $\theta = 135°$. For $k \in J$,

$$-2 + 2i = 2\sqrt{2}[\cos(135° + k \cdot 360°) + i\sin(135° + k \cdot 360°)].$$

For $k = 0$,

$$-2 + 2i = 2\sqrt{2}(\cos 135° + i\sin 135°).$$

Example Represent $3(\cos 60° + i\sin 60°)$ graphically and express the number in the form $a + bi$.

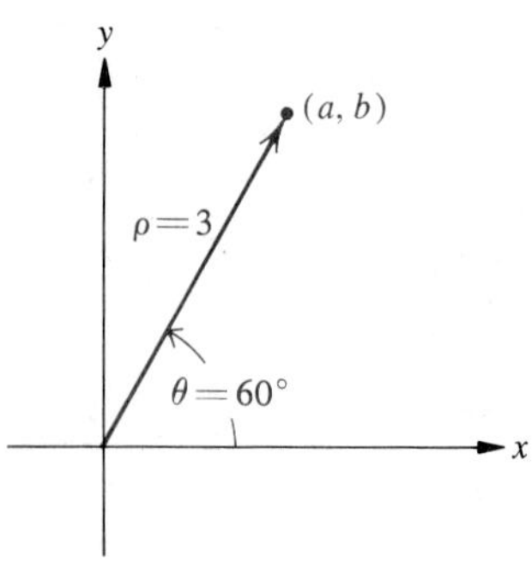

Solution Since

$$a = \rho\cos\theta = 3\cos 60° = 3\left(\frac{1}{2}\right) = \frac{3}{2},$$

and

$$b = \rho\sin\theta = 3\sin 60° = 3\left(\frac{\sqrt{3}}{2}\right) = \frac{3\sqrt{3}}{2},$$

we have

$$3(\cos 60° + i\sin 60°) = \frac{3}{2} + \frac{3\sqrt{3}}{2}i.$$

Products and Quotients

Products and quotients of complex numbers expressed in trigonometric form can be found quite readily. If

$$z_1 = \rho_1(\cos\theta_1 + i\sin\theta_1) \quad \text{and} \quad z_2 = \rho_2(\cos\theta_2 + i\sin\theta_2),$$

then

$$\begin{aligned} z_1 \cdot z_2 &= \rho_1(\cos\theta_1 + i\sin\theta_1)\cdot\rho_2(\cos\theta_2 + i\sin\theta_2) \\ &= \rho_1\rho_2[(\cos\theta_1\cos\theta_2 - \sin\theta_1\sin\theta_2) \\ &\qquad + i(\cos\theta_1\sin\theta_2 + \sin\theta_1\cos\theta_2)]. \end{aligned}$$

Substituting $\cos(\theta_1 + \theta_2)$ for $\cos\theta_1\cos\theta_2 - \sin\theta_1\sin\theta_2$ and then $\sin(\theta_1 + \theta_2)$ for $\cos\theta_1\sin\theta_2 + \sin\theta_1\cos\theta_2$, we have

$$\mathbf{z_1 \cdot z_2 = \rho_1\rho_2[\cos(\theta_1 + \theta_2) + i\sin(\theta_1 + \theta_2)].} \tag{1}$$

It can also be shown in a similar way that

$$\mathbf{\frac{z_1}{z_2} = \frac{\rho_1}{\rho_2}[\cos(\theta_1 - \theta_2) + i\sin(\theta_1 - \theta_2)], \quad z_2 \neq (0, 0).} \tag{2}$$

Example Express $6(\cos 75° + i\sin 75°)\cdot 3(\cos 15° + i\sin 15°)$ in the form $a + bi$.

Solution From Equation (1),

$$\begin{aligned} &6(\cos 75° + i\sin 75°)\cdot 3(\cos 15° + i\sin 15°) \\ &\qquad = 6\cdot 3[\cos(75° + 15°) + i\sin(75° + 15°)] \\ &\qquad = 18(\cos 90° + i\sin 90°) = 18(0 + i) = 18i. \end{aligned}$$

Example Express $\dfrac{6(\cos 75° + i\sin 75°)}{3(\cos 15° + i\sin 15°)}$ in the form $a + bi$.

Solution From Equation (2),

$$\begin{aligned} \frac{6(\cos 75° + i\sin 75°)}{3(\cos 15° + i\sin 15°)} &= \frac{6}{3}[\cos(75° - 15°) + i\sin(75° - 15°)] \\ &= 2(\cos 60° + i\sin 60°) \\ &= 2\left(\frac{1}{2} + i\frac{\sqrt{3}}{2}\right) = 1 + \sqrt{3}i. \end{aligned}$$

EXERCISE SET 6.4

A

Represent each complex number in trigonometric form, using the angle with the least nonnegative measure.

Example $\sqrt{3} - i$

Solution From Definition 6.8, $\rho = |\sqrt{3} - i| = \sqrt{3+1} = 2$. From Definition 6.9, $\cos\theta = \sqrt{3}/2$ and $\sin\theta = -1/2$. Hence θ is in Quadrant IV and $\theta = 330°$. Therefore,

$$\sqrt{3} - i = 2[\cos(330° + k \cdot 360°) + i\sin(330° + k \cdot 360°)],\ k \in J.$$

For $k = 0$, we obtain

$$\sqrt{3} - i = 2(\cos 330° + i \sin 330°).$$

1. $-2 + 2\sqrt{3}i$ **2.** $6 + 6i$ **3.** $-5\sqrt{3} - 5i$ **4.** $3\sqrt{3} + 3i$

5. $2 - 2i$ **6.** $-4 - 4i$ **7.** -8 **8.** $-9i$

Represent each complex number graphically and write as a + bi.

Example $6(\cos 315° + i \sin 315°)$

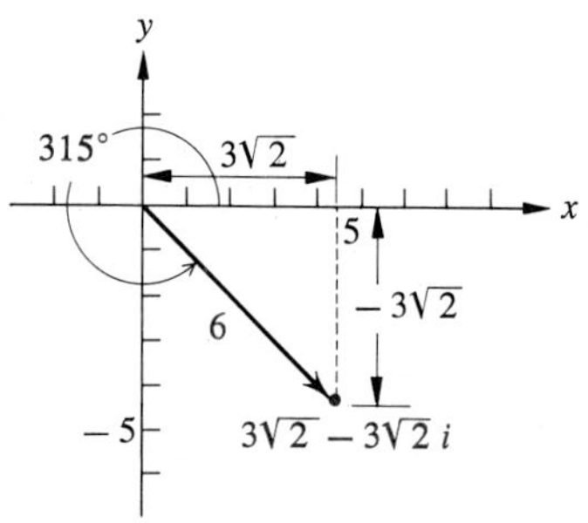

Solution Since

$$a = \rho\cos\theta = 6\left(\frac{\sqrt{2}}{2}\right) = 3\sqrt{2},$$

and

$$b = \rho\sin\theta = 6\left(\frac{-\sqrt{2}}{2}\right) = -3\sqrt{2},$$

we have

$$a + bi = 3\sqrt{2} - 3\sqrt{2}i.$$

9. $2(\cos 240° + i \sin 240°)$ **10.** $2(\cos 90° + i \sin 90°)$

11. $2(\cos 135° + i \sin 135°)$ **12.** $3(\cos 30° + i \sin 30°)$

13. $\cos(-45°) + i \sin(-45°)$ **14.** $\cos(-210°) + i \sin(-210°)$

15. $\frac{1}{2}(\cos 150° + i \sin 150°)$ **16.** $\frac{3}{4}(\cos 330° + i \sin 330°)$

Express each product in the form a + bi.

Example $2(\cos 140^\circ + i \sin 140^\circ) \cdot 5(\cos 70^\circ + i \sin 70^\circ)$

Solution From Equation (1) in this section,

$$\begin{aligned} 2(\cos 140^\circ + i \sin 140^\circ) \cdot 5(\cos 70^\circ + i \sin 70^\circ) &= 2 \cdot 5[\cos(140^\circ + 70^\circ) + i \sin(140^\circ + 70^\circ)] \\ &= 10(\cos 210^\circ + i \sin 210^\circ) \\ &= 10\left[\frac{-\sqrt{3}}{2} + i\left(-\frac{1}{2}\right)\right] = -5\sqrt{3} - 5i. \end{aligned}$$

17. $5(\cos 100^\circ + i \sin 100^\circ) \cdot 2(\cos 35^\circ + i \sin 35^\circ)$

18. $3(\cos 130^\circ + i \sin 130^\circ) \cdot \frac{2}{3}(\cos 140^\circ + i \sin 140^\circ)$

19. $4(\cos 85^\circ + i \sin 85^\circ) \cdot \frac{3}{4}(\cos 245^\circ + i \sin 245^\circ)$

20. $4(\cos 20^\circ + i \sin 20^\circ) \cdot \frac{1}{2}(\cos 40^\circ + i \sin 40^\circ)$

21. $2(\cos 40^\circ + i \sin 40^\circ) \cdot (\cos 95^\circ + i \sin 95^\circ)$

22. $2(\cos 5^\circ + i \sin 5^\circ) \cdot 3(\cos 40^\circ + i \sin 40^\circ)$

Express each quotient in the form a + bi.

Example $\dfrac{6(\cos 85^\circ + i \sin 85^\circ)}{2(\cos 40^\circ + i \sin 40^\circ)}$

Solution From Equation (2) in this section,

$$\begin{aligned} \frac{6(\cos 85^\circ + i \sin 85^\circ)}{2(\cos 40^\circ + i \sin 40^\circ)} &= \frac{6}{2}[\cos(85^\circ - 40^\circ) + i \sin(85^\circ - 40^\circ)] \\ &= 3(\cos 45^\circ + i \sin 45^\circ) \\ &= \frac{3}{\sqrt{2}} + \frac{3}{\sqrt{2}}i. \end{aligned}$$

23. $\dfrac{8(\cos 260^\circ + i \sin 260^\circ)}{2(\cos 80^\circ + i \sin 80^\circ)}$

24. $\dfrac{6(\cos 215^\circ + i \sin 215^\circ)}{3(\cos 80^\circ + i \sin 80^\circ)}$

25. $\dfrac{2(\cos 255^\circ + i \sin 255^\circ)}{5(\cos 15^\circ + i \sin 15^\circ)}$

26. $\dfrac{2(\cos 280^\circ + i \sin 280^\circ)}{3(\cos 70^\circ + i \sin 70^\circ)}$

27. $\dfrac{3(\cos 70^\circ + i \sin 70^\circ)}{\cos 130^\circ + i \sin 130^\circ}$

28. $\dfrac{\cos 19^\circ + i \sin 19^\circ}{4(\cos 259^\circ + i \sin 259^\circ)}$

29. Write $[2(\cos 20° + i \sin 20°)]^3$ in the form $a + bi$.

30. Write $\left(\frac{1}{\sqrt{2}} + \frac{1}{\sqrt{2}} i\right)^3$ in the form $a + bi$.

31. Show that if

$$a + bi = \rho(\cos \theta + i \sin \theta),$$

then

$$(a + bi)^2 = \rho^2(\cos 2\theta + i \sin 2\theta).$$

32. Use the result of Exercise 31 to show that if

$$a + bi = \rho(\cos \theta + i \sin \theta),$$

then

$$(a + bi)^3 = \rho^3(\cos 3\theta + i \sin 3\theta).$$

33. Use the result of Exercise 32 to show that if

$$a + bi = \rho(\cos \theta + i \sin \theta),$$

then

$$(a + bi)^4 = \rho^4(\cos 4\theta + i \sin 4\theta).$$

34. Using the results of Exercises 31–33, what can you conjecture about $(a + bi)^n$, $n \in N$, if $a + bi = \rho(\cos \theta + i \sin \theta)$?

35. Show that $\cos \theta + i \sin \theta$ and $\cos(-\theta) + i \sin(-\theta)$ are conjugates of each other.

36. Show that $\cos \theta + i \sin \theta$ and $\cos(-\theta) + i \sin(-\theta)$ are reciprocals of each other.

6.5 DeMoivre's Theorem; Powers and Roots

Consider the complex number $z = a + bi = \rho(\cos \theta + i \sin \theta)$. From Equation (1) in Section 6.4,

$$\begin{aligned}[\rho(\cos \theta + i \sin \theta)]^2 &= \rho(\cos \theta + i \sin \theta) \cdot \rho(\cos \theta + i \sin \theta) \\ &= \rho^2[\cos(\theta + \theta) + i \sin(\theta + \theta)] \\ &= \rho^2(\cos 2\theta + i \sin 2\theta);\end{aligned}$$

$$\begin{aligned}[\rho(\cos \theta + i \sin \theta)]^3 &= \rho^2(\cos 2\theta + i \sin 2\theta) \cdot \rho(\cos \theta + i \sin \theta) \\ &= \rho^3[\cos(2\theta + \theta) + i \sin(2\theta + \theta)] \\ &= \rho^3(\cos 3\theta + i \sin 3\theta).\end{aligned}$$

It can be shown that similar results are valid for each such power for $n \in N$. That is, if $z = \rho(\cos\theta + i\sin\theta)$ and $n \in N$, then

$$\mathbf{z^n = \rho^n(\cos n\theta + i \sin n\theta).} \tag{1}$$

This statement is known as **DeMoivre's theorem.** The proof of the statement and extensions of it which are presented in this section involve the process of mathematical induction and will not be shown. However, the informal argument above should make the validity of Equation (1) plausible.

Example Express $(\sqrt{3} + i)^5$ in the form $a + bi$.

Solution From Definition 6.8,

$$\rho = |\sqrt{3} + i| = \sqrt{3 + 1} = 2.$$

By Definition 6.9, $\cos\theta = \sqrt{3}/2$ and $\sin\theta = 1/2$. Therefore θ is in Quadrant I and $\theta = 30°$. Thus,

$$\begin{aligned}(\sqrt{3} + i)^5 &= [2(\cos 30° + i \sin 30°)]^5\\ &= 2^5(\cos 5 \cdot 30° + i \sin 5 \cdot 30°)\\ &= 32(\cos 150° + i \sin 150°)\\ &= 32\left(-\frac{\sqrt{3}}{2} + \frac{1}{2}i\right) = -16\sqrt{3} + 16i.\end{aligned}$$

Equation (1) was stated for powers with natural number exponents. If this statement is to be valid for integral exponents, consistent meanings must be assigned to z^0 and z^{-n} for $n \in N$.

Substituting 0 for n in Equation (1), we have

$$\begin{aligned}z^0 = \rho^0(\cos 0 \cdot \theta + i \sin 0 \cdot \theta) &= 1 \cdot (\cos 0 + i \sin 0)\\ &= 1 \cdot (1 + 0) = 1.\end{aligned}$$

Substituting $-n$ for n in Equation (1), we have

$$z^{-n} = \rho^{-n}[\cos(-n\theta) + i \sin(-n\theta)].$$

Since $\cos(-n\theta) = \cos n\theta$ and $\sin(-n\theta) = -\sin n\theta$, we have

$$z^{-n} = \frac{1}{\rho^n} \cdot (\cos n\theta - i \sin n\theta).$$

Multiplying the numerator and the denominator of the right-hand member by the conjugate of the numerator yields

$$\begin{aligned} z^{-n} &= \frac{1}{\rho^n} \cdot \frac{(\cos n\theta - i \sin n\theta) \cdot (\cos n\theta + i \sin n\theta)}{\cos n\theta + i \sin n\theta} \\ &= \frac{1}{\rho^n} \cdot \frac{\cos^2 n\theta + \sin^2 n\theta}{\cos n\theta + i \sin n\theta} \\ &= \frac{1}{\rho^n} \cdot \frac{1}{\cos n\theta + i \sin n\theta} = \frac{1}{\rho^n(\cos n\theta + i \sin n\theta)}. \end{aligned}$$

Substituting z^n for $\rho^n(\cos n\theta + i \sin n\theta)$ yields

$$z^{-n} = \frac{1}{z^n}.$$

Thus, if we want Equation (1) to be valid for every exponent $n \in J$, z^0 and z^{-n} must be given the following interpretations:

Definition 6.10 *For all* $z \neq 0 + 0i$,

I. $\mathbf{z^0 = 1 + 0i = 1.}$

II. $\mathbf{z^{-n} = \dfrac{1}{z^n}, \quad n \in N.}$

DeMoivre's theorem [Equation (1)] can now be applied to powers with $n \in J$.

Example Express $(-\sqrt{3} + i)^{-3}$ in the form $a + bi$.

Solution From Definitions 6.8 and 6.9,

$$\rho = |-\sqrt{3} + i| = \sqrt{3 + 1} = 2;$$

since

$$\cos\theta = -\frac{\sqrt{3}}{2} \quad \text{and} \quad \sin\theta = \frac{1}{2},$$

θ is in Quadrant II and $\theta = 150°$. Thus,

$$(-\sqrt{3} + i)^{-3} = [2(\cos 150° + i \sin 150°)]^{-3},$$

and from DeMoivre's theorem we have

$$(-\sqrt{3}+i)^{-3} = \frac{1}{2^3}[\cos(-3\cdot 150°) + i\sin(-3\cdot 150°)]$$

$$= \frac{1}{8}[\cos(-450°) + i\sin(-450°)]$$

$$= \frac{1}{8}[0 - 1\cdot i] = -\frac{1}{8}i.$$

Another extension of DeMoivre's theorem is possible for rational number exponents $1/n$, where $n \in N$. First we define an nth root of z in the same way that an nth root of a real number a is defined, when such a number exists.

Definition 6.11 *For $z \in C$, $n \in N$, the number ω is an nth root of z if $\omega^n = z$.*

We can now state another extension of DeMoivre's theorem applicable to roots of complex numbers.

If $n \in N$ and $z = \rho(\cos\theta + i\sin\theta)$, $z \neq 0$, then

$$\boldsymbol{\rho^{1/n}\left[\cos\left(\frac{\theta + k\cdot 360°}{n}\right) + i\sin\left(\frac{\theta + k\cdot 360°}{n}\right)\right], \quad k \in J} \tag{2}$$

specifies all nth roots of z. It can be shown that all complex numbers (and hence all real numbers) have n nth roots.

Using this extension of DeMoivre's theorem, we can now find n distinct complex nth roots for each $z \in C$, where $z \neq 0$. For $k = 0$, one nth root is

$$\rho^{1/n}\left(\cos\frac{\theta}{n} + i\sin\frac{\theta}{n}\right),$$

where θ is the angle with the least positive measure. This root is called the **principal *n*th root** of z.

Example Express each of the five fifth roots of $z = \sqrt{2} + \sqrt{2}i$ in trigonometric form.

Solution From Definition 6.8,

$$\rho = |\sqrt{2} + \sqrt{2}i| = \sqrt{2+2} = 2;$$

(*continued*)

since, from Definition 6.9,

$$\cos\theta = \frac{\sqrt{2}}{2} \quad \text{and} \quad \sin\theta = \frac{\sqrt{2}}{2},$$

the argument θ is in Quadrant I and $\theta = 45°$. Thus, in trigonometric form,

$$z = 2(\cos 45° + i \sin 45°),$$

and by Equation (2),

$$2^{1/5}\left[\cos\left(\frac{45° + k \cdot 360°}{5}\right) + i \sin\left(\frac{45° + k \cdot 360°}{5}\right)\right], \quad k \in J,$$

is a fifth root of z. Taking $k = 0, 1, 2, 3,$ and 4 in turn yields the roots

$$2^{1/5}(\cos 9° + i \sin 9°),$$
$$2^{1/5}(\cos 81° + i \sin 81°),$$
$$2^{1/5}(\cos 153° + i \sin 153°),$$
$$2^{1/5}(\cos 225° + i \sin 225°),$$
$$2^{1/5}(\cos 297° + i \sin 297°).$$

Notice that the substitution of any other integer for k will simply produce one of these five complex numbers. For example, if $k = 5$,

$$2^{1/5}[\cos(9 + 360)° + i \sin(9 + 360)°] = 2^{1/5}(\cos 9° + i \sin 9°).$$

EXERCISE SET 6.5

A

Express each power in the form $a + bi$.

Example $(\sqrt{3} + i)^{-2}$

Solution From Definition 6.8,

$$\rho = |\sqrt{3} + i| = \sqrt{(\sqrt{3})^2 + (1)^2} = \sqrt{4} = 2;$$

since, from Definition 6.9,

$$\cos\theta = \frac{\sqrt{3}}{2} \quad \text{and} \quad \sin\theta = \frac{1}{2},$$

the argument θ is in Quadrant I and $\theta = 30°$. Thus,

$$(\sqrt{3} + i)^{-2} = [2(\cos 30° + i \sin 30°)]^{-2}.$$

From DeMoivre's theorem,

$$\begin{aligned}(\sqrt{3} + i)^{-2} &= \frac{1}{2^2}[\cos(-2 \cdot 30°) + i \sin(-2 \cdot 30°)] \\ &= \frac{1}{4}[\cos(-60°) + i \sin(-60°)] \\ &= \frac{1}{4}\left(\frac{1}{2} - \frac{\sqrt{3}}{2} i\right) \\ &= \frac{1}{8} - \frac{\sqrt{3}}{8} i.\end{aligned}$$

1. $[3(\cos 20° + i \sin 20°)]^3$

2. $[2(\cos 42° + i \sin 42°)]^5$

3. $(-1 + i)^6$

4. $(2\sqrt{3} - 2i)^5$

5. $\left(\frac{1}{4} + \frac{\sqrt{3}}{4} i\right)^3$

6. $\left(-\frac{\sqrt{3}}{2} + \frac{1}{2} i\right)^6$

7. $[9(\cos 315° + i \sin 315°)]^{-2}$

8. $[8(\cos 120° + i \sin 120°)]^{-3}$

9. $(2 - 2i)^{-3}$

10. $(\sqrt{3} - i)^{-3}$

11. $(-1 - \sqrt{3}i)^{-5}$

12. $(-2 - 2i)^{-4}$

In Exercises 13–18, find the roots indicated and express each of them in trigonometric form.

Example The two square roots of $\frac{\sqrt{3}}{2} - \frac{1}{2} i$

Solution

$$\rho = \left|\frac{\sqrt{3}}{2} - \frac{1}{2} i\right| = \sqrt{\frac{3}{4} + \frac{1}{4}} = 1.$$

Thus, $\cos \theta = \sqrt{3}/2$ and $\sin \theta = -1/2$, and θ is in Quadrant IV. Hence $\theta = 330°$ and

$$\frac{\sqrt{3}}{2} - \frac{1}{2} i = \cos(330° + k \cdot 360°) + i \sin(330° + k \cdot 360°).$$

(continued)

The square roots of $\frac{\sqrt{3}}{2} - \frac{1}{2}i$, using DeMoivre's theorem, are

$$1^{1/2}\left[\cos\left(\frac{330^\circ + k \cdot 360}{2}\right) + i\sin\left(\frac{330^\circ + k \cdot 360^\circ}{2}\right)\right],$$

where $k = 0$ and $k = 1$. For $k = 0$, we have

$$\cos 165^\circ + i \sin 165^\circ,$$

and for $k = 1$, we have

$$\cos 345^\circ + i \sin 345^\circ.$$

13. The two square roots of

a. $\cos 80^\circ + i \sin 80^\circ$ **b.** $\frac{1}{2} + \frac{\sqrt{3}}{2}i$

14. The two square roots of

a. $9(\cos 140^\circ + i \sin 140^\circ)$ **b.** $-4i$

15. The three cube roots of

a. $8(\cos 120^\circ + i \sin 120^\circ)$ **b.** $\frac{1}{2} + \frac{\sqrt{3}}{2}i$

16. The three cube roots of

a. $2(\cos 75^\circ + i \sin 75^\circ)$ **b.** $-27i$

17. The four fourth roots of

a. $\cos 240^\circ + i \sin 240^\circ$ **b.** $\frac{1}{2} - \frac{\sqrt{3}}{2}i$

18. The four fourth roots of

a. $81(\cos 60^\circ + i \sin 60^\circ)$ **b.** $-16i$

Solve each equation for all $x \in C$.

19. $x^3 + 1 = 0$. *Hint:* $x^3 = -1 = -1 + 0i$.
Thus, x is a cube root of $-1 + 0i$.

20. $x^5 + 1 = 0$

21. $x^3 - 1 = 0$

22. $x^5 - 1 = 0$

23. $x^3 = \frac{\sqrt{3}}{4} - \frac{1}{4}i$

24. $x^5 = 12 - 5i$

B

25. Graph each of the following on a separate Cartesian coordinate system; make a conjecture about the graphs of all nth roots of 1.

a. The two square roots of 1 **b.** The three cube roots of 1
c. The four fourth roots of 1 **d.** The five fifth roots of 1
e. The six sixth roots of 1 **f.** The eight eighth roots of 1

26. Use the results of Exercise 21 to show that the reciprocal of any cube root of unity is itself a cube root of unity.

27. Use DeMoivre's theorem and Definition 6.2 to find an identity for $\cos 3\theta$ in terms of $\cos \theta$.
Hint: $(\cos \theta + i \sin \theta)^3 = \cos 3\theta + i \sin 3\theta$.

28. Use the hint in Exercise 27 to find an identity for $\sin 3\theta$ in terms of $\sin \theta$.

6.6 Polar Coordinates

In Chapter 2 we graphed functions using Cartesian coordinates, say (3, 5), which locate a point in the plane 3 units from the y-axis and 5 units from the x-axis. There are other ways to designate a point in the plane, however. One method is to use "polar coordinates."

Graphing Ordered Pairs (ρ, θ)

The location of a point A in the plane can also be specified by ρ (rho), the distance (a positive number) from the pole O to A and the measure of an angle θ* that the ray through point A makes with a fixed ray called the **polar axis** (Figure 6.4 on page 278). If we extend the replacement set of ρ to include the negative real numbers and the replacement set of θ to include the set of all angles, we can specify point A by many ordered pairs of the form (ρ, θ). The two components of each of these ordered pairs are called **polar coordinates** of A. For example, in Figure 6.4a, if (ρ, θ) are polar coordinates of A, then so are $[\rho, (\theta + k \cdot 360)°]$, $k \in J$. Also, if $-\rho$ $(\rho > 0)$ denotes the directed distance from O to A along the negative extension of the ray OA', as shown in Figure 6.4b, then $[-\rho, (\theta + 180 + k \cdot 360)°]$ are also acceptable coordinates of A. For example, in Figure 6.5, the point having polar coordinates (4, 30°) also has polar coordinates (4, 390°) and (4, −330°) for positive values of ρ, and (−4, 210°) for a negative value of ρ. Of course, there are infinitely many other possible polar coordinates for this point.

If point A is at the pole, then $\rho = 0$, and the coordinates of A are $(0, \theta)$, where θ is an arbitrary angle, not necessarily zero. For example, (0, 90°) and (0, 60°) are both coordinates of the pole.

* The measure of θ can be expressed in degrees or radians.

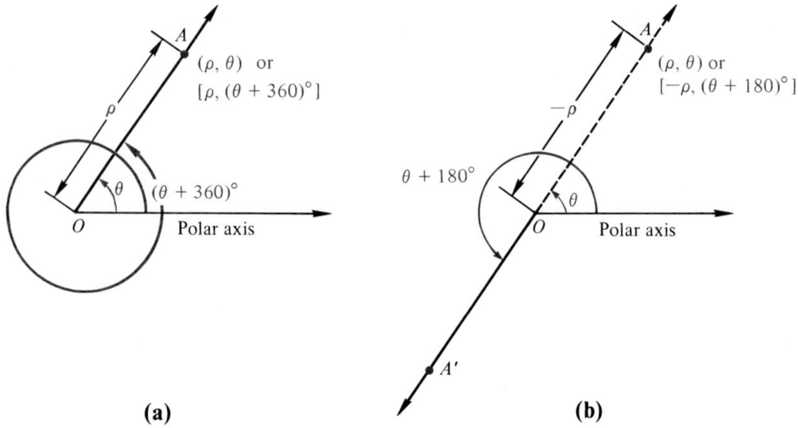

Figure 6.4

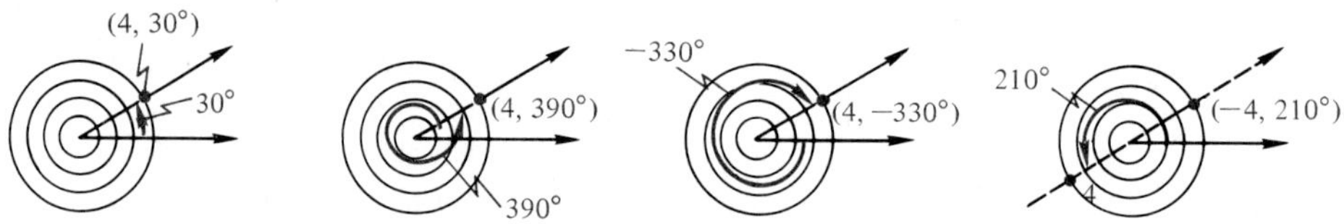

Figure 6.5

Cartesian and Polar Coordinates

If the pole coincides with the origin, and the polar axis coincides with the positive x-axis, as in Figure 6.6, then for any point on the terminal ray of angle θ with Cartesian coordinates x and y, $\cos\theta = x/\rho$ and $\sin\theta = y/\rho$. Hence, the Cartesian and polar coordinates can be related as follows.

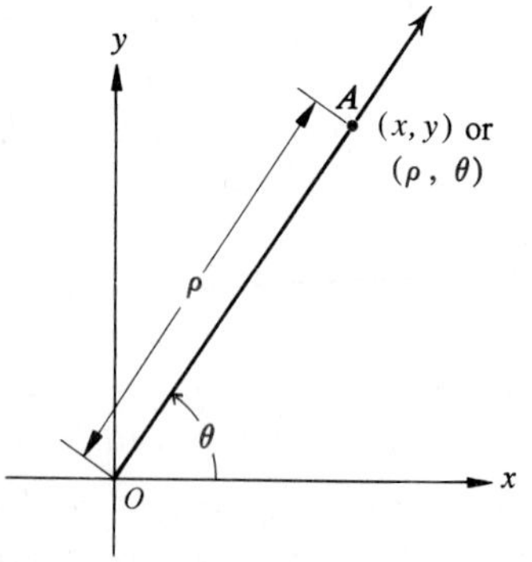

Figure 6.6

$$\left.\begin{aligned} x &= \rho\cos\theta \\ y &= \rho\sin\theta \end{aligned}\right\} \qquad (1)$$

$$\left.\begin{aligned} \rho &= \pm\sqrt{x^2 + y^2} \\ \cos\theta &= \frac{x}{\rho} \\ \sin\theta &= \frac{y}{\rho}, \quad \rho \neq 0 \end{aligned}\right\} \qquad (2)$$

Equations (1) can be used to find the rectangular coordinates for a point if the polar coordinates are known, and Equations (2) can be used to find the polar coordinates for a point if the Cartesian coordinates are known. The positive or negative radical expression is chosen depending on the particular problem.

Example Find the Cartesian coordinates of the point with polar coordinates (3, 60°).

Solution From Equations (1), it follows that

$$x = 3 \cos 60^\circ$$
$$= 3 \cdot \frac{1}{2} = \frac{3}{2}$$

and

$$y = 3 \sin 60^\circ$$
$$= 3 \cdot \frac{\sqrt{3}}{2} = \frac{3\sqrt{3}}{2}.$$

The point and the corresponding ordered pairs are shown in the figure.

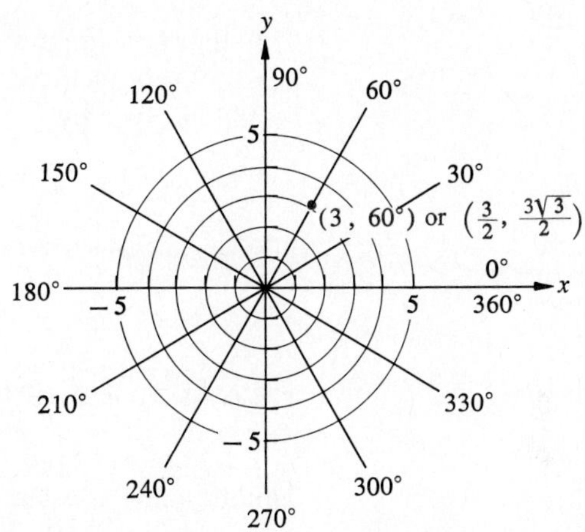

Example Find an approximation for the polar coordinates of the point with Cartesian coordinates (5, −3), where $0^\circ < \theta < 360^\circ$ and $\rho > 0$. Graph the point.

Solution From Equations (2) and since $\rho > 0$,

$$\rho = \sqrt{x^2 + y^2} = \sqrt{25 + 9} = \sqrt{34} \approx 5.831;$$
$$\cos \theta = \frac{5}{5.831} \approx 0.8575.$$

Since the point is in Quadrant IV, then $\theta \approx 329.0^\circ$. Therefore, the pair of polar coordinates is $(\sqrt{34}, 329^\circ)$. The point and the corresponding ordered pairs are shown in the figure on page 280.

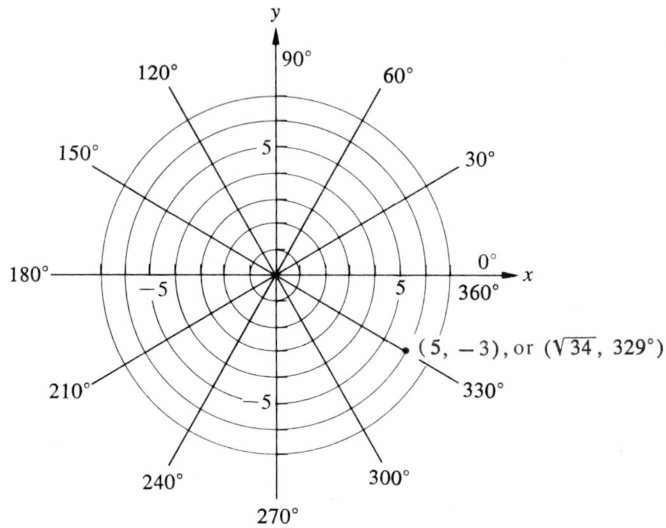

In the above example, θ was restricted to values between 0° and 360° and $\rho > 0$. Without these restrictions, the infinite pairs of polar coordinates would be given by

$$[\sqrt{34}, (329^\circ + k \cdot 360^\circ)]$$

and

$$[-\sqrt{34}, (329^\circ - 180^\circ + k \cdot 360^\circ)].$$

Rewriting Equations

Equations (1) and (2) on page 278 can also be used to transform an equation in Cartesian form to polar form, and vice versa.

Example Rewrite $x^2 + y^2 - 4x = 0$ in polar form.

Solution Substituting ρ^2 for $x^2 + y^2$ and $\rho \cos \theta$ for x in

$$x^2 + y^2 - 4x = 0$$

yields

$$\rho^2 - 4\rho \cos \theta = 0,$$

from which

$$\rho(\rho - 4 \cos \theta) = 0.$$

Hence,

$$\rho = 0 \qquad \text{or} \qquad \rho - 4 \cos \theta = 0.$$

Note that the graph of $\rho = 0$ is the origin and also that the graph of $\rho = 4 \cos \theta$ includes the origin as one of its points (for example, for $\theta = 90°$, we have $\rho = 0$). Hence the equation $\rho = 4 \cos \theta$ corresponds to $x^2 + y^2 - 4x = 0$, and it is not necessary to write $\rho = 0$ as part of the answer.

Example Rewrite $\rho(1 - 3 \cos \theta) = 2$ in Cartesian form.

Solution Substituting $\pm\sqrt{x^2 + y^2}$ for ρ and x/ρ, or $x/\pm\sqrt{x^2 + y^2}$, for $\cos \theta$, we have

$$\pm\sqrt{x^2 + y^2}\left[1 - 3\left(\frac{x}{\pm\sqrt{x^2 + y^2}}\right)\right] = 2$$

from which

$$\pm\sqrt{x^2 + y^2} - 3x = 2,$$

$$\pm\sqrt{x^2 + y^2} = 3x + 2.$$

Squaring each member of this equation, we have

$$x^2 + y^2 = 9x^2 + 12x + 4,$$

from which

$$8x^2 - y^2 + 12x + 4 = 0.$$

EXERCISE SET 6.6

A

Find four additional sets of polar coordinates ($-360° < \theta \leq 360°$) for each point whose coordinates are given.

1. $(3, 380°)$ **2.** $(6, 840°)$ **3.** $(5, -450°)$

4. $(4, -540°)$ **5.** $(-4, 420°)$ **6.** $(-5, -600°)$

Find the Cartesian coordinates of each point with polar coordinates as given.

7. $\left(4, \frac{\pi}{6}^R\right)$ **8.** $\left(6, \frac{2\pi}{3}^R\right)$ **9.** $\left(3, -\frac{3\pi}{4}^R\right)$

10. $\left(0, \frac{13\pi}{3}^R\right)$ **11.** $\left(-3, \frac{5\pi}{6}^R\right)$ **12.** $\left(-4, \frac{\pi}{4}^R\right)$

Find the pair of polar coordinates (θ in radians) using an angle θ such that $0^R < \theta < 2\pi^R$ and $\rho > 0$ for each point with Cartesian coordinates as given

13. $(4, 0)$ **14.** $(0, -3)$ **15.** $(2, 2)$

16. $(-1, -\sqrt{3})$ **17.** $\left(-\frac{\sqrt{3}}{2}, \frac{1}{2}\right)$ **18.** $\left(\frac{1}{2}, -\frac{\sqrt{3}}{2}\right)$

Transform each Cartesian equation to an equation in polar form.

Example $y = 3x$

Solution Substituting $\rho \sin \theta$ for y and $\rho \cos \theta$ for x yields

$$\rho \sin \theta = 3 \rho \cos \theta$$

Multiplying each member by $1/(\rho \cos \theta)$, $\rho \neq 0$, $\theta \neq 90° + k \cdot 180°$, yields

$$\frac{\sin \theta}{\cos \theta} = 3,$$

$$\tan \theta = 3.$$

Notice that for values of θ for which the equation is satisfied, $\tan \theta$ is a constant, namely 3. Thus the graph of the equation is a straight line with slope 3.

19. $x = 4$ **20.** $y = -3$ **21.** $x^2 = 8y$

22. $y^2 = 4x$ **23.** $x^2 + y^2 = 16$ **24.** $4x^2 + y^2 = 4$

25. $x^2 + y^2 - 3y = 0$ **26.** $x^2 + y^2 + 4x = 0$

Transform each polar equation to an equation in Cartesian form.

Example $\rho = 2 \sec \theta$

Solution Substituting $1/\cos \theta$ for $\sec \theta$ yields

$$\rho = \frac{2}{\cos \theta},$$

$$\rho \cos \theta = 2.$$

Because $\rho \cos \theta = x$, the Cartesian form of the equation is $x = 2$.

27. $\rho = 6$ **28.** $\rho = 9$ **29.** $\rho = 4 \sin \theta$

30. $\rho = 3 \cos \theta$ **31.** $\rho = 2 \csc \theta$ **32.** $\rho = -4 \sec \theta$

33. $\rho = \dfrac{4}{1 - \sin \theta}$ **34.** $\rho = \dfrac{4}{1 + 2 \cos \theta}$

6.7 Graphing Equations in Polar Form

A solution to an equation in two variables x and y can be represented graphically by considering those ordered pairs (x, y) which satisfy the equation as the coordinates of points in the Cartesian plane. Thus, each point on the graph of an equation represents a solution to the equation. In a similar fashion, a solution to an equation in two variables (usually ρ and θ in this case) can be viewed as the *polar* coordinates of a point in the plane. The set of solutions to the equation would then be represented as a graph in the polar coordinate system.

θ	$\sin \theta$	ρ or $4 \sin \theta$
0°	0	0
30°	0.50	2.0
45°	0.71	2.8
60°	0.87	3.5
90°	1.00	4.0
120°	0.87	3.5
135°	0.71	2.8
150°	0.50	2.0
180°	0	0

Example Graph $\rho = 4 \sin \theta$ in the interval $0° \leq \theta < 360°$.

Solution Some ordered pairs with commonly used angles 30°, 45°, etc., are first obtained. The tabulation of the data in the arrangement shown is sometimes helpful. The approximations shown for some values of $\sin \theta$ are sufficient for our purpose. These ordered pairs (ρ, θ) are shown in figure a.

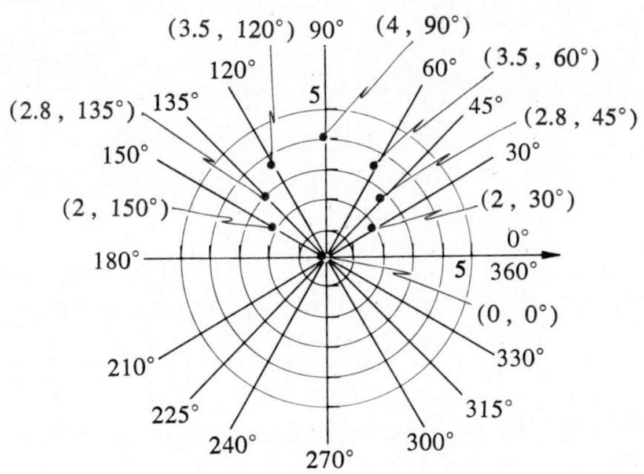

a.

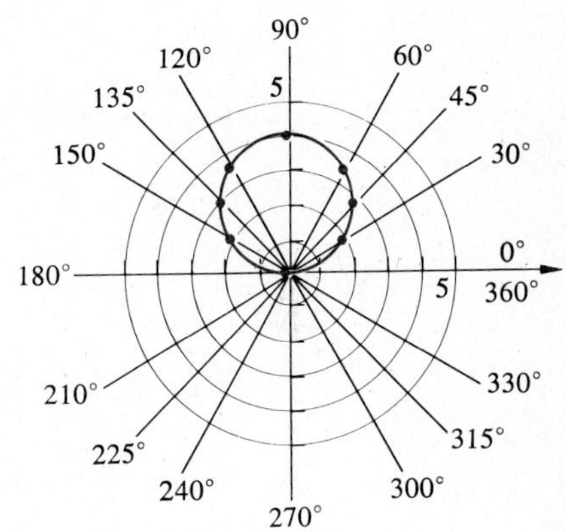

b.

The completed curve is obtained for $0° \leq \theta < 180°$. For values of θ where $180° \leq \theta < 360°$, the components of ordered pairs (ρ, θ) are also coordinates of points on the curve shown in figure b.

Example Graph $\rho = \dfrac{3}{\sin \theta + \cos \theta}$.

Solution First obtain some ordered pairs as shown in the table on page 284 and in figure a. The graph, a straight line, is shown in figure b.

(continued)

θ	$\sin\theta$	$\cos\theta$	$\sin\theta+\cos\theta$	$\rho=\dfrac{3}{\sin\theta+\cos\theta}$
0°	0	1.00	1.00	3.0
30°	0.50	0.87	1.37	2.2
45°	0.71	0.71	1.42	2.1
60°	0.87	0.50	1.37	2.2
90°	1.00	0	1.00	3.0
120°	0.87	−0.50	0.37	8.1
135°	0.71	−0.71	0	undef.

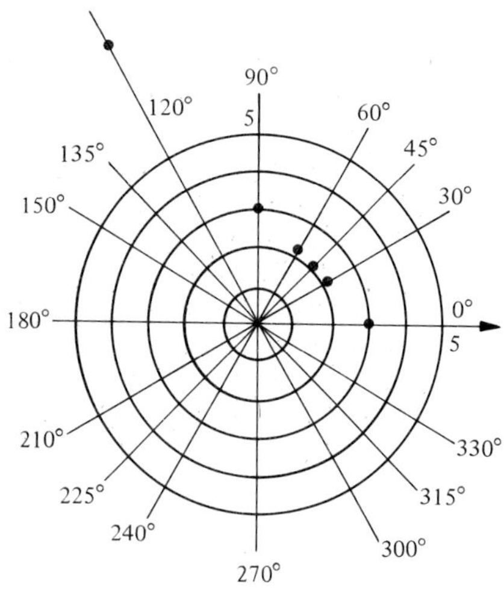

a.

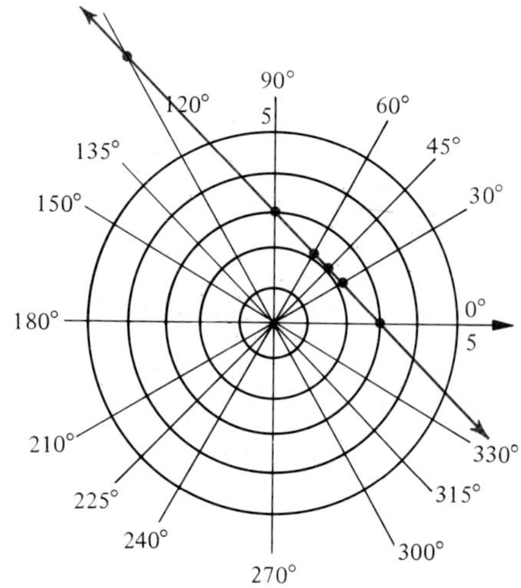

b.

For some common curves, the polar form of an equation is more useful than the rectangular form; furthermore, the graph may be simpler to sketch from the polar representation.

Excluded Values of θ

By analyzing a polar equation, we can sometimes determine what its graph looks like. For example, if

$$\rho^2 = 9\cos 2\theta \tag{1}$$

and we select $\theta = 60°$, then

$$\rho^2 = 9\cos 120° = \frac{-9}{2},$$

and $\rho = \pm\sqrt{-9/2}$, an imaginary number. Therefore, 60° must be

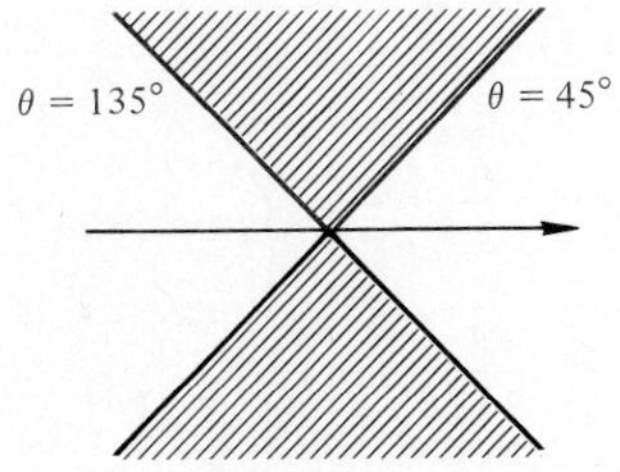

Figure 6.7

excluded as a value of θ. In fact, for all θ such that

$$(90° + k \cdot 360°) < 2\theta < (270° + k \cdot 360°),$$

or

$$(45° + k \cdot 180°) < \theta < (135° + k \cdot 180°), \ k \in J,$$

it follows that $\cos 2\theta < 0$, and hence these values of θ must be excluded (see Figure 6.7).

Hence, in order to graph $\rho^2 = 9 \cos 2\theta$, it is necessary to select values of θ that are not excluded. Some ordered pairs (ρ, θ) that are solutions of the equation are shown in Table 6.1, and its graph is shown in Figure 6.8. The graph is called a **lemniscate.**

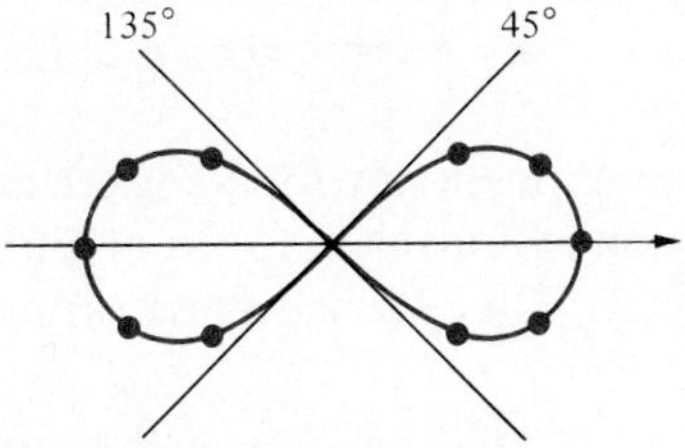

Figure 6.8

TABLE 6.1

θ	2θ	$\cos 2\theta$	ρ^2 or $9 \cos 2\theta$	ρ
0°	0°	1	9	±3
15°	30°	$\frac{\sqrt{3}}{2}$	$\frac{9\sqrt{3}}{2}$	±2.8
30°	60°	$\frac{1}{2}$	$\frac{9}{2}$	±2.1
45°	90°	0	0	0
(45° < θ < 135° excluded)				
135°	270°	0	0	0
150°	300°	$\frac{1}{2}$	$\frac{9}{2}$	±2.1
165°	330°	$\frac{\sqrt{3}}{2}$	$\frac{9\sqrt{3}}{2}$	±2.8
180°	360°	1	9	±3

Tangent Lines at the Pole

If, for some particular value of θ, say α, the value of ρ becomes 0, then the line $\theta = \alpha$ is tangent to the graph at the pole. For example, in the equation $\rho^2 = 9 \cos 2\theta$ for the lemniscate above, $\rho = 0$ when

$$\cos 2\theta = 0,$$

from which

$$2\theta = \frac{\pi}{2} \quad \text{or} \quad 2\theta = \frac{3\pi}{2},$$

$$\theta = \frac{\pi}{4} \quad \text{or} \quad \theta = \frac{3\pi}{4}.$$

Thus, the graphs of these equations are two lines tangent to the graph at the pole as shown in Figure 6.8.

Symmetry

The graph of an equation in polar form is often symmetric with respect to a line or a point. In particular, considerations of symmetry with respect to the polar axis ($\theta = 0°$), the vertical line through the pole, sometimes called the vertical axis ($\theta = 90°$), and the pole ($\rho = 0$) are helpful when graphing equations in polar form.

Testing an equation for each of these symmetries involves making certain replacements of ρ and θ. If the new equation is equivalent to the original, then the graph has that particular symmetry. Table 6.2 lists these various

TABLE 6.2

Symmetry with respect to		**Replacement for** ρ	**Replacement for** θ	**Geometric representation**
Polar axis ($\theta = 0°$)	(a)		$-\theta$	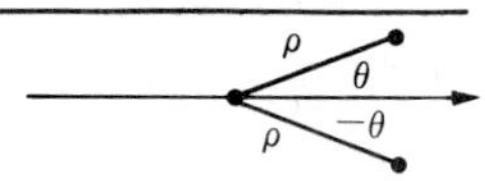
	(b)	$-\rho$	$\pi - \theta$	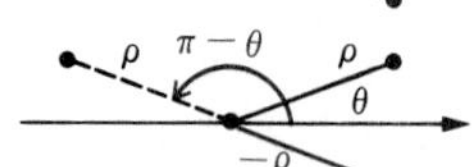
Vertical axis ($\theta = 90°$)	(c)		$\pi - \theta$	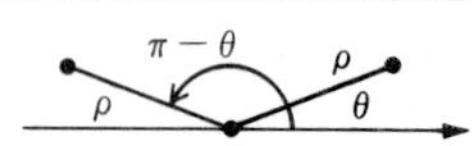
	(d)	$-\rho$	$-\theta$	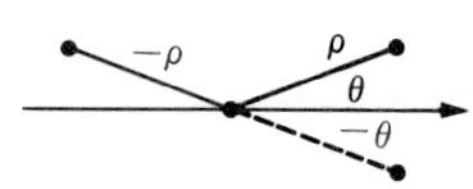
Pole ($\rho = 0$)	(e)		$\pi + \theta$	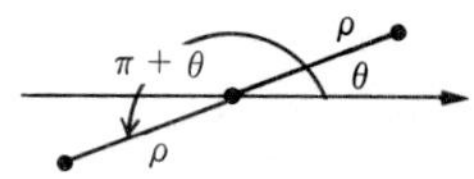
	(f)	$-\rho$		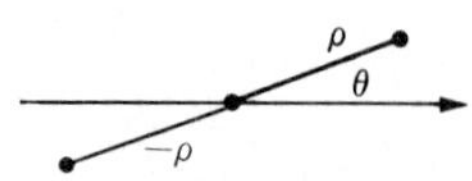

replacements. Note that for each possible symmetry, there are two tests. If either test is successful, the graph has that particular symmetry.

All of the tests in Table 6.2 are successful when applied to the equation $\rho^2 = 9 \cos 2\theta$ that we considered above. To illustrate we will check (a), (d), and (f). Thus, for test (a), if θ is replaced by $-\theta$.

$$\rho^2 = 9 \cos 2(-\theta) = 9 \cos(-2\theta)$$
$$= 9 \cos 2\theta.$$

Then for test (d), replacing ρ with $-\rho$ and θ with $-\theta$,

$$(-\rho)^2 = 9 \cos 2(-\theta)$$

or

$$\rho^2 = 9 \cos 2\theta.$$

Finally, for test (f), replacing ρ with $-\rho$,

$$(-\rho)^2 = 9 \cos 2\theta$$

or

$$\rho^2 = 9 \cos 2\theta.$$

Hence, as indicated in Figure 6.8, the graph is symmetric with respect to the polar axis, the vertical axis, and the pole.

Example Test the equation $\rho = 1 + \cos \theta$ for the symmetries in Table 6.2.

Test (a): Replacing θ with $-\theta$, we have

$$\rho = 1 + \cos(-\theta)$$
$$= 1 + \cos \theta.$$

The new equation is equivalent to the original, hence the graph is symmetric with respect to the polar axis.

Test (c): Replacing θ with $\pi - \theta$, we have

$$\rho = 1 + \cos(\pi - \theta)$$
$$= 1 - \cos \theta.$$

This is not equivalent to the original equation, so we try the second test.

(continued)

Test (d): Replacing ρ with $-\rho$, and θ with $-\theta$, we have

$$-\rho = 1 + \cos(-\theta)$$
$$= 1 + \cos\theta.$$

Since this equation also is not equivalent to the original, we cannot conclude that the graph is symmetric with respect to the vertical axis.

Test (e): Replacing θ with $\pi + \theta$, we have

$$\rho = 1 + \cos(\pi + \theta)$$
$$= 1 - \cos\theta,$$

which is not equivalent to the original equation. Hence, we try the second test.

Test (f): Replacing ρ with $-\rho$, we have

$$-\rho = 1 + \cos\theta.$$

This again is not equivalent to the original equation, so we cannot conclude that the graph is symmetric with respect to the pole.

θ	$\cos\theta$	ρ or $1 + \cos\theta$
0°	1	2
30°	$\frac{\sqrt{3}}{2}$	1.87
60°	$\frac{1}{2}$	1.5
90°	0	1
120°	$-\frac{1}{2}$	$\frac{1}{2}$
150°	$-\frac{\sqrt{3}}{2}$	0.13
180°	-1	0

We have shown only that the graph is symmetric with respect to the polar axis. Now, using selected values of θ between 0° and 180°, we first graph the equation in this interval. Then, using the fact that the graph is symmetric with respect to the polar axis from test (a), we sketch the entire graph. The graph is known as a **cardioid.**

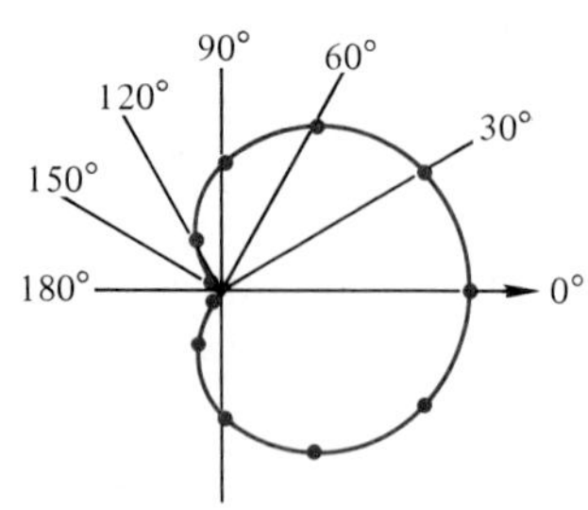

With a little practice, tests in Table 6.2 can be done quickly. We can usually facilitate graphing an equation in polar coordinates by applying two or three of the tests.

Several additional interesting types of curves that are graphs of polar equations are shown in Figure 6.9.

LIMACONS

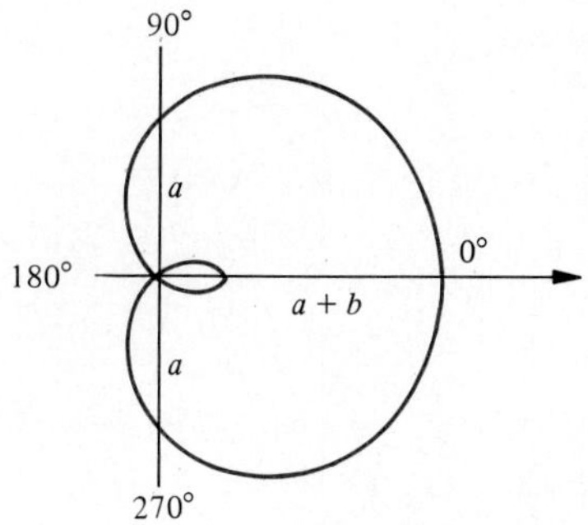

$\rho = a + b \cos \theta, \ a < b$

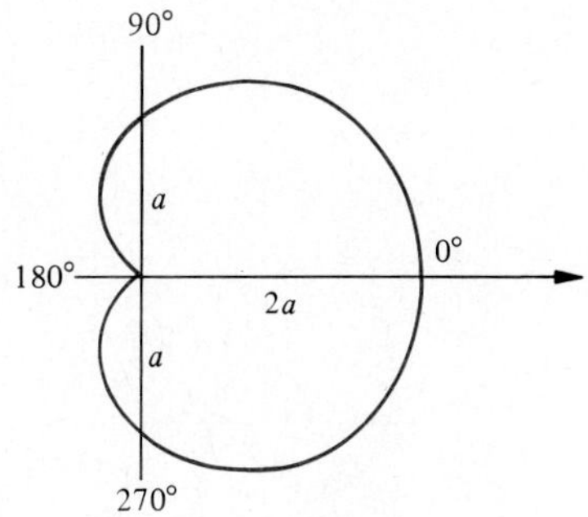

$\rho = a + a \cos \theta$

Cardioid

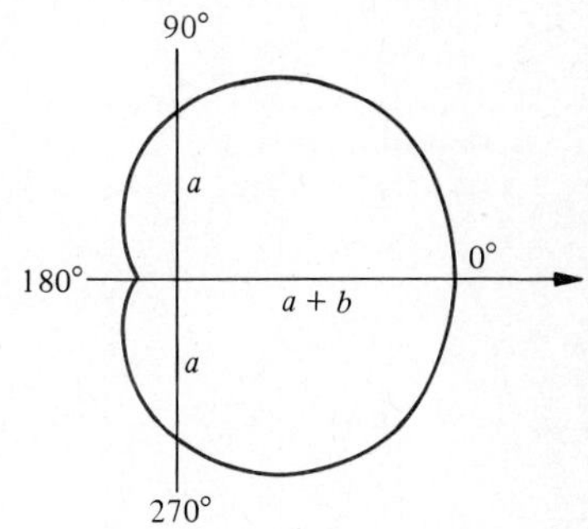

$\rho = a + b \cos \theta, \ a > b$

LEMNISCATES

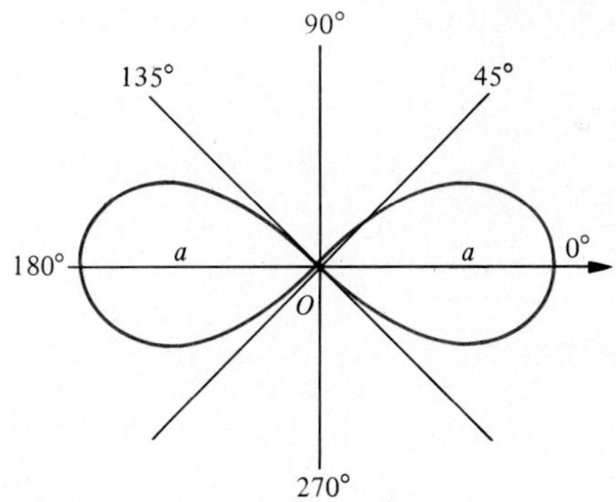

$\rho^2 = a^2 \cos 2\theta$

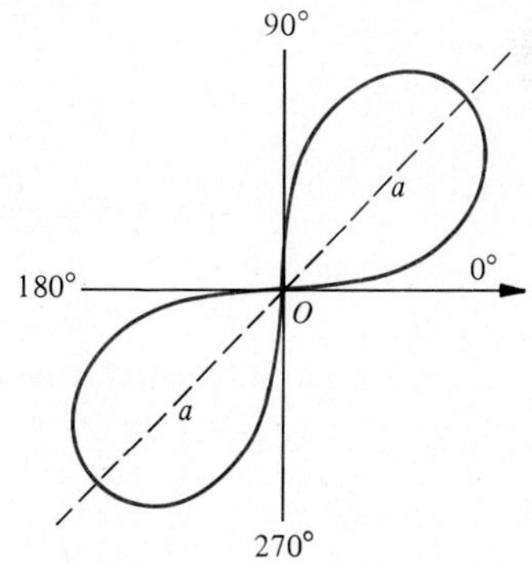

$\rho^2 = a^2 \sin 2\theta$

THREE-LEAVED ROSE

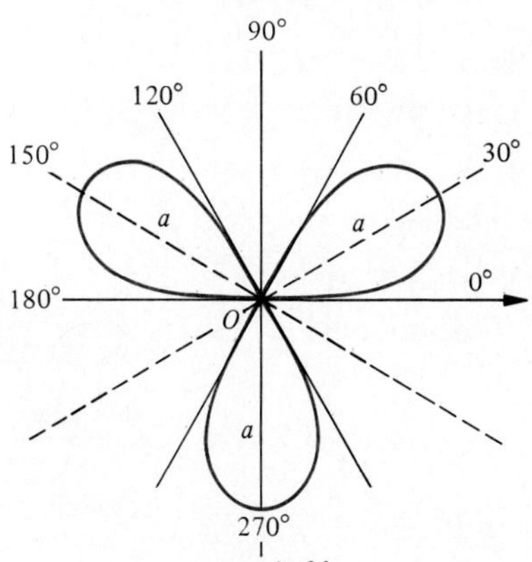

$\rho = a \sin 3\theta$

FOUR-LEAVED ROSE

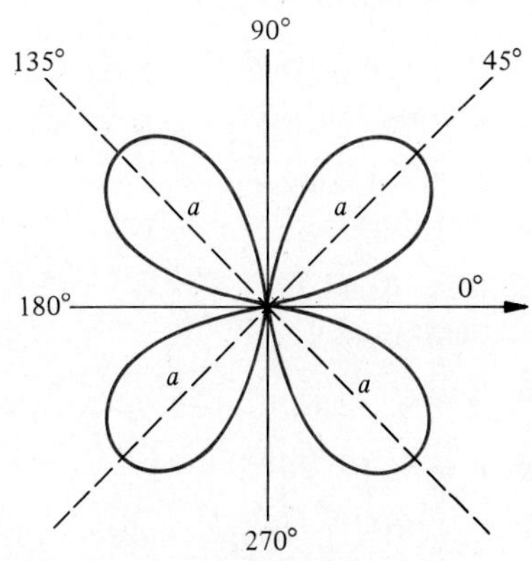

$\rho = a \sin 2\theta$

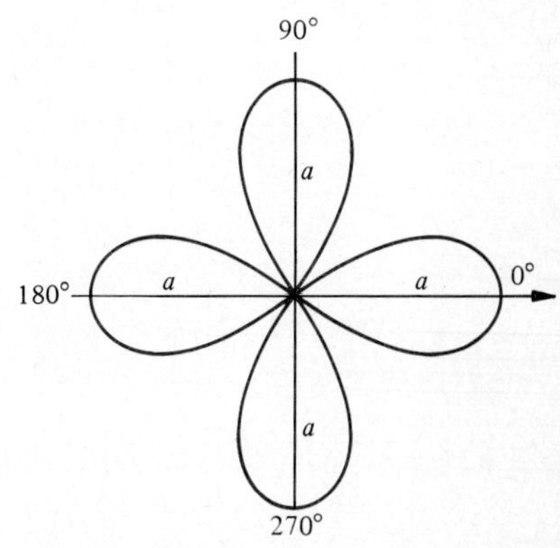

$\rho = a \cos 2\theta$

Figure 6.9 **(continued)**

SPIRAL OF ARCHIMEDES — HYPERBOLIC SPIRAL — LOGARITHMIC SPIRAL

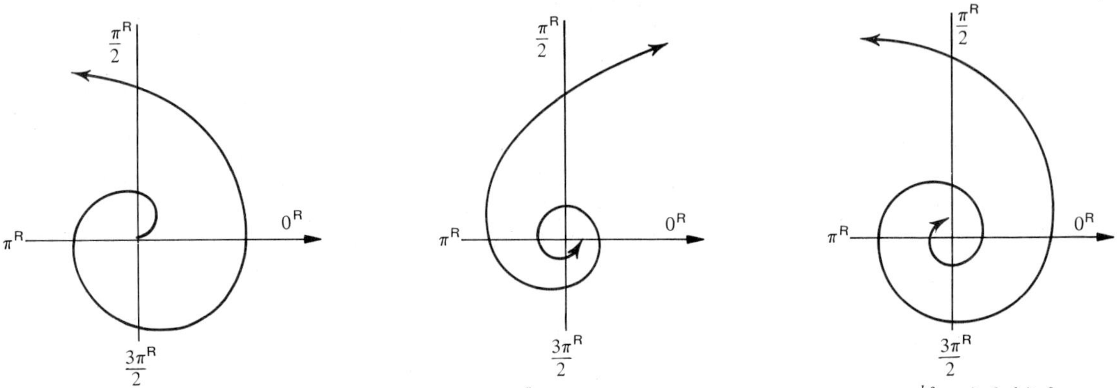

$\rho = a\theta,\ a > 0,\ \theta > 0$ — $\rho = \frac{a}{\theta},\ a > 0,\ \theta > 0$ — $\rho = a^{k\theta},\ a > 0, k > 0$

Figure 6.9 (continued)

EXERCISE SET 6.7

A

Graph each equation.

1. $\rho = 4 \cos \theta$ (circle)

2. $\rho = 3 \sin \theta$ (circle)

3. $\rho = \cos 3\theta$ (three-leaved rose)

4. $\rho = 2 \sin 3\theta$ (three-leaved rose)

5. $\rho = \sin 2\theta$ (four-leaved rose)

6. $\rho = 2 \cos 2\theta$ (four-leaved rose)

7. $\rho = 1 + 2 \cos \theta$ (limaçon)

8. $\rho = 3 + 3 \cos \theta$ (cardioid)

9. $\rho = 2 + 2 \sin \theta$ (cardioid)

10. $\rho = 2 + \cos \theta$ (limaçon)

11. $\rho^2 = 4 \cos 2\theta$ (lemniscate)

12. $\rho^2 = 4 \sin 2\theta$ (lemniscate)

13. $\rho = 4$ *Hint:* Consider the equation $\rho = 0 \cdot \theta + 4$.

14. $\rho = 5$

15. $\theta = 30°$ *Hint:* Consider the equation $\theta = 0 \cdot \rho + 30°$.

16. $\theta = 45°$

B

Graph each equation.

17. $\rho = \frac{1}{\theta}$, $\theta > 0$
(hyperbolic spiral)

18. $\rho = 2\theta$, $\theta > 0$
(spiral of Archimedes)

Graph each of the given pairs of equations and find coordinates for the points of intersection.

19. $\rho = \sin \theta$
$\rho = \cos \theta$

20. $\rho = 1 + \sin \theta$
$\rho = 1$

21. $\rho = 1 + \sin \theta$
$\rho = 1 + \cos \theta$

22. $\rho = 2 \sin \theta$
$\rho = 2 \sin 2\theta$

23. Solve the system in Exercise 19 analytically. Explain why the coordinates of the origin are not in the solution set of the system.

24. Solve the system in Exercise 20 analytically.

25. Solve the system in Exercise 21 analytically.

26. Solve the system in Exercise 22 analytically.

27. Show that the distance between the points with coordinates (ρ_1, θ_1) and (ρ_2, θ_2) is given by $d = \sqrt{\rho_1{}^2 + \rho_2{}^2 - 2\rho_1\rho_2 \cos(\theta_1 - \theta_2)}$.

28. Find an equation in polar form for the circle with center at $(a, 0^R)$ and radius of length a. *Hint:* Recall that an angle inscribed in a semicircle is a right angle.

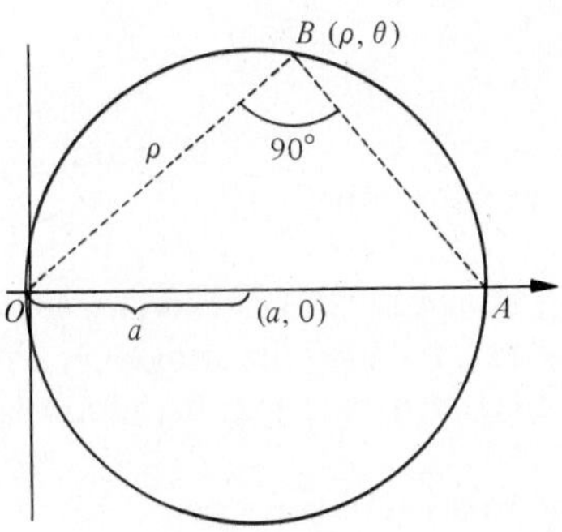

Ex. 28

Chapter Summary

6.1 A **complex number** is defined by

$$z = a + bi, \quad a, b \in R \quad \text{and} \quad i^2 = -1.$$

If $z_1 = a_1 + b_1 i$ and $z_2 = a_2 + b_2 i$, then

$$z_1 = z_2 \text{ if and only if } a_1 = a_2 \text{ and } b_1 = b_2.$$

Sums and products of complex numbers are found by treating $a + bi$ as a binomial and replacing i^2 with -1.

A complex number $a + bi$ can be identified with a **real number** if $b = 0$; if $b \neq 0$, the number is an **imaginary number,** and for the special case where $a = 0$, the number is a **pure imaginary number.**

For all $b \in R,\ b > 0$,

$$\sqrt{-b} = \sqrt{bi} = i\sqrt{b}.$$

Hence, for a and b positive real numbers,

$$\sqrt{-a}\sqrt{-b} = i^2\sqrt{ab} = -\sqrt{ab}.$$

6.2 Similar to the corresponding properties of real numbers:

$$z_1 - z_2 = z_1 + (-z_2);$$

$$\frac{z_1}{z_2} = z_1 \cdot \frac{1}{z_2} \quad (z_2 \neq 0).$$

If $z_2, z_3 \neq 0$, then

$$\frac{z_1}{z_2} = \frac{z_1 z_3}{z_2 z_3}.$$

If $z = a + bi$, then the **conjugate** of z is $\bar{z} = a - bi$. The product of $a + bi$ and its conjugate equals $a^2 + b^2$, a real number.

6.3 A plane on which complex numbers are graphed is called a **complex plane.** The x-axis is called the **real axis** and the y-axis is called the **imaginary axis.**

Complex numbers can be added graphically by the parallelogram law applicable to geometric vectors.

The **absolute value** or **modulus** of the complex number z is

$$\rho = |z| = |a + bi| = |(a, b)| = \sqrt{a^2 + b^2}.$$

The modulus is the distance from the origin to the graph of (a, b). An **argument** or **amplitude** of the complex number $z = a + bi = (a, b)$, denoted by $\mathbf{arg}(\boldsymbol{a} + \boldsymbol{bi})$ or $\mathbf{arg}(\boldsymbol{a}, \boldsymbol{b})$, is an angle θ such that

$$\cos\theta = \frac{a}{\rho} \quad \text{and} \quad \sin\theta = \frac{b}{\rho} \quad (a^2 + b^2 \neq 0).$$

If θ is an argument of $a + bi$, then $(\theta + k \cdot 360)^\circ,\ k \in J,$ is also an argument of $a + bi$. The angle θ with the least positive measure is called the **principal argument.**

6.4 A nonzero complex number $a + bi$ can be expressed in **trigonometric form** as

$$a + bi = \rho[\cos(\theta + k \cdot 360^\circ) + i\sin(\theta + k \cdot 360^\circ)],\ k \in J.$$

If $z_1 = \rho_1(\cos\theta_1 + i\sin\theta_1)$ and $z_2 = \rho_2(\cos\theta_2 + i\sin\theta_2)$, then

$$z_1 \cdot z_2 = \rho_1 \cdot \rho_2[\cos(\theta_1 + \theta_2) + i\sin(\theta_1 + \theta_2)],$$

$$\frac{z_1}{z_2} = \frac{\rho_1}{\rho_2}[\cos(\theta_1 - \theta_2) + i\sin(\theta_1 - \theta_2)].$$

6.5 If $z = \rho(\cos\theta + i\sin\theta)$ and $n \in N$, then

$$z^n = \rho^n(\cos n\theta + i\sin n\theta).$$

This is known as **DeMoivre's theorem.**

If $z \neq 0 + 0i$, then

$$z^0 = 1 + 0i = 1 \qquad \text{and} \qquad z^{-n} = \frac{1}{z^n},\ n \in N.$$

If $n \in N$ and $z = \rho(\cos\theta + i\sin\theta)$, then

$$\rho^{1/n}\left[\cos\left(\frac{\theta + k\cdot 360^\circ}{n}\right) + i\sin\left(\frac{\theta + k\cdot 360^\circ}{n}\right)\right],\quad k \in J$$

is an ***n*th root of** z. For $k = 0$, the nth root

$$\rho^{1/n}\left(\cos\frac{\theta}{n} + i\sin\frac{\theta}{n}\right)$$

is called the **principal root.**

6.6 The components of an ordered pair (ρ, θ°) or (ρ, θ^R) that locate a point A in the plane are called the **polar coordinates** of A. Each such ordered pair specifies a unique point; however, each point in the plane corresponds to infinitely many ordered pairs of polar coordinates.

Cartesian and polar coordinates are related as follows.

$$x = \rho\cos\theta \qquad \rho = \pm\sqrt{x^2 + y^2},$$

$$y = \rho\sin\theta \qquad \cos\theta = \frac{x}{\rho} = \frac{x}{\pm\sqrt{x^2 + y^2}},$$

$$\sin\theta = \frac{y}{\rho} = \frac{y}{\pm\sqrt{x^2 + y^2}} \qquad [(x, y) \neq (0, 0)].$$

6.7 Polar equations can be graphed by obtaining a number of ordered pairs

(ρ, θ). The notions of excluded values, tangent lines, and symmetry can facilitate the procedures of graphing such equations.

A list of symbols introduced in this chapter is shown inside the front cover, and important properties are shown inside the back cover.

Review Exercises

A

6.1 *Find real numbers x and y for which each statement is true.*

1. $9x - 6yi = -3 + 4i$

2. $-8x - 5i = -2y - 5x^2 i$

Write each expression in the form a + bi.

3. $(-4 - i) + (-4 + 9i)$

4. $(-3 + i) + (3 - 2i)$

5. $(8 + 6i)(-2 + 5i)$

6. $(5 + 3i)(2 - 4i)$

Write each expression as one of the complex numbers i, −1, −i, or 1.

7. i^{13}

8. i^{23}

Write each expression in the form a + bi.

9. $\sqrt{-5}(3 - \sqrt{-2})$

10. $(3 - \sqrt{-5})(3 + \sqrt{-5})$

11. Show that $3i$ and $-3i$ satisfy the equation $x^2 + 9 = 0$.

12. Show that $1 + 3i$ and $1 - 3i$ satisfy $x^2 - 2x + 10 = 0$.

6.2 *Express each difference in the form a + bi.*

13. $(-7 + 2i) - (8 + 3i)$

14. $(6 - 9i) - (3 - 4i)$

Express each quotient in the form a + bi.

15. $\dfrac{8 + 3i}{2 + 4i}$

16. $\dfrac{-7 + i}{6 - 10i}$

Write each expression in the form bi or a + bi.

17. $\dfrac{-2}{\sqrt{-6}}$

18. $\dfrac{4}{2 + \sqrt{-5}}$

6.3 *Express each complex number as an ordered pair.*

19. $-4 + i$

20. $2\sqrt{3} - 2i$

Express each complex number in the form $a + bi$.

21. $(9, -6)$

22. $(3, 4)$

Graph each complex number z, its negative $-z$, and its conjugate $\bar{z}$.

23. $-3 + 5i$

24. $(-2, -6)$

Represent each sum graphically and check the results by algebraic methods.

25. $(-4 + 3i) + (8 - 2i)$

26. $(-1, -8) + (6, 7)$

Find the absolute value and the principal argument of each complex number.

27. $4 + 2i$

28. $(-2, -5)$

6.4 *Represent each complex number in trigonometric form, using the angle with least nonnegative measure.*

29. $1 + i$

30. $-3 + 3\sqrt{3}i$

Represent each complex number graphically and write it in the form $a + bi$.

31. $7(\cos 120° + i \sin 120°)$

32. $2[\cos(-315°) + i \sin(-315°)]$

Express each product in the form $a + bi$.

33. $3(\cos 200° + i \sin 200°) \cdot 2(\cos 25° + i \sin 25°)$

34. $\sqrt{3}(\cos 50° + i \sin 50°) \cdot \sqrt{12}[\cos(-20°) + i \sin(-20°)]$

Express each quotient in the form $a + bi$.

35. $\dfrac{8(\cos 309° + i \sin 309°)}{9(\cos 129° + i \sin 129°)}$

36. $\dfrac{12[\cos(-315°) + i \sin(-315°)]}{3(\cos 270° + i \sin 270°)}$

6.5 *Express each power in the form $a + bi$.*

37. $[3(\cos 15° + i \sin 15°)]^4$

38. $(\sqrt{3} - i)^3$

39. $[4(\cos 240° + i \sin 240°)]^{-2}$

40. $(4 - 4i)^{-3}$

41. Find the two square roots of $\frac{\sqrt{3}}{2} + \frac{1}{2}i$.

42. Find the three cube roots of $125(\cos 330° + i \sin 330°)$.

Solve each equation, $x \in C$.

43. $x^4 + 1 = 0$

44. $x^5 = -5 - 5\sqrt{3}i$

6.6 *Find four additional sets of polar coordinates* $(-360° < \theta \leq 360°)$ *for each point.*

45. $(2, 610°)$

46. $(-3, 390°)$

Find the Cartesian coordinates for each point whose polar coordinates are given.

47. $\left(3, \frac{\pi}{3}^{R}\right)$

48. $\left(-2, \frac{3\pi}{4}^{R}\right)$

Find the pair of polar coordinates (θ *in radians*) *using an angle* θ *such that* $0^R < \theta < 2\pi^R$ *and* $\rho > 0$ *for each point whose Cartesian coordinates are given.*

49. $(2, -2)$

50. $(-\sqrt{3}, -1)$

Transform each equation to an equation in polar form.

51. $x = 3$

52. $x^2 + 3y^2 = 5$

Transform each equation to an equation in Cartesian form.

53. $\rho = 5 \sin \theta$

54. $\rho = \frac{2}{1 - \cos \theta}$

6.7 *Graph each equation.*

55. $\rho = 1 + \sin \theta$

56. $\rho = 1 - \sin \theta$

57. $\rho = \cos 2\theta$

58. $\rho = 2 \cos 3\theta$

A ALGEBRA AND GEOMETRY REVIEW

In Appendix A we review the basic ideas of algebra and geometry, using terminology and notation that are needed in the development of various topics in the text.

A.1 Set Notation

Recall that a set is simply a collection of "things." Each "thing" in the set is called a **member** or **element** of the set. The membership can be described by listing the names of the members in braces, { }, or by stating a rule that identifies the members in the set.

The symbol ∈ is used to denote membership in a set. For example, we can write

$$3 \in \{3, 4, 5\}$$

to represent the statement "3 is an element of {3, 4, 5}."

Definition A.1 *Two sets are* **equal** *if and only if they have the same members.*

Customarily, the slant bar, /, drawn through symbols for relations, indicates negation. Thus,

$$\{3, 4, 5\} = \{5, 3, 4\} \qquad \text{but} \qquad \{3, 4, 5\} \neq \{4, 5, 6\}.$$

Sets are also designated by means of capital letters, A, B, C, R, etc.

Example If $A = \{2, 3, 4\}$ and $B = \{4, 5, 6\}$, then

$$3 \in A \quad \text{but} \quad 3 \notin B.$$

An unspecified element of a set is usually denoted by a lowercase italic letter such as a, b, c, x, and y or by a lowercase letter from the Greek alphabet such as α, β, γ, and θ. Such a symbol is called a **variable;** a symbol used to denote a specific element is called a **constant.** Variables are used in conjunction with braces in another symbolism that is often useful in discussing sets. This symbolism, illustrated by

$$\{x \mid x \in A \quad \text{and} \quad x \in B\}$$

and read "the set of all x such that x is an element of A and x is an element of B," is called **set-builder notation.** The vertical line, |, is read "such that." Set-builder notation names a variable and states conditions on the variable in order for it to be an element of the set.

Example If $A = \{2, 3, 4\}$ and $B = \{4, 5, 6\}$, then

$$\{x \mid x \in A \quad \textbf{or} \quad x \in B\} = \{2, 3, 4, 5, 6\} \tag{1}$$

because 2, 3, 4, 5, 6 are elements of *one or the other* of the two sets, whereas

$$\{x \mid x \in A \quad \textbf{and} \quad x \in B\} = \{4\} \tag{2}$$

because 4 is an element of *both* sets.

The preceding example illustrates the following two operations on sets.

Definition A.2 *The* **union** *of two sets A and B is the set of all elements that belong to A or to B or to both.*

The symbol $\cup$ is used to denote this operation; $A \cup B$ is read "the union of A and B."

Example If $A = \{2, 3, 4\}$ and $B = \{4, 5, 6\}$, then

$$A \cup B = \{2, 3, 4, 5, 6\}. \tag{3}$$

Notice that each element in $A \cup B$ is listed only once. Since the symbol 4 denotes only one number, it would be redundant to list the name of the element twice and to write $A \cup B = \{2, 3, 4, 4, 5, 6\}$. A second useful operation on sets is the following.

Definition A.3 *The* **intersection** *of two sets A and B is the set of all elements that belong to both A and B.*

The symbol $\cap$ is used to denote this operation; $A \cap B$ is read "the intersection of A and B."

Example If $A = \{2, 3, 4\}$ and $B = \{4, 5, 6\}$, then

$$A \cap B = \{4\}. \tag{4}$$

Notice from Equations (1) and (3) that the operation of union is associated with the word *or*, and from Equations (2) and (4) that the operation of intersection is associated with the word *and*.

The set that contains no elements is called the **null set** or the **empty set** and is denoted by the symbol $\varnothing$.

Example The intersection of two sets is a set. Therefore, if $C = \{2, 3, 4\}$ and $D = \{5, 6, 7\}$, we can write

$$C \cap D = \{2, 3, 4\} \cap \{5, 6, 7\} = \varnothing.$$

Two sets such as C and D, which do not have any elements in common, are said to be **disjoint.**

A.2 The Set of Real Numbers

Recall from your study of algebra that each real number can be associated with one and only one point on a number line and each point on the line can be associated with one and only one real number. Several examples are shown in Figure A.1. The real number corresponding to a point on the line is called the **coordinate** of the point, and the point is called the **graph** of the number. A number whose graph is to the right of the graph of zero is **positive,** and a number whose graph is to the left is **negative.**

The following sets of numbers are contained in the set of real numbers R. The capital letters shown are the ones that are often associated with the respective sets, and are the symbols we use in this text.

1. The set N of **natural numbers,** whose elements are 1, 2, 3, 4, ...

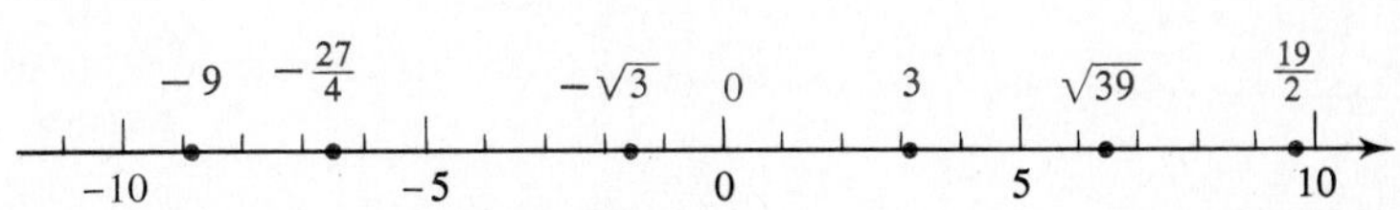

Figure A.1

2. The set J of **integers,** whose elements consist of the natural numbers, their negatives, and zero. Included are such elements as -24, $-9, 0, 7$, and 314.
3. The set Q of **rational numbers,** whose elements are all those numbers that can be represented in the form $\frac{a}{b}$, or a/b, where a and b are integers and b is not zero. Included are such elements as $-\frac{1}{3}, 0, \frac{1}{4}$, and 6.
4. The set H of **irrational numbers,** whose elements are those real numbers that are not rational numbers. An irrational number cannot be represented in the form a/b, where a and b are integers. Some examples are $\sqrt{2}$, $-\sqrt[3]{12}$, and π.

Figure A.2 summarizes the relationships among these sets of numbers.

If *every* element of a set A is also an element of a set B, then A is said to be a **subset** of B. Thus the sets H and Q can be viewed as subsets of R, the set J as a subset of Q or R, and the set N as a subset of J, Q, or R.

Radical notation is useful to represent some irrational numbers.

Definition A.4 *For all* $a \in R$ *where* a *is positive or zero,* $\sqrt{a}$ *is the nonnegative number such that*

$$\sqrt{a} \cdot \sqrt{a} = a.$$

Examples

a. $\sqrt{9}$ is the positive number such that $\sqrt{9}\sqrt{9} = 9$.

b. For $x > 0$, $\sqrt{x}$ is the positive number such that $\sqrt{x}\sqrt{x} = x$.

Every irrational number can be approximated to any desired degree of accuracy by a rational number. For example, by using the symbol $\approx$ for "approximately equal to," we write $\sqrt{2} \approx 1.414$, $\sqrt{3} \approx 1.732$, $\sqrt{5} \approx 2.236$,

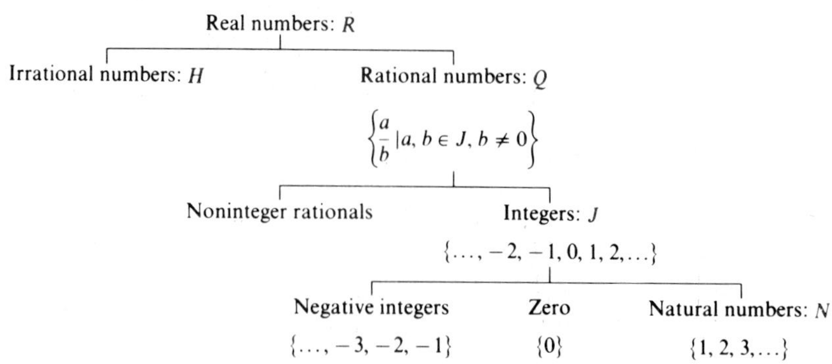

Figure A.2

etc. Table I in Appendix C gives rational number approximations for some square roots that are irrational numbers to three decimal places.

If the elements of a set can be paired with the numerals 1, 2, 3, ..., n for some natural number n so that each element is paired with exactly one numeral and each numeral is paired with exactly one element, then the set is said to be a **finite set.** A set that is not the empty set and is not finite is called an **infinite set.** Some infinite sets can be designated by using three dots and some of the members listed to establish a pattern as shown in Figure A.2.

Often we wish to consider only the magnitude of a real number while ignoring its sign. The special notation $|a|$ is used for this purpose. To extend the notation for all $a \in R$, we define $|0| = 0$.

Definition A.5 *For all* $a \in R$,

if a *is positive or zero, then* $|a| = a$;

if a *is negative, then* $|a| = -a$.

$|a|$ *is called the* **absolute value** *of* a.

Examples a. $|4| = 4$ b. $|-4| = -(-4) = 4$ c. $|0| = 0$

An ordering of the elements of the set of real numbers can be established by the following.

Definition A.6 *For all* $a, b \in R$, b *is* **less than** a *if for some positive real number* c,

$$b + c = a.$$

For such conditions, a *is said to be* **greater than** b.

Examples

a. 4 is less than 7 because there is a *positive number* c (equal to 3) such that $4 + c = 7$.

b. -9 is less than -5 because there is a *positive number* c (equal to 4) such that $-9 + c = -5$.

The number line shown in Figure A.3 is very helpful in visualizing whether one real number is less than or greater than a second real number.

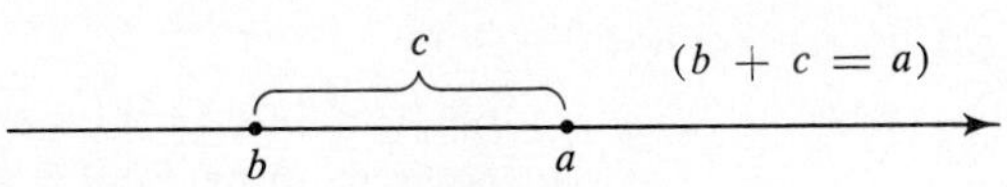

Figure A.3

If the graph of b lies to the *left* of the graph of a, then b is less than a. The following symbols are used in connection with the property of order.

$<$ read "is less than";

$\leq$ read "is less than or equal to";

$>$ read "is greater than";

$\geq$ read "is greater than or equal to."

Observe that $b < a$ and $a > b$ are equivalent statements.

Examples

a. $4 < 7$ and $7 > 4$

b. $-9 < -5$ and $-5 > -9$

c. $x < 7$ is equivalent to $7 > x$

In your study of arithmetic and algebra, you applied the operations of addition and multiplication to elements of the set of real numbers. You also made assumptions, called **axioms,** about these operations and studied the logical consequences of these axioms. For your reference, we include a list of some of these axioms and properties on page 303.

A.3 Equations in Two Variables; Relations and Functions

Recall from your study of algebra that the ordered pair (2, 9) is called a **solution** of the equation $y = x + 7$ because the replacement of x by the first component 2 and the replacement of y by the second component 9 results in a true statement, $9 = 2 + 7$. The ordered pair (2, 9) is said to *satisfy* the equation. The set of all such ordered pairs that satisfy an equation is called the **solution set** of the equation.

The equation $y = x + 7$ is not a true statement for all ordered pairs of real numbers. For example, (3, 5) is not a solution because $5 \neq 3 + 7$. Such equations are called **conditional equations.** On the other hand, an equation such as

$$x + y = y + x$$

is a true statement for every replacement of x and y for which both members have meaning. Such an equation is call an **identity.** The left-hand member and the right-hand member of identities are called **equivalent expressions.**

For $a, b, c, d \in R$:

Equality Axioms

1. $a = a$	Reflexive law
2. If $a = b$, then $b = a$	Symmetric law
3. If $a = b$ and $b = c$, then $a = c$	Transitive law
4. If $a = b$, then a may be substituted for b or b for a in any expression.	Substitution law

Order Axioms

1. Exactly one of the following is true: $a < b$, $a = b$, or $a > b$.	Trichotomy law
2. If $a < b$ and $b < c$, then $a < c$.	Transitive law
3. If $a, b > 0$, then $a + b > 0$ and $a \cdot b > 0$	Closure for positive numbers

Axioms for Operations

1. $a + b \in R$	Closure law for addition
2. $a + b = b + a$	Commutative law of addition
3. $(a + b) + c = a + (b + c)$	Associative law of addition
4. $a + 0 = 0 + a = a$	Identity law of addition
5. $a + (-a) = (-a) + a = 0$	Additive inverse law
6. $a \cdot b \in R$	Closure law for multiplication
7. $a \cdot b = b \cdot a$	Commutative law of multiplication
8. $(a \cdot b) \cdot c = a \cdot (b \cdot c)$	Associative law of multiplication
9. $a \cdot 1 = 1 \cdot a = a$	Identity law of multiplication
10. $a \cdot \frac{1}{a} = \frac{1}{a} \cdot a = 1 \ (a \neq 0)$	Multiplicative inverse law
11. $a \cdot (b + c) = a \cdot b + a \cdot c$	Distributive law

Some Properties

1. $a \cdot 0$
2. $-(-a) = a$
3. $-a = -1 \cdot a$
4. If $a = b$, then $a + c = b + c$
5. If $a + c = b + c$, then $a = b$
6. If $a = b$, then $a \cdot c = b \cdot c$
7. If $a \cdot c = b \cdot c$, then $a = b \ (c \neq 0)$
8. $a - b = a + (-b)$
9. $\frac{a}{b} = \frac{c}{d}$ if and only if $a \cdot d = b \cdot c \ (b, d \neq 0)$
10. $\frac{a}{b} = \frac{a \cdot c}{b \cdot c} \ (b, c \neq 0)$
11. $\frac{a}{c} + \frac{b}{c} = \frac{a + b}{c} \ (c \neq 0)$
12. $\frac{a}{b} \cdot \frac{c}{d} = \frac{a \cdot c}{b \cdot d} \ (b, d \neq 0)$
13. $\frac{a}{b} = a \cdot \frac{1}{b} \ (b \neq 0)$
14. $\frac{a}{b} \div \frac{c}{d} = \frac{a}{b} \cdot \frac{d}{c} \ (b, c, d \neq 0)$
15. $\frac{a}{b} = \frac{-a}{-b} = -\frac{a}{-b} = -\frac{-a}{b}$;

    $\frac{-a}{b} = \frac{a}{-b} = -\frac{a}{b} = -\frac{-a}{-b} \ (b \neq 0)$
16. $ab = 0$ if and only if $a = 0$ or $b = 0$

To show that an equation is not an identity, it is necessary only to find one replacement for the variables that will make the equation false. Such a procedure is referred to as showing a **counterexample.**

Example We can show by counterexample that $x^2 - 3y = 3y - x^2$ is not an identity by substituting some arbitrary values for the variables. If we use 1 for x and 1 for y, the left-hand member of the equation equals $1^2 - 3 \cdot 1$, or -2, and the right-hand member equals $3 \cdot 1 - 1^2$, or 2. Hence,

$$x^2 - 3y \neq 3y - x^2$$

for the ordered pair (1, 1), and the equation is not an identity.

Example We can show that the equation

$$(x - y)^2 + 2xy = x^2 + y^2$$

is an identity by simplifying the left-hand member

$$\begin{aligned}(x - y)^2 + 2xy &= x^2 - 2xy + y^2 + 2xy \\ &= x^2 + y^2.\end{aligned}$$

Thus, the left-hand member and the right-hand member are equivalent, and the equation is an identity.

Sets of ordered pairs play an important role in mathematics and are given a special name.

Definition A.7 *A* **relation** *is a set of ordered pairs.*

The set of first components in a set of ordered pairs is called the **domain** of the relation and the set of second components is called the **range.**

Examples

a. $\{(5, 8), (6, 9), (7, 10)\}$ is a relation with domain $\{5, 6, 7\}$ and range $\{8, 9, 10\}$.

b. $\{(5, 8), (5, 9), (7, 10)\}$ is a relation with domain $\{5, 7\}$ and range $\{8, 9, 10\}$.

The set of ordered pairs exhibited in Example a is a special kind of relation in which each element in the domain is paired with only one element in the range. Such relations are given a special name.

Definition A.8 *A* **function** *is a relation in which each element in the domain is associated with only one element in the range.*

Note that the relation in Example b above is not a function because 5, an element in the domain, is associated with more than one element in the range.

The relations above were exhibited by listing the members in the set of ordered pairs. They can also be described by equations that specify which ordered pairs are in the relation and which are not; those that are in the relation are elements of the solution set. For example,

$$\{(x, y) \mid y = x + 3,\ x \in \{5, 6, 7\}\}$$

also specifies the function in Example a above.

(Domain elements)

(5, 8), (6, 9), (7, 10)

(Range elements)

The elements in the range, 8, 9, and 10, are obtained by replacing x in the equation $y = x + 3$ with 5, 6, and 7 in turn and solving for y. This function consists of three ordered pairs only.

Example Given the function

$$\{(x, y) \mid y = x - 3, \quad x \in \{1, 2, 3\}\}:$$

1. The domain is $\{1, 2, 3\}$.
2. Replacing x with each element of the domain in turn, you can find the associated values of y to be -2, -1, and 0, respectively. Thus, the range is $\{-2, -1, 0\}$.
3. The function is $\{(1, -2), (2, -1), (3, 0)\}$.

If the domain of a function is the set of real numbers R, rather than a finite set, the function consists of an *infinite set* of ordered pairs and obviously not all its elements can be listed. *If the domain is not specified, we shall assume that it includes all real numbers for which both members of the defining equation are defined.*

Examples

a. The domain of the function defined by $y = x + 3$ is the set R because $x + 3$, and therefore y, are defined for all real number replacements for x.

(continued)

b. The domain of the function defined by $y = 3/(x - 4)$ is the set R with the restriction that $x \neq 4$ because $x - 4 = 0$ and y is undefined for $x = 4$.

Functions are sometimes designated by names (linear, quadratic, etc.) or by means of a single symbol, generally the symbol f. If the discussion includes a consideration of more than one function, other letters such as g, h, F, P, etc., are used.

The symbol $f(x)$ (read "f of x" or "the value of f at x") stands for the element in the range of f associated or paired with the element x in the domain. Thus, we can discuss functions defined by an equation such as

$$y = x - 2,$$

$$f(x) = x - 2,$$

$$g(x) = x - 2,$$

etc.

where the symbols $f(x)$, $g(x)$, etc., are playing exactly the same role as y. The variable x represents an element in the domain, and y, $f(x)$, $g(x)$, etc., represent an element in the range.

The notation $f(x)$ is especially useful because, by replacing x with a specific real number a in the domain, the notation $f(a)$ will then denote the paired, or corresponding, element in the range of the function.

Example If $f(x) = 2x - 5$,

$$f(3) = 2(3) - 5 = 6 - 5 = 1,$$

$$f(0) = 2(0) - 5 = 0 - 5 = -5,$$

$$f(-3) = 2(-3) - 5 = -6 - 5 = -11,$$

$$f(a) = 2(a) - 5 = 2a - 5.$$

A.4 Graphs of Relations and Functions

As you probably recall from your study of algebra, a correspondence exists between ordered pairs of real numbers and the points on a geometric plane, sometimes called the **real plane.** If two number lines are drawn perpendicular to each other at their origins as in Figure A. 4, they form a rectangular coordinate system. These perpendicular number lines are called **axes** (singular, axis). The horizontal line is usually called the **x-axis,** and the vertical line is usually called the **y-axis.** The x-axis and the y-axis divide the plane ino four regions, called **quadrants,** as shown.

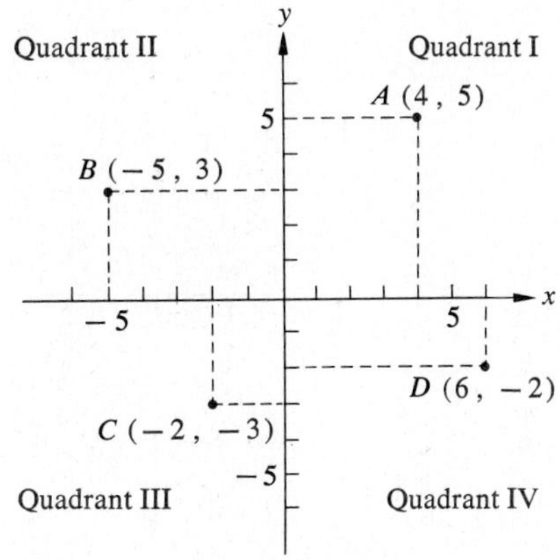

Figure A.4

For a given rectangular coordinate system, any point in the plane can be associated with a unique ordered pair of real numbers. For example, the point A in Quadrant I, in Figure A.4, can be associated with the ordered pair (4, 5), where 4 represents the distance of the point to the right of the y-axis and 5 represents the distance of the point above the x-axis. The points B, C, and D correspond to the ordered pairs $(-5, 3)$, $(-2, -3)$, and $(6, -2)$, respectively.

The *components* of an ordered pair associated with a point in the plane are called the **coordinates** of the point. The first component is called the **abscissa** of the point; the second component is called the **ordinate** of the point. The *point* is called the **graph** of the ordered pair. An ordered pair is sometimes used as a name for the corresponding point. Thus, we occasionally speak of the point (2, 3) or the point $(4, -3)$ or, in general, the point (x, y).

Recall that in Section A.3, a function was defined in part as a set of ordered pairs. Consequently, a function—or at least a part of a function—can now be displayed as a set of points on the plane. The phrase "part of a function" is used here since the domain or range or both may be the infinite set of real numbers, which cannot be shown on a finite line. For example, consider the function

$$y = x - 4.$$

If arbitrary values are assigned to x, say -3, 0, 3, and 6, and the corresponding values for y are computed, the four ordered pairs

$$(-3, -7),\quad (0, -4),\quad (3, -1),\quad \text{and}\quad (6, 2)$$

are obtained, which are solutions to the equation. These points can be located on a coordinate system as shown in Figure A.5a. These points

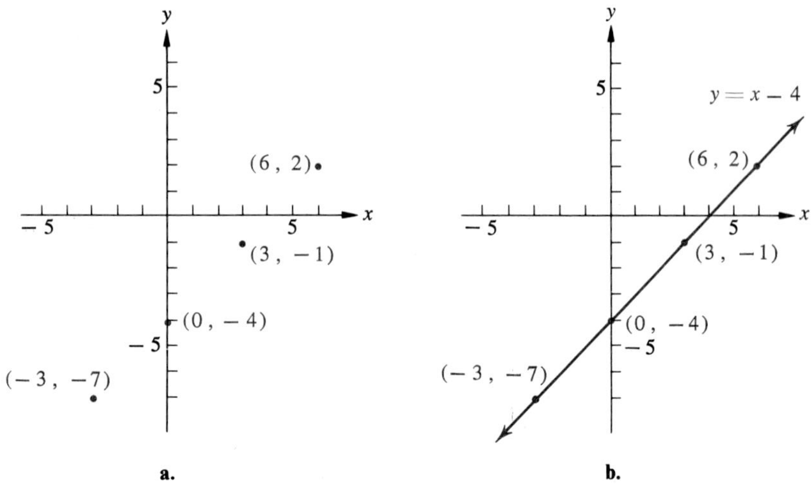

Figure A.5

appear to lie on a straight line, and in fact they do. It can be shown that the coordinates of any point on the line (Figure A.5b) constitute a solution of the first-degree equation in two variables, $y = x - 4$ and that, conversely, every solution of $y = x - 4$ corresponds to a point on this line. The line is referred to as the *graph of the function* $y = x - 4$ or, alternatively, as the *graph of the solution set of the equation* or simply as the *graph of the equation.* Obviously, only part of the line can be displayed, so an arrowhead is placed at each end to indicate that the graph, or line, continues in both directions indefinitely. Furthermore, since the graphs of first-degree equations

$$y = mx + b \qquad \text{or} \qquad ax + by = c$$

are straight lines, it is only necessary to find two ordered pairs in the solution set to sketch the graph. Two pairs that are easy to find are $(0, b)$ and $(a, 0)$, in which the numbers a and b are called the **intercepts.**

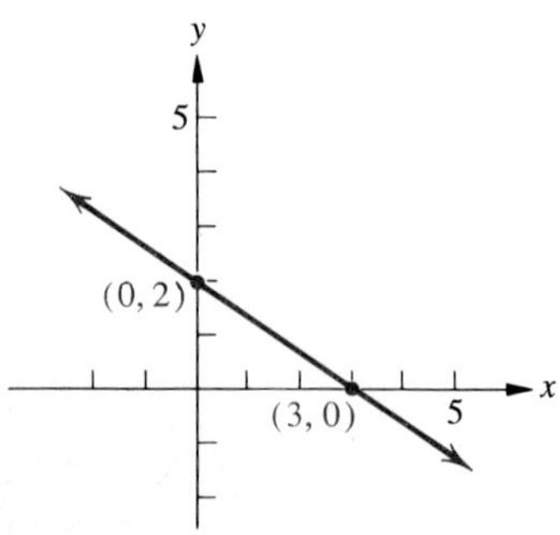

Example The graph of the function $2x + 3y = 6$ can be obtained by first finding the x and y intercepts.

If $y = 0$, then $x = 3$ and (3, 0) is a solution.

If $x = 0$, then $y = 2$ and (0, 2) is a solution.

The x and y intercepts are 3 and 2, respectively. See the graph.

First-degree equations are also called **linear equations,** and the functions defined by these equations are called **linear functions.**

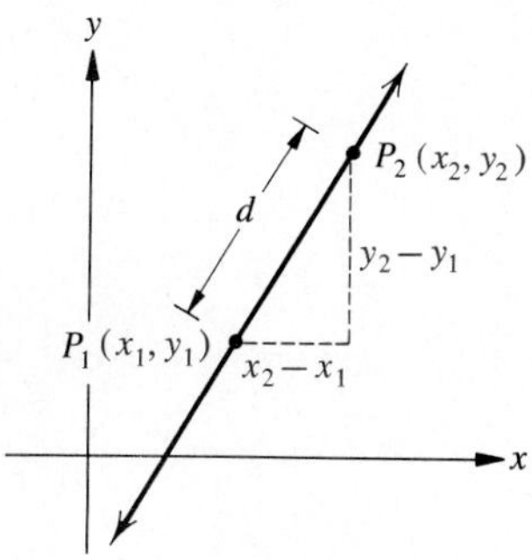

Figure A.6

For any two points P_1 and P_2 on a line, the set of points containing P_1 and P_2 and all points lying between P_1 and P_2 is called a **line segment.** We shall designate the line segment by $\overline{P_1P_2}$. A line segment has the important property of length. If we designate the endpoints of a line segment $\overline{P_1P_2}$ with the pairs of coordinates (x_1, y_1) and (x_2, y_2), as in Figure A.6, the length, or measure, of the line segment, which we shall designate by P_1P_2 or some variable such as d, can be determined by an application of the Pythagorean theorem (see Section A.8). From the right triangle shown, we observe that the lengths of the perpendicular sides are $y_2 - y_1$ and $x_2 - x_1$; the square of the distance between P_1 and P_2 is given by

$$d^2 = (x_2 - x_1)^2 + (y_2 - y_1)^2.$$

By considering only the positive square root of the right-hand member, we have that

$$\mathbf{d = \sqrt{(x_2 - x_1)^2 + (y_2 - y_1)^2}.}$$

Example The distance between the pair of points (1, 3) and (5, 6) is given by

$$d = \sqrt{(5-1)^2 + (6-3)^2}$$
$$= \sqrt{16 + 9} = \sqrt{25} = 5.$$

As with first-degree equations in two variables, solutions of higher-degree equations in two variables are also ordered pairs, which can be found by arbitrarily assigning values to x and finding the associated values of y. For example, in the equation

$$y = x^2 + 2,$$

If $x = -2$, then

$$y = (-2)^2 + 2 = 6,$$

and $(-2, 6)$ is a solution. Similarly, by replacing x with $-1, 0, 1$, and 2, we obtain $(-1, 3)$, $(0, 2)$, $(1, 3)$, and $(2, 6)$ as additional solutions. Plotting these points on the plane, we have the graph in Figure A.7a. Clearly, these points do not lie on a straight line. By plotting additonal solutions of $y = x^2 + 2$, such as

$$\left(-\frac{3}{2}, \frac{17}{4}\right), \left(-\frac{1}{2}, \frac{9}{4}\right), \left(\frac{1}{2}, \frac{9}{4}\right), \text{ and } \left(\frac{3}{2}, \frac{17}{4}\right),$$

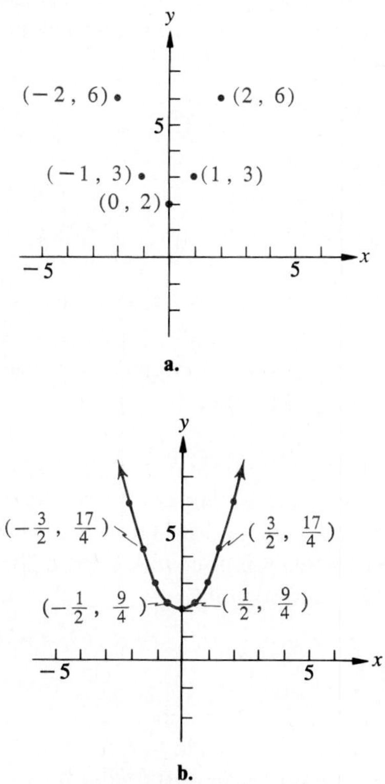

Figure A.7

we have the additional points in Figure A.7b. These points can now be

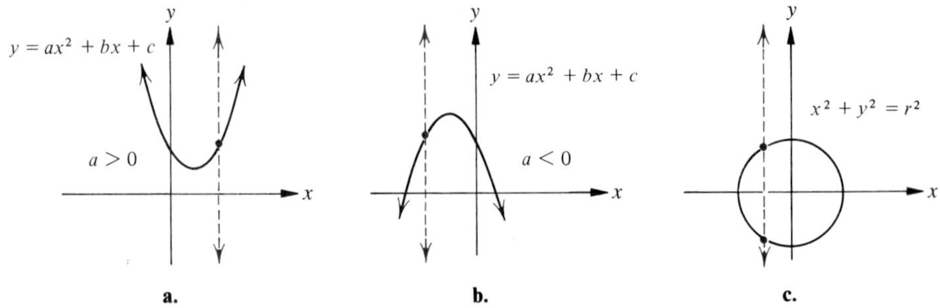

Figure A.8

connected in sequence from left to right by a smooth curve as shown. We assume that this curve is a good approximation to the graph of $y = x^2 + 2$.

More generally, the graph of any equation of the form

$$y = ax^2 + bx + c \quad (a, b, c, x \in R,\ a \neq 0),$$

is called a **parabola.** Notice that such an equation defines a function, since with each x, an equation of this form will associate one and only one y. Such functions are called **quadratic functions.**

The graph of

$$x^2 + y^2 = r^2,$$

where r is a real number greater than zero, is a **circle** with center at the origin and radius of length r. However, although this equation defines a relation, it does not define a function.

Whether or not a relation is also a function can be determined by inspecting its graph. For example, consider the graphs of equations $y = ax^2 + bx + c$ and $x^2 + y^2 = r^2$ in Figure A.8. Imagine a vertical line moving across each graph from left to right. Since the line cuts the graph of $y = ax^2 + bx + c$ at *only one point at each position, this equation defines a function;* that is, for each x in this relation, there is only one y. Since the line cuts the graph of $x^2 + y^2 = r^2$ at *more than one point at certain positions, the equation does not define a function,* although it does define a relation. There are ordered pairs in the relation which have the same first components and different second components.

A.5 Zeros of Functions; Equations in One Variable

In any function, the element(s) in the domain that are paired with the element 0 in the range are called **zeros of the function.**

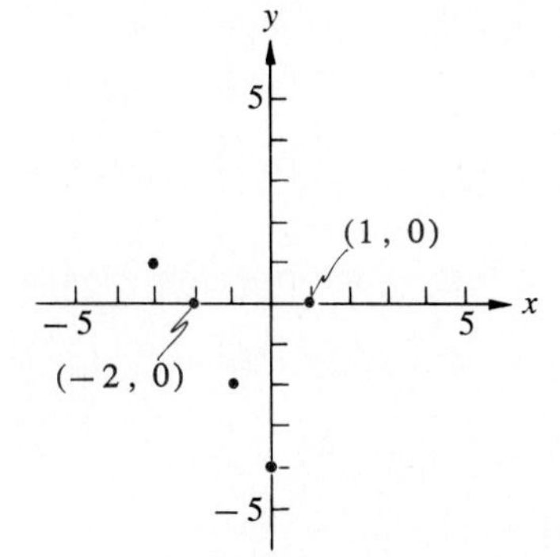

Figure A.9

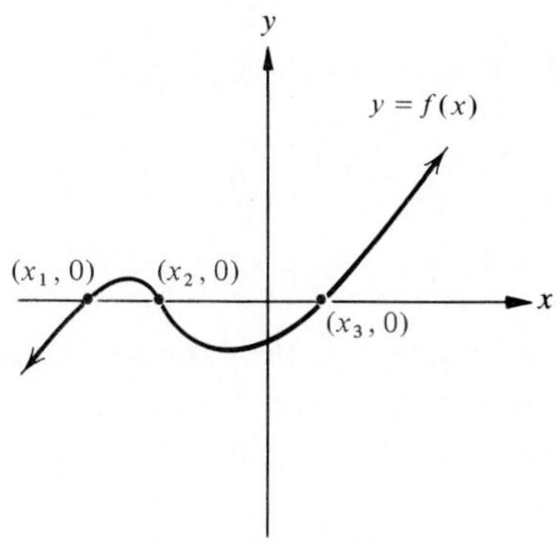

Figure A.10

Example The zeros of

$$f = \{(-3, 1), (-2, 0), (-1, -2), (0, -4), (1, 0)\}$$

are -2 and 1 because each of these first components is paired with a second component 0.

The graph of each zero in the above example lies on the x-axis, as shown in Figure A.9.

For any function

$$y = f(x),$$

the zeros of the function are values of x (x_1, x_2, x_3 in Figure A.10) for which

$$y = f(x) = 0.$$

These values of x are the solutions of the equation in one variable, $f(x) = 0$, and the set of such numbers is the solution set of the equation.

Note that the ordered pairs $(x_1, 0)$, $(x_2, 0)$ and $(x_3, 0)$ are the coordinates of the points of intersection of the graph of $y = f(x)$ and the x-axis. Thus, there are three different names for a single idea.

1. The zeros of $y = f(x)$.
2. The x intercepts of the graph of $y = f(x)$.
3. Solutions of $f(x) = 0$.

In many simple cases, such as

$$x - 2 = 0,$$

the solution set of an equation in one variable is evident by inspection. However, this is not the case for most first-degree equations. Recall from your work in algebra that solutions of first-degree equations in one variable can usually be found by generating **equivalent equations** (a set of equations with the same solution set) until you obtain an equation whose solution set is evident by inspection.

In general, if each member of an equation is multiplied by the same nonzero number, or if the same quantity is added to both members, the result is an equation equivalent to the original equation.

Solutions of quadratic equations in one variable can be obtained by various methods. One of these is *solution by factoring*, which depends upon the zero factor principle:

A product is equal to zero if and only if one or more of its factors equals zero.

Example We can solve $x^2 + x = 30$, by first writing the equation equivalently as

$$x^2 + x - 30 = 0, \tag{1}$$

in which form the right-hand member is zero. Factoring the left-hand member yields

$$(x + 6)(x - 5) = 0. \tag{2}$$

The left-hand member equals zero for values of x for which

$$x + 6 = 0 \quad \text{or} \quad x - 5 = 0 \quad \text{or both.}$$

By inspection, the solution set of (2) and, therefore, of (1) is $\{-6, 5\}$.

Quadratic equations of the form

$$x^2 = a \ (a \geq 0)$$

can be solved by a method called *extraction of roots*. If the equation has a solution, then x must be a square root of a (see Definition A.4). Since each positive real number a has two square roots, the solution set, for $a > 0$, is $\{\sqrt{a}, -\sqrt{a}\}$. If $a = 0$, the solution set is $\{0\}$.

Example The method of extraction of roots can be used to find the solution set of the equation $4x^2 = 24$. Multiplying each member by 1/4 yields

$$x^2 = 6.$$

Hence, $x = \sqrt{6}$ or $x = -\sqrt{6}$ and the solution set is $\{\sqrt{6}, -\sqrt{6}\}$.

From the method of extraction of roots applied to the equation

$$(x - a)^2 = b \ (b \geq 0).$$

we obtain

$$x - a = \sqrt{b} \quad \text{and} \quad x - a = -\sqrt{b},$$

from which the solution set $\{a + \sqrt{b}, a - \sqrt{b}\}$ is obtained. If $b = 0$, the solution set is $\{a\}$. From this it follows that we can find the solution set of any quadratic equation by first rewriting the equation in the form

$$(x - a)^2 = b.$$

This procedure of solving a quadratic equation is called *completing the square*.

Applying this procedure to the general quadratic equation

$$ax^2 + bx + c = 0 \ (a \neq 0)$$

results in the formula

$$x = \frac{-b \pm \sqrt{b^2 - 4ac}}{2a},$$

called the **quadratic formula,** which gives us the solutions of any quadratic equation in the form $ax^2 + bx + c = 0$ expressed in terms of a, b, and c. The solution set is

$$\left\{\frac{-b + \sqrt{b^2 - 4ac}}{2a}, \frac{-b - \sqrt{b^2 - 4ac}}{2a}\right\}.$$

If $a, b, c \in R$, the elements in the solution set are real numbers if and only if $b^2 - 4ac \geq 0$.

Example The equation $x^2 - 5x = 6$ can be solved by using the quadratic formula as follows. First, rewrite the equation as

$$x^2 - 5x - 6 = 0.$$

Then, replacing a with 1, b with -5, and c with -6 in the quadratic formula yields

$$x = \frac{-(-5) \pm \sqrt{(-5)^2 - 4(1)(-6)}}{2(1)},$$

from which

$$x = \frac{5 + \sqrt{25 + 24}}{2} = 6 \quad \text{or} \quad x = \frac{5 - \sqrt{25 + 24}}{2} = -1,$$

and the solution set is $\{6, -1\}$.

A.6 Inverse of a Function

If the components of each ordered pair in a given relation are interchanged, the resulting relation is called the **inverse** of the original relation.

Example $\{(1, 3), (5, 7), (11, 13)\}$ and $\{(3, 1), (7, 5), (13, 11)\}$ are inverse relations.

The inverse of a relation $\mathscr{R}$ is denoted by $\mathscr{R}^{-1}$ (read "$\mathscr{R}$ inverse" or "the inverse of $\mathscr{R}$"). It is evident from the above example that the domain and range of $\mathscr{R}^{-1}$ are the range and domain, respectively, of $\mathscr{R}$. We have used the script $\mathscr{R}$ for a relation so that it should not be confused with the R that represents the set of real numbers. Furthermore, it should be understood that -1 in the inverse notation symbol $\mathscr{R}^{-1}$ is not an exponent.

The inverse of the relation

$$\mathscr{R}: y = 3x - 5 \tag{1}$$

can be obtained by replacing x with y and y with x. Hence, the inverse relation of (1) is given by

$$\mathscr{R}^{-1}: x = 3y - 5 \tag{2}$$

or, when y is expressed in terms of x in the defining equation,

$$\mathscr{R}^{-1}: y = \frac{1}{3}(x + 5). \tag{2'}$$

Notice in the above example that $y = 3x - 5$ and $x = 3y - 5$ from (1) and (2) define inverse relations. The domains and ranges of the relations are interchanged because the variables have been interchanged. However, $x = 3y - 5$ from (2) and $y = \frac{1}{3}(x + 5)$ from (2′) are equivalent equations and define the same relation.

The graphs of inverse relations are always located symmetrically with respect to the graph of the linear equation $y = x$. To see this, notice the graphs of the ordered pairs (a, b) and (b, a) in Figure A.11. They are the same distance from, but on opposite sides of, the line that is the graph of the equation $y = x$. The graph of $y = x$ serves as a reflecting line, or mirror, for these points.

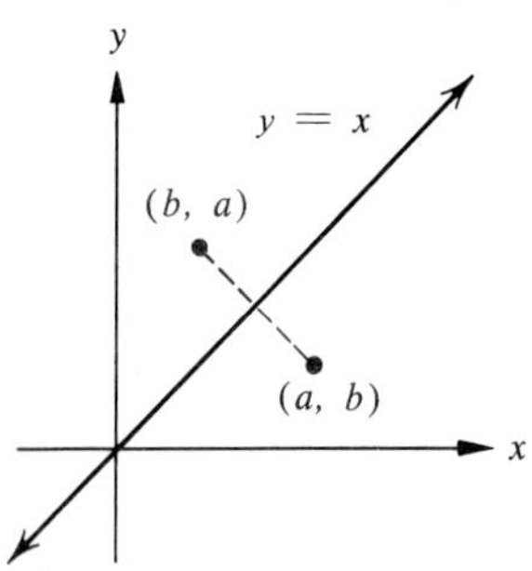

Figure A.11

Using the example above, Figure A.12 shows the graphs of

$$y = 3x - 5$$

and its inverse,

$$x = 3y - 5 \qquad \text{or} \qquad y = \frac{1}{3}(x + 5),$$

together with the graph of $y = x$.

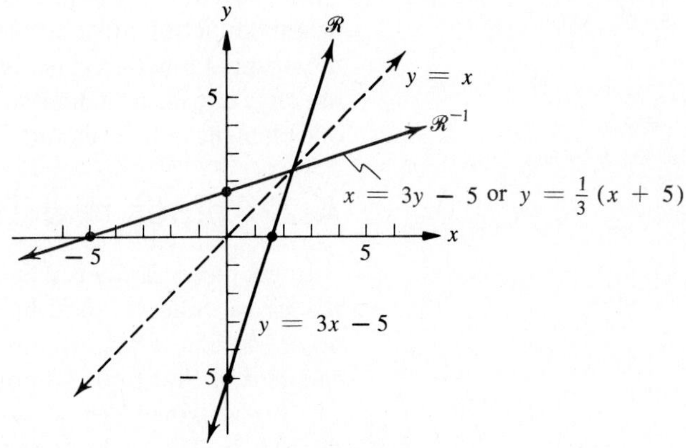

Figure A.12

Because every function is a relation, *every function has an inverse, but the inverse is not always a function.* For example, consider the function

$$y = x^2. \tag{3}$$

Its graph and the graph of its inverse,

$$x = y^2 \qquad \text{or} \qquad y = \pm\sqrt{x}. \tag{4}$$

are shown in Figure A.13. Since for all but one value of $x(x = 0)$, the inverse specified by (4) associates two different values of y (one positive and one negative) in the range, the inverse of (3) is not a function. Observe in Figure A.13 that the graphs of these relations are symmetric with respect to the graph of $y = x$.

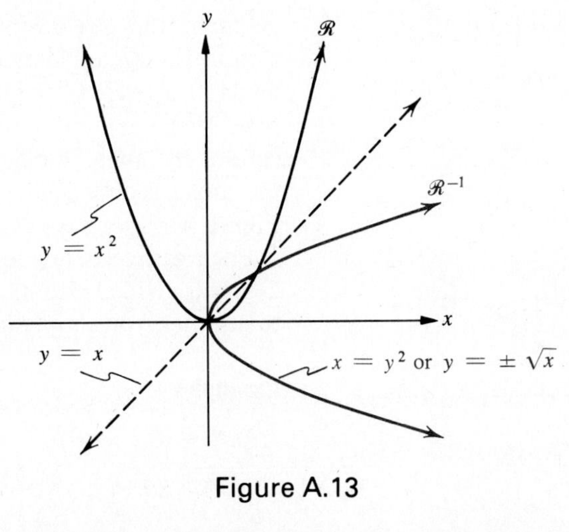

Figure A.13

In order for a function to have an inverse that is also a function, the original function must be *one-to-one;* that is, each element in the domain of the original function must be associated with one and only one element in its range and each element in its range must be associated with one and only one element in its domain.

A.7 Angles and Their Measure

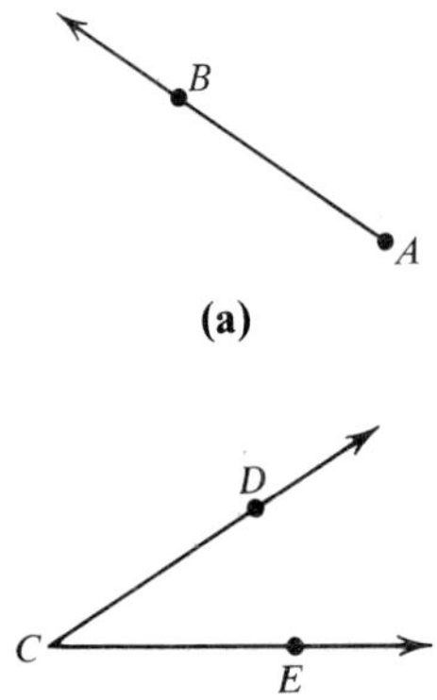

Figure A.14

In geometry, a **ray** is a figure such as that shown in Figure A.14a, where the arrow indicates that the ray extends infinitely far in one direction. The point A (points are usually named with capital letters) in A.14a is the **endpoint** of that ray, and point c is the endpoint of the two rays in A.14b.

A ray is named by first naming its endpoint and then one other point on the ray. In Figure A.14a the ray is named by $\overrightarrow{AB}$ (read "ray AB"), and in A.14b the rays are named by $\overrightarrow{CD}$ and $\overrightarrow{CE}$.

An **angle** is the union of two rays having a common endpoint called the **vertex** of the angle. The rays are the **sides** of the angle. In Figure A.14b, an angle is formed by $\overrightarrow{CD}$ and $\overrightarrow{CE}$, where $\overrightarrow{CD}$ and $\overrightarrow{CE}$ are the sides, and point C is the vertex.

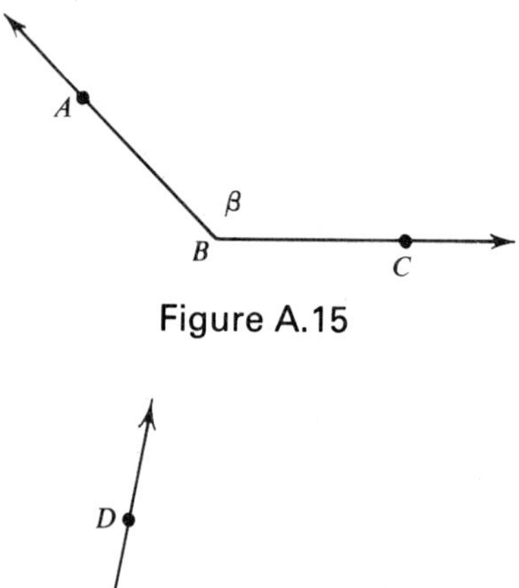

Figure A.15

An angle can be named in several ways. For example, if a point is the vertex of only one angle, as in Figure A.15, then that letter can be used to name the angle; in this case, $\angle B$. Also, an angle can be named using a single letter placed "inside" the angle, such that β (beta), or by using three letters, where the middle letter names the vertex, such as $\angle ABC$ or $\angle CBA$.

If a point is the vertex of more than one angle, as in Figure A.16, each angle can also be named either by using three letters, such as $\angle DEF$, $\angle DEG$, or $\angle GEF$, or by using a single letter "inside" the angle, such as α (alpha), or γ (gamma). However, since three angles have the vertex E, "$\angle E$" cannot be used since it would not clearly name any one of them.

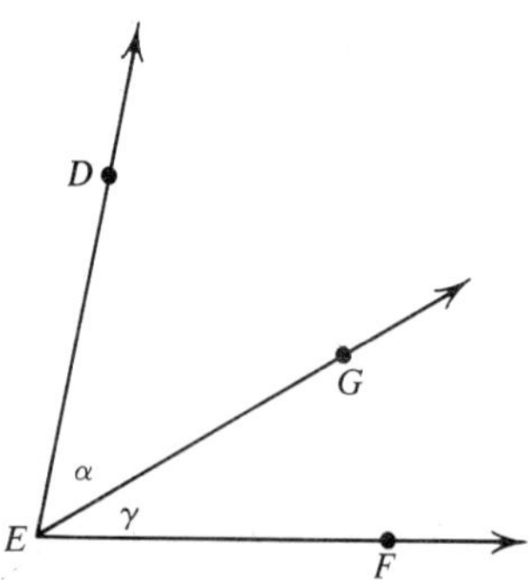

Figure A.16

Angles can be measured by various units of measurement, but in geometry they are commonly measured in **degrees.** Such measures can be assigned to an angle by using a circle whose center is at the vertex of the angle. If a circle is divided into 360 arcs of equal length (Figure A.17 on page 317), the number of arcs intercepted by a central angle α is the measure of the angle in degree units and is designated by $\alpha°$ (the measure of the angle is independent of the radius of the measuring circle). From Figure A.17, we have that $\alpha = 40°$.

Each degree is divided into sixty minutes (60′) and each minute of a degree is divided into sixty seconds (60″). To convert from tenths of degrees to minutes, you simply multiply the number of tenths by 60.

Examples

a. $0.4° = (0.4 \times 60)'$

$= 24'$

b. $35.2° = 35° + (0.2 \times 60)'$

$= 35°\ 12'$

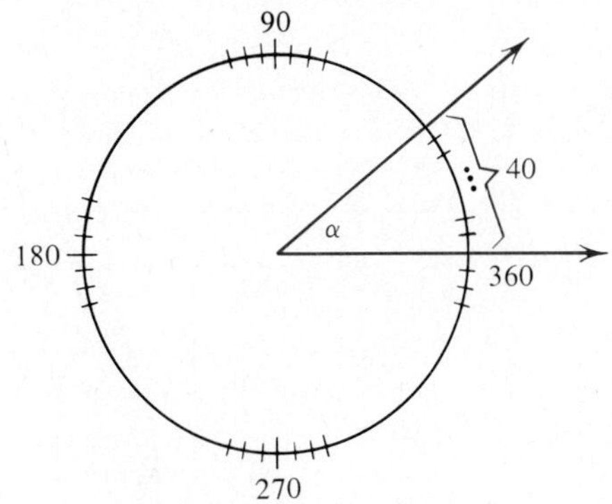

Figure A.17

Conversely, to convert from minutes to tenths of degrees, simply divide the number of minutes by 60.

Examples

a. $36' = \left(\frac{36}{60}\right)^\circ$
$= 0.6^\circ$

b. $75^\circ\ 18' = 75^\circ + \left(\frac{18}{60}\right)^\circ$
$= 75.3^\circ$

A **protractor,** usually in the shape of a semicircle, as shown in Figure A.18, with 180 equally spaced marks on the semicircle, can be used to measure angles. If an angle is placed with its vertex at the center of the semicircle with one side passing through the 0° mark, the measure of the angle can be read directly.

Examples In Figure A.18,

$$\angle ABC = 50^\circ,$$

$$\angle ABE = 110^\circ,$$

and

$$\angle DBE = 70^\circ.$$

Also

$$\angle CBE = 110^\circ - 50^\circ = 60^\circ.$$

If the measure of an angle is less than 90°, the angle is an **acute angle;** if more than 90° and less than 180°, it is an **obtuse angle;** if equal to 90°, it is a **right angle.** If two lines, or rays, form a right angle, the lines, or rays, are **perpendicular,** indicated by the symbol $\perp$. Thus $\overrightarrow{DE} \perp \overrightarrow{FG}$ is read "ray DE is perpendicular to ray FG."

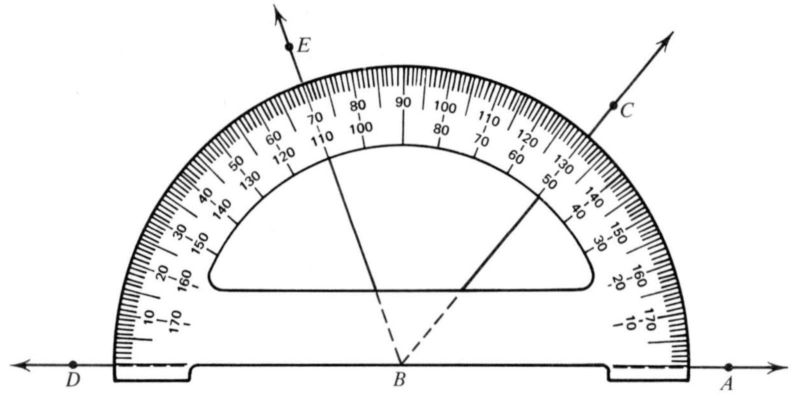

Figure A.18

Examples In the figure, α is an acute angle, β is an obtuse angle, and γ is a right angle. Also, $\overrightarrow{AC} \perp \overrightarrow{AB}$.

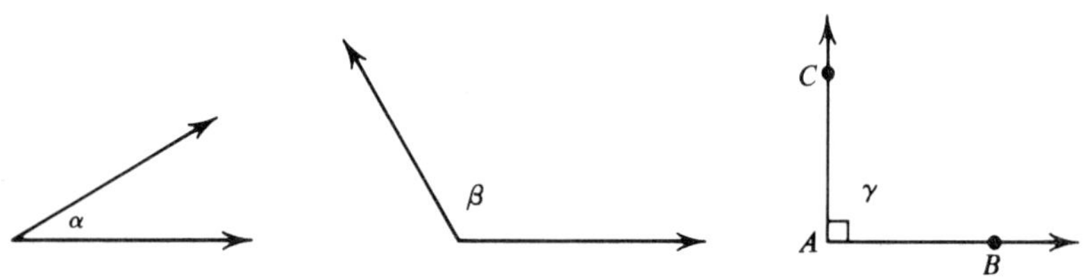

If the sum of the measures of two angles is 90°, the angles are **complementary;** if the sum is 180°, the angles are **supplementary.**

Examples

a. If $\angle D = 32°$, $\angle E = 148°$, and $\angle F = 58°$, then $\angle D$ and $\angle F$ are complementary, and $\angle D$ and $\angle E$ are supplementary.

b. In the figure, angles α and β are complementary, and angles α and θ (theta) are supplementary.

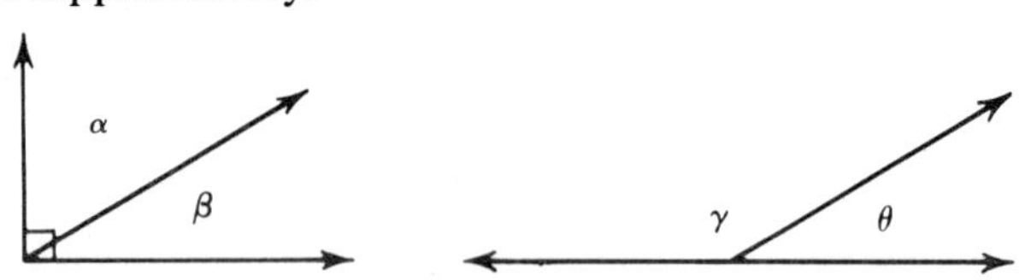

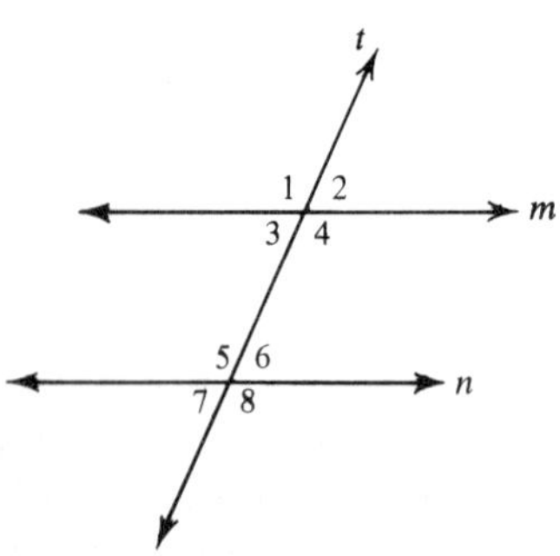

Figure A.19

Frequently, pairs of supplementary angles are introduced through the use of parallel lines. In Figure A.19, in which lines m and n are intersected by a transversal, line t, angles 1, 2, 7, and 8 are called **exterior angles** and angles 3, 4, 5, and 6 are called **interior angles.** Interior angles on opposite sides of the transversal are called **alternate interior angles** and exterior angles on opposite sides of the transversal are called **alternate exterior angles.** For example, pairs of angles 3 and 6, and 4 and 5, are alternate interior angles and pairs of angles 1 and 8, and 7 and 2 are alternate exterior angles.

Angles that are on the same side of the transversal and also in the same relative position in relation to the parallel lines are called **corresponding angles.** For example, pairs of angles 1 and 5; 2 and 6; 3 and 7; and 4 and 8 are corresponding angles.

If lines m and n are parallel, it follows that alternate interior angles are equal and alternate exterior angles are equal, as are corresponding angles. Also interior angles on the same side of the transversal are supplementary, as are exterior angles on the same side of the transversal.

A.8 Triangles

A **triangle** is a figure formed by three line segments joining three points that are not in the same line. The line segments are the **sides,** and the three points are the **vertices** (plural of **vertex**) of the triangle. The letters assigned to the vertices can be used to name a triangle.

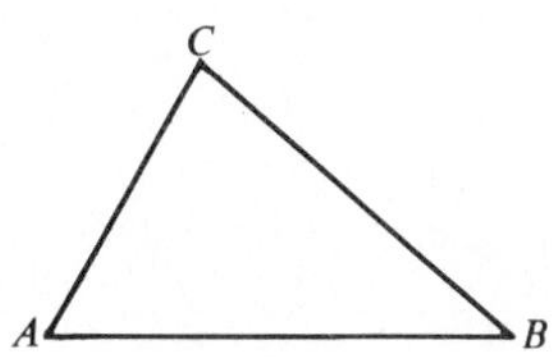

Example In the figure, the points A, B, and C are the vertices of $\triangle ABC$ (read "triangle ABC"), which can also be named by $\triangle ACB$, $\triangle BAC$, $\triangle BCA$, $\triangle CAB$, or $\triangle CBA$. The sides are $\overline{AB}$, $\overline{BC}$, and $\overline{CA}$; their lengths (positive real numbers) can be represented by AB, BC, and CA, respectively.

If each angle of a triangle is acute, the triangle is an **acute triangle;** if one angle is obtuse, it is an **obtuse triangle;** if one angle is a right angle, it is a **right triangle.** In a right triangle, the side opposite the right angle is the **hypotenuse,** and the other sides are the **legs.**

Examples In the figure, $\triangle DEF$ is an acute triangle, $\triangle KGH$ is an obtuse triangle, and $\triangle PQR$ is a right triangle. In $\triangle PQR$, $\overline{PQ}$ and $\overline{PR}$ are the legs, and $\overline{RQ}$ is the hypotenuse.

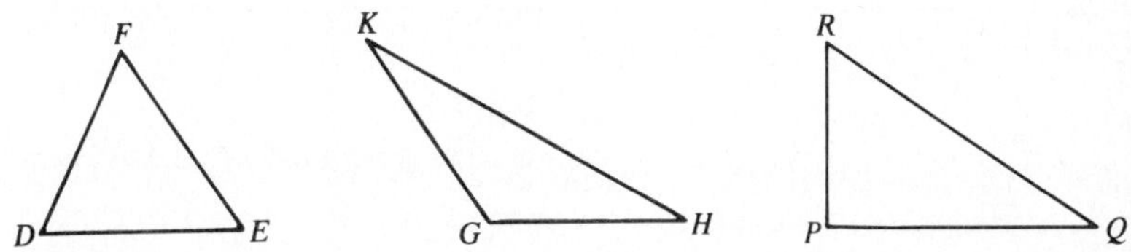

Triangles also have names related to the relative lengths of the sides. If the three sides are of different lengths, the triangle is a **scalene triangle;** if two sides have equal lengths, it is an **isosceles triangle;** if all three sides have equal lengths, it is an **equilateral triangle.**

Examples In the figure, $\triangle ABC$ is scalene, $\triangle DEF$ is isosceles $(FD = FE)$, $\triangle HKL$ is equilateral $(HK = KL = LH)$. Note, further, that $\triangle ABC$ is an obtuse triangle, whereas $\triangle DEF$ and $\triangle HKL$ are both acute triangles.

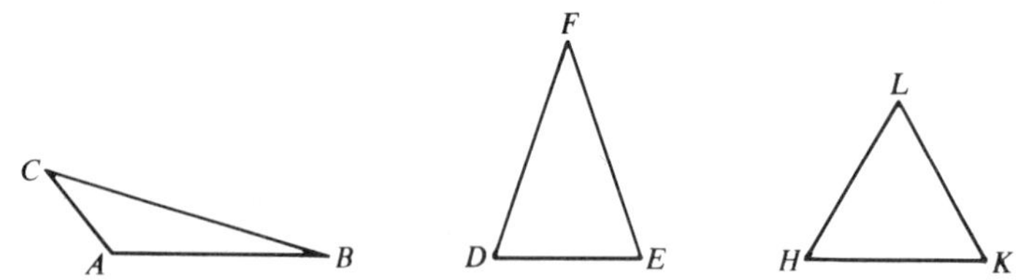

An **altitude** of a triangle is the length of a segment from any vertex of a triangle perpendicular to the opposite side. Sometimes the segment itself is called the altitude.

Examples In the figure, in $\triangle ABC$, h, or $\overline{CD}$, is the altitude to AB. In $\triangle EFG$, GH is the altitude from G, and $\overline{FL}$ is the altitude from F.

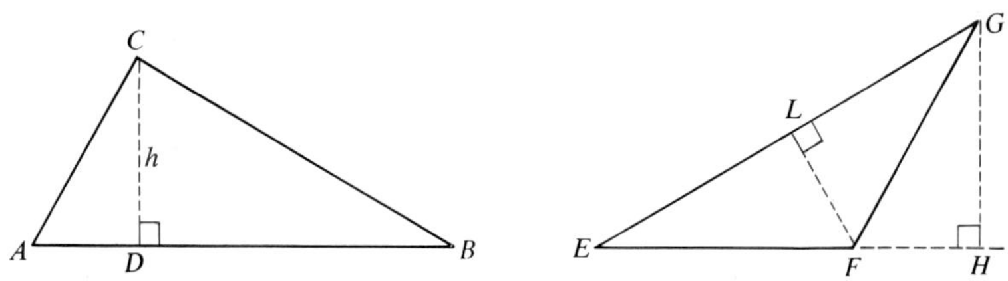

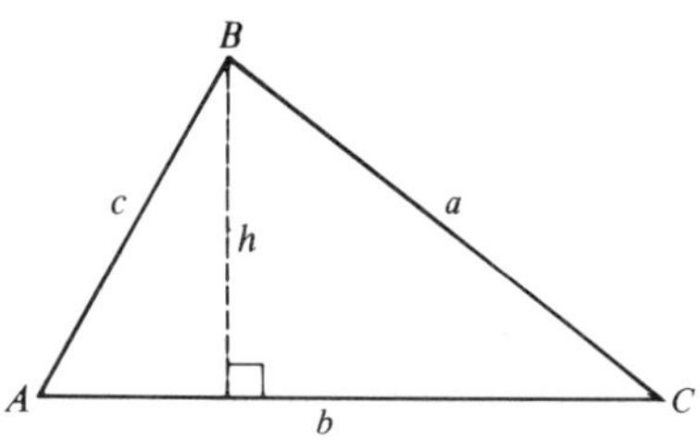

Figure A.20

The following properties hold for any triangle (see Figure A.20).

1. The perimeter is $P = a + b + c$.
2. The area is $\mathscr{A} = \frac{1}{2}bh$.
3. $\angle A + \angle B + \angle C = 180°$.

As in Properties 1 and 2, lowercase letters are often used to indicate the lengths of the sides opposite the angles named by the corresponding capital letters.

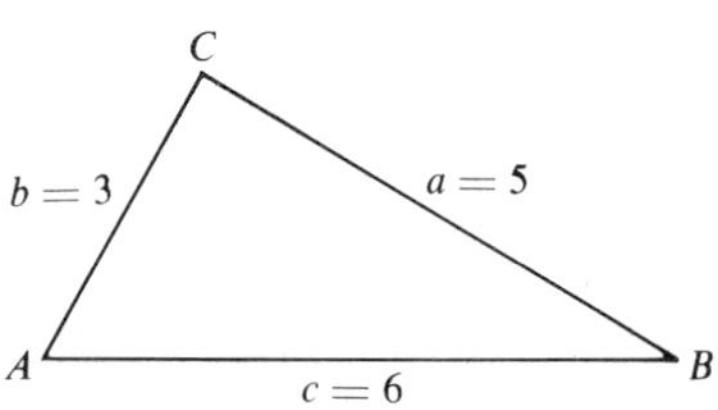

a.

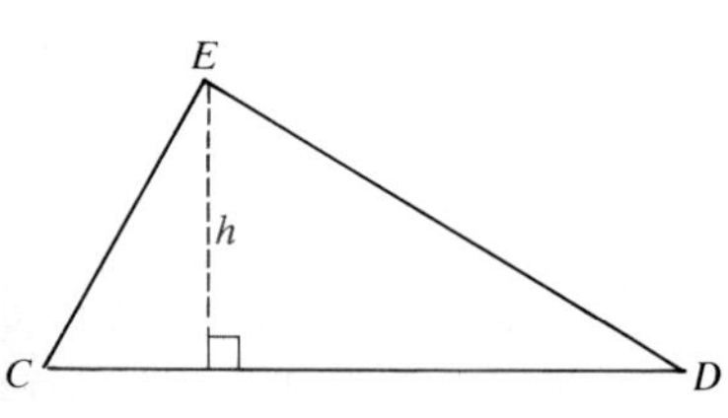

b.

Examples

a. The perimeter of $\triangle ABC$ is

$$P = 3 + 5 + 6 = 14$$

b. If $h = 6$ and $CD = 10$, the area of $\triangle CDE$ is

$$\mathscr{A} = \frac{1}{2}(10)(6) = 30.$$

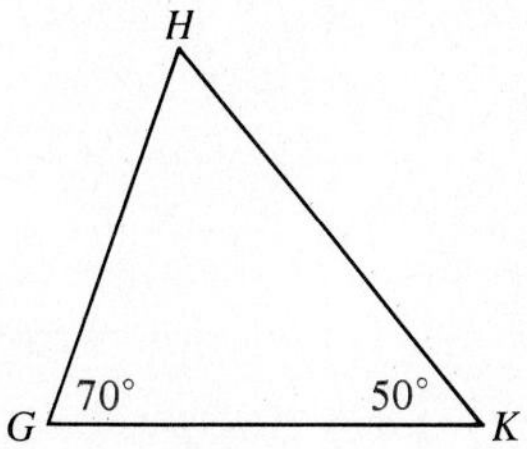

c.

c. If $\angle G = 70°$ and $\angle K = 50°$,

$$70° + 50° + \angle H = 180°,$$

from which we have that $\angle H = 60°$.

In addition to Properties 1, 2, and 3, the following properties also apply to certain special triangles.

4. In an isosceles triangle, the angles opposite the equal sides are equal.

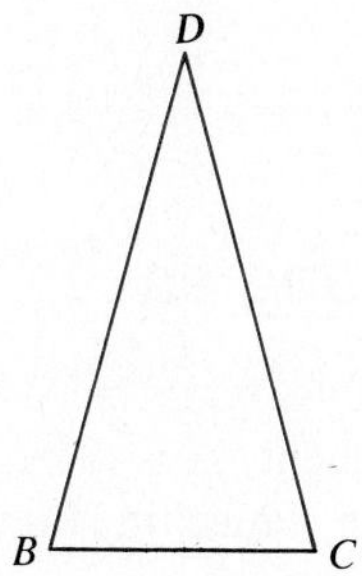

Example If $BD = CD$, then $\angle B = \angle C$.

5. If a triangle is equilateral, then (see Figure A.21)

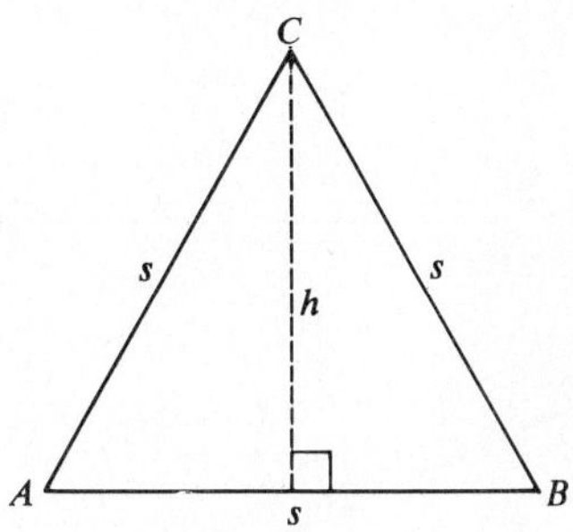

Figure A.21

$$\angle A = \angle B = \angle C = 60°, \quad h = \frac{s}{2}\sqrt{3}, \quad \text{and} \quad \mathscr{A} = \frac{\sqrt{3}}{4}s^2.$$

Example If $s = 12$, then

$$h = \frac{12}{2}\sqrt{3} = 6\sqrt{3} \quad \text{and} \quad \mathscr{A} = \frac{\sqrt{3}}{4}(12)^2 = 36\sqrt{3}.$$

6. In a right triangle ($\angle C = 90°$),
   (i) $a^2 + b^2 = c^2$ (the Pythagorean theorem).
   (ii) If $a = b$, then $\angle A = \angle B = 45°$, and $c = a\sqrt{2}$.
   (iii) If one acute angle is 30°, and the length of the opposite side is a, then the other acute angle is 60°, the length of the hypotenuse is $2a$, and the length of the longer leg is $a\sqrt{3}$.

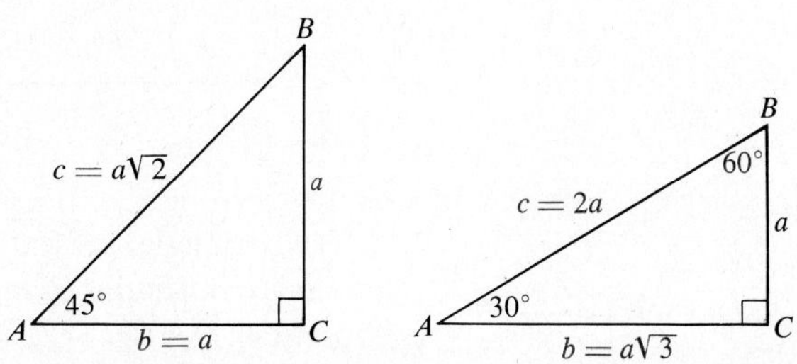

Examples Using the measurements of the appropriate triangle, we obtain

a. $c^2 = 3^2 + 4^2 = 9 + 16 = 25$; $c = 5$

b. $\angle A = \angle B = 45°$; $c = 5\sqrt{2}$

c. $\angle A = 30°$; $c = 2 \cdot 4 = 8$; $b = 4\sqrt{3}$

(continued)

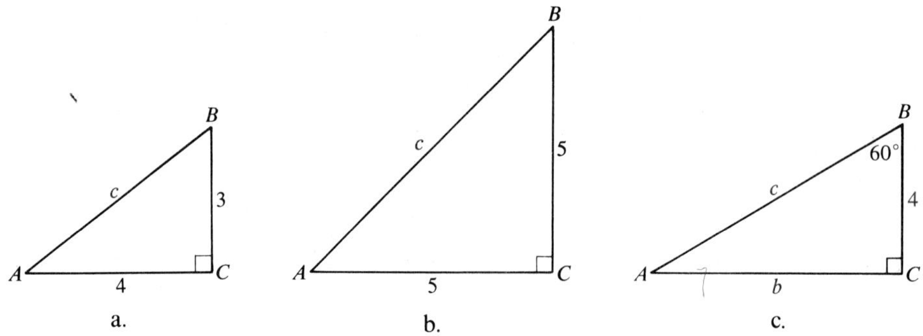

A.9 Congruent and Similar Triangles

If two triangles are the same size and shape, they are said to be **congruent triangles.** In Figure A.22, the triangles are congruent. This is indicated by use of the symbol ≅ (read "is congruent to"). Thus, in Figure A.22, $\triangle PQR \cong \triangle STV$. From this it follows that

$$\angle P = \angle S, \ p = s,$$
$$\angle Q = \angle T, \ q = t,$$
$$\angle R = \angle V, \ r = v.$$

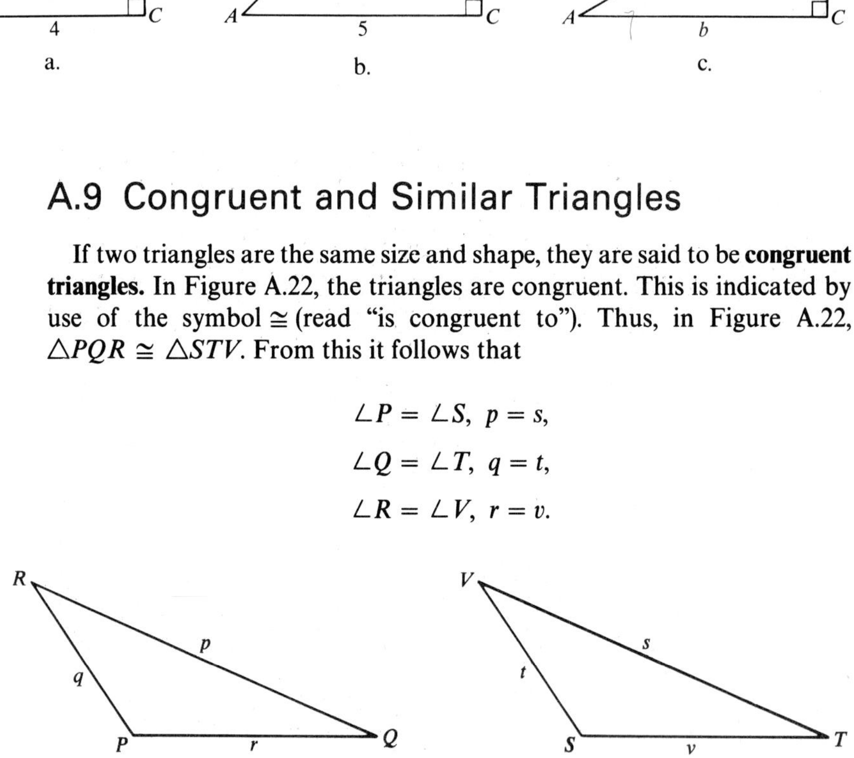

Figure A.22

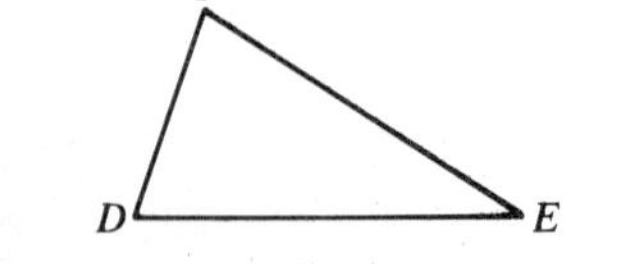

Figure A.23

If two triangles are the same shape, but not necessarily the same size, they are said to be **similar triangles.** In Figure A.23, the triangles are similar. This is written $\triangle ABC \sim \triangle DEF$, where "$\sim$" is read "is similar to." In two similar triangles, the corresponding angles are equal—that is, $\angle A = \angle D$, $\angle B = \angle E$, $\angle C = \angle F$; and the corresponding sides are proportional—that is,

$$\frac{AC}{DF} = \frac{AB}{DE} = \frac{BC}{EF}.$$

Example Given that $\triangle RST \sim \triangle ADF$, we can find AD as follows. Since the triangles are similar, we have

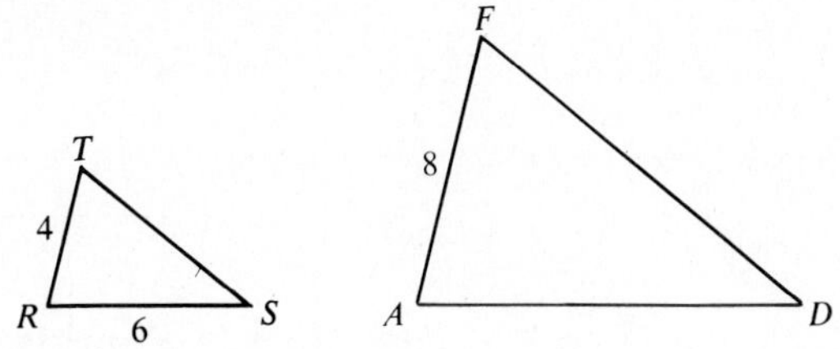

$$\frac{AD}{RS} = \frac{AF}{RT},$$

$$\frac{AD}{6} = \frac{8}{4},$$

from which

$$AD = 12.$$

The following properties can be used to prove that two triangles are similar.

1. two angles of one triangle are equal, respectively, to two angles of the other, or
2. one angle of one triangle is equal to one angle of the other, and the including sides are proportional, or
3. three sides of one triangle are proportional to three sides of the other.

Examples Using the measurements in the figures, we obtain

a. $\angle P = \angle M$ and $\angle Q = \angle N$;

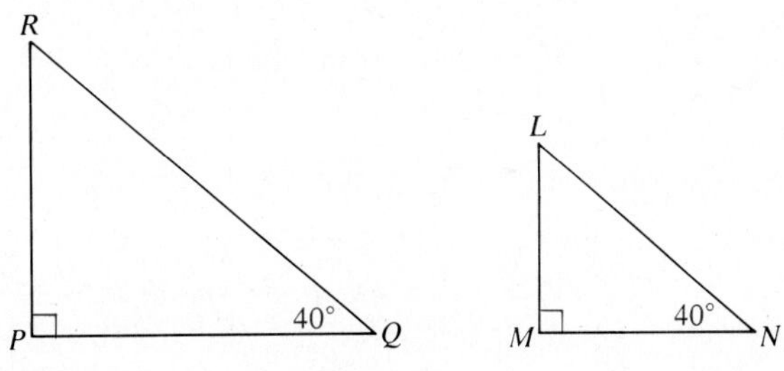

therefore, by Property 1, $\triangle PQR \sim \triangle MNL$.

b. $\angle A = \angle D$ and $\dfrac{AB}{DE} = \dfrac{AC}{DF}$, because $\dfrac{9}{6} = \dfrac{6}{4}$;

(continued)

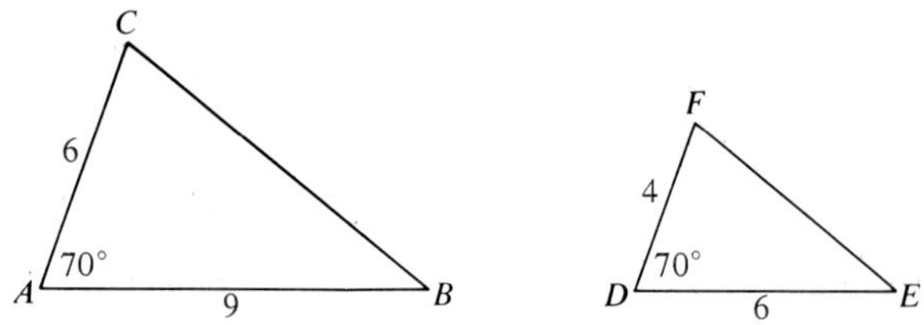

therefore, by Property 2, $\triangle ABC \sim \triangle DEF$.

c. $\dfrac{GK}{SW} = \dfrac{KH}{WT} = \dfrac{GH}{ST}$ because $\dfrac{8}{4} = \dfrac{10}{5} = \dfrac{12}{6}$;

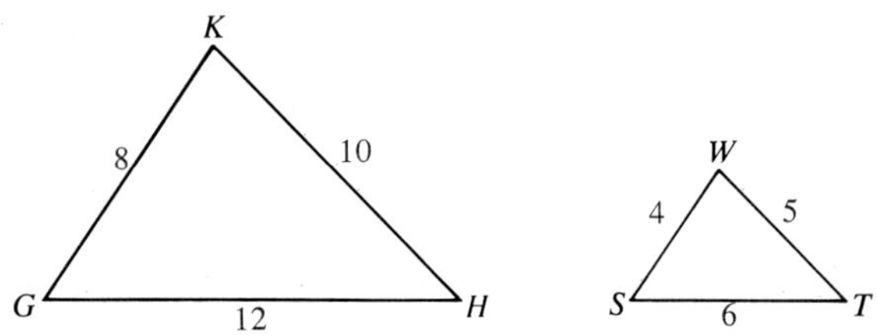

therefore, by Property 3, $\triangle GHK \sim \triangle STW$.

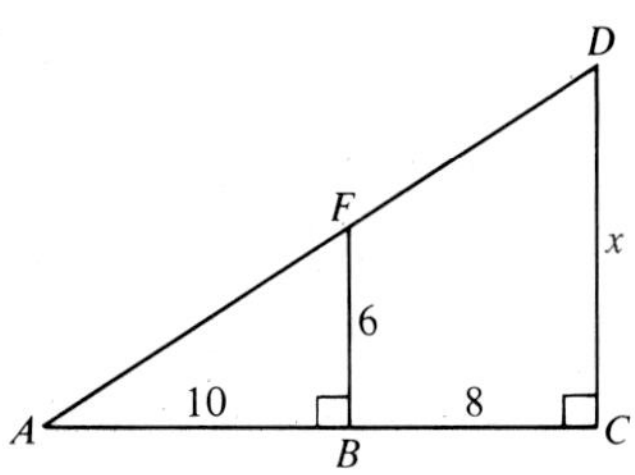

Example To find the length x in the figure, we proceed as follows. $\angle A = \angle A$ and $\angle ABF = \angle ACD$; hence, by Property 1, we have $\triangle ABF \sim \triangle ACD$. Also, $AC = AB + BC = 18$. Because the triangles are similar, the corresponding sides are proportional. Thus,

$$\frac{DC}{FB} = \frac{AC}{AB} \quad \text{and} \quad \frac{x}{6} = \frac{18}{10},$$

from which

$$x = \frac{6 \cdot 18}{10} = \frac{54}{5}.$$

A list of symbols introduced in Appendix A is shown inside the front cover.

B ROUNDING-OFF PROCEDURES

Figure B.1 shows the names of selected place values for the decimal number system: *whole number places* (to the left of the decimal point) are named to thousands; *decimal places* (to the right of the decimal point) are named to ten-thousandths. Familiarity with these place values is sufficient for our work.

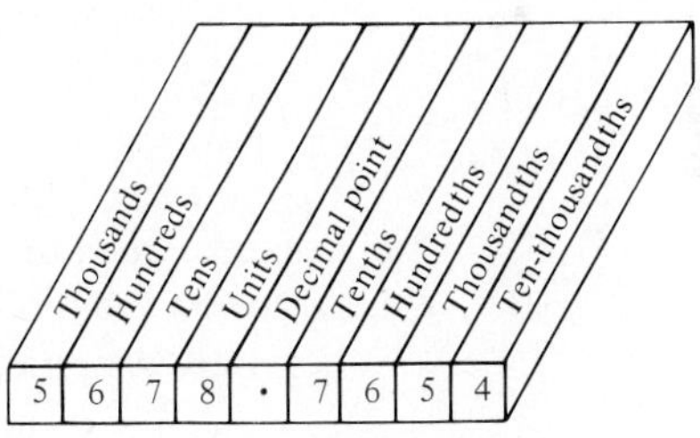

Figure B. 1

Frequently the result of a computation includes more digits than may be needed for a particular purpose. For example, the result of a calculation that involves the cost of an item might be $32.527. The actual cost, to the nearest cent, is $32.53. We say that 32.527 has been "rounded off to the nearest hundredth" (or "to the nearest cent"). Note that 32.527 is "nearer" to 32.53 than it is to 32.52.

The process shown below indicates the rounding-off process most commonly used. In this procedure, the *round-off digit* is the last digit to be retained; the *test digit* is the digit immediately to the right of the round-off digit. For example, to round off 318.04591 to the nearest *hundredth*, underline the round-off digit 4 in the hundredths place, then proceed as follows:

round-off digit ↓

$$318.0\underline{4}591$$

↑ test digit

If the test digit is 5 or more, the round-off digit is increased by 1, but if the test digit is less than 5, the round-off digit is not changed. Here the test digit is 5, hence 1 is added to the round-off digit.

$$318.0\overset{5}{\not4}591$$

The round-off digit is in a decimal place, so all digits to the right of the round-off digit are dropped.

$$318.0\overset{5}{\not4}\not5\not9\not1 \longrightarrow 318.05$$

Thus, 318.04591 is written as 318.05 to the nearest hundredth.

To round off 318.04591 to the nearest *hundred*, we have

round-off digit ↓

$$\underline{3}18.04591$$

↑ test digit

Here the test digit is less than 5, so the round-off digit is not changed. Also, because the round-off digit is to the left of the decimal point, the digits following are replaced with zeros, and all digits to the right of the decimal point are dropped. Thus, to the nearest hundred, we have

$$\underline{3}18.04591 \longrightarrow 3\overset{00}{\not1\not8}.\not0\not4\not5\not9\not1 \longrightarrow 300$$

Examples In each case the round-off digit is underlined.

a. Round off to the nearest ten.

$$3\underline{1}8.04591 \longrightarrow 3\overset{20}{\underline{\not1}\not8}.\not0\not4\not5\not9\not1 \longrightarrow 320$$

b. Round off to the nearest tenth.

$$318.\underline{0}4591 \longrightarrow 318.\underline{0}\not4\not5\not9\not1 \longrightarrow 318.0$$

c. Round off to the nearest thousandth:

$$318.04\underline{5}91 \longrightarrow 318.04\overset{6}{\underline{\not{5}}}\not{9}\not{1} \longrightarrow 318.046$$

Round-off instructions sometimes specify a given number of decimal places.

Examples

a. Round off to two decimal places.

$$21.5\underline{9}71 \longrightarrow 21.\overset{60}{\not{5}\underline{\not{9}}}\not{7}\not{1} \longrightarrow 21.60$$

b. Round off to three decimal places.

$$21.59\underline{7}1 \longrightarrow 21.59\underline{7}\not{1} \longrightarrow 21.597$$

Exact and Approximate Numbers

Any number arrived at by a counting process is an **exact number.** For example, if we say "60 students are enrolled in a class," the number 60 is exact. Exact numbers may also arise by definition. In the statement, "One hour equals 60 minutes," the number 60 is exact. On the other hand, a number that is arrived at by some process of measurement is an **approximate number.** Thus, if the distance between two cities, measured by driving from one city to the other, is given as 60 miles, this distance is an approximation to the actual distance, and the number 60 is an approximate number. The actual distance may be a little more or a little less than 60 miles.

Examples

a. "One dime is equal to 10 cents," the number 10 is *exact*; 10 is defined to be the number of cents in a dime.

b. "This sheet of postage stamps contains 10 stamps," the number 10 is *exact*; 10 is associated with the concept of counting.

c. "The reading shown on a voltmeter is 10 volts," the number 10 is *approximate*; 10 is a *measure* of voltage.

Significant Digits

When working with approximate numbers, the term **significant digits** is used to refer to those digits in a number that have meaning. The following rules can be used to decide which digits of an approximate number are significant.

SIGNIFICANT DIGITS OF APPROXIMATE NUMBERS

1. Nonzero digits (1, 2, 3, 4, 5, 6, 7, 8, 9) are always significant.
2. Zeros that are preceded and followed by nonzero significant digits are always significant.
3. When a number has no nonzero digits on the left of the decimal point, zeros between the decimal point and the first nonzero digit are not significant.
4. Final zeros on the right of a decimal point are significant.
5. Final zeros on a whole number are assumed to be not significant unless further information is available.

In the examples illustrating each rule, the significant digits are underlined.

Examples

a. $\underline{62}.\underline{18}$ has four significant digits. (Rule 1)

b. $0.0\underline{42}$ has two significant digits. (Rules 1 and 3)

c. $\underline{23}.\underline{0017}$ has six significant digits. (Rule 2)

d. $\underline{4005}$ has four significant digits. (Rule 2)

e. $\underline{7}.\underline{200}$ has four significant digits. (Rule 4)

f. $0.0\underline{160}$ has three significant digits. (Rules 3 and 4)

g. In the statement "Hawaii is 2400 miles west of Los Angeles," without further information on how the measurement was made, assume that $\underline{24}00$ has only two significant digits. (Rule 5)

h. If a measurement is given as 2400 miles to *the nearest mile*, then the number $\underline{2400}$ has four significant digits. (Rule 5)

Numbers can be rounded off to a specified number of significant digits by using the same procedure that is used to round off a number to a specified place value.

Examples

a. Round off to two significant digits.

$$\underline{58}.603 \longrightarrow \underline{5\overset{9}{\not8}}.\not6\not0\not3 \longrightarrow 59$$

b. Round off to four significant digits.

$$\underline{58}.\underline{603} \longrightarrow \underline{58}.\underline{60}\not3 \longrightarrow 58.60$$

C USING TABLES

In previous sections we obtained approximations to powers, square roots, reciprocals, and trigonometric function values using a scientific calculator. Approximations for these values can also be obtained by using tables that have been prepared for this purpose.

Table I in this appendix lists approximations to powers, square roots, and reciprocals of some natural numbers and can be used if a calculator is not available. Tables II and III list trigonometric function values.

Function Values for Acute Angles

Table II lists rational number approximations for function values for angles whose measures are given in radian units in intervals of 0.01^{R} and corresponding measures in degrees. (See extract of Table II.)

PART OF TABLE II

θ radians or Real Number x	**θ degrees**	**$\sin\theta$ or $\sin x$**	**$\cos\theta$ or $\cos x$**	**$\tan\theta$ or $\tan x$**	**$\csc\theta$ or $\csc x$**	**$\sec\theta$ or $\sec x$**	**$\cot\theta$ or $\cot x$**
0.45	25° 47′	0.4350	0.9004	0.4831	2.299	1.111	2.070
.46	26° 21′	.4439	.8961	.4954	2.253	1.116	2.018
.47	26° 56′	.4529	.8916	.5080	2.208	1.122	1.969
.48	27° 30′	.4618	.8870	.5206	2.166	1.127	1.921
.49	28° 04′	.4706	.8823	.5334	2.125	1.133	1.875
0.50	28° 39′	0.4794	0.8776	0.5463	2.086	1.139	1.830
.51	29° 13′	.4882	.8727	.5594	2.048	1.146	1.788
.52	29° 48′	.4969	.8678	.5726	2.013	1.152	1.747
.53	30° 22′	.5055	.8628	.5859	1.978	1.159	1.707
.54	30° 56′	.5141	.8577	.5994	1.945	1.166	1.668

Examples Find approximations for

a. $\cos 0.47^{\text{R}}$ b. $\cot 0.53^{\text{R}}$

Solutions From Table II,

a. $\cos 0.47^{\text{R}} \approx 0.8916$ b. $\cot 0.53^{\text{R}} \approx 1.707$

Table III lists rational number approximations for function values of selected angles over the interval $0° \leq \alpha \leq 90°$. Observe that the table reads from top to bottom in the left margin with function values identified at the tops of the columns. It reads from bottom to top in the right margin with function values identified at the bottoms of the columns. (See extract of Table III.)

Function values are, in general, irrational numbers, and the four-place decimals shown in the table are mostly rational number approximations.

Examples Find approximations for

a. $\sin 30° 12'$ b. $\cot 59.5°$

Solutions From Table III,

a. $\sin 30° 12' \approx 0.5030$ b. $\cot 59.5° \approx 0.5890$

Notice that in both tables the function values $\sin \theta$, $\tan \theta$, and $\sec \theta$ increase, while $\cos \theta$, $\cot \theta$, and $\csc \theta$ decrease, for increasing measures of θ where $0° \leq \theta \leq 90°$.

Tables II and III consist of function values for angles with the specified measures. To approximate function values for angle measures not listed in the tables, we can read the entry in the table for that angle measure nearest the angle measure with which we are concerned. Thus, from Table II,

$$\sin 0.514^{\text{R}} \approx \sin 0.51^{\text{R}} \approx 0.4882.$$

PART OF TABLE III

θ deg	deg min	sin θ	cos θ	tan θ	csc θ	sec θ	cot θ		
30.0	30 0	0.5000	0.8660	0.5774	2.0000	1.1547	1.7321	60 0	60.0
30.1	30 6	0.5015	0.8652	0.5797	1.9940	1.1559	1.7251	59 54	59.9
30.2	30 12	0.5030	0.8643	0.5820	1.9880	1.1570	1.7182	59 48	59.8
30.3	30 18	0.5045	0.8634	0.5844	1.9821	1.1582	1.7113	59 42	59.7
30.4	30 24	0.5060	0.8625	0.5867	1.9762	1.1594	1.7045	59 36	59.6
30.5	30 30	0.5075	0.8616	0.5890	1.9703	1.1606	1.6977	59 30	59.5
30.6	30 36	0.5090	0.8607	0.5914	1.9645	1.1618	1.6909	59 24	59.4
30.7	30 42	0.5105	0.8599	0.5938	1.9587	1.1630	1.6842	59 18	59.3
30.8	30 48	0.5120	0.8590	0.5961	1.9530	1.1642	1.6775	59 12	59.2
		cos θ	sin θ	cot θ	sec θ	csc θ	tan θ	deg min	θ deg

Function Values for All Angles

Because the trigonometric tables are restricted to acute angles, we have to use a reference angle in conjunction with these tables to find approximations for function values for angles that are not acute. We will use the same procedure that we used to find function values for integral multiples of 30°, 45°, and 60° in Section 1.4.

Examples Find an approximation for each function value.

a. csc 334° 36′ b. sec 128° 12′.

Solutions A sketch showing the reference angle is helpful.

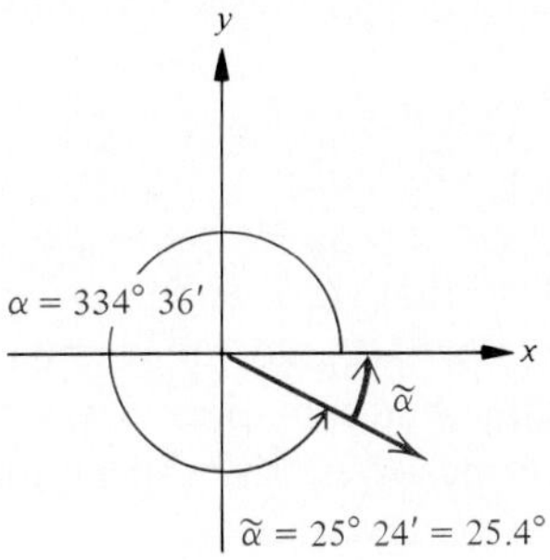

a. $$\tilde{\alpha} = 360° - 334°\,36' = 25°\,24'.$$

Also, since α is in Quadrant IV, csc 334° 36′ is negative. Therefore, using Table II, we obtain

$$\csc 334°\,36' = -\csc \tilde{\alpha}$$
$$= -\csc 25°\,24' \approx -2.3314.$$

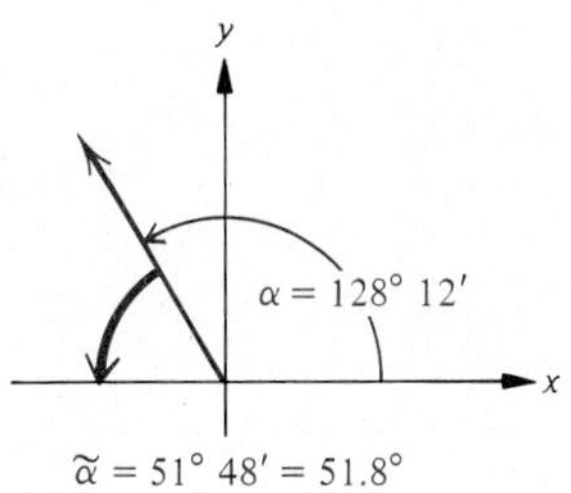

b. $$\tilde{\alpha} = 180° - 128°\,12' = 51°\,48'.$$

Since 128° 12′ is in Quadrant II, sec 128° 12′ is negative. Therefore, using Table II, we obtain

$$\sec 128°\,12' = -\sec 51°\,48' = -1.6171.$$

When the measure of an angle is given in radians, and a reference angle is needed, we have to use an approximation for π. We shall use 3.14.

Examples Find an approximation for each function value.

a. $\cos 1.95^{R}$ b. $\cot 5.36^{R}$

Solutions A sketch showing the reference angle is helpful.

a. Using $\pi = 3.14$, we have

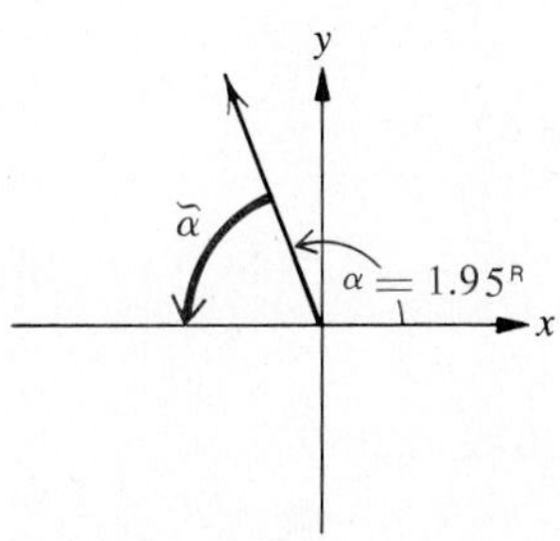

$$\tilde{\alpha} \approx (3.14 - 1.95)^{R} = 1.19^{R}.$$

Since α is in Quadrant II, $\cos 1.95^{R}$ is negative. Therefore, using Table III, we obtain

$$\cos 1.95^{R} = -\cos \tilde{\alpha}$$
$$\approx -\cos 1.19^{R} \approx -0.3717.$$

(continued)

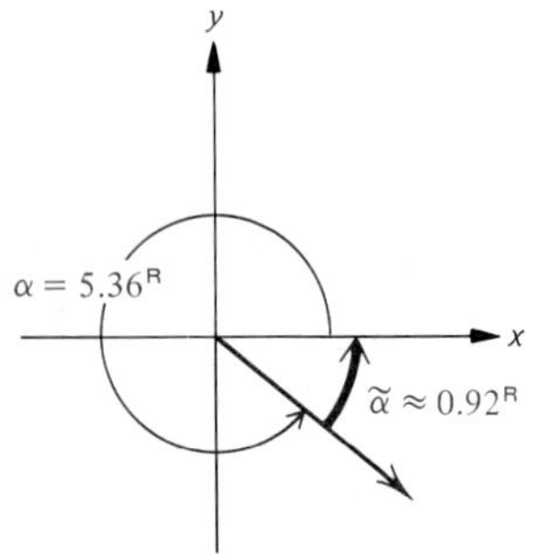

b. Using 6.28 for 2π, we obtain

$$\tilde{\alpha} \approx 6.28^{\mathrm{R}} - 5.36^{\mathrm{R}} = 0.92^{\mathrm{R}}.$$

Since 5.36^{R} is in Quadrant IV, cot 5.36^{R} is negative. Hence, using Table II, we obtain

$$\cot 5.36^{\mathrm{R}} \approx -\cot 0.92^{\mathrm{R}} \approx -0.7615.$$

The answers in the last example were obtained using $\pi \approx 3.14$. Different results will be obtained with other approximations for π such as 3.142 or 3.1416. In fact, if a calculator is used to find an approximation in Example a above, we obtain

$$\cos 1.95^{\mathrm{R}} \approx -0.3701808 \approx -0.3702,$$

a significant difference from the value -0.3717 that we obtained using the table with $\pi \approx 3.14$. The calculator value is a more accurate approximation because it is programmed for a more accurate approximation to π.

Function values for angles with negative measures can also be found using the procedures we used for angles with positive measures.

Examples Find an approximation for each function value.

a. $\sin(-34^\circ\ 12')$ b. $\cos(-422^\circ\ 30')$

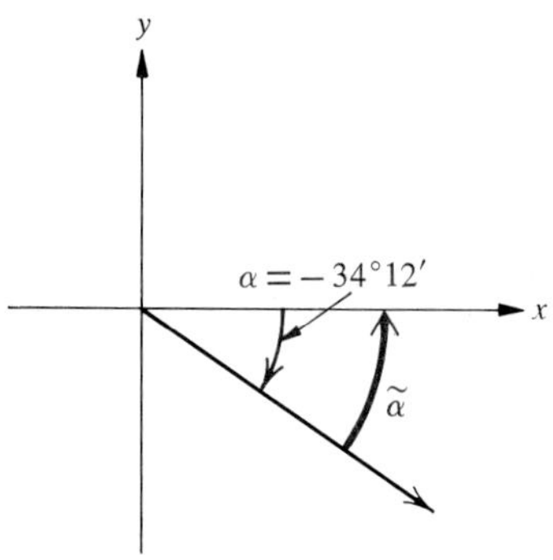

Solutions

a. The reference angle is

$$\tilde{\alpha} = 34^\circ\ 12'.$$

Because $-34^\circ\ 12'$ is in Quadrant IV, we have, using Table III,

$$\sin(-34^\circ\ 12') = -\sin 34^\circ\ 12'$$
$$= -\sin 34.2^\circ \approx -0.5621.$$

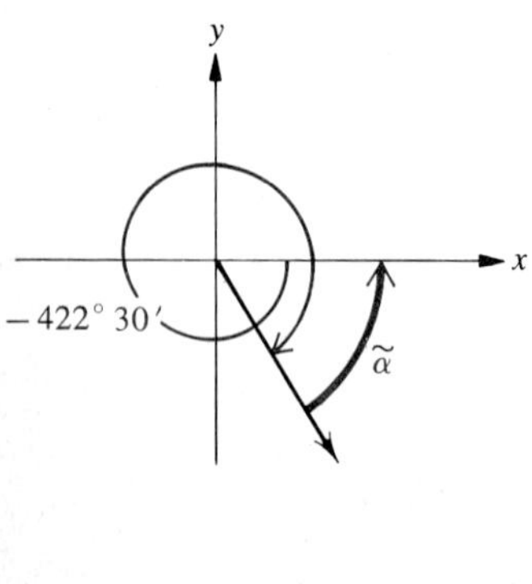

b. The reference angle is

$$\tilde{\alpha} = |360^\circ - 422^\circ\ 30'| = 62^\circ\ 30'.$$

Because $-422^\circ\ 30'$ is in Quadrant IV, $\cos(-422^\circ\ 30')$ is positive, and we have, using Table III,

$$\cos(-422^\circ\ 30') = \cos 62^\circ\ 30' \approx 0.4617.$$

TABLE I Powers, roots, and reciprocals

N	N^2	$\sqrt{N}$	$\sqrt{10N}$	$1/N$	N	N^2	$\sqrt{N}$	$\sqrt{10N}$	$1/N$
1	1	1.0 00	3.1 62	1.00 00	51	2 601	7.1 41	22.5 83	.019 61
2	4	1.4 14	4.4 72	.500 00	52	2 704	7.2 11	22.8 04	.019 23
3	9	1.7 32	5.4 77	.333 33	53	2 809	7.2 80	23.0 22	.018 87
4	16	2.0 00	6.3 25	.250 00	54	2 916	7.3 48	23.2 38	.018 52
5	25	2.2 36	7.0 71	.200 00	55	3 025	7.4 16	23.4 52	.018 18
6	36	2.4 49	7.7 46	.166 67	56	3 136	7.4 83	23.6 64	.017 86
7	49	2.6 46	8.3 67	.142 86	57	3 249	7.5 50	23.8 75	.017 54
8	64	2.8 28	8.9 44	.125 00	58	3 364	7.6 16	24.0 83	.017 24
9	81	3.0 00	9.4 87	.111 11	59	3 481	7.6 81	24.2 90	.016 95
10	100	3.1 62	10.0 00	.100 00	**60**	3 600	7.7 46	24.4 95	.016 67
11	121	3.3 17	10.4 88	.090 91	61	3 721	7.8 10	24.6 98	.016 39
12	144	3.4 64	10.9 54	.083 33	62	3 844	7.8 74	24.9 00	.016 13
13	169	3.6 06	11.4 02	.076 92	63	3 969	7.9 37	25.1 00	.015 87
14	196	3.7 42	11.8 32	.071 43	64	4 096	8.0 00	25.2 98	.015 62
15	225	3.8 73	12.2 47	.066 67	65	4 225	8.0 62	25.4 95	.015 38
16	256	4.0 00	12.6 49	.062 50	66	4 356	8.1 24	25.6 90	.015 15
17	289	4.1 23	13.0 38	.058 82	67	4 489	8.1 85	25.8 84	.014 93
18	324	4.2 43	13.4 16	.055 56	68	4 624	8.2 46	26.0 77	.014 71
19	361	4.3 59	13.7 84	.052 63	69	4 761	8.3 07	26.2 68	.014 49
20	400	4.4 72	14.1 42	.050 00	**70**	4 900	8.3 67	26.4 58	.014 29
21	441	4.5 83	14.4 91	.047 62	71	5 041	8.4 26	26.6 46	.014 08
22	484	4.6 90	14.8 32	.045 45	72	5 184	8.4 85	26.8 33	.013 89
23	529	4.7 96	15.1 66	.043 48	73	5 329	8.5 44	27.0 19	.013 70
24	576	4.8 99	15.4 92	.041 67	74	5 476	8.6 02	27.2 03	.013 51
25	625	5.0 00	15.8 11	.040 00	75	5 625	8.6 60	27.3 86	.013 33
26	676	5.0 99	16.1 25	.038 46	76	5 776	8.7 18	27.5 68	.013 16
27	729	5.1 96	16.4 32	.037 04	77	5 929	8.7 75	27.7 49	.012 99
28	784	5.2 92	16.7 33	.035 71	78	6 084	8.8 32	27.9 28	.012 82
29	841	5.3 85	17.0 29	.034 48	79	6 241	8.8 88	28.1 07	.012 66
30	900	5.4 77	17.3 21	.033 33	**80**	6 400	8.9 44	28.2 84	.012 50
31	961	5.5 68	17.6 07	.032 26	81	6 561	9.0 00	28.4 60	.012 35
32	1 024	5.6 57	17.8 89	.031 25	82	6 724	9.0 55	28.6 36	.012 20
33	1 089	5.7 45	18.1 66	.030 30	83	6 889	9.1 10	28.8 10	.012 05
34	1 156	5.8 31	18.4 39	.029 41	84	7 056	9.1 65	28.9 83	.011 90
35	1 225	5.9 16	18.7 08	.028 57	85	7 225	9.2 20	29.1 55	.011 76
36	1 296	6.0 00	18.9 74	.027 78	86	7 396	9.2 74	29.3 26	.011 63
37	1 369	6.0 83	19.2 35	.027 03	87	7 569	9.3 27	29.4 96	.011 49
38	1 444	6.1 64	19.4 94	.026 32	88	7 744	9.3 81	29.6 65	.011 36
39	1 521	6.2 45	19.7 48	.025 64	89	7 921	9.4 34	29.8 33	.011 24
40	1 600	6.3 25	20.0 00	.025 00	**90**	8 100	9.4 87	30.0 00	.011 11
41	1 681	6.4 03	20.2 48	.024 39	91	8 281	9.5 39	30.1 66	.010 99
42	1 764	6.4 81	20.4 94	.023 81	92	8 464	9.5 92	30.3 32	.010 87
43	1 849	6.5 57	20.7 36	.023 26	93	8 649	9.6 44	30.4 96	.010 75
44	1 936	6.6 33	20.9 76	.022 73	94	8 836	9.6 95	30.6 59	.010 64
45	2 025	6.7 08	21.2 13	.022 22	95	9 025	9.7 47	30.8 22	.010 53
46	2 116	6.7 82	21.4 48	.021 74	96	9 216	9.7 98	30.9 84	.010 42
47	2 209	6.8 56	21.6 79	.021 28	97	9 409	9.8 49	31.1 45	.010 31
48	2 304	6.9 28	21.9 09	.020 83	98	9 604	9.8 99	31.3 05	.010 20
49	2 401	7.0 00	22.1 36	.020 41	99	9 801	9.9 50	31.4 64	.010 10
50	2 500	7.0 71	22.3 61	.020 00	**100**	10 000	10.0 00	31.6 23	.010 00
N	N^2	$\sqrt{N}$	$\sqrt{10N}$	$1/N$	N	N^2	$\sqrt{N}$	$\sqrt{10N}$	$1/N$

TABLE II Values of circular and trigonometric functions (radians)

θ radians or Real Number x	θ degrees	sin θ or sin x	cos θ or cos x	tan θ or tan x	csc θ or csc x	sec θ or sec x	cot θ or cot x
0.00	0° 00′	0.0000	1.000	0.0000	No value	1.000	No value
.01	0° 34′	.0100	1.000	.0100	100.0	1.000	100.0
.02	1° 09′	.0200	0.9998	.0200	50.00	1.000	49.99
.03	1° 43′	.0300	0.9996	.0300	33.34	1.000	33.32
.04	2° 18′	.0400	0.9992	.0400	25.01	1.001	24.99
0.05	2° 52′	0.0500	0.9988	0.0500	20.01	1.001	19.98
.06	3° 26′	.0600	.9982	.0601	16.68	1.002	16.65
.07	4° 01′	.0699	.9976	.0701	14.30	1.002	14.26
.08	4° 35′	.0799	.9968	.0802	12.51	1.003	12.47
.09	5° 09′	.0899	.9960	.0902	11.13	1.004	11.08
0.10	5° 44′	0.0998	0.9950	0.1003	10.02	1.005	9.967
.11	6° 18′	.1098	.9940	.1104	9.109	1.006	9.054
.12	6° 53′	.1197	.9928	.1206	8.353	1.007	8.293
.13	7° 27′	.1296	.9916	.1307	7.714	1.009	7.649
.14	8° 01′	.1395	.9902	.1409	7.166	1.010	7.096
0.15	8° 36′	0.1494	0.9888	0.1511	6.692	1.011	6.617
.16	9° 10′	.1593	.9872	.1614	6.277	1.013	6.197
.17	9° 44′	.1692	.9856	.1717	5.911	1.015	5.826
.18	10° 19′	.1790	.9838	.1820	5.586	1.016	5.495
.19	10° 53′	.1889	.9820	.1923	5.295	1.018	5.200
0.20	11° 28′	0.1987	0.9801	0.2027	5.033	1.020	4.933
.21	12° 02′	.2085	.9780	.2131	4.797	1.022	4.692
.22	12° 36′	.2182	.9759	.2236	4.582	1.025	4.472
.23	13° 11′	.2280	.9737	.2341	4.386	1.027	4.271
.24	13° 45′	.2377	.9713	.2447	4.207	1.030	4.086
0.25	14° 19′	0.2474	0.9689	0.2553	4.042	1.032	3.916
.26	14° 54′	.2571	.9664	.2660	3.890	1.035	3.759
.27	15° 28′	.2667	.9638	.2768	3.749	1.038	3.613
.28	16° 03′	.2764	.9611	.2876	3.619	1.041	3.478
.29	16° 37′	.2860	.9582	.2984	3.497	1.044	3.351
0.30	17° 11′	0.2955	0.9553	0.3093	3.384	1.047	3.233
.31	17° 46′	.3051	.9523	.3203	3.278	1.050	3.122
.32	18° 20′	.3146	.9492	.3314	3.179	1.053	3.018
.33	18° 54′	.3240	.9460	.3425	3.086	1.057	2.920
.34	19° 29′	.3335	.9428	.3537	2.999	1.061	1.827
0.35	20° 03′	0.3429	0.9394	0.3650	2.916	1.065	2.740
.36	20° 38′	.3523	.9359	.3764	2.839	1.068	2.657
.37	21° 12′	.3616	.9323	.3879	2.765	1.073	2.578
.38	21° 46′	.3709	.9287	.3994	2.696	1.077	2.504
.39	22° 21′	.3802	.9249	.4111	2.630	1.081	2.433
0.40	22° 55′	0.3894	0.9211	0.4228	2.568	1.086	2.365
.41	23° 29′	.3986	.9171	.4346	2.509	1.090	2.301
.42	24° 04′	.4078	.9131	.4466	2.452	1.095	2.239
.43	24° 38′	.4169	.9090	.4586	2.399	1.100	2.180
.44	25° 13′	.4259	.9048	.4708	2.348	1.105	2.124
0.45	25° 47′	0.4350	0.9004	0.4831	2.299	1.111	2.070

TABLE II (*continued*)

θ radians or Real Number x	θ degrees	sin θ or sin x	cos θ or cos x	tan θ or tan x	csc θ or csc x	sec θ or sec x	cot θ or cot x
0.45	25° 47′	0.4350	0.9004	0.4831	2.299	1.111	2.070
.46	26° 21′	.4439	.8961	.4954	2.253	1.116	2.018
.47	26° 56′	.4529	.8916	.5080	2.208	1.122	1.969
.48	27° 30′	.4618	.8870	.5206	2.166	1.127	1.921
.49	28° 04′	.4706	.8823	.5334	2.125	1.133	1.875
0.50	28° 39′	0.4794	0.8776	0.5463	2.086	1.139	1.830
.51	29° 13′	.4882	.8727	.5594	2.048	1.146	1.788
.52	29° 48′	.4969	.8678	.5726	2.013	1.152	1.747
.53	30° 22′	.5055	.8628	.5859	1.978	1.159	1.707
.54	30° 56′	.5141	.8577	.5994	1.945	1.166	1.668
0.55	31° 31′	0.5227	0.8525	0.6131	1.913	1.173	1.631
.56	32° 05′	.5312	.8473	.6269	1.883	1.180	1.595
.57	32° 40′	.5396	.8419	.6410	1.853	1.188	1.560
.58	33° 14′	.5480	.8365	.6552	1.825	1.196	1.526
.59	33° 48′	.5564	.8309	.6696	1.797	1.203	1.494
0.60	34° 23′	0.5646	0.8253	0.6841	1.771	1.212	1.462
.61	34° 57′	.5729	.8196	.6989	1.746	1.220	1.431
.62	35° 31′	.5810	.8139	.7139	1.721	1.229	1.401
.63	36° 06′	.5891	.8080	.7291	1.697	1.238	1.372
.64	36° 40′	.5972	.8021	.7445	1.674	1.247	1.343
0.65	37° 15′	0.6052	0.7961	0.7602	1.652	1.256	1.315
.66	37° 49′	.6131	.7900	.7761	1.631	1.266	1.288
.67	38° 23′	.6210	.7838	.7923	1.610	1.276	1.262
.68	38° 58′	.6288	.7776	.8087	1.590	1.286	1.237
.69	39° 32′	.6365	.7712	.8253	1.571	1.297	1.212
0.70	40° 06′	0.6442	0.7648	0.8423	1.552	1.307	1.187
.71	40° 41′	.6518	.7584	.8595	1.534	1.319	1.163
.72	41° 15′	.6594	.7518	.8771	1.517	1.330	1.140
.73	41° 50′	.6669	.7452	.8949	1.500	1.342	1.117
.74	42° 24′	.6743	.7385	.9131	1.483	1.354	1.095
0.75	42° 58′	0.6816	0.7317	0.9316	1.467	1.367	1.073
.76	43° 33′	.6889	.7248	.9505	1.452	1.380	1.052
.77	44° 07′	.6961	.7179	.9697	1.436	1.393	1.031
.78	44° 41′	.7033	.7109	.9893	1.422	1.407	1.011
.79	45° 16′	.7104	.7038	1.009	1.408	1.421	0.9908
0.80	45° 50′	0.7174	0.6967	1.030	1.394	1.435	0.9712
.81	46° 25′	.7243	.6895	1.050	1.381	1.450	.9520
.82	46° 59′	.7311	.6822	1.072	1.368	1.466	.9331
.83	47° 33′	.7379	.6749	1.093	1.355	1.482	.9146
.84	48° 08′	.7446	.6675	1.116	1.343	1.498	.8964
0.85	48° 42′	0.7513	0.6600	1.138	1.331	1.515	0.8785
.86	49° 16′	.7578	.6524	1.162	1.320	1.533	.8609
.87	49° 51′	.7643	.6448	1.185	1.308	1.551	.8437
.88	50° 25′	.7707	.6372	1.210	1.297	1.569	.8267
.89	51° 00′	.7771	.6294	1.235	1.287	1.589	.8100
0.90	51° 34′	0.7833	0.6216	1.260	1.277	1.609	0.7936
.91	52° 08′	.7895	.6137	1.286	1.267	1.629	.7774
.92	52° 43′	.7956	.6058	1.313	1.257	1.651	.7615
.93	53° 17′	.8016	.5978	1.341	1.247	1.673	.7458
.94	53° 51′	.8076	.5898	1.369	1.238	1.696	.7303
0.95	54° 26′	0.8134	0.5817	1.398	1.229	1.719	0.7151

II

TABLE II (*continued*)

θ radians or Real Number x	**θ degrees**	**sin θ or sin x**	**cos θ or cos x**	**tan θ or tan x**	**csc θ or csc x**	**sec θ or sec x**	**cot θ or cot x**
0.95	54° 26′	0.8134	0.5817	1.398	1.229	1.719	0.7151
.96	55° 00′	.8192	.5735	1.428	1.221	1.744	.7001
.97	55° 35′	.8249	.5653	1.459	1.212	1.769	.6853
.98	56° 09′	.8305	.5570	1.491	1.204	1.795	.6707
.99	56° 43′	.8360	.5487	1.524	1.196	1.823	.6563
1.00	57° 18′	0.8415	0.5403	1.557	1.188	1.851	0.6421
1.01	57° 52′	.8468	.5319	1.592	1.181	1.880	.6281
1.02	58° 27′	.8521	.5234	1.628	1.174	1.911	.6142
1.03	59° 01′	.8573	.5148	1.665	1.166	1.942	.6005
1.04	59° 35′	.8624	.5062	1.704	1.160	1.975	.5870
1.05	60° 10′	0.8674	0.4976	1.743	1.153	2.010	0.5736
1.06	60° 44′	.8724	.4889	1.784	1.146	2.046	.5604
1.07	61° 18′	.8772	.4801	1.827	1.140	2.083	.5473
1.08	61° 53′	.8820	.4713	1.871	1.134	2.122	.5344
1.09	62° 27′	.8866	.4625	1.917	1.128	2.162	.5216
1.10	63° 02′	0.8912	0.4536	1.965	1.122	2.205	0.5090
1.11	63° 36′	.8957	.4447	2.014	1.116	2.249	.4964
1.12	64° 10′	.9001	.4357	2.066	1.111	2.295	.4840
1.13	64° 45′	.9044	.4267	2.120	1.106	2.344	.4718
1.14	65° 19′	.9086	.4176	2.176	1.101	2.395	.4596
1.15	65° 53′	0.9128	0.4085	2.234	1.096	2.448	0.4475
1.16	66° 28′	.9168	.3993	2.296	1.091	2.504	.4356
1.17	67° 02′	.9208	.3902	2.360	1.086	2.563	.4237
1.18	67° 37′	.9246	.3809	2.427	1.082	2.625	.4120
1.19	68° 11′	.9284	.3717	2.498	1.077	2.691	.4003
1.20	68° 45′	0.9320	0.3624	2.572	1.073	2.760	0.3888
1.21	69° 20′	.9356	.3530	2.650	1.069	2.833	.3773
1.22	69° 54′	.9391	.3436	2.733	1.065	2.910	.3659
1.23	70° 28′	.9425	.3342	2.820	1.061	2.992	.3546
1.24	71° 03′	.9458	.3248	2.912	1.057	3.079	.3434
1.25	71° 37′	0.9490	0.3153	3.010	1.054	3.171	0.3323
1.26	72° 12′	.9521	.3058	3.113	1.050	3.270	.3212
1.27	72° 46′	.9551	.2963	3.224	1.047	3.375	.3102
1.28	73° 20′	.9580	.2867	3.341	1.044	3.488	.2993
1.29	73° 55′	.9608	.2771	3.467	1.041	3.609	.2884
1.30	74° 29′	0.9636	0.2675	3.602	1.038	3.738	0.2776
1.31	75° 03′	.9662	.2579	3.747	1.035	3.878	.2669
1.32	75° 38′	.9687	.2482	3.903	1.032	4.029	.2562
1.33	76° 12′	.9711	.2385	4.072	1.030	4.193	.2456
1.34	76° 47′	.9735	.2288	4.256	1.027	4.372	.2350
1.35	77° 21′	0.9757	0.2190	4.455	1.025	4.566	0.2245
1.36	77° 55′	.9779	.2092	4.673	1.023	4.779	.2140
1.37	78° 30′	.9799	.1994	4.913	1.021	5.014	.2035
1.38	79° 04′	.9819	.1896	5.177	1.018	5.273	.1931
1.39	79° 38′	.9837	.1798	5.471	1.017	5.561	.1828
1.40	80° 13′	0.9854	0.1700	5.798	1.015	5.883	0.1725
1.41	80° 47′	.9871	.1601	6.165	1.013	6.246	.1622
1.42	81° 22′	.9887	.1502	6.581	1.011	6.657	.1519
1.43	81° 56′	.9901	.1403	7.055	1.010	7.126	.1417
1.44	82° 30′	.9915	.1304	7.602	1.009	7.667	.1315
1.45	83° 05′	0.9927	0.1205	8.238	1.007	8.299	0.1214

TABLE II *(continued)*

θ radians or Real Number x	θ degrees	sin θ or sin x	cos θ or cos x	tan θ or tan x	csc θ or csc x	sec θ or sec x	cot θ or cot x
1.45	83° 05′	0.9927	0.1205	8.238	1.007	8.299	0.1214
1.46	83° 39′	.9939	.1106	8.989	1.006	9.044	.1113
1.47	84° 13′	.9949	.1006	9.887	1.005	9.938	.1001
1.48	84° 48′	.9959	.0907	10.98	1.004	11.03	.0910
1.49	85° 22′	.9967	.0807	12.35	1.003	12.39	.0810
1.50	85° 57′	0.9975	0.0707	14.10	1.003	14.14	0.0709
1.51	86° 31′	.9982	.0608	16.43	1.002	16.46	.0609
1.52	87° 05′	.9987	.0508	19.67	1.001	19.69	.0508
1.53	87° 40′	.9992	.0408	24.50	1.001	24.52	.0408
1.54	88° 14′	.9995	.0308	32.46	1.000	32.48	.0308
1.55	88° 49′	0.9998	0.0208	48.08	1.000	48.09	0.0208
1.56	89° 23′	.9999	.0108	92.62	1.000	92.63	.0108
1.57	89° 57′	1.000	.0008	1256	1.000	1256	.0008

II

TABLE III Values of trigonometric functions

θ deg	deg min	sin θ	cos θ	tan θ	csc θ	sec θ	cot θ		
0.0	0 0	0.0000	1.0000	0.0000	No value	1.0000	No value	90 0	90.0
0.1	0 6	0.0017	1.0000	0.0017	572.96	1.0000	572.96	89 54	89.9
0.2	0 12	0.0035	1.0000	0.0035	286.48	1.0000	286.48	89 48	89.8
0.3	0 18	0.0052	1.0000	0.0052	190.99	1.0000	190.98	89 42	89.7
0.4	0 24	0.0070	1.0000	0.0070	143.24	1.0000	143.24	89 36	89.6
0.5	0 30	0.0087	1.0000	0.0087	114.59	1.0000	114.59	89 30	89.5
0.6	0 36	0.0105	0.9999	0.0105	95.495	1.0001	95.490	89 24	89.4
0.7	0 42	0.0122	0.9999	0.0122	81.853	1.0001	81.847	89 18	89.3
0.8	0 48	0.0140	0.9999	0.0140	71.622	1.0001	71.615	89 12	89.2
0.9	0 54	0.0157	0.9999	0.0157	63.665	1.0001	63.657	89 6	89.1
1.0	1 0	0.0175	0.9998	0.0175	57.299	1.0002	57.290	89 0	89.0
1.1	1 6	0.0192	0.9998	0.0192	52.090	1.0002	52.081	88 54	88.9
1.2	1 12	0.0209	0.9998	0.0209	47.750	1.0002	47.740	88 48	88.8
1.3	1 18	0.0227	0.9997	0.0227	44.077	1.0003	44.066	88 42	88.7
1.4	1 24	0.0244	0.9997	0.0244	40.930	1.0003	40.917	88 36	88.6
1.5	1 30	0.0262	0.9997	0.0262	38.202	1.0003	38.188	88 30	88.5
1.6	1 36	0.0279	0.9996	0.0279	35.815	1.0004	35.801	88 24	88.4
1.7	1 42	0.0297	0.9996	0.0297	33.708	1.0004	33.694	88 18	88.3
1.8	1 48	0.0314	0.9995	0.0314	31.836	1.0005	31.821	88 12	88.2
1.9	1 54	0.0332	0.9995	0.0332	30.161	1.0005	30.145	88 6	88.1
2.0	2 0	0.0349	0.9994	0.0349	28.654	1.0006	28.636	88 0	88.0
2.1	2 6	0.0366	0.9993	0.0367	27.290	1.0007	27.271	87 54	87.9
2.2	2 12	0.0384	0.9993	0.0384	26.050	1.0007	26.031	87 48	87.8
2.3	2 18	0.0401	0.9992	0.0402	24.918	1.0008	24.898	87 42	87.7
2.4	2 24	0.0419	0.9991	0.0419	23.880	1.0009	23.859	87 36	87.6
2.5	2 30	0.0436	0.9990	0.0437	22.926	1.0010	22.904	87 30	87.5
2.6	2 36	0.0454	0.9990	0.0454	22.044	1.0010	22.022	87 24	87.4
2.7	2 42	0.0471	0.9989	0.0472	21.229	1.0011	21.205	87 18	87.3
2.8	2 48	0.0488	0.9988	0.0489	20.471	1.0012	20.446	87 12	87.2
2.9	2 54	0.0506	0.9987	0.0507	19.766	1.0013	19.740	87 6	87.1
3.0	3 0	0.0523	0.9986	0.0524	19.107	1.0014	19.081	87 0	87.0
3.1	3 6	0.0541	0.9985	0.0542	18.492	1.0015	18.464	86 54	86.9
3.2	3 12	0.0558	0.9984	0.0559	17.914	1.0016	17.886	86 48	86.8
3.3	3 18	0.0576	0.9983	0.0577	17.372	1.0017	17.343	86 42	86.7
3.4	3 24	0.0593	0.9982	0.0594	16.862	1.0018	16.832	86 36	86.6
3.5	3 30	0.0610	0.9981	0.0612	16.380	1.0019	16.350	86 30	86.5
3.6	3 36	0.0628	0.9980	0.0629	15.926	1.0020	15.895	86 24	86.4
3.7	3 42	0.0645	0.9979	0.0647	15.496	1.0021	15.464	86 18	86.3
3.8	3 48	0.0663	0.9978	0.0664	15.089	1.0022	15.056	86 12	86.2
3.9	3 54	0.0680	0.9977	0.0682	14.703	1.0023	14.669	86 6	86.1
4.0	4 0	0.0698	0.9976	0.0699	14.336	1.0024	14.301	86 0	86.0
4.1	4 6	0.0715	0.9974	0.0717	13.987	1.0026	13.951	85 54	85.9
4.2	4 12	0.0732	0.9973	0.0734	13.654	1.0027	13.617	85 48	85.8
4.3	4 18	0.0750	0.9972	0.0752	13.337	1.0028	13.300	85 42	85.7
4.4	4 24	0.0767	0.9971	0.0769	13.035	1.0030	12.996	85 36	85.6
4.5	4 30	0.0785	0.9969	0.0787	12.746	1.0031	12.706	85 30	85.5
4.6	4 36	0.0802	0.9968	0.0805	12.469	1.0032	12.429	85 24	85.4
4.7	4 42	0.0819	0.9966	0.0822	12.204	1.0034	12.163	85 18	85.3
4.8	4 48	0.0837	0.9965	0.0840	11.951	1.0035	11.909	85 12	85.2
4.9	4 54	0.0854	0.9963	0.0857	11.707	1.0037	11.665	85 6	85.1
		cos θ	sin θ	cot θ	sec θ	csc θ	tan θ	deg min	θ deg

TABLE III (*continued*)

θ deg	deg	min	sin θ	cos θ	tan θ	csc θ	sec θ	cot θ			
5.0	5	0	0.0872	0.9962	0.0875	11.474	1.0038	11.430	85	0	85.0
5.1	5	6	0.0889	0.9960	0.0892	11.249	1.0040	11.205	84	54	84.9
5.2	5	12	0.0906	0.9959	0.0910	11.034	1.0041	10.988	84	48	84.8
5.3	5	18	0.0924	0.9957	0.0928	10.826	1.0043	10.780	84	42	84.7
5.4	5	24	0.0941	0.9956	0.0945	10.626	1.0045	10.579	84	36	84.6
5.5	5	30	0.0958	0.9954	0.0963	10.433	1.0046	10.385	84	30	84.5
5.6	5	36	0.0976	0.9952	0.0981	10.248	1.0048	10.199	84	24	84.4
5.7	5	42	0.0993	0.9951	0.0998	10.069	1.0050	10.019	84	18	84.3
5.8	5	48	0.1011	0.9949	0.1016	9.8955	1.0051	9.8448	84	12	84.2
5.9	5	54	0.1028	0.9947	0.1033	9.7283	1.0053	9.6768	84	6	84.1
6.0	6	0	0.1045	0.9945	0.1051	9.5668	1.0055	9.5144	84	0	84.0
6.1	6	6	0.1063	0.9943	0.1069	9.4105	1.0057	9.3573	83	54	83.9
6.2	6	12	0.1080	0.9942	0.1086	9.2593	1.0059	9.2052	83	48	83.8
6.3	6	18	0.1097	0.9940	0.1104	9.1129	1.0061	9.0579	83	42	83.7
6.4	6	24	0.1115	0.9938	0.1122	8.9711	1.0063	8.9152	83	36	83.6
6.5	6	30	0.1132	0.9936	0.1139	8.8337	1.0065	8.7769	83	30	83.5
6.6	6	36	0.1149	0.9934	0.1157	8.7004	1.0067	8.6428	83	24	83.4
6.7	6	42	0.1167	0.9932	0.1175	8.5711	1.0069	8.5126	83	18	83.3
6.8	6	48	0.1184	0.9930	0.1192	8.4457	1.0071	8.3863	83	12	83.2
6.9	6	54	0.1201	0.9928	0.1210	8.3238	1.0073	8.2636	83	6	83.1
7.0	7	0	0.1219	0.9925	0.1228	8.2055	1.0075	8.1444	83	0	83.0
7.1	7	6	0.1236	0.9923	0.1246	8.0905	1.0077	8.0285	82	54	82.9
7.2	7	12	0.1253	0.9921	0.1263	7.9787	1.0079	7.9158	82	48	82.8
7.3	7	18	0.1271	0.9919	0.1281	7.8700	1.0082	7.8062	82	42	82.7
7.4	7	24	0.1288	0.9917	0.1299	7.7642	1.0084	7.6996	82	36	82.6
7.5	7	30	0.1305	0.9914	0.1317	7.6613	1.0086	7.5958	82	30	82.5
7.6	7	36	0.1323	0.9912	0.1334	7.5611	1.0089	7.4947	82	24	82.4
7.7	7	42	0.1340	0.9910	0.1352	7.4635	1.0091	7.3962	82	18	82.3
7.8	7	48	0.1357	0.9907	0.1370	7.3684	1.0093	7.3002	82	12	82.2
7.9	7	54	0.1374	0.9905	0.1388	7.2757	1.0096	7.2066	82	6	82.1
8.0	8	0	0.1392	0.9903	0.1405	7.1853	1.0098	7.1154	82	0	82.0
8.1	8	6	0.1409	0.9900	0.1423	7.0972	1.0101	7.0264	81	54	81.9
8.2	8	12	0.1426	0.9898	0.1441	7.0112	1.0103	6.9395	81	48	81.8
8.3	8	18	0.1444	0.9895	0.1459	6.9273	1.0106	6.8548	81	42	81.7
8.4	8	24	0.1461	0.9893	0.1477	6.8454	1.0108	6.7720	81	36	81.6
8.5	8	30	0.1478	0.9890	0.1495	6.7655	1.0111	6.6912	81	30	81.5
8.6	8	36	0.1495	0.9888	0.1512	6.6874	1.0114	6.6122	81	24	81.4
8.7	8	42	0.1513	0.9885	0.1530	6.6111	1.0116	6.5350	81	18	81.3
8.8	8	48	0.1530	0.9882	0.1548	6.5366	1.0119	6.4596	81	12	81.2
8.9	8	54	0.1547	0.9880	0.1566	6.4637	1.0122	6.3859	81	6	81.1
9.0	9	0	0.1564	0.9877	0.1584	6.3925	1.0125	6.3138	81	0	81.0
9.1	9	6	0.1582	0.9874	0.1602	6.3228	1.0127	6.2432	80	54	80.9
9.2	9	12	0.1599	0.9871	0.1620	6.2547	1.0130	6.1742	80	48	80.8
9.3	9	18	0.1616	0.9869	0.1638	6.1880	1.0133	6.1066	80	42	80.7
9.4	9	24	0.1633	0.9866	0.1655	6.1227	1.0136	6.0405	80	36	80.6
9.5	9	30	0.1650	0.9863	0.1673	6.0589	1.0139	5.9758	80	30	80.5
9.6	9	36	0.1668	0.9860	0.1691	5.9963	1.0142	5.9124	80	24	80.4
9.7	9	42	0.1685	0.9857	0.1709	5.9351	1.0145	5.8502	80	18	80.3
9.8	9	48	0.1702	0.9854	0.1727	5.8751	1.0148	5.7894	80	12	80.2
9.9	9	54	0.1719	0.9851	0.1745	5.8164	1.0151	5.7297	80	6	80.1
			cos θ	sin θ	cot θ	sec θ	csc θ	tan θ	deg	min	θ deg

TABLE III *(continued)*

θ deg	deg min	sin θ	cos θ	tan θ	csc θ	sec θ	cot θ		
10.0	10 0	0.1736	0.9848	0.1763	5.7588	1.0154	5.6713	80 0	80.0
10.1	10 6	0.1754	0.9845	0.1781	5.7023	1.0157	5.6140	79 54	79.9
10.2	10 12	0.1771	0.9842	0.1799	5.6470	1.0161	5.5578	79 48	79.8
10.3	10 18	0.1788	0.9839	0.1817	5.5928	1.0164	5.5027	79 42	79.7
10.4	10 24	0.1805	0.9836	0.1835	5.5396	1.0167	5.4486	79 36	79.6
10.5	10 30	0.1822	0.9833	0.1853	5.4874	1.0170	5.3955	79 30	79.5
10.6	10 36	0.1840	0.9829	0.1871	5.4362	1.0174	5.3435	79 24	79.4
10.7	10 42	0.1857	0.9826	0.1890	5.3860	1.0177	5.2924	79 18	79.3
10.8	10 48	0.1874	0.9823	0.1908	5.3367	1.0180	5.2422	79 12	79.2
10.9	10 54	0.1891	0.9820	0.1926	5.2883	1.0184	5.1929	79 6	79.1
11.0	11 0	0.1908	0.9816	0.1944	5.2408	1.0187	5.1446	79 0	79.0
11.1	11 6	0.1925	0.9813	0.1962	5.1942	1.0191	5.0970	78 54	78.9
11.2	11 12	0.1942	9.9810	0.1980	5.1484	1.0194	5.0504	78 48	78.8
11.3	11 18	0.1959	0.9806	0.1998	5.1034	1.0198	5.0045	78 42	78.7
11.4	11 24	0.1977	0.9803	0.2016	5.0593	1.0201	4.9595	78 36	78.6
11.5	11 30	0.1994	0.9799	0.2035	5.0159	1.0205	4.9152	78 30	78.5
11.6	11 36	0.2011	0.9796	0.2053	4.9732	1.0209	4.8716	78 24	78.4
11.7	11 42	0.2028	0.9792	0.2071	4.9313	1.0212	4.8288	78 18	78.3
11.8	11 48	0.2045	0.9789	0.2089	4.8901	1.0216	4.7867	78 12	78.2
11.9	11 54	0.2062	0.9785	0.2107	4.8496	1.0220	4.7453	78 6	78.1
12.0	12 0	0.2079	0.9781	0.2126	4.8097	1.0223	4.7046	78 0	78.0
12.1	12 6	0.2096	0.9778	0.2144	4.7706	1.0227	4.6646	77 54	77.9
12.2	12 12	0.2113	0.9774	0.2162	4.7321	1.0231	4.6252	77 48	77.8
12.3	12 18	0.2130	0.9770	0.2180	4.6942	1.0235	4.5864	77 42	77.7
12.4	12 24	0.2147	0.9767	0.2199	4.6569	1.0239	4.5483	77 36	77.6
12.5	12 30	0.2164	0.9763	0.2217	4.6202	1.0243	4.5107	77 30	77.5
12.6	12 36	0.2181	0.9759	0.2235	4.5841	1.0247	4.4737	77 24	77.4
12.7	12 42	0.2198	0.9755	0.2254	4.5486	1.0251	4.4374	77 18	77.3
12.8	12 48	0.2215	0.9751	0.2272	4.5137	1.0255	4.4015	77 12	77.2
12.9	12 54	0.2232	0.9748	0.2290	4.4793	1.0259	4.3662	77 6	77.1
13.0	13 0	0.2250	0.9744	0.2309	4.4454	1.0263	4.3315	77 0	77.0
13.1	13 6	0.2267	0.9740	0.2327	4.4121	1.0267	4.2972	76 54	76.9
13.2	13 12	0.2284	0.9736	0.2345	4.3792	1.0271	4.2635	76 48	76.8
13.3	13 18	0.2300	0.9732	0.2364	4.3469	1.0276	4.2303	76 42	76.7
13.4	13 24	0.2317	0.9728	0.2382	4.3150	1.0280	4.1976	76 36	76.6
13.5	13 30	0.2334	0.9724	0.2401	4.2837	1.0284	4.1653	76 30	76.5
13.6	13 36	0.2351	0.9720	0.2419	4.2528	1.0288	4.1335	76 24	76.4
13.7	13 42	0.2368	0.9715	0.2438	4.2223	1.0293	4.1022	76 18	76.3
13.8	13 48	0.2385	0.9711	0.2456	4.1923	1.0297	4.0713	76 12	76.2
13.9	13 54	0.2402	0.9707	0.2475	4.1627	1.0302	4.0408	76 6	76.1
14.0	14 0	0.2419	0.9703	0.2493	4.1336	1.0306	4.0108	76 0	76.0
14.1	14 6	0.2436	0.9699	0.2512	4.1048	1.0311	3.9812	75 54	75.9
14.2	14 12	0.2453	0.9694	0.2530	4.0765	1.0315	3.9520	75 48	75.8
14.3	14 18	0.2470	0.9690	0.2549	4.0486	1.0320	3.9232	75 42	75.7
14.4	14 24	0.2487	0.9686	0.2568	4.0211	1.0324	3.8947	75 36	75.6
14.5	14 30	0.2504	0.9681	0.2586	3.9939	1.0329	3.8667	75 30	75.5
14.6	14 36	0.2521	0.9677	0.2605	3.9672	1.0334	3.8391	75 24	75.4
14.7	14 42	0.2538	0.9673	0.2623	3.9408	1.0338	3.8118	75 18	75.3
14.8	14 48	0.2554	0.9668	0.2642	3.9147	1.0343	3.7849	75 12	75.2
14.9	14 54	0.2571	0.9664	0.2661	3.8890	1.0348	3.7583	75 6	75.1
		cos θ	sin θ	cot θ	sec θ	csc θ	tan θ	deg min	θ deg

TABLE III (*continued*)

θ deg	deg min	sin θ	cos θ	tan θ	csc θ	sec θ	cot θ		
15.0	15 0	0.2588	0.9659	0.2679	3.8637	1.0353	3.7321	75 0	75.0
15.1	15 6	0.2605	0.9655	0.2698	3.8387	1.0358	3.7062	74 54	74.9
15.2	15 12	0.2622	0.9650	0.2717	3.8140	1.0363	3.6806	74 48	74.8
15.3	15 18	0.2639	0.9646	0.2736	3.7897	1.0367	3.6554	74 42	74.7
15.4	15 24	0.2656	0.9641	0.2754	3.7657	1.0372	3.6305	74 36	74.6
15.5	15 30	0.2672	0.9636	0.2773	3.7420	1.0377	3.6059	74 30	74.5
15.6	15 36	0.2689	0.9632	0.2792	3.7186	1.0382	3.5816	74 24	74.4
15.7	15 42	0.2706	0.9627	0.2811	3.6955	1.0388	3.5576	74 18	74.3
15.8	15 48	0.2723	0.9622	0.2830	3.6727	1.0393	3.5339	74 12	74.2
15.9	15 54	0.2740	0.9617	0.2849	3.6502	1.0398	3.5105	74 6	74.1
16.0	16 0	0.2756	0.9613	0.2867	3.6280	1.0403	3.4874	74 0	74.0
16.1	16 6	0.2773	0.9608	0.2886	3.6060	1.0408	3.4646	73 54	73.9
16.2	16 12	0.2790	0.9603	0.2905	3.5843	1.0413	3.4420	73 48	73.8
16.3	16 18	0.2807	0.9598	0.2924	3.5629	1.0419	3.4197	73 42	73.7
16.4	16 24	0.2823	0.9593	0.2943	3.5418	1.0424	3.3977	73 36	73.6
16.5	16 30	0.2840	0.9588	0.2962	3.5209	1.0429	3.3759	73 30	73.5
16.6	16 36	0.2857	0.9583	0.2981	3.5003	1.0435	3.3544	73 24	74.4
16.7	16 42	0.2874	0.9578	0.3000	3.4800	1.0440	3.3332	73 18	73.3
16.8	16 48	0.2890	0.9573	0.3019	3.4598	1.0446	3.3122	73 12	73.2
16.9	16 54	0.2907	0.9568	0.3038	3.4399	1.0451	3.2914	73 6	73.1
17.0	17 0	0.2924	0.9563	0.3057	3.4203	1.0457	3.2709	73 0	73.0
17.1	17 6	0.2940	0.9558	0.3076	3.4009	1.0463	3.2506	72 54	72.9
17.2	17 12	0.2957	0.9553	0.3096	3.3817	1.0468	3.2305	72 48	72.8
17.3	17 18	0.2974	0.9548	0.3115	3.3628	1.0474	3.2106	72 42	72.7
17.4	17 24	0.2990	0.9542	0.3134	3.3440	1.0480	3.1910	72 36	72.6
17.5	17 30	0.3007	0.9537	0.3153	3.3255	1.0485	3.1716	72 30	72.5
17.6	17 36	0.3024	0.9532	0.3172	3.3072	1.0491	3.1524	72 24	72.4
17.7	17 42	0.3040	0.9527	0.3191	3.2891	1.0497	3.1334	72 18	72.3
17.8	17 48	0.3057	0.9521	0.3211	3.2712	1.0503	3.1146	72 12	72.2
17.9	17 54	0.3074	0.9516	0.3230	3.2536	1.0509	3.0961	72 6	72.1
18.0	18 0	0.3090	0.9511	0.3249	3.2361	1.0515	3.0777	72 0	72.0
18.1	18 6	0.3107	0.9505	0.3268	3.2188	1.0521	3.0595	71 54	71.9
18.2	18 12	0.3123	0.9500	0.3288	3.2017	1.0527	3.0415	71 48	71.8
18.3	18 18	0.3140	0.9494	0.3307	3.1848	1.0533	3.0237	71 42	71.7
18.4	18 24	0.3156	0.9489	0.3327	3.1681	1.0539	3.0061	71 36	71.6
18.5	18 30	0.3173	0.9483	0.3346	3.1515	1.0545	2.9887	71 30	71.5
18.6	18 36	0.3190	0.9478	0.3365	3.1352	1.0551	2.9714	71 24	71.4
18.7	18 42	0.3206	0.9472	0.3385	3.1190	1.0557	2.9544	71 18	71.3
18.8	18 48	0.3223	0.9466	0.3404	3.1030	1.0564	2.9375	71 12	71.2
18.9	18 54	0.3239	0.9461	0.3424	3.0872	1.0570	2.9208	71 6	71.1
19.0	19 0	0.3256	0.9455	0.3443	2.0716	1.0576	2.9042	71 0	71.0
19.1	19 6	0.3272	0.9449	0.3463	3.0561	1.0583	2.8878	70 54	70.9
19.2	19 12	0.3289	0.9444	0.3482	3.0407	1.0589	2.8716	70 48	70.8
19.3	19 18	0.3305	0.9438	0.3502	3.0256	1.0595	2.8556	70 42	70.7
19.4	19 24	0.3322	0.9432	0.3522	3.0106	1.0602	2.8397	70 36	70.6
19.5	19 30	0.3338	0.9426	0.3541	2.9957	1.0608	2.8239	70 30	70.5
19.6	19 36	0.3355	0.9421	0.3561	2.9811	1.0615	2.8083	70 24	70.4
19.7	19 42	0.3371	0.9415	0.3581	2.9665	1.0622	2.7929	70 18	70.3
19.8	19 48	0.3387	0.9409	0.3600	2.9521	1.0628	2.7776	70 12	70.2
19.9	19 54	0.3404	0.9403	0.3620	2.9379	1.0635	2.7625	70 6	70.1
		cos θ	sin θ	cot θ	sec θ	csc θ	tan θ	deg min	θ deg

TABLE III (*continued*)

θ deg	**deg**	**min**	**sin θ**	**cos θ**	**tan θ**	**csc θ**	**sec θ**	**cot θ**			
20.0	20	0	0.3420	0.9397	0.3640	2.9238	1.0642	2.7475	70	0	70.0
20.1	20	6	0.3437	0.9391	0.3659	2.9099	1.0649	2.7326	69	54	69.9
20.2	20	12	0.3453	0.9385	0.3679	2.8960	1.0655	2.7179	69	48	69.8
20.3	20	18	0.3469	0.9379	0.3699	2.8824	1.0662	2.7034	69	42	69.7
20.4	20	24	0.3486	0.9373	0.3719	2.8688	1.0669	2.6889	69	36	69.6
20.5	20	30	0.3502	0.9367	0.3739	2.8555	1.0676	2.6746	69	30	69.5
20.6	20	36	0.3518	0.9361	0.3759	2.8422	1.0683	2.6605	69	24	69.4
20.7	20	42	0.3535	0.9354	0.3779	2.8291	1.0690	2.6464	69	18	69.3
20.8	20	48	0.3551	0.9348	0.3799	2.8161	1.0697	2.6325	69	12	69.2
20.9	20	54	0.3567	0.9342	0.3819	2.8032	1.0704	2.6187	69	6	69.1
21.0	21	0	0.3584	0.9336	0.3839	2.7904	1.0711	2.6051	69	0	69.0
21.1	21	6	0.3600	0.9330	0.3859	2.7778	1.0719	2.5916	68	54	68.9
21.2	21	12	0.3616	0.9323	0.3879	2.7653	1.0726	2.5782	68	48	68.8
21.3	21	18	0.3633	0.9317	0.3899	2.7529	1.0733	2.5649	68	42	68.7
21.4	21	24	0.3649	0.9311	0.3919	2.7407	1.0740	2.5517	68	36	68.6
21.5	21	30	0.3665	0.9304	0.3939	2.7285	1.0748	2.5386	68	30	68.5
21.6	21	36	0.3681	0.9298	0.3959	2.7165	1.0755	2.5257	68	24	68.4
21.7	21	42	0.3697	0.9291	0.3979	2.7046	1.0763	2.5129	68	18	68.3
21.8	21	48	0.3714	0.9285	0.4000	2.6927	1.0770	2.5002	68	12	68.2
21.9	21	54	0.3730	0.9278	0.4020	2.6811	1.0778	2.4876	68	6	68.1
22.0	22	0	0.3746	0.9272	0.4040	2.6695	1.0785	2.4751	68	0	68.0
22.1	22	6	0.3762	0.9265	0.4061	2.6580	1.0793	2.4627	67	54	67.9
22.2	22	12	0.3778	0.9259	0.4081	2.6466	1.0801	2.4504	67	48	67.8
22.3	22	18	0.3795	0.9252	0.4101	2.6354	1.0808	2.4383	67	42	67.7
22.4	22	24	0.3811	0.9245	0.4122	2.6242	1.0816	2.4262	67	36	67.6
22.5	22	30	0.3827	0.9239	0.4142	2.6131	1.0824	2.4142	67	30	67.5
22.6	22	36	0.3843	0.9232	0.4163	2.6022	1.0832	2.4023	67	24	67.4
22.7	22	42	0.3859	0.9225	0.4183	2.5913	1.0840	2.3906	67	18	67.3
22.8	22	48	0.3875	0.9219	0.4204	2.5805	1.0848	2.3789	67	12	67.2
22.9	22	54	0.3891	0.9212	0.4224	2.5699	1.0856	2.3673	67	6	67.1
23.0	23	0	0.3907	0.9205	0.4245	2.5593	1.0864	2.3559	67	0	67.0
23.1	23	6	0.3923	0.9198	0.4265	2.5488	1.0872	2.3445	66	54	66.9
23.3	23	12	0.3939	0.9191	0.4286	2.5384	1.0880	2.3332	66	48	66.8
23.3	23	18	0.3955	0.9184	0.4307	2.5282	1.0888	2.3220	66	42	66.7
23.4	23	24	0.3971	0.9178	0.4327	2.5180	1.0896	2.3109	66	36	66.6
23.5	23	30	0.3987	0.9171	0.4348	2.5078	1.0904	2.2998	66	30	66.5
23.6	23	36	0.4003	0.9164	0.4369	2.4978	1.0913	2.2889	66	24	66.4
23.7	23	42	0.4019	0.9157	0.4390	2.4879	1.0921	2.2781	66	18	66.3
23.8	23	48	0.4035	0.9150	0.4411	2.4780	1.0929	2.2673	66	12	66.2
23.9	23	54	0.4051	0.9143	0.4431	2.4683	1.0938	2.2566	66	6	66.1
24.0	24	0	0.4067	0.9135	0.4452	2.4586	1.0946	2.2460	66	0	66.0
24.1	24	6	0.4083	0.9128	0.4473	2.4490	1.0955	2.2355	65	54	65.9
24.2	24	12	0.4099	0.9121	0.4494	2.4395	1.0963	2.2251	65	48	65.8
24.3	24	18	0.4115	0.9114	0.4515	2.4301	1.0972	2.2148	65	42	65.7
24.4	24	24	0.4131	0.9107	0.4536	2.4207	1.0981	2.2045	65	36	65.6
24.5	24	30	0.4147	0.9100	0.4557	2.4114	1.0989	2.1943	65	30	65.5
24.6	24	36	0.4163	0.9092	0.4578	2.4022	1.0998	2.1842	65	24	65.4
24.7	24	42	0.4179	0.9085	0.4599	2.3931	1.1007	2.1742	65	18	65.3
24.8	24	48	0.4195	0.9078	0.4621	2.3841	1.1016	2.1642	65	12	65.2
24.9	24	54	0.4210	0.9070	0.4642	2.3751	1.1025	2.1543	65	6	65.1
			cos θ	**sin θ**	**cot θ**	**sec θ**	**csc θ**	**tan θ**	**deg**	**min**	**θ deg**

TABLE III (*continued*)

θ deg	deg min	sin θ	cos θ	tan θ	csc θ	sec θ	cot θ		
25.0	25 0	0.4226	0.9063	0.4663	2.3662	1.1034	2.1445	65 0	65.0
25.1	25 6	0.4242	0.9056	0.4684	2.3574	1.1043	2.1348	64 54	64.9
25.2	25 12	0.4258	0.9048	0.4706	2.3486	1.1052	2.1251	64 48	64.8
25.3	25 18	0.4274	0.9041	0.4727	2.3400	1.1061	2.1155	64 42	64.7
25.4	25 24	0.4289	0.9033	0.4748	2.3314	1.1070	2.1060	64 36	64.6
25.5	25 30	0.4305	0.9026	0.4770	2.3228	1.1079	2.0965	64 30	64.5
25.6	25 36	0.4321	0.9018	0.4791	2.3144	1.1089	2.0872	64 24	64.4
25.7	25 42	0.4337	0.9011	0.4813	2.3060	1.1098	2.0778	64 18	64.3
25.8	25 48	0.4352	0.9003	0.4834	2.2976	1.1107	2.0686	64 12	64.2
25.9	25 54	0.4368	0.8996	0.4856	2.2894	1.1117	2.0594	64 6	64.1
26.0	26 0	0.4384	0.8988	0.4877	2.2812	1.1126	2.0503	64 0	64.0
26.1	26 6	0.4399	0.8980	0.4899	2.2730	1.1136	2.0413	63 54	63.9
26.2	26 12	0.4415	0.8973	0.4921	2.2650	1.1145	2.0323	63 48	63.8
26.3	26 18	0.4431	0.8965	0.4942	2.2570	1.1155	2.0233	63 42	63.7
26.4	26 24	0.4446	0.8957	0.4964	2.2490	1.1164	2.0145	63 36	63.6
26.5	26 30	0.4462	0.8949	0.4986	2.2412	1.1174	2.0057	63 30	63.5
26.6	26 36	0.4478	0.8942	0.5008	2.2333	1.1184	1.9970	63 24	63.4
26.7	26 42	0.4493	0.8934	0.5029	2.2256	1.1194	1.9883	63 18	63.3
26.8	26 48	0.4509	0.8926	0.5051	2.2179	1.1203	1.9797	63 12	63.2
26.9	26 54	0.4524	0.8918	0.5073	2.2103	1.1213	1.9711	63 6	63.1
27.0	27 0	0.4540	0.8910	0.5095	2.2027	1.1223	1.9626	63 0	63.0
27.1	27 6	0.4555	0.8902	0.5117	2.1952	1.1233	1.9542	62 54	62.9
27.2	27 12	0.4571	0.8894	0.5139	2.1877	1.1243	1.9458	62 48	62.8
27.3	27 18	0.4586	0.8886	0.5161	2.1803	1.1253	1.9375	62 42	62.7
27.4	27 24	0.4602	0.8878	0.5184	2.1730	1.1264	1.9292	62 36	62.6
27.5	27 30	0.4617	0.8870	0.5206	2.1657	1.1274	1.9210	62 30	62.5
27.6	27 36	0.4633	0.8862	0.5228	2.1584	1.1284	1.9128	62 24	62.4
27.7	27 42	0.4648	0.8854	0.5250	2.1513	1.1294	1.9047	62 18	62.3
27.8	27 48	0.4664	0.8846	0.5272	2.1441	1.1305	1.8967	62 12	62.2
27.9	27 54	0.4679	0.8838	0.5295	2.1371	1.1315	1.8887	62 6	62.1
28.0	28 0	0.4695	0.8829	0.5317	2.1301	1.1326	1,8807	62 0	62.0
28.1	28 6	0.4710	0.8821	0.5339	2.1231	1.1336	1.8728	61 54	61.9
28.2	28 12	0.4726	0.8813	0.5362	2.1162	1.1347	1.8650	61 48	61.8
28.3	28 18	0.4741	0.8805	0.5384	2.1093	1.1357	1.8572	61 42	61.7
28.4	28 24	0.4756	0.8796	0.5407	2.1025	1.1368	1.8495	61 36	61.6
28.5	28 30	0.4772	0.8788	0.5430	2.0957	1.1379	1.8418	61 30	61.5
28.6	28 36	0.4787	0.8780	0.5452	2.0890	1.1390	1.8341	61 24	61.4
28.7	28 42	0.4802	0.8771	0.5475	2.0824	1.1401	1.8265	61 18	61.3
28.8	28 48	0.4818	0.8763	0.5498	2.0758	1.1412	1.8190	61 12	61.2
28.9	28 54	0.4833	0.8755	0.5520	2.0692	1.1423	1.8115	61 6	61.1
29.0	29 0	0.4848	0.8746	0.5543	2.0627	1.1434	1.8040	61 0	61.0
29.1	29 6	0.4863	0.8738	0.5566	2.0562	1.1445	1.7966	60 54	60.9
29.2	29 12	0.4879	0.8729	0.5589	2.0598	1.1456	1.7893	60 48	60.8
29.3	29 18	0.4894	0.8721	0.5612	2.0434	1.1467	1.7820	60 42	60.7
29.4	29 24	0.4909	0.8712	0.5635	2.0371	1.1478	1.7747	60 36	60.6
29.5	29 30	0.4924	0.8704	0.5658	2.0308	1.1490	1.7675	60 30	60.5
29.6	29 36	0.4939	0.8695	0.5681	2.0245	1.1501	1.7603	60 24	60.4
29.7	29 42	0.4955	0.8686	0.5704	2.0183	1.1512	1.7532	60 18	60.3
29.8	29 48	0.4970	0.8678	0.5727	2.0122	1.1524	1.7461	60 12	60.2
29.9	29 54	0.4985	0.8669	0.5750	2.0061	1.1535	1.7391	60 6	60.1
		cos θ	sin θ	cot θ	sec θ	csc θ	tan θ	deg min	θ deg

III

TABLE III (*continued*)

θ deg	deg min	sin θ	cos θ	tan θ	csc θ	sec θ	cot θ		
30.0	30 0	0.5000	0.8660	0.5774	2.0000	1.1547	1.7321	60 0	60.0
30.1	30 6	0.5015	0.8652	0.5797	1.9940	1.1559	1.7251	59 54	59.9
30.2	30 12	0.5030	0.8643	0.5820	1.9880	1.1570	1.7182	59 48	59.8
30.3	30 18	0.5045	0.8634	0.5844	1.9821	1.1582	1.7113	59 42	59.7
30.4	30 24	0.5060	0.8625	0.5867	1.9762	1.1594	1.7045	59 36	59.6
30.5	30 30	0.5075	0.8616	0.5890	1.9703	1.1606	1.6977	59 30	59.5
30.6	30 36	0.5090	0.8607	0.5914	1.9645	1.1618	1.6909	59 24	59.4
30.7	30 42	0.5105	0.8599	0.5938	1.9587	1.1630	1.6842	59 18	59.3
30.8	30 48	0.5120	0.8590	0.5961	1.9530	1.1642	1.6775	59 12	59.2
30.9	30 54	0.5135	0.8581	0.5985	1.9473	1.1654	1.6709	59 6	59.1
31.0	31 0	0.5150	0.8572	0.6009	1.9416	1.1666	1.6643	59 0	59.0
31.1	31 6	0.5165	0.8563	0.6032	1.9360	1.1679	1.6577	58 54	58.9
31.2	31 12	0.5180	0.8554	0.6056	1.9304	1.1691	1.6512	58 48	58.8
31.3	31 18	0.5195	0.8545	0.6080	1.9249	1.1703	1.6447	58 42	58.7
31.4	31 24	0.5210	0.8536	0.6104	1.9194	1.1716	1.6383	58 36	58.6
31.5	31 30	0.5225	0.8526	0.6128	1.9139	1.1728	1.6319	58 30	58.5
31.6	31 36	0.5240	0.8517	0.6152	1.9084	1.1741	1.6255	58 24	58.4
31.7	31 42	0.5255	0.8508	0.6176	1.9031	1.1753	1.6191	58 18	58.3
31.8	31 48	0.5270	0.8499	0.6200	1.8977	1.1766	1.6128	58 12	58.2
31.9	31 54	0.5284	0.8490	0.6224	1.8924	1.1779	1.6066	58 6	58.1
32.0	32 0	0.5299	0.8480	0.6249	1.8871	1.1792	1.6003	58 0	58.0
32.1	32 6	0.5314	0.8471	0.6273	1.8818	1.1805	1.5941	57 54	57.9
32.2	32 12	0.5329	0.8462	0.6297	1.8766	1.1818	1.5880	57 48	57.8
32.3	32 18	0.5344	0.8453	0.6322	1.8714	1.1831	1.5818	57 42	57.7
32.4	32 24	0.5358	0.8443	0.6346	1.8663	1.1844	1.5757	57 36	57.6
32.5	32 30	0.5373	0.8434	0.6371	1.8612	1.1857	1.5697	57 30	57.5
32.6	32 36	0.5388	0.8425	0.6395	1.8561	1.1870	1.5637	57 24	57.4
32.7	32 42	0.5402	0.8415	0.6420	1.8510	1.1883	1.5577	57 18	57.3
32.8	32 48	0.5417	0.8406	0.6445	1.8460	1.1897	1.5517	57 12	57.2
32.9	32 54	0.5432	0.8396	0.6469	1.8410	1.1910	1.5458	57 6	57.1
33.0	33 0	0.5446	0.8387	0.6494	1.8361	1.1924	1.5399	57 0	57.0
33.1	33 6	0.5461	0.8377	0.6519	1.8312	1.1937	1.5340	56 54	56.9
33.2	33 12	0.5476	0.8368	0.6544	1.8263	1.1951	1.5282	56 48	56.8
33.3	33 18	0.5490	0.8358	0.6569	1.8214	1.1964	1.5224	56 42	56.7
33.4	33 24	0.5505	0.8348	0.6594	1.8166	1.1978	1.5166	56 36	56.6
33.5	33 30	0.5519	0.8339	0.6619	1.8118	1.1992	1.5108	56 30	56.5
33.6	33 36	0.5534	0.8329	0.6644	1.8070	1.2006	1.5051	56 24	56.4
33.7	33 42	0.5548	0.8320	0.6669	1.8023	1.2020	1.4994	56 18	56.3
33.8	33 48	0.5563	0.8310	0.6694	1.7976	1.2034	1.4938	56 12	56.2
33.9	33 54	0.5577	0.8300	0.6720	1.7929	1.2048	1.4882	56 6	56.1
34.0	34 0	0.5592	0.8290	0.6745	1.7883	1.2062	1.4826	56 0	56.0
34.1	34 6	0.5606	0.8281	0.6771	1.7837	1.2076	1.4770	55 54	55.9
34.2	34 12	0.5621	0.8271	0.6796	1.7791	1.2091	1.4715	55 48	55.8
34.3	34 18	0.5635	0.8261	0.6822	1.7745	1.2105	1.4659	55 42	55.7
34.4	34 24	0.5650	0.8251	0.6847	1.7700	1.2120	1.4605	55 36	55.6
34.5	34 30	0.5664	0.8241	0.6873	1.7655	1.2134	1.4550	55 30	55.5
34.6	34 36	0.5678	0.8231	0.6899	1.7610	1.2149	1.4496	55 24	55.4
34.7	34 42	0.5693	0.8221	0.6924	1.7566	1.2163	1.4442	55 18	55.3
34.8	34 48	0.5707	0.8211	0.6950	1.7522	1.2178	1.4388	55 12	55.2
34.9	34 54	0.5721	0.8202	0.6976	1.7478	1.2193	1.4335	55 6	55.1
		cos θ	sin θ	cot θ	sec θ	csc θ	tan θ	deg min	θ deg

TABLE III (*continued*)

θ deg	deg	min	sin θ	cos θ	tan θ	csc θ	sec θ	cot θ	deg	min	θ deg
35.0	35	0	0.5736	0.8192	0.7002	1.7434	1.2208	1.4281	55	0	55.0
35.1	35	6	0.5750	0.8181	0.7028	1.7391	1.2223	1.4229	54	54	54.9
35.2	35	12	0.5764	0.8171	0.7054	1.7348	1.2238	1.4176	54	48	54.8
35.3	35	18	0.5779	0.8161	0.7080	1.7305	1.2253	1.4124	54	42	54.7
35.4	35	24	0.5793	0.8151	0.7107	1.7263	1.2268	1.4071	54	36	54.6
35.5	35	30	0.5807	0.8141	0.7133	1.7221	1.2283	1.4019	54	30	54.5
35.6	35	36	0.5821	0.8131	0.7159	1.7179	1.2299	1.3968	54	24	54.4
35.7	35	42	0.5835	0.8121	0.7186	1.7137	1.2314	1.3916	54	18	54.3
35.8	35	48	0.5850	0.8111	0.7212	1.7095	1.2329	1.3865	54	12	54.2
35.9	35	54	0.5864	0.8100	0.7239	1.7054	1.2345	1.3814	54	6	54.1
36.0	36	0	0.5878	0.8090	0.7265	1.7013	1.2361	1.3764	54	0	54.0
36.1	36	6	0.5892	0.8080	0.7292	1.6972	1.2376	1.3713	53	54	53.9
36.2	36	12	0.5906	0.8070	0.7319	1.6932	1.2392	1.3663	53	48	53.8
36.3	36	18	0.5920	0.8059	0.7346	1.6892	1.2408	1.3613	53	42	53.7
36.4	36	24	0.5934	0.8049	0.7373	1.6852	1.2424	1.3564	53	36	53.6
36.5	36	30	0.5948	0.8039	0.7400	1.6812	1.2440	1.3514	53	30	53.5
36.6	36	36	0.5962	0.8028	0.7427	1.6772	1.2456	1.3465	53	24	53.4
36.7	36	42	0.5976	0.8018	0.7454	1.6733	1.2472	1.3416	53	18	53.3
36.8	36	48	0.5990	0.8007	0.7481	1.6694	1.2489	1.3367	53	12	53.2
36.9	36	54	0.6004	0.7997	0.7508	1.6655	1.2505	1.3319	53	6	53.1
37.0	37	0	0.6018	0.7986	0.7536	1.6616	1.2521	1.3270	53	0	53.0
37.1	37	6	0.6032	0.7976	0.7563	1.6578	1.2538	1.3222	52	54	52.9
37.2	37	12	0.6046	0.7965	0.7590	1.6540	1.2554	1.3175	52	48	52.8
37.3	37	18	0.6060	0.7955	0.7618	1.6502	1.2571	1.3127	52	42	52.7
37.4	37	24	0.6074	0.7944	0.7646	1.6464	1.2588	1.3079	52	36	52.6
37.5	37	30	0.6088	0.7934	0.7673	1.6427	1.2605	1.3032	52	30	52.5
37.6	37	36	0.6101	0.7923	0.7701	1.6390	1.2622	1.2985	52	24	52.4
37.7	37	42	0.6115	0.7912	0.7729	1.6353	1.2639	1.2938	52	18	52.3
37.8	37	48	0.6129	0.7902	0.7757	1.6316	1.2656	1.2892	52	12	52.2
37.9	37	54	0.6143	0.7891	0.7785	1.6279	1.2673	1.2846	52	6	52.1
38.0	38	0	0.6157	0.7880	0.7813	1.6243	1.2690	1.2799	52	0	52.0
38.1	38	6	0.6170	0.7869	0.7841	1.6207	1.2708	1.2753	51	54	51.9
38.2	38	12	0.6184	0.7859	0.7869	1.6171	1.2725	1.2708	51	48	51.8
38.3	38	18	0.6198	0.7848	0.7898	1.6135	1.2742	1.2662	51	42	51.7
38.4	38	24	0.6211	0.7837	0.7926	1.6099	1.2760	1.2617	51	36	51.6
38.5	38	30	0.6225	0.7826	0.7954	1.6064	1.2778	1.2572	51	30	51.5
38.6	38	36	0.6239	0.7815	0.7983	1.6029	1.2796	1.2527	51	24	51.4
38.7	38	42	0.6252	0.7804	0.8012	1.5994	1.2813	1.2482	51	18	51.3
38.8	38	48	0.6266	0.7793	0.8040	1.5959	1.2831	1.2437	51	12	51.2
38.9	38	54	0.6280	0.7782	0.8069	1.5925	1.2849	1.2393	51	6	51.1
39.0	39	0	0.6293	0.7771	0.8098	1.5890	1.2868	1.2349	51	0	51.0
39.1	39	6	0.6307	0.7760	0.8127	1.5856	1.2886	1.2305	50	54	50.9
39.2	39	12	0.6320	0.7749	0.8156	1.5822	1.2904	1.2261	50	48	50.8
39.3	39	18	0.6334	0.7738	0.8185	1.5788	1.2923	1.2218	50	42	50.7
39.4	39	24	0.6347	0.7727	0.8214	1.5755	1.2941	1.2174	50	36	50.6
39.5	39	30	0.6361	0.7716	0.8243	1.5721	1.2960	1.2131	50	30	50.5
39.6	39	36	0.6374	0.7705	0.8273	1.5688	1.2978	1.2088	50	24	50.4
39.7	39	42	0.6388	0.7694	0.8302	1.5655	1.2997	1.2045	50	18	50.3
39.8	39	48	0.6401	0.7683	0.8332	1.5622	1.3016	1.2002	50	12	50.2
39.9	39	54	0.6414	0.7672	0.8361	1.5590	1.3035	1.1960	50	6	50.1
			cos θ	sin θ	cot θ	sec θ	csc θ	tan θ	deg	min	θ deg

III

TABLE III (*continued*)

θ deg	deg min	sin θ	cos θ	tan θ	csc θ	sec θ	cot θ		
40.0	40 0	0.6428	0.7660	0.8391	1.5557	1.3054	1.1918	50 0	50.0
40.1	40 6	0.6441	0.7649	0.8421	1.5525	1.3073	1.1875	49 54	49.9
40.2	40 12	0.6455	0.7638	0.8451	1.5493	1.3092	1.1833	49 48	49.8
40.3	40 18	0.6468	0.7627	0.8481	1.5461	1.3112	1.1792	49 42	49.7
40.4	40 24	0.6481	0.7615	0.8511	1.5429	1.3131	1.1750	49 36	49.6
40.5	40 30	0.6494	0.7604	0.8541	1.5398	1.3151	1.1708	49 30	49.5
40.6	40 36	0.6508	0.7593	0.8571	1.5366	1.3171	1.1667	49 24	49.4
40.7	40 42	0.6521	0.7581	0.8601	1.5335	1.3190	1.1626	49 18	49.3
40.8	40 48	0.6534	0.7570	0.8632	1.5304	1.3210	1.1585	49 12	49.2
40.9	40 54	0.6547	0.7559	0.8662	1.5273	1.3230	1.1544	49 6	49.1
41.0	41 0	0.6561	0.7547	0.8693	1.5243	1.3250	1.1504	49 0	49.0
41.1	41 6	0.6574	0.7536	0.8724	1.5212	1.3270	1.1463	48 54	48.9
41.2	41 12	0.6587	0.7524	0.8754	1.5182	1.3291	1.1423	48 48	48.8
41.3	41 18	0.6600	0.7513	0.8785	1.5151	1.3311	1.1383	48 42	48.7
41.4	41 24	0.6613	0.7501	0.8816	1.5121	1.3331	1.1343	48 36	48.6
41.5	41 30	0.6626	0.7490	0.8847	1.5092	1.3352	1.1303	48 30	48.5
41.6	41 36	0.6639	0.7478	0.8878	1.5062	1.3373	1.1263	48 24	48.4
41.7	41 42	0.6652	0.7466	0.8910	1.5032	1.3393	1.1224	48 18	48.3
41.8	41 48	0.6665	0.7455	0.8941	1.5003	1.3414	1.1184	48 12	48.2
41.9	41 54	0.6678	0.7443	0.8972	1.4974	1.3435	1.1145	48 6	48.1
42.0	42 0	0.6691	0.7431	0.9004	1.4945	1.3456	1.1106	48 0	48.0
42.1	42 6	0.6704	0.7420	0.9036	1.4916	1.3478	1.1067	47 54	47.9
42.2	42 12	0.6717	0.7408	0.9067	1.4887	1.3499	1.1028	47 48	47.8
42.3	42 18	0.6730	0.7396	0.9099	1.4859	1.3520	1.0990	47 42	47.7
42.4	42 24	0.6743	0.7385	0.9131	1.4830	1.3542	1.0951	47 36	47.6
42.5	42 30	0.6756	0.7373	0.9163	1.4802	1.3563	1.0913	47 30	47.5
42.6	42 36	0.6769	0.7361	0.9195	1.4774	1.3585	1.0875	47 24	47.4
42.7	42 42	0.6782	0.7349	0.9228	1.4746	1.3607	1.0837	47 18	47.3
42.8	42 48	0.6794	0.7337	0.9260	1.4718	1.3629	1.0799	47 12	47.2
42.9	42 54	0.6807	0.7325	0.9293	1.4690	1.3651	1.0761	47 6	47.1
43.0	43 0	0.6820	0.7314	0.9325	1.4663	1.3673	1.0724	47 0	47.0
43.1	43 6	0.6833	0.7302	0.9358	1.4635	1.3696	1.0686	46 54	46.9
43.2	43 12	0.6845	0.7290	0.9391	1.4608	1.3718	1.0649	46 48	46.8
43.3	43 18	0.6858	0.7278	0.9424	1.4581	1.3741	1.0612	46 42	46.7
43.4	43 24	0.6871	0.7266	0.9457	1.4554	1.3763	1.0575	46 36	46.6
43.5	43 30	0.6884	0.7254	0.9490	1.4527	1.3786	1.0538	46 30	46.5
43.6	43 36	0.6896	0.7242	0.9523	1.4501	1.3809	1.0501	46 24	46.4
43.7	43 42	0.6909	0.7230	0.9556	1.4474	1.3832	1.0464	46 18	46.3
43.8	43 48	0.6921	0.7218	0.9590	1.4448	1.3855	1.0428	46 12	46.2
43.9	43 54	0.6934	0.7206	0.9623	1.4422	1.3878	1.0392	46 6	46.1
44.0	44 0	0.6947	0.7193	0.9657	1.4396	1.3902	1.0355	46 0	46.0
44.1	44 6	0.6959	0.7181	0.9691	1.4370	1.3925	1.0319	45 54	45.9
44.2	44 12	0.6972	0.7169	0.9725	1.4344	1.3949	1.0283	45 48	45.8
44.3	44 18	0.6984	0.7157	0.9759	1.4318	1.3972	1.0247	45 42	45.7
44.4	44 24	0.6997	0.7145	0.9793	1.4293	1.3996	1.0212	45 36	45.6
44.5	44 30	0.7009	0.7133	0.9827	1.4267	1.4020	1.0176	45 30	45.5
44.6	44 36	0.7022	0.7120	0.9861	1.4242	1.4044	1.0141	45 24	45.4
44.7	44 42	0.7034	0.7108	0.9896	1.4217	1.4069	1.0105	45 18	45.3
44.8	44 48	0.7046	0.7096	0.9930	1.4192	1.4093	1.0070	45 12	45.2
44.9	44 54	0.7059	0.7083	0.9965	1.4167	1.4118	1.0035	45 6	45.1
45.0	45 0	0.7071	0.7071	1.0000	1.4142	1.4142	1.0000	45 0	45.0
		cos θ	sin θ	cot θ	sec θ	csc θ	tan θ	deg min	θ deg

ANSWERS TO ODD-NUMBERED EXERCISES

Exercise Set 1.1

1. II	**3.** IV	**5.** I	**7.** IV	**9.** II	**11.** IV
13. II	**15.** IV	**17.** II	**19.** IV	**21.** III	**23.** III
25. I	**27.** IV	**29.** 4/5	**31.** $-\sqrt{3}/2$		
33. a. $\sqrt{3}/2$ **b.** $2/\sqrt{3}$ **c.** 2		**35.** III	**37.** IV	**39.** II	

Exercise Set 1.2

1. $s = 0, (1, 0)$; $s = \pi/2, (0, 1)$; $s = \pi, (-1, 0)$; $s = 3\pi/2, (0, -1)$

3. 1/4	**5.** 1	**7.** 3/4	**9.** 0	**11.** $1/\sqrt{2}$	**13.** $2/\sqrt{3}$
15. 0	**17.** 1	**19.** 0.9968	**21.** 0.3429	**23.** 0.7518	**25.** 0.2522
27. 0.2838	**29.** 0.3830	**31.** 0.7420	**33.** 1.5747	**35.** 0.6768	**37.** 2.2081
39. 13.5007	**43.** 0.0100	**45.** 0.8776	**47.** 0.8415		

Exercise Set 1.3

1. $x = 0, (1, 0)$; $x = \pi/4, (1/\sqrt{2}, 1/\sqrt{2})$; $x = \pi/2, (0, 1)$; $x = 3\pi/4, (-1/\sqrt{2}, 1/\sqrt{2})$; $x = \pi, (-1, 0)$; $x = 5\pi/4, (-1/\sqrt{2}, -1/\sqrt{2})$; $x = 3\pi/2, (0, -1)$; $x = 7\pi/4, (1/\sqrt{2}, -1/\sqrt{2})$

3.

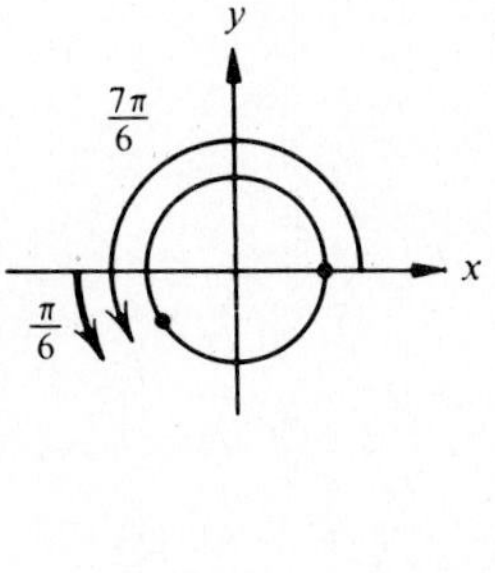

5.

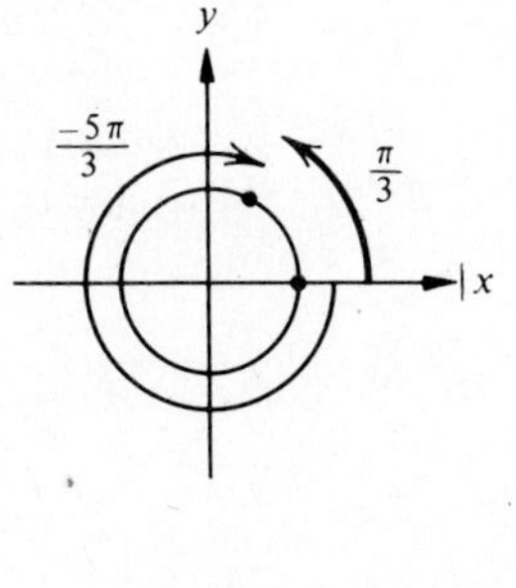

7.

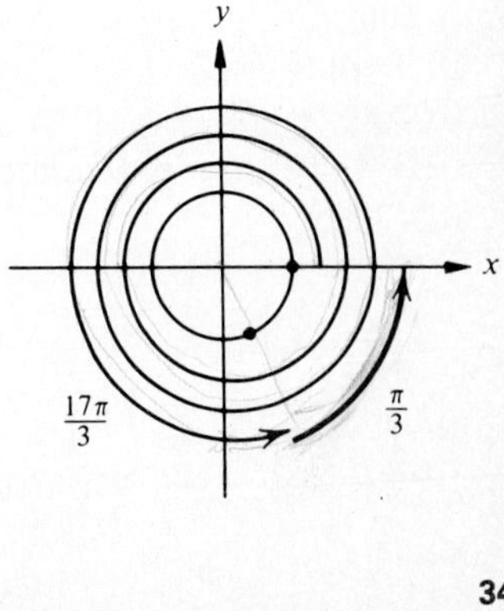

9.

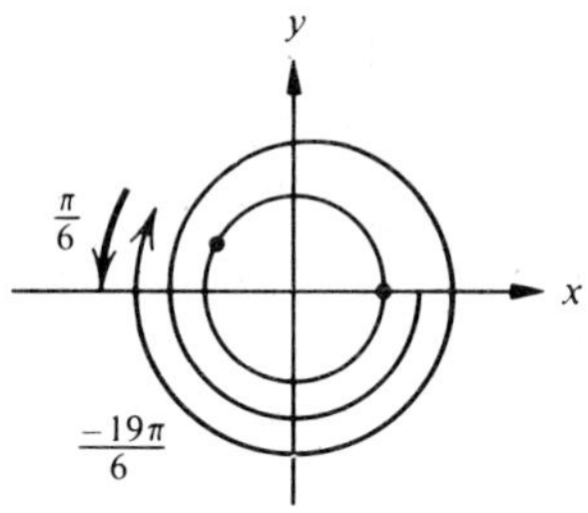

11.

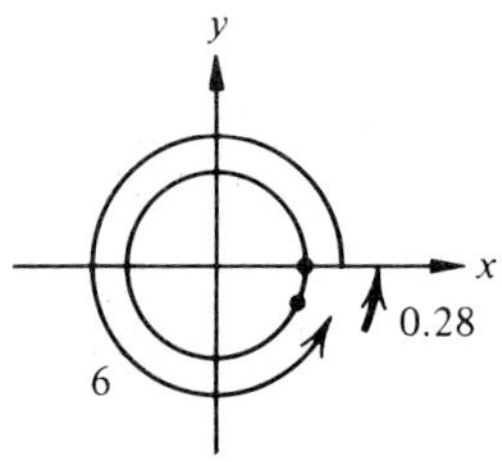

13.

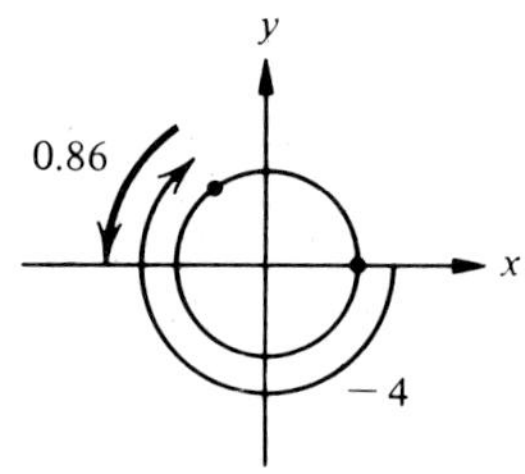

15. $-1/2$ **17.** $1/2$ **19.** $-\sqrt{3}/2$ **21.** $-1/2$ **23.** $1/\sqrt{2}$ **25.** $1/\sqrt{2}$
27. $-1/\sqrt{2}$ **29.** $\sqrt{3}/2$ **31.** 0.9819 **33.** 0.1994 **35.** -0.2182 **37.** 0.9664
39. -0.5564 **41.** -0.7584 **43.** 1.7469 **45.** -1.6771 **47.** -0.4120 **49.** -3.2487

Exercise Set 1.4

1. 1 **3.** $\sqrt{3}$ **5.** $-1/\sqrt{3}$ **7.** 2 **9.** $\sqrt{2}$ **11.** $-\sqrt{2}$
13. $-\sqrt{2}$ **15.** 2 **17.** 2 **19.** $1/\sqrt{3}$ **21.** $1/\sqrt{3}$ **23.** $-\sqrt{3}$
25. $\sqrt{3}+\sqrt{2}$ **27.** -1 **29.** $\frac{2\sqrt{6}}{3}+1$ **31.** 0.5206 **33.** 2.2081 **35.** -4.9131
37. -4.5823 **39.** -3.7591 **41.** -1.2035 **43.** 3.3528 **45.** 0.3387 **47.** -1.4907
49. II **51.** III **53.** II **55.** -1 **57.** $-2/\sqrt{3}$ **59.** $-3/4$
61. $\frac{1}{\cos x \sin x}$ **63.** $\cos^2 x$ **65.** $\frac{\cos^3 x}{\sin^2 x}$ **67.** $\frac{1}{\cos x}$ **69.** $\frac{\cos x}{\sin x}$
73. $\tan x = AT/OA = AT/1 = AT$
75. $\csc x = OP/PQ = OC/OB$ (similar triangles) $= OC/1 = OC$

Exercise Set 1.5

1. $\cos\left(\frac{3\pi}{4}+2\pi\right)$ **3.** $\tan\left(\frac{\pi}{4}+\pi\right)$ **5.** $\sin\left(\frac{5\pi}{4}+2\pi\right)$
7. $\tan\left(\frac{3\pi}{4}+3\pi\right)$ **9.** $\sin\left(\frac{7\pi}{6}+2\pi\right)$ **11.** $\sin\left(\frac{\pi}{3}+4\pi\right)$
13. 10 centimeters; -10 centimeters
15. -8 centimeters; 0 centimeter
17. 22 centimeters
19. $40/\sqrt{3}$ centimeters
21. -106 volts
23. -68.5 volts
25. 81.1 volts
27. -0.0294 ampere
29. 0.0000888 ampere
31. 0.000444 ampere
33. 0 dynes per square centimeter
35. 0 dynes per square centimeter
37. 0.242 feet

Review Exercises

1. II **2.** I **3.** IV **4.** II **5.** IV **6.** II
7. III **8.** II **9.** $-3/5$ **10.** $4/3$ **11.** $-5/4$ **12.** $-5/3$

13. $-\sqrt{2}/4$ **14.** 2 **15.** 0.5735 **16.** 0.2764 **17.** 0.8820 **18.** 0.2092
19. 3.72 **20.** -0.4122 **21.** 0.2079 **22.** 0.1437

23.

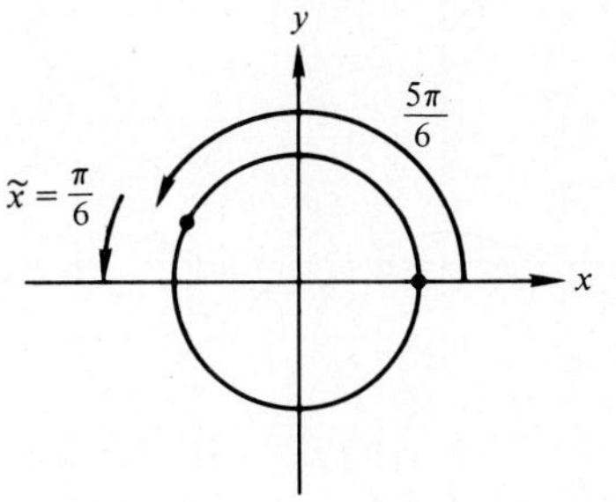

24.

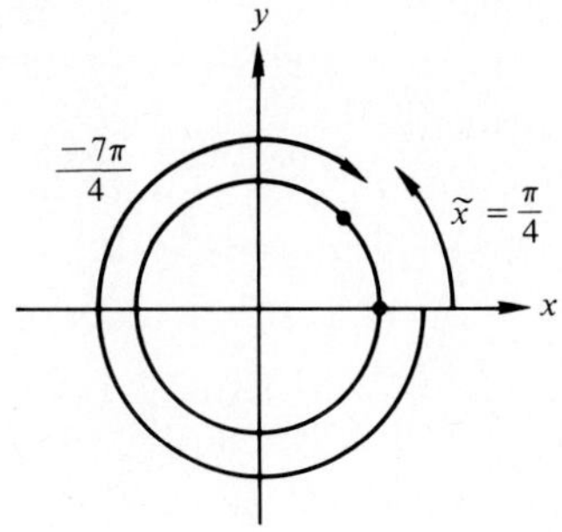

25.

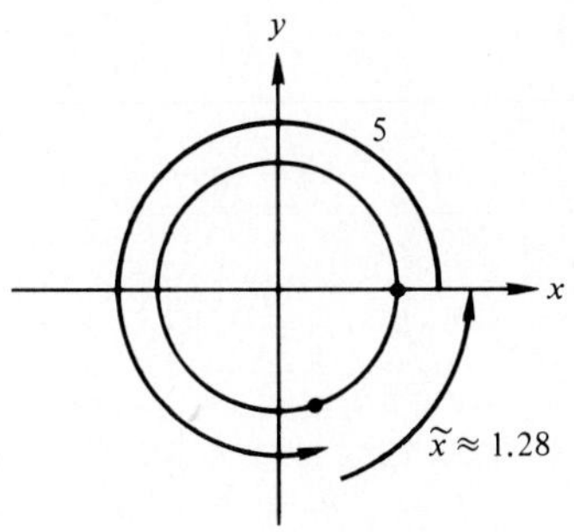

26.

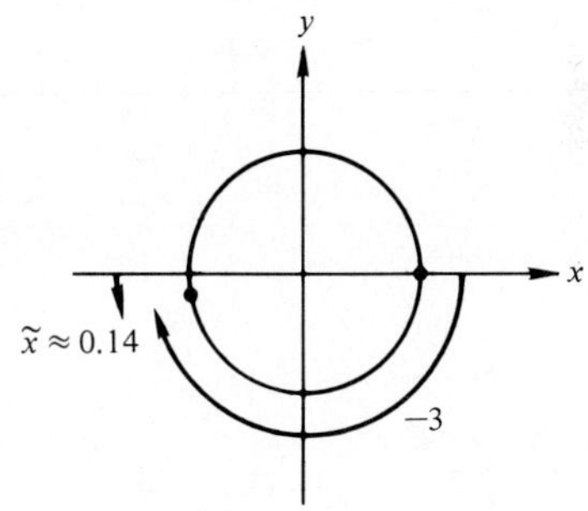

27. $1/\sqrt{2}$ **28.** $-\sqrt{3}/2$ **29.** $1/2$ **30.** $-1/\sqrt{2}$ **31.** -0.1304
32. -0.2474 **33.** $-1/\sqrt{3}$ **34.** $-1/\sqrt{3}$ **35.** $-\sqrt{2}$ **36.** $2/\sqrt{3}$
37. -2 **38.** $2/\sqrt{3}$ **39.** 1 **40.** $-\sqrt{3}$ **41.** $3\sqrt{2}+2$
42. $1/2+\sqrt{3}$ **43.** 0.9131 **44.** 1.7953 **45.** -1.2475 **46.** -11.0811
47. 0.0382 **48.** -0.3114 **49.** I **50.** III **51.** $2/\sqrt{3}$
52. $-1/\sqrt{3}$ **53.** $\dfrac{\sin x}{\cos^2 x}$ **54.** $\dfrac{\cos x}{\sin^2 x}$ **55.** $\cos\left(\dfrac{2\pi}{3}+2\pi\right)$ **56.** $\sin\left(\dfrac{\pi}{4}+4\pi\right)$
57. 4 centimeters **58.** 0 centimeter **59.** 0 volt **60.** 0.0259 ampere

Exercise Set 2.1

1.

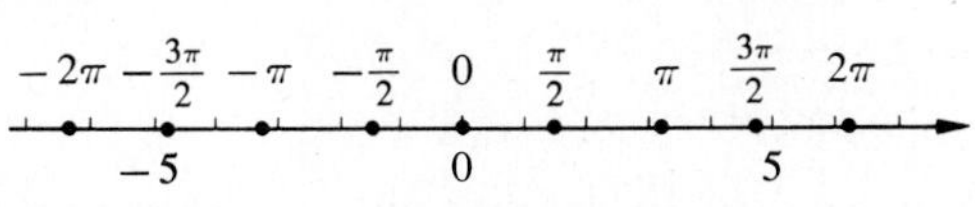

3.

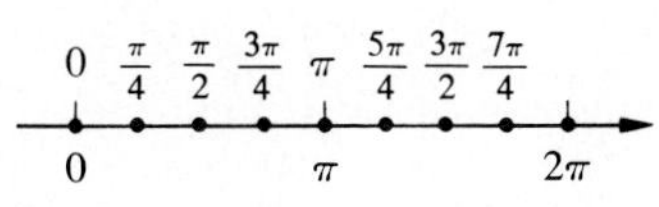

5.

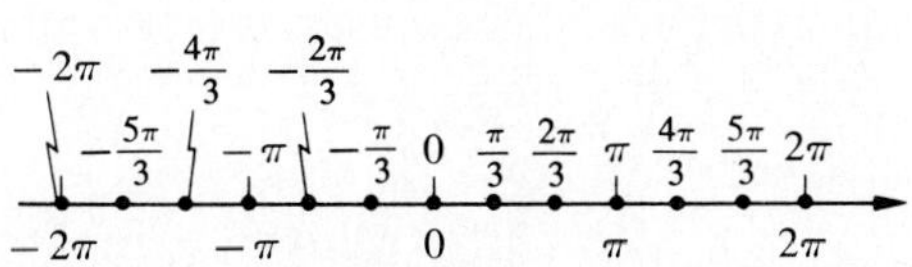

7. Zeros: $\{x \mid x = k\pi,\ k \in J\}$;
amplitude: 3

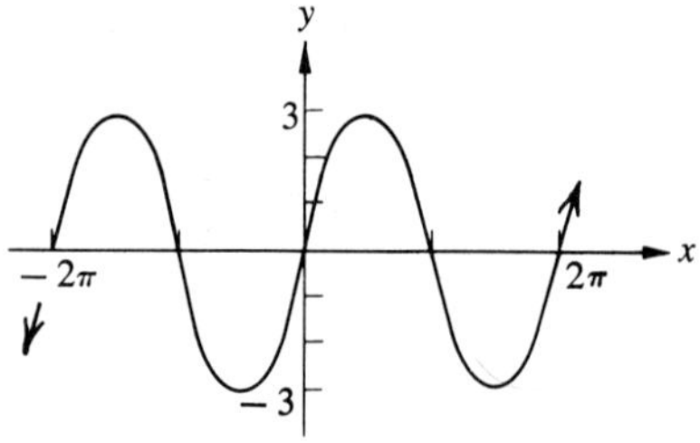

9. Zeros: $\{x \mid x = \pi/2 + k\pi,\ k \in J\}$;
amplitude: 1/2

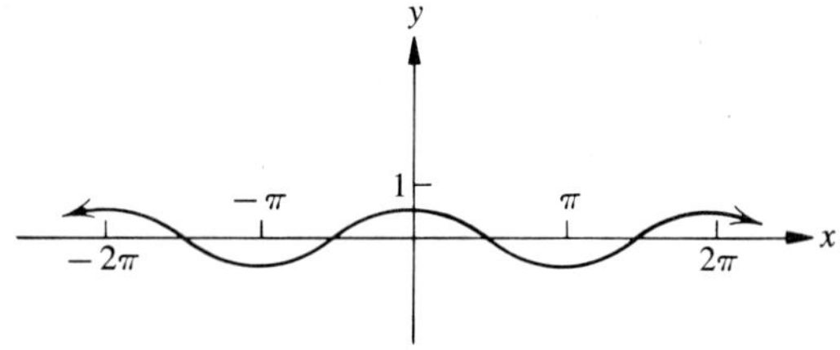

11. Zeros: $\{x \mid x = k\pi,\ k \in J\}$:
amplitude: 1

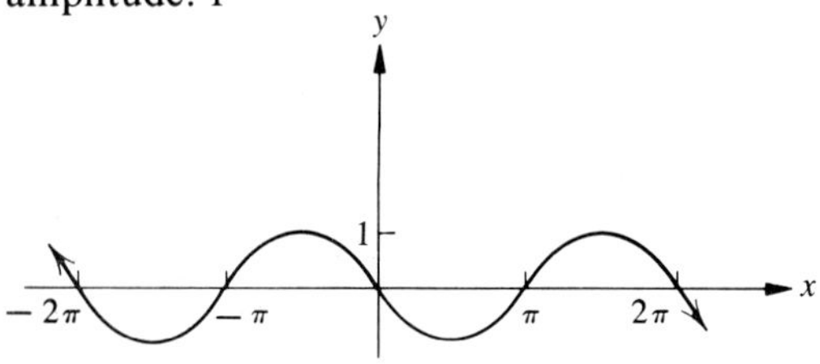

13. Zeros: $\{x \mid x = k\pi/2,\ k \in J\}$;
amplitude: 1

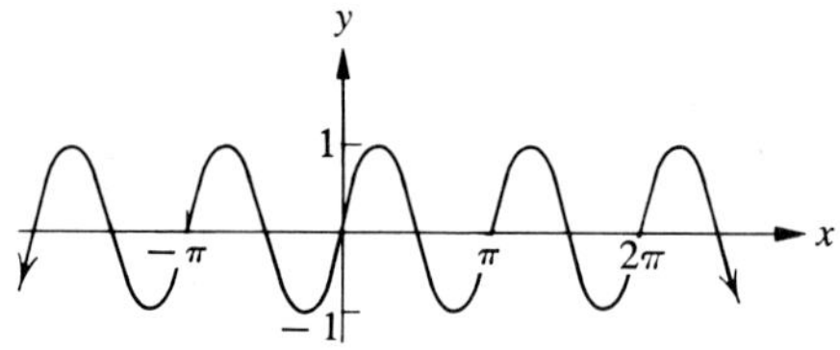

15. Zeros: $\{x \mid x = \pi + k \cdot 2\pi,\ k \in J\}$;
amplitude: 1

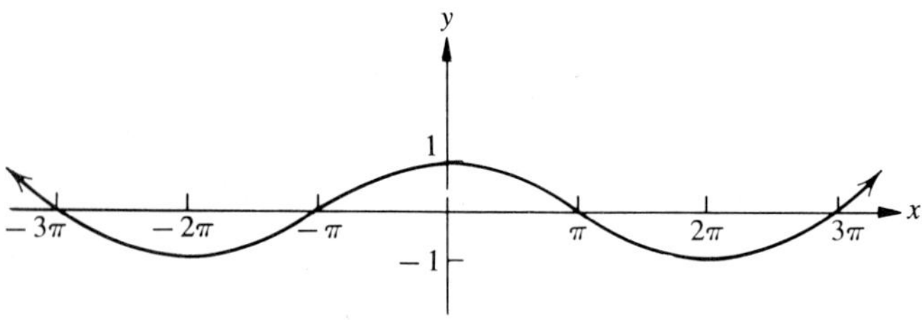

17. Zeros: $\{x \mid x = \pi/2 + k\pi,\ k \in J\}$;
amplitude: 1

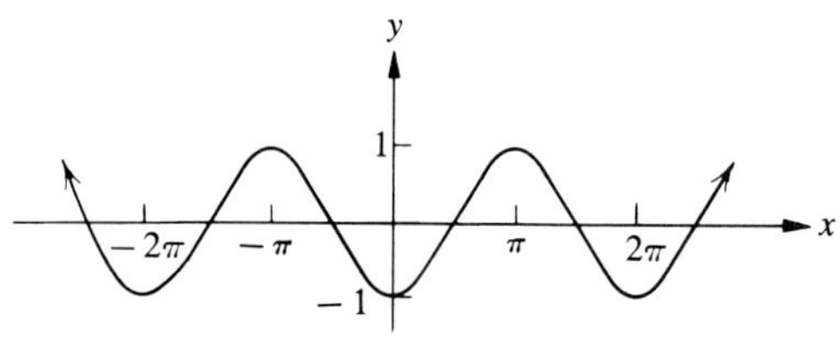

19.

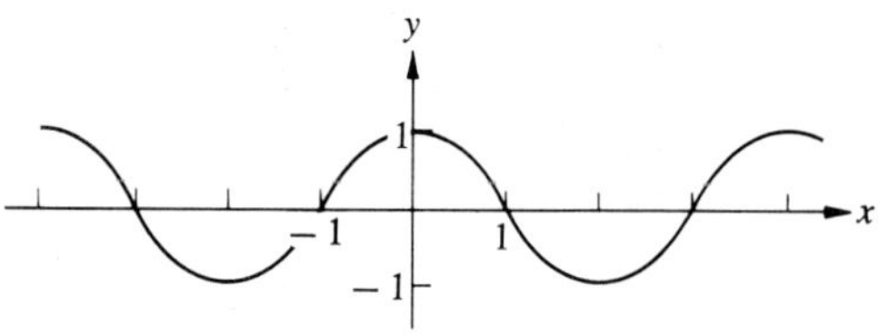

21.

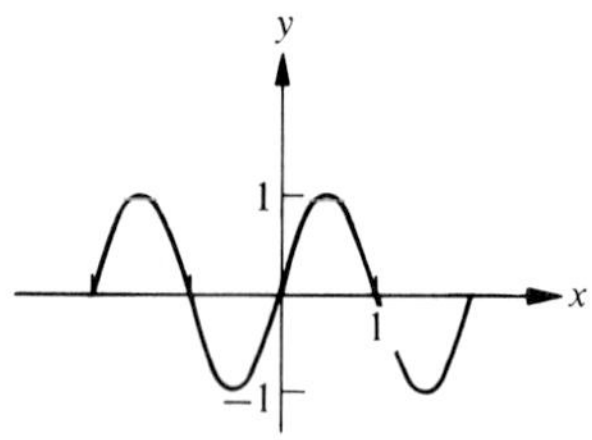

23.

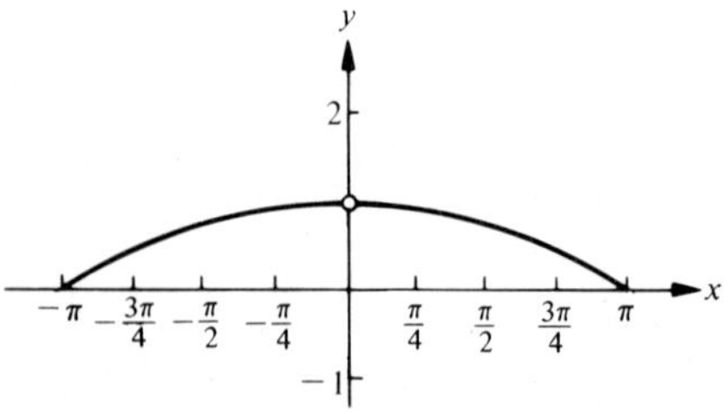

As x approaches zero from the left or right, $\dfrac{\sin x}{x}$ approaches a value of 1. The graph is not periodic because there is no number a ($a \neq 0$) such that $\dfrac{\sin(x + a)}{x + a} = \dfrac{\sin x}{x}$.

25.

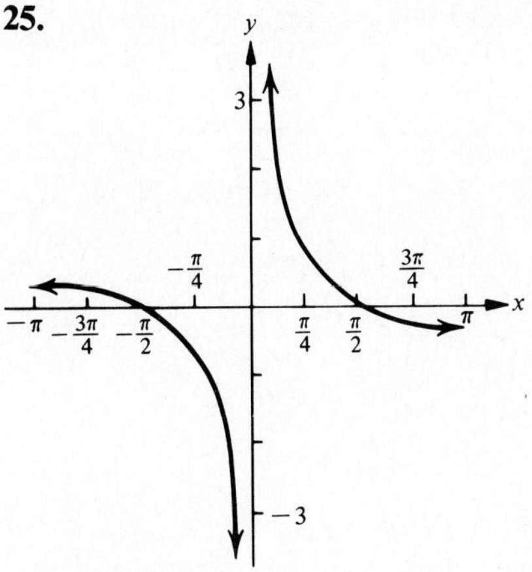

As x approaches zero from the left, $\frac{\cos x}{x}$ is negative and its absolute value becomes greater. As x approaches zero from the right, $\frac{\cos x}{x}$ is positive and its absolute value becomes greater. The graph is not periodic because there is no number $a (a \neq 0)$ such that $\frac{\cos(x + a)}{x + a} = \frac{\cos x}{x}$.

27. Period is 1.

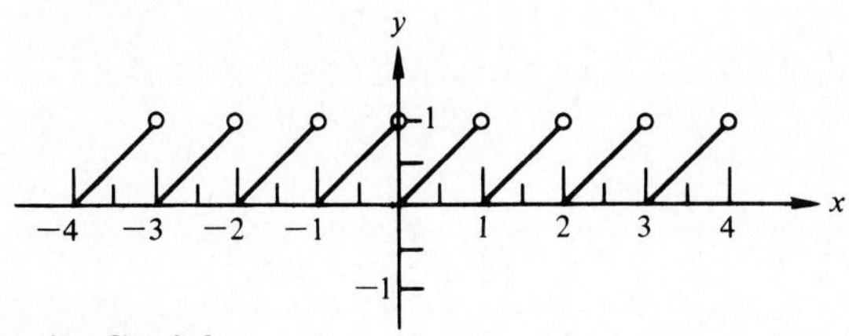

29. Period is 1/2.

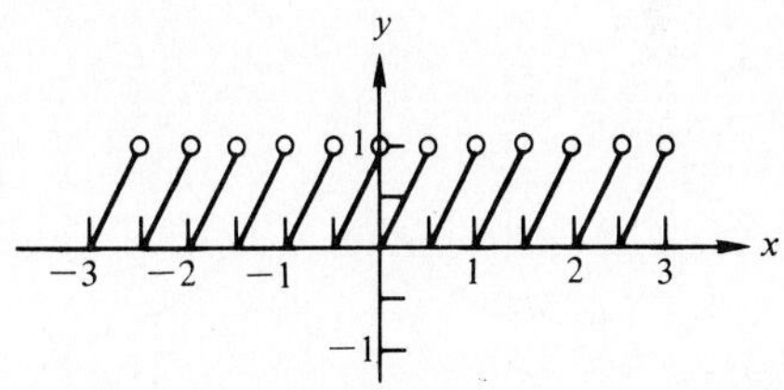

Exercise Set 2.2

1. Zeros: $\{-2\pi, -\pi, 0, \pi, 2\pi\}$; amplitude: 3

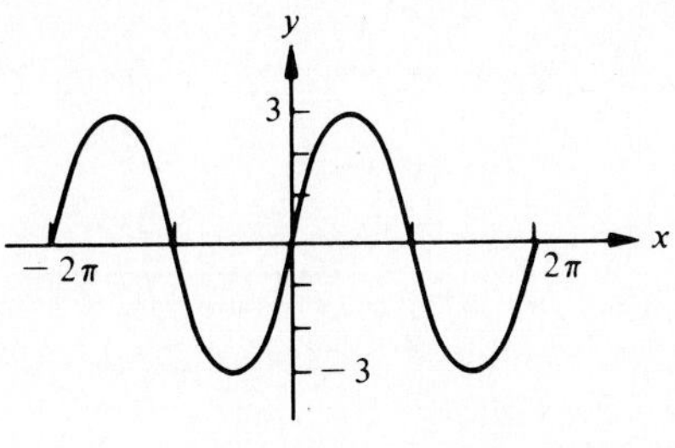

3. Zeros: $\left\{-\frac{3\pi}{2}, -\frac{\pi}{2}, \frac{\pi}{2}, \frac{3\pi}{2}\right\}$; amplitude: 2

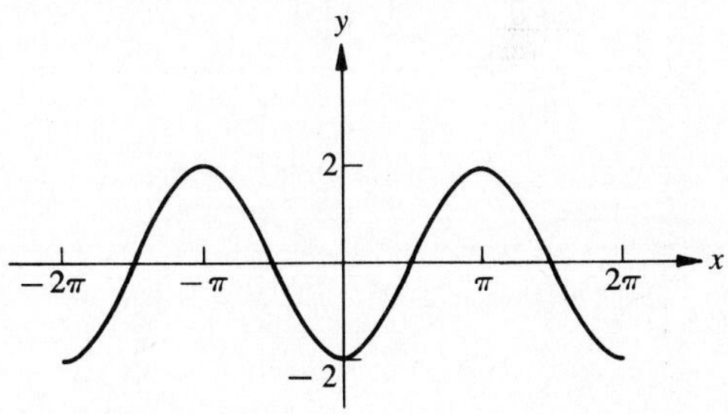

5. Zeros: $\{-2\pi, -\pi, 0, \pi, 2\pi\}$; amplitude: 1/2

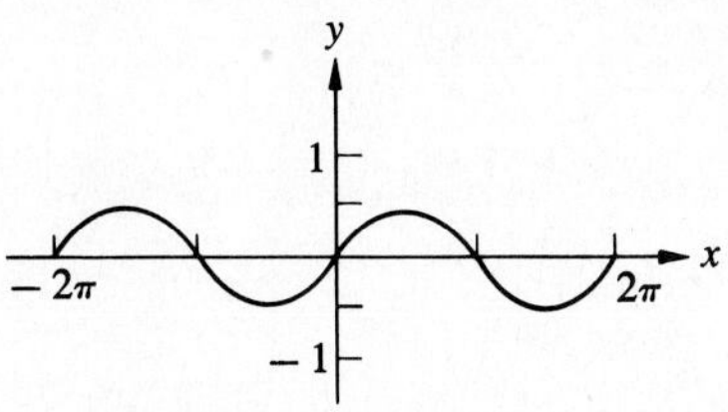

7. Zeros: $\left\{-\frac{7\pi}{4}, -\frac{5\pi}{4}, -\frac{3\pi}{4}, -\frac{\pi}{4}, \frac{\pi}{4}, \frac{3\pi}{4}, \frac{5\pi}{4}, \frac{7\pi}{4}\right\}$; amplitude: 1

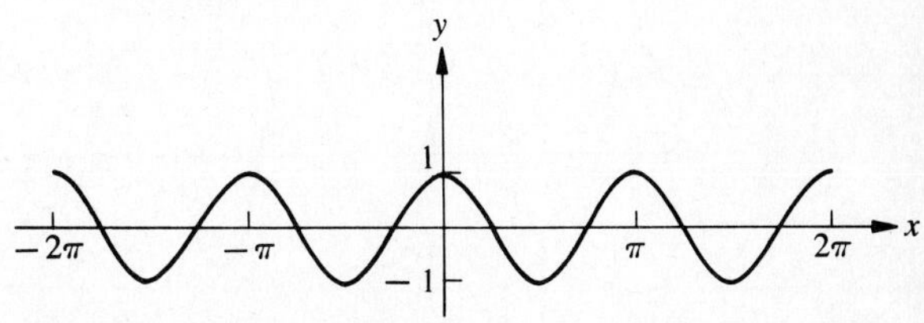

9. Zeros: $\{x \mid -2\pi, 0, 2\pi\}$; amplitude: 1

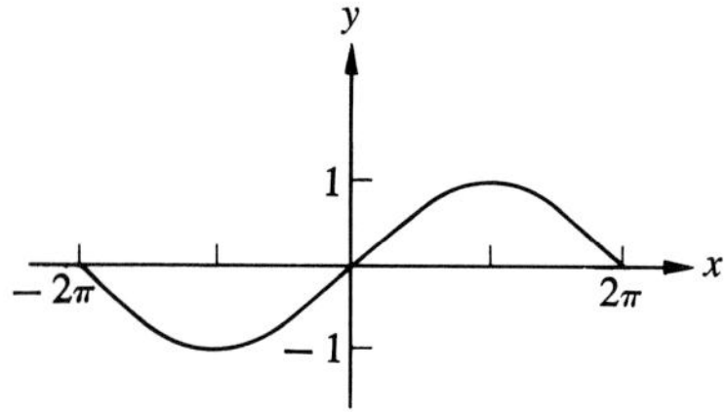

11. Zeros: $\left\{-2\pi, -\frac{3\pi}{2}, -\pi, -\frac{\pi}{2}, 0, \frac{\pi}{2}, \pi, \frac{3\pi}{2}, 2\pi\right\}$;
amplitude: 2

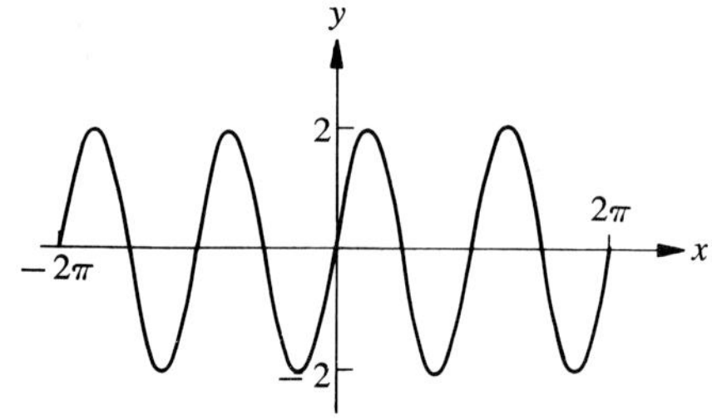

13. Zeros: $\left\{-\frac{3\pi}{2}, -\frac{\pi}{2}, \frac{\pi}{2}, \frac{3\pi}{2}\right\}$;
amplitude: 1; phase shift: π

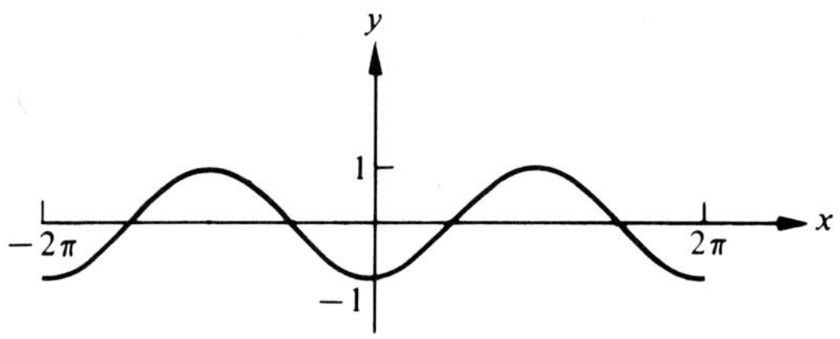

15. Zeros: $\left\{-\frac{7\pi}{4}, -\frac{3\pi}{4}, \frac{\pi}{4}, \frac{5\pi}{4}\right\}$;
amplitude: 2; phase shift: $\pi/4$

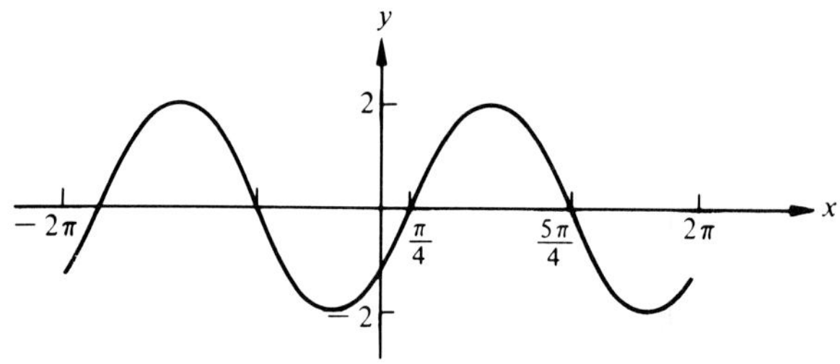

17. Zeros: $\left\{-\frac{4\pi}{3}, -\frac{\pi}{3}, \frac{2\pi}{3}, \frac{5\pi}{3}\right\}$; amplitude: 1/2; phase shift: $\pi/3$

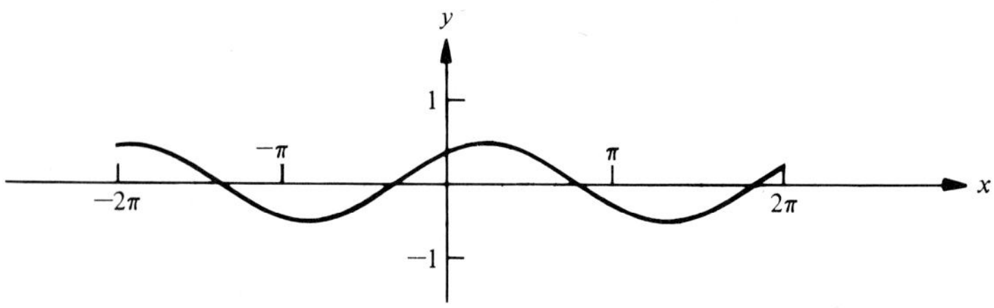

19. Zeros: $\left\{-2\pi, -\frac{3\pi}{2}, -\pi, -\frac{\pi}{2}, 0, \frac{\pi}{2}, \pi, \frac{3\pi}{2}, 2\pi\right\}$; amplitude: 2; phase shift: $\frac{\pi}{4}$

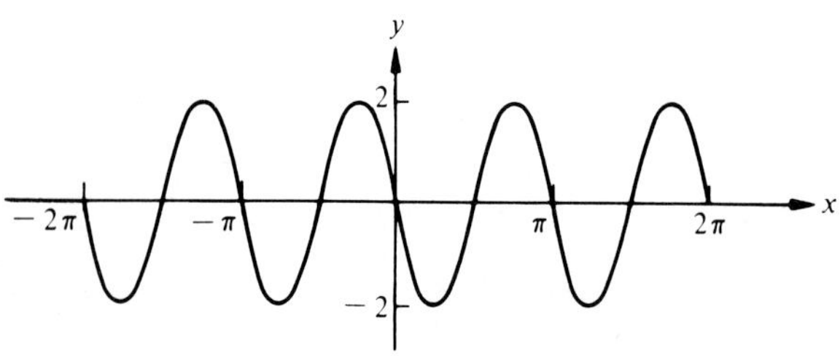

21. Zeros: $\{-\pi, \pi\}$; amplitude: 1/2; phase shift: π

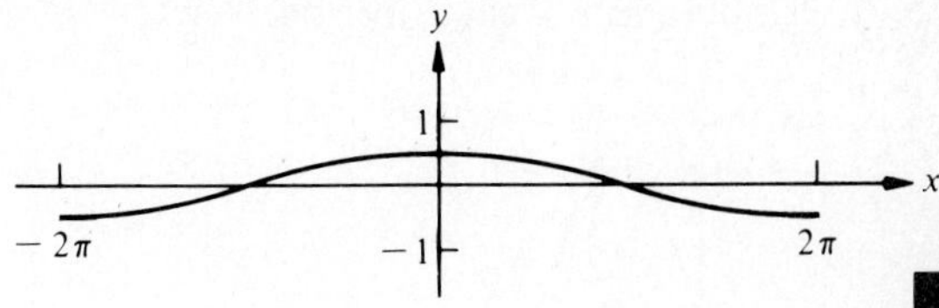

23. Zeros: $\left\{-\frac{\pi}{2}\right\}$; amplitude: 3; phase shift: π

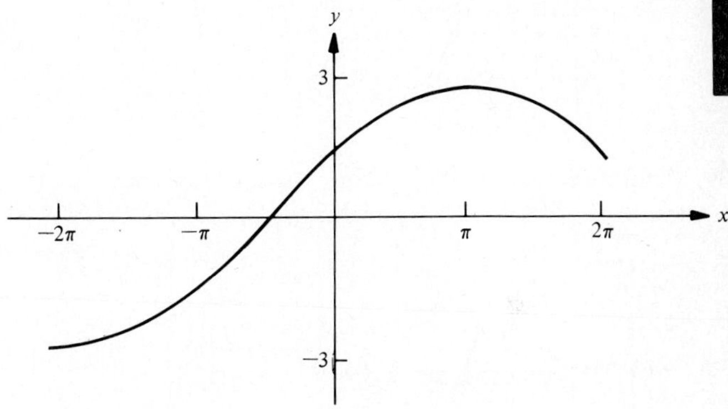

25. Zeros: $\left\{\frac{1}{2}, \frac{3}{2}, \frac{5}{2}, \frac{7}{2}\right\}$; amplitude: 1

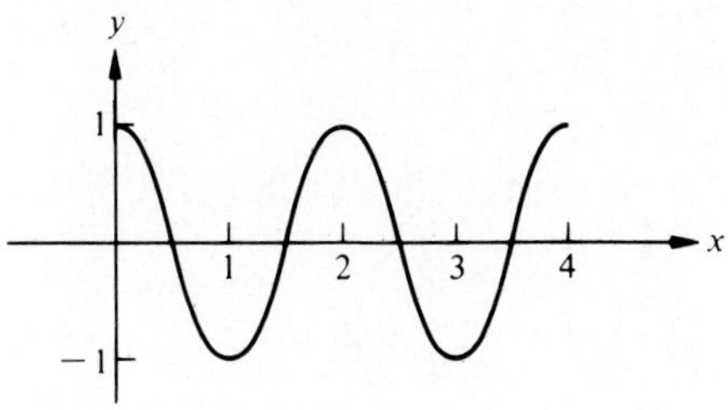

27. Zeros: $\{0, 3, 6, 9, 12\}$; amplitude: 1/2

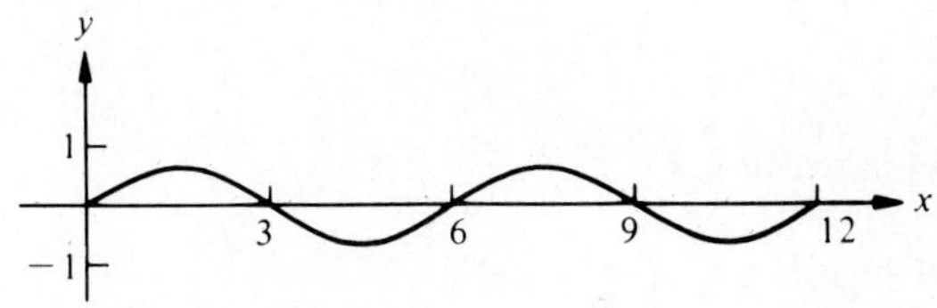

29.

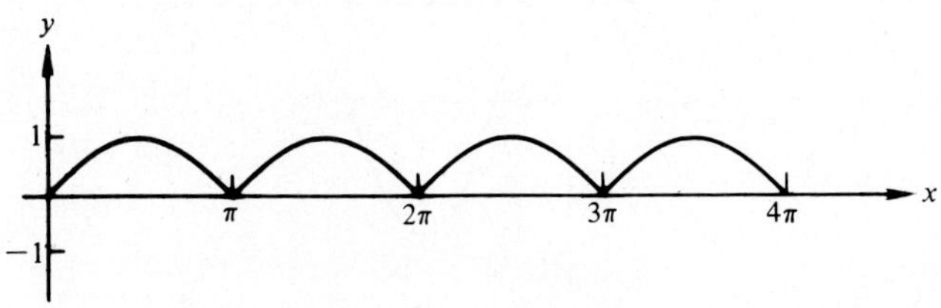

31.

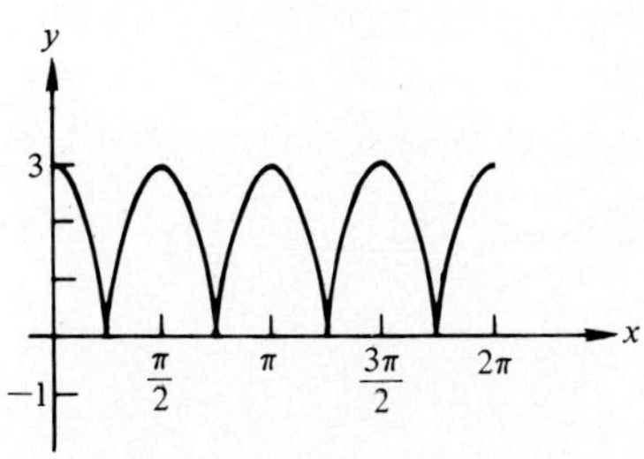

33.

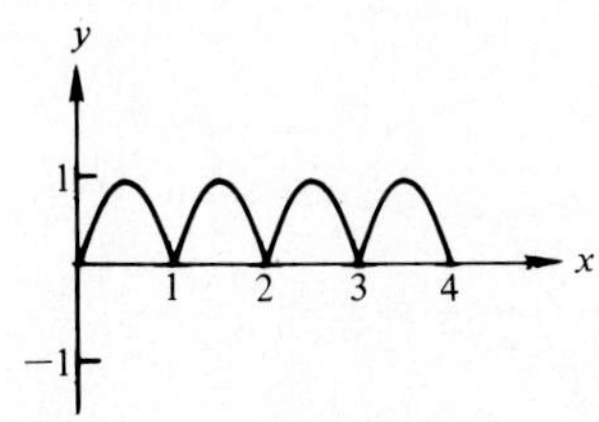

35. **a.** Amplitude: 20 cm.; period: 1 sec.

b.

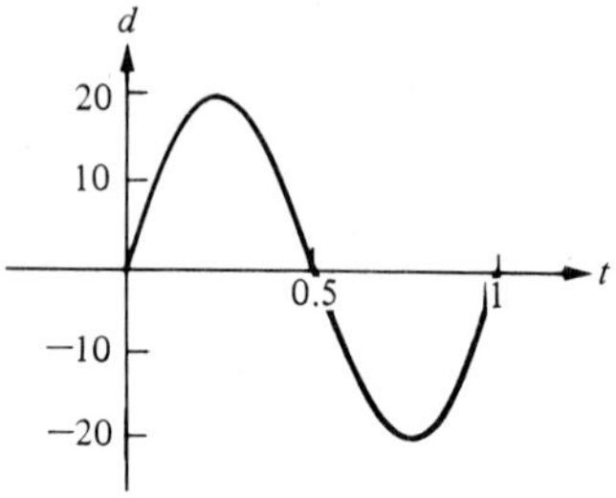

37. **a.** Amplitude: 4.2 volts; period: 0.025 sec.

b.

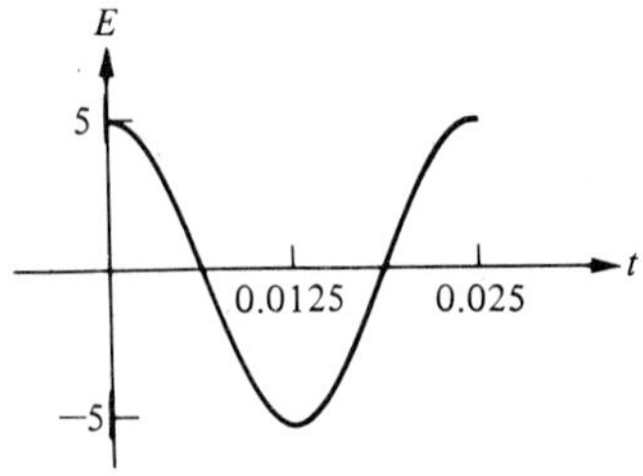

39. Amplitude: 10 dynes/sq. cm; period: 0.01 sec.

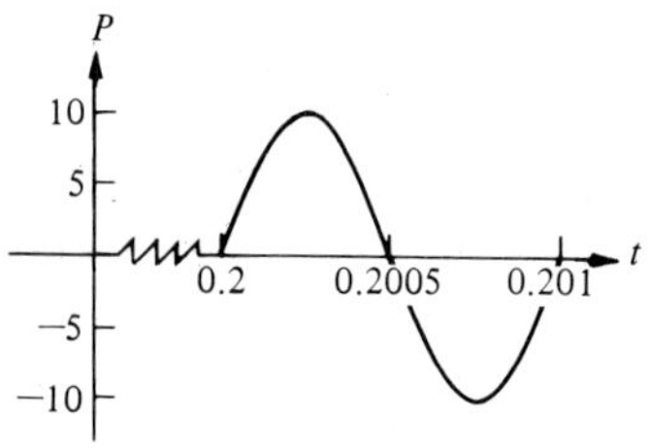

Exercise Set 2.3

1.

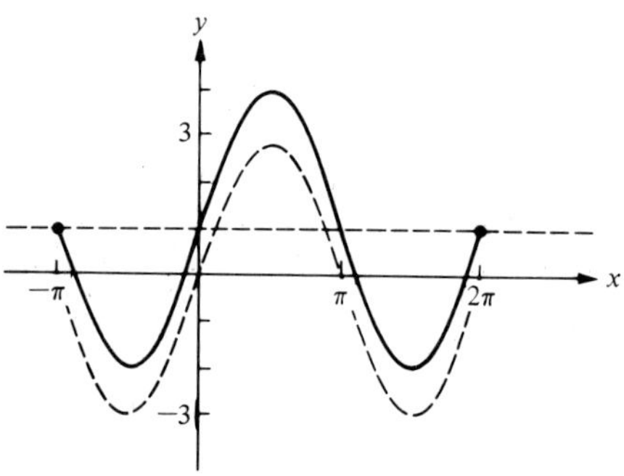

3.

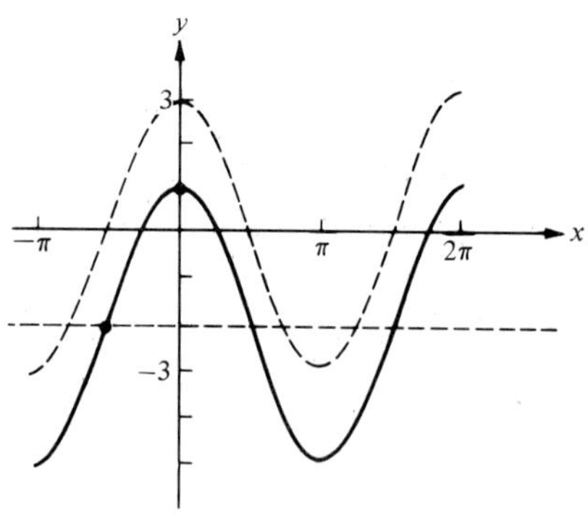

5.

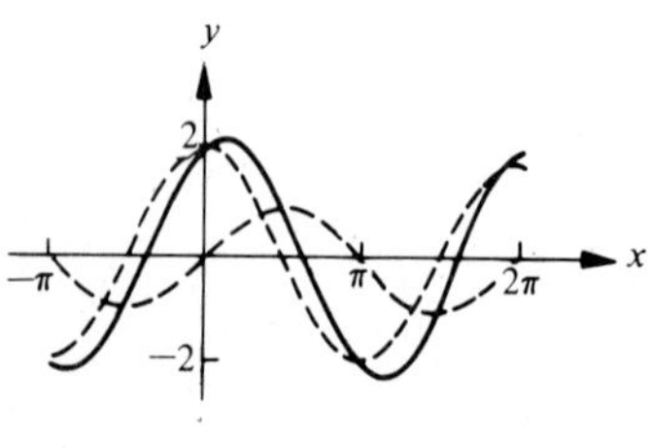

7.

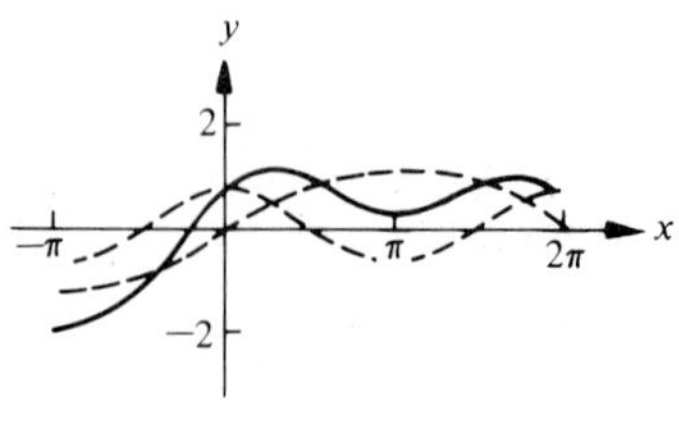

9.

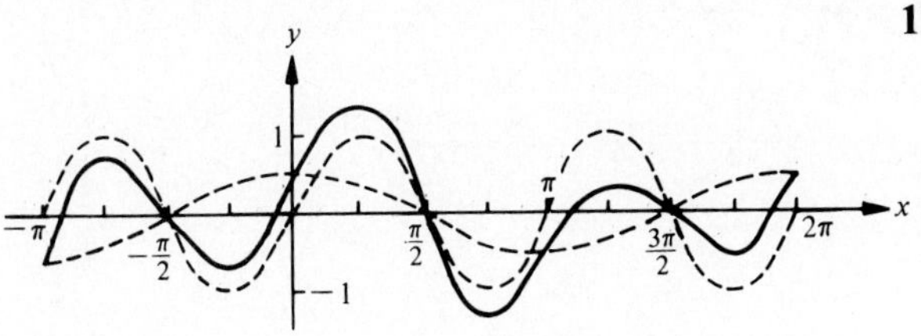

11.

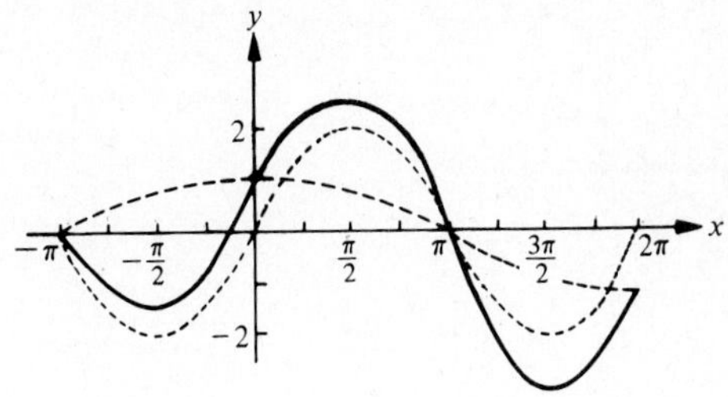

13.

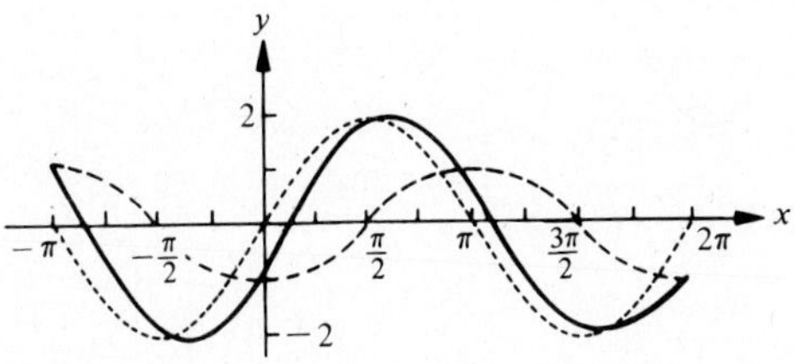

15.

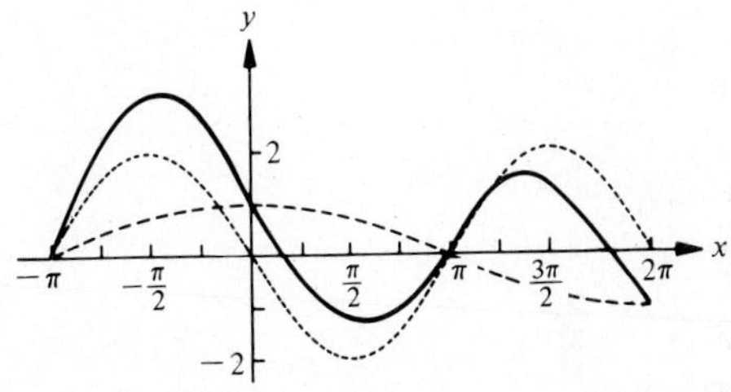

17.

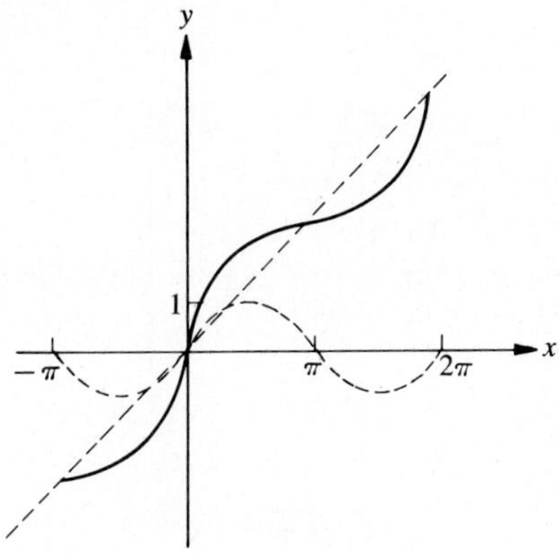

19.

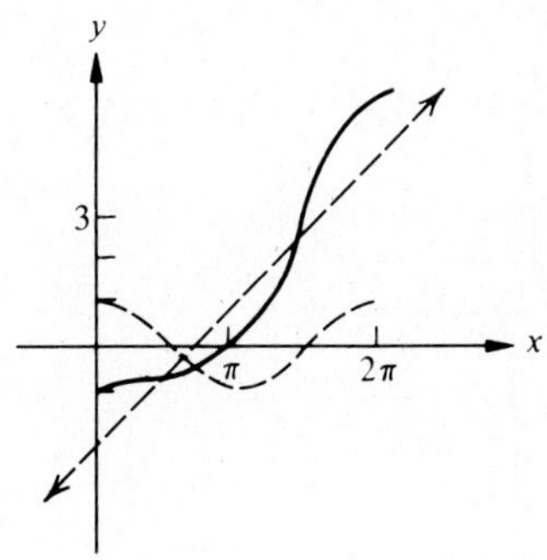

21.

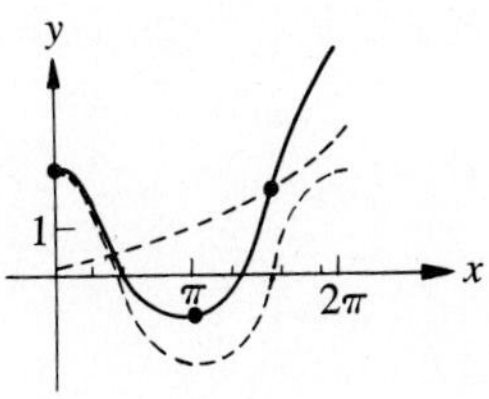

23.

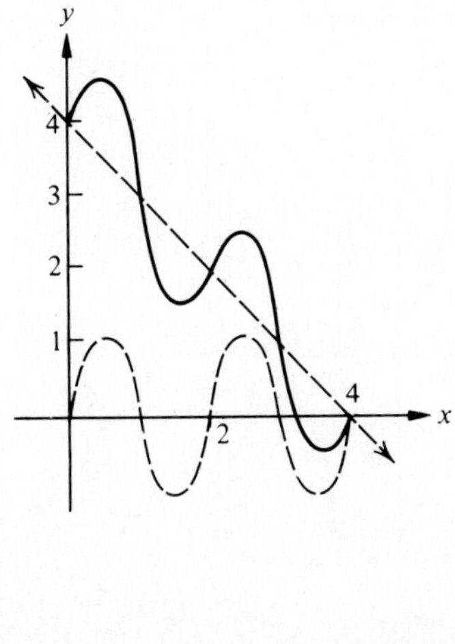

25.

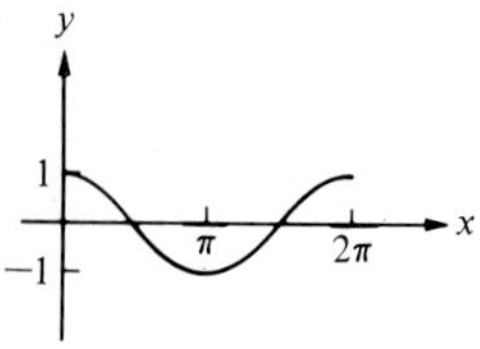

27.

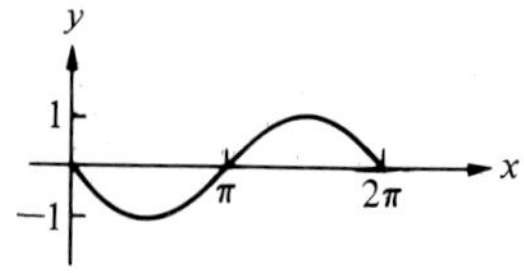

29.

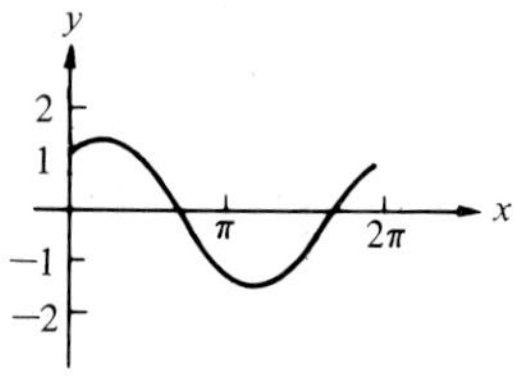

31.

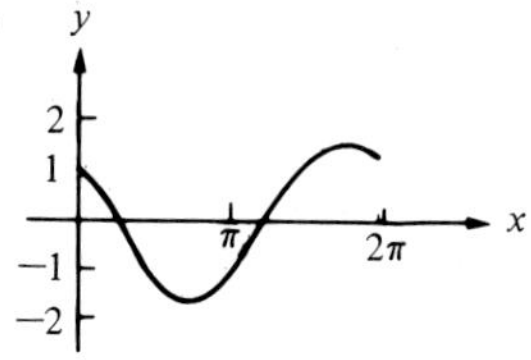

Exercise Set 2.4

1.

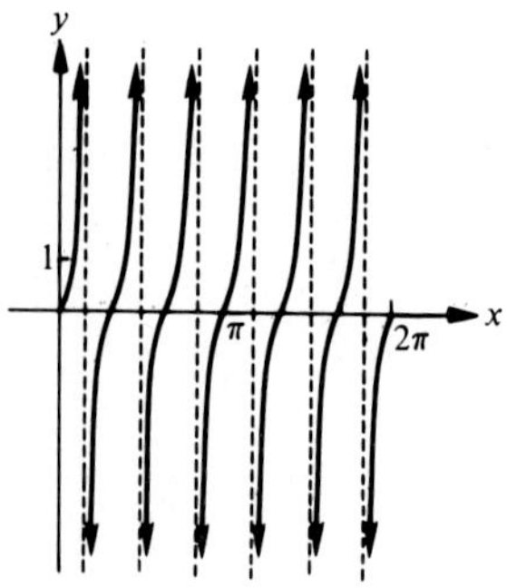

a. $\{y \mid y \in R\}$
b. $x = \pi/6 + k \cdot \pi/3,\ k \in J$
c. $\{x \mid x = k \cdot \pi/3,\ k \in J\}$

3.

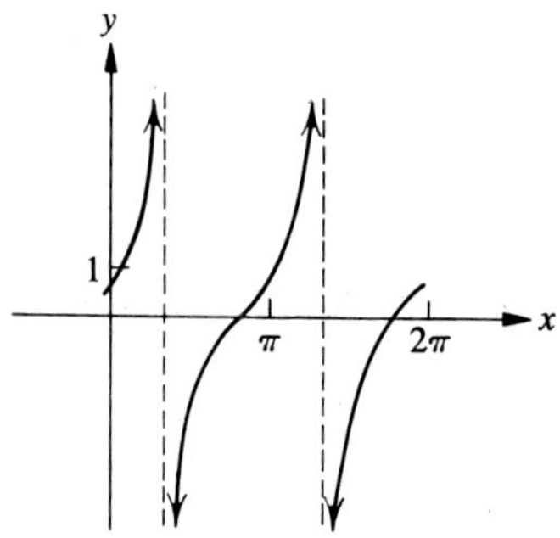

a. $\{y \mid y \in R\}$
b. $x = \pi/3 + k\pi,\ k \in J$
c. $\{x \mid x = 5\pi/6 + k\pi,\ k \in J\}$

5.

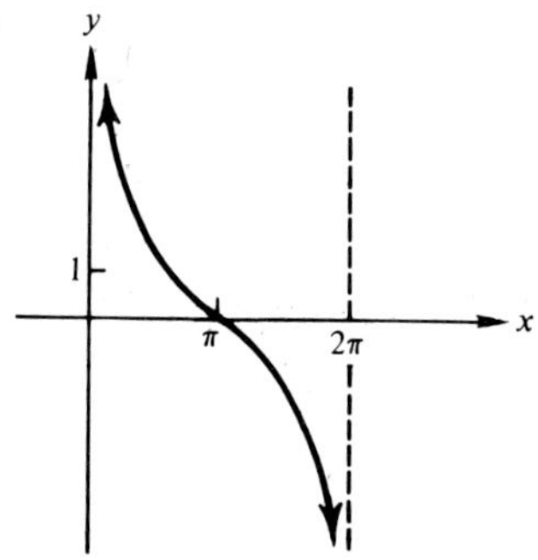

a. $\{y \mid y \in R\}$
b. $x = k \cdot 2\pi,\ k \in J$
c. $\{x \mid x = \pi + k \cdot 2\pi,\ k \in J\}$

7.

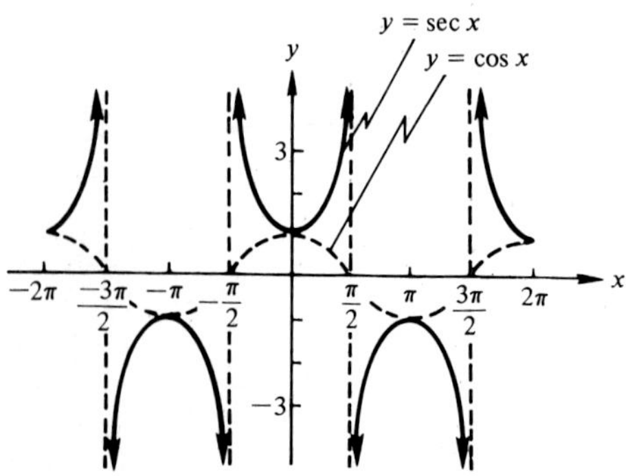

9. $x = \pi/2 + k \cdot \pi,\ k \in J$

11.

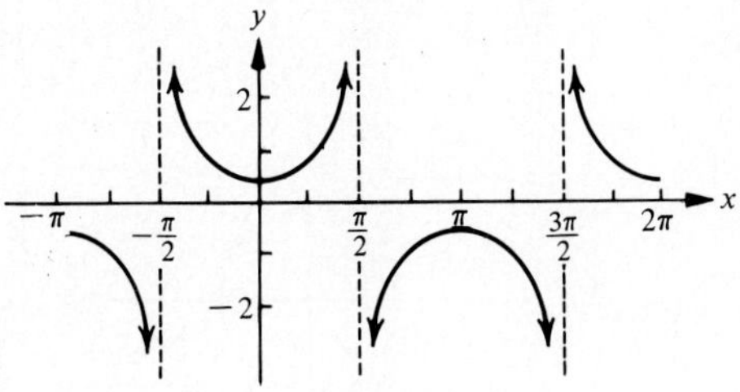

13.

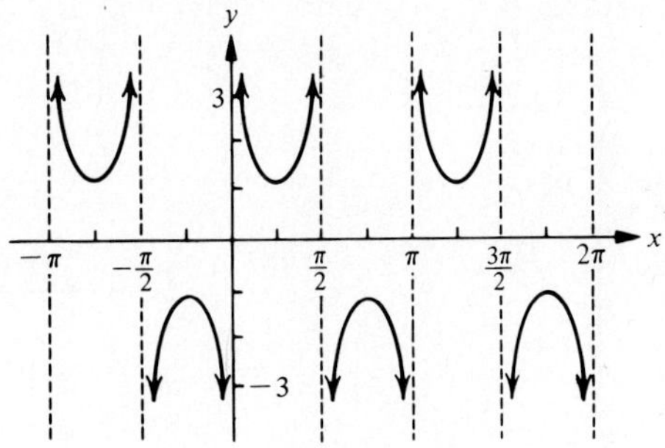

15.

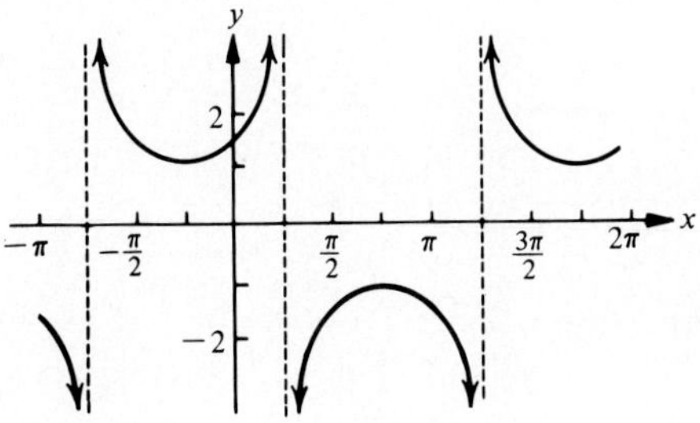

Exercise Set 2.5

1. $\{(0, 1), (6.3, 1)\}$

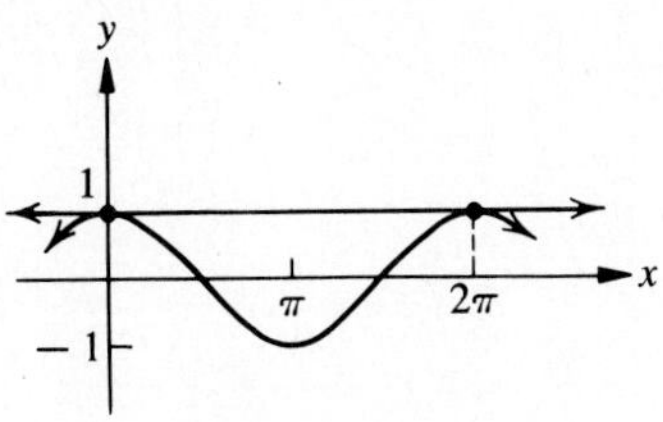

3. $\{(2.1, 0.5)\}$

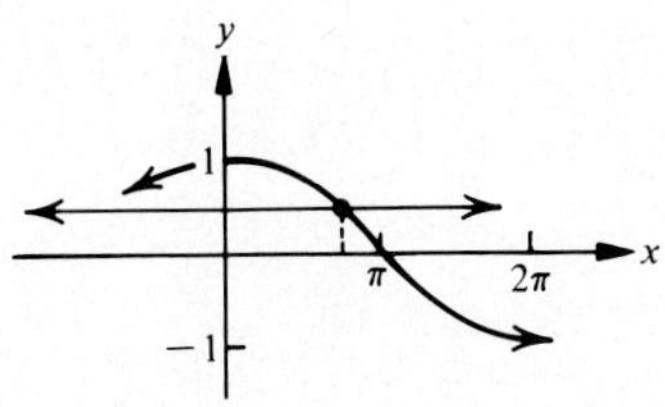

5. $\{(3\pi/4, -1), (7\pi/4, -1)\}$

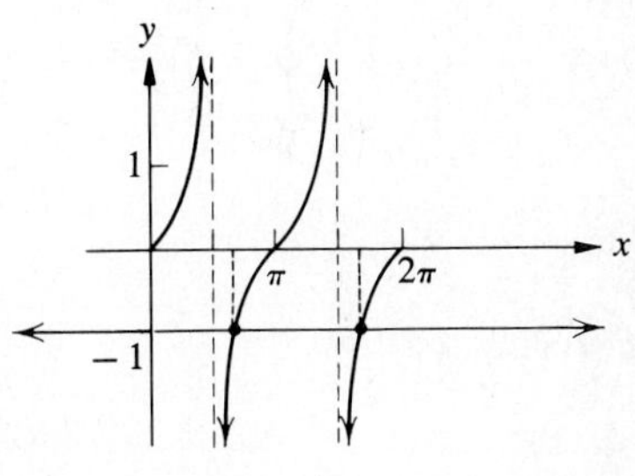

7. $\{0, 1.3, \pi, 5.0, 2\pi\}$

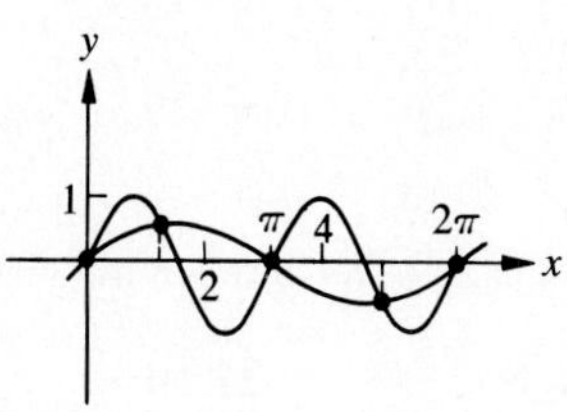

9. $\{0.7, 2.6, 3.7, 5.6\}$

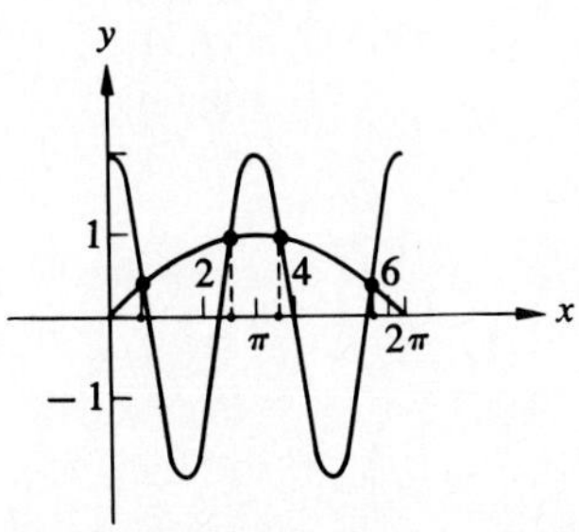

11. $2x - 3y = 21$

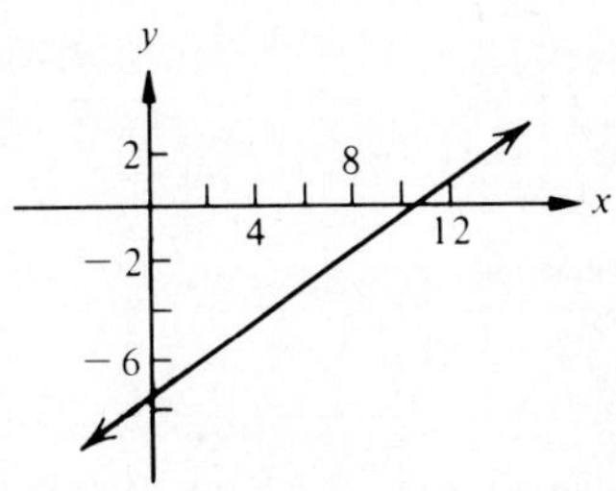

2

13. $x^2 + y^2 = 4$

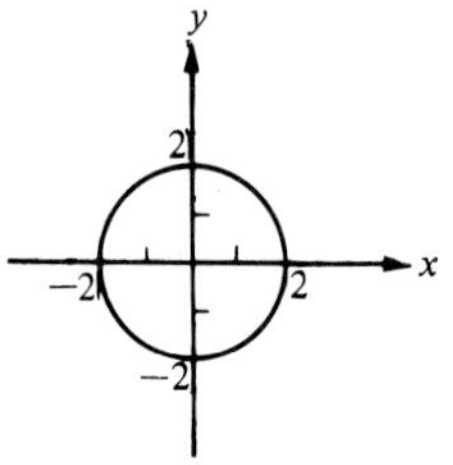

15. $x^2 + y = 1, \ y \geq 0$

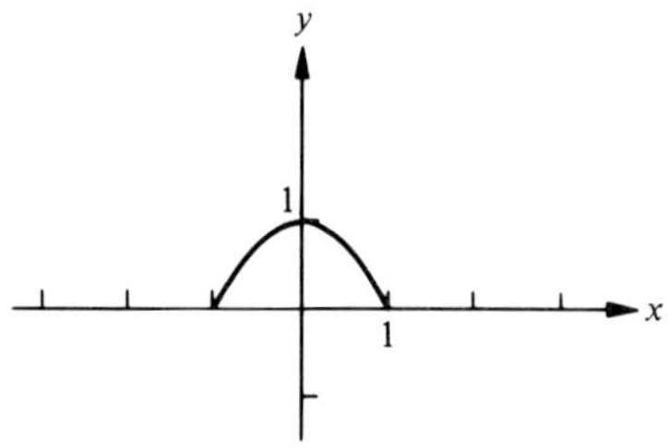

17. $x^2 + y^2 = 1; \ x \geq 0$

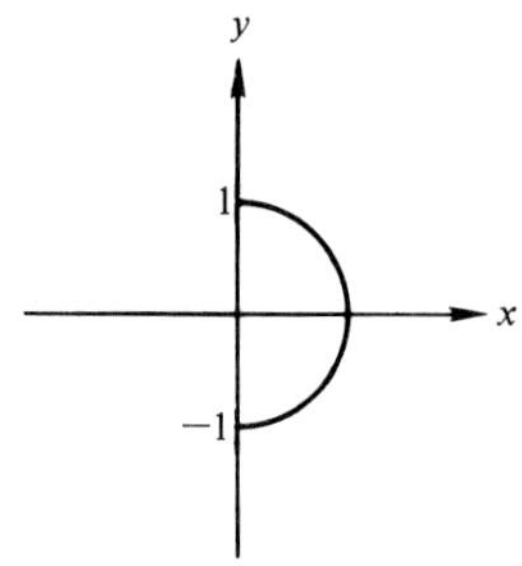

19. $y = \dfrac{8}{x^2 + 4}$

21.

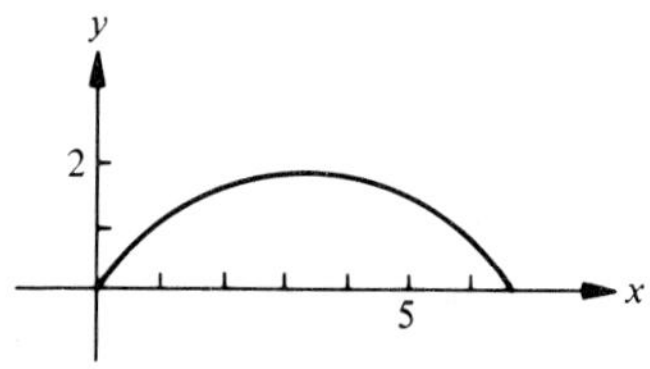

23. $xy = 1; \ |x| \geq 1, |y| \leq 1$

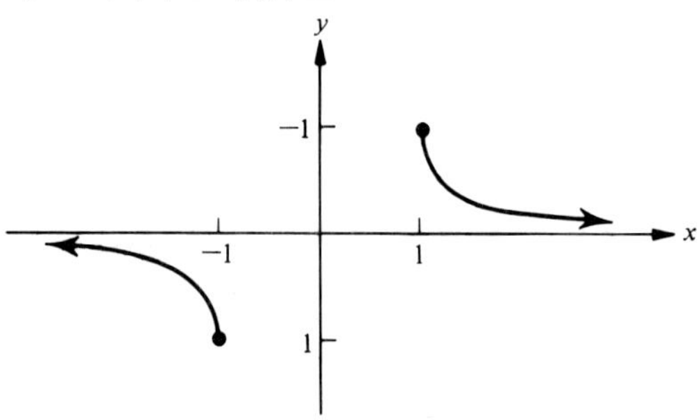

25. $\dfrac{(x-h)^2}{a^2} + \dfrac{(y-k)^2}{b^2} = 1$

Exercise Set 2.6

1. $\cos y = \sqrt{3}/2, \ 0 \leq y \leq \pi; \ \pi/6$
3. $\sin y = 1, \ -\pi/2 \leq y \leq \pi/2; \ \pi/2$
5. $\cot y = 1/\sqrt{3}, \ -\pi/2 \leq y \leq \pi/2, \ y \neq 0; \ \pi/3$
7. $\pi/4$
9. $\pi/4$
11. $\pi/3$
13. 0
15. $\pi/6$
17. $\pi/4$
19. $-\pi/6$
21. $-\pi/3$
23. $2\pi/3$
25. 0.52
27. 0.46
29. 0.36
31. 1.24
33. 1.45
35. 1.19
37. −0.50
39. 2.92
41. −1.45

Exercise Set 2.7

1. $\sqrt{3}/2$
3. 0
5. $-1/\sqrt{2}$
7. $\pi/4$
9. $\pi/6$
11. 0
13. π
15. 1.3983
17. 0.9018
19. 0.6285
21. 1.1728
23. 0.3641
25. $x = \mathrm{Cos}^{-1} \dfrac{3}{4}$
27. $x = \dfrac{1}{3} \mathrm{Sin}^{-1} \dfrac{1}{4}$
29. $x = \dfrac{1}{2} \mathrm{Tan}^{-1} \dfrac{5}{2}$

31. $x = \frac{1}{3}\,\mathrm{Cos}^{-1}\,0.62$ **33.** $x = \frac{1}{3}\,\mathrm{Tan}^{-1}\,\frac{y}{2}$ **35.** $x = 2\,\mathrm{Sin}^{-1}\,\frac{y}{4}$

37. x **39.** $3x$ **41.** $2x$

43.

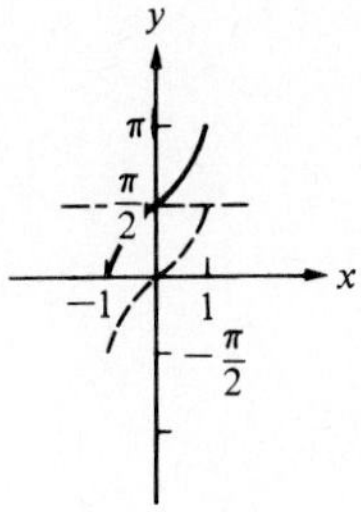

45.

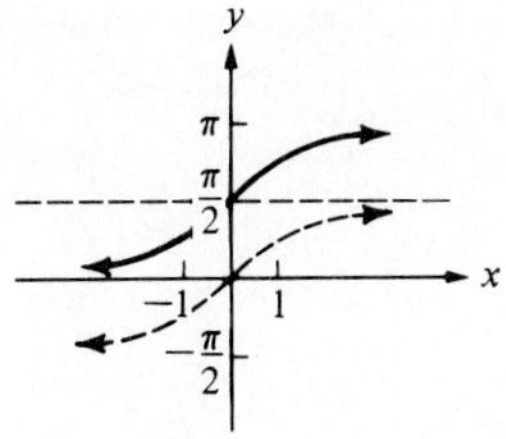

Review Exercises

1.

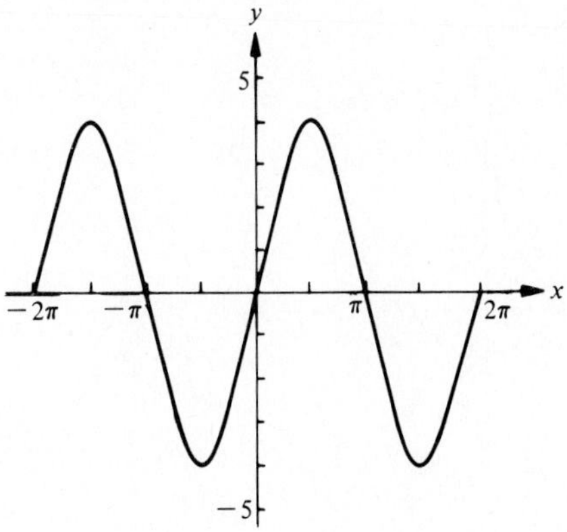

a. amplitude: 4
b. period: 2π
c. no phase shift

2.

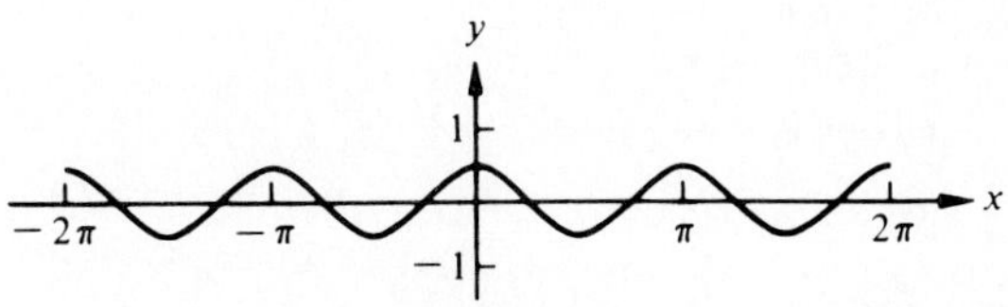

a. amplitude: $1/2$
b. period: π
c. no phase shift

3.

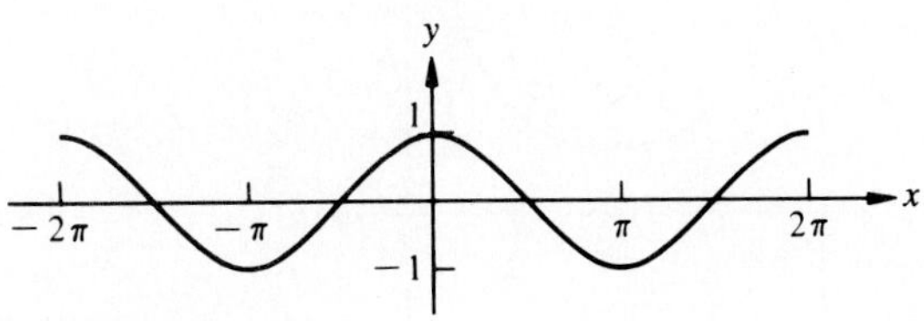

a. amplitude: 1
b. period: 2π
c. phase shift: $\pi/2$ to the left

4.

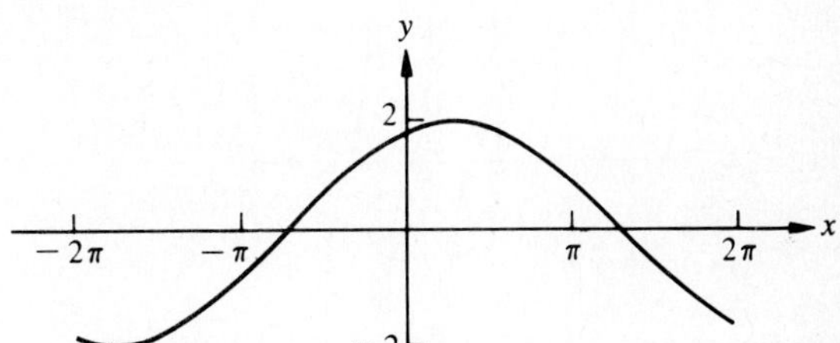

a. amplitude: 2
b. period: 4π
c. phase shift: $\pi/3$ to the right

5.

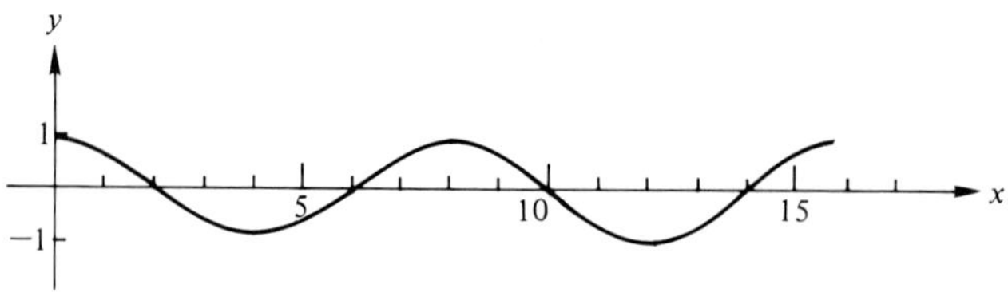

6.

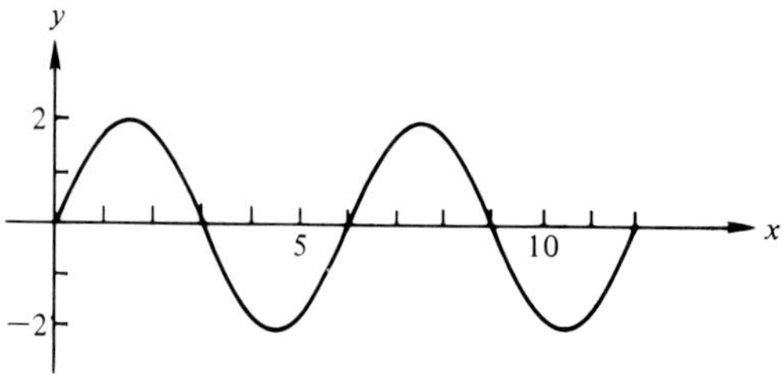

7.

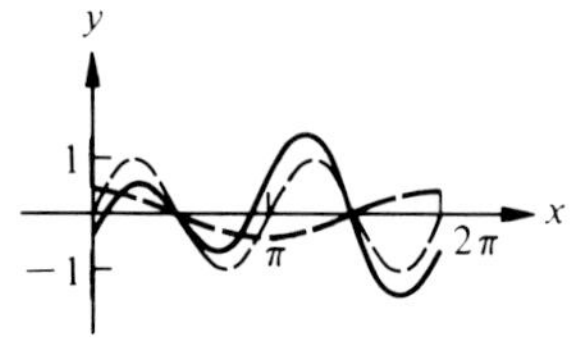

8.

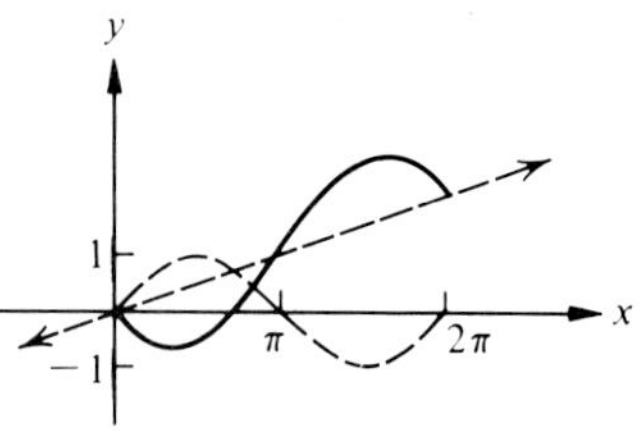

9.

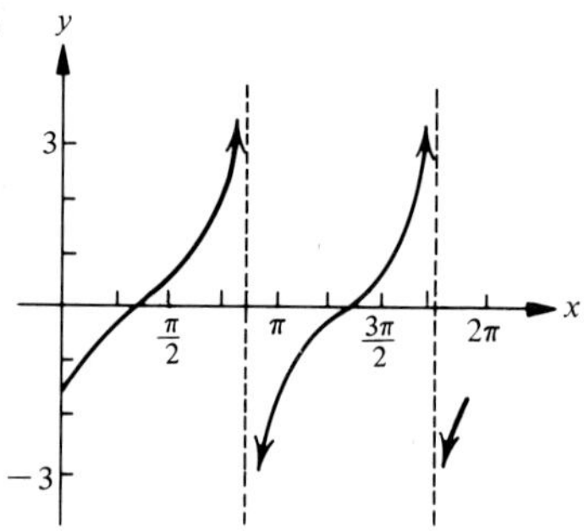

a. range: $\{y \mid y \in R\}$
b. asymptotes: $x = 5\pi/6 + k \cdot \pi,\ k \in J$
c. zeros: $\{x \mid x = \pi/3 + k \cdot \pi,\ k \in J\}$

10.

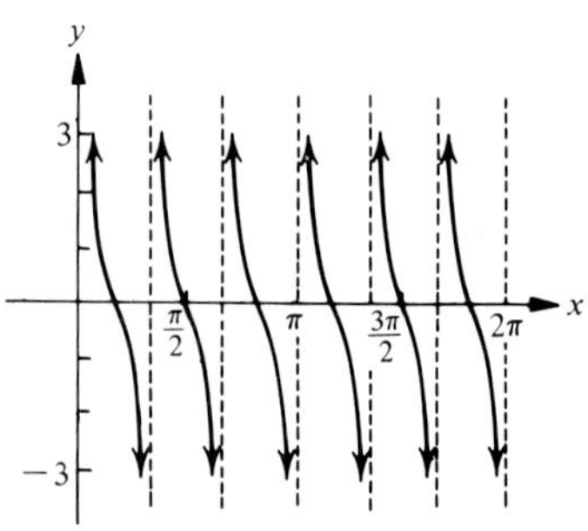

a. range: $\{y \mid y \in R\}$
b. asymptotes: $x = k \cdot (\pi/3),\ k \in J$
c. zeros: $\{x \mid x = \pi/6 + k \cdot (\pi/3),\ k \in J\}$

11.

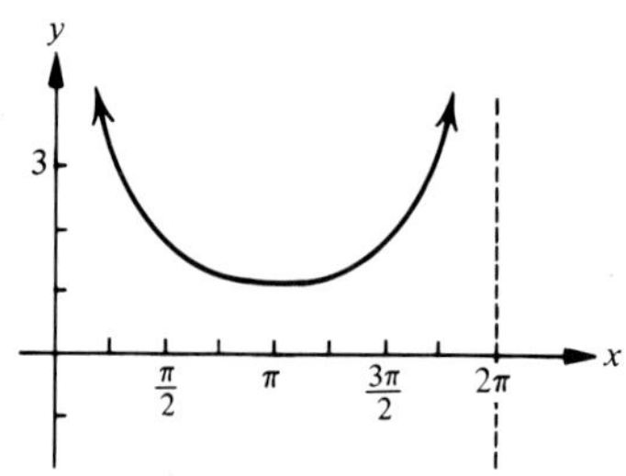

a. range: $\{y \mid y \leq -1 \text{ or } y \geq 1\}$
b. asymptotes: $x = k \cdot 2\pi,\ k \in J$
c. no zeros

12.

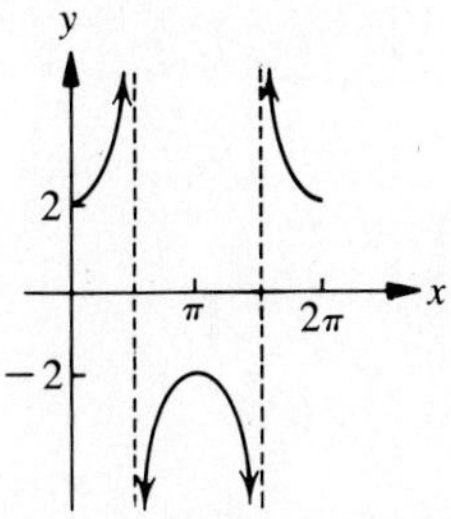

a. range: $\{y \mid y \leq -2 \text{ or } y \geq 2\}$
b. asymptotes: $x = \pi/2 + k \cdot \pi,\ k \in J$
c. no zeros

13.

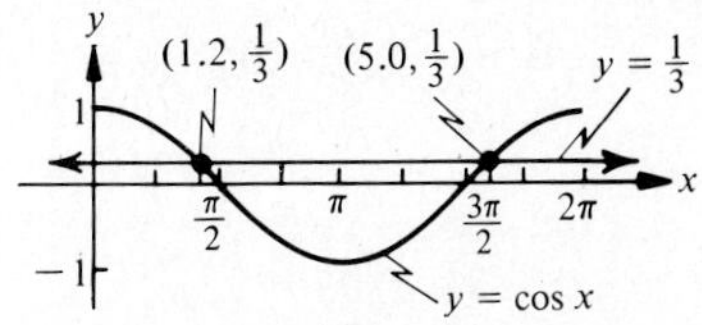

Solution set is $\{(1.2, \frac{1}{3}), (5.0, \frac{1}{3})\}$

14.

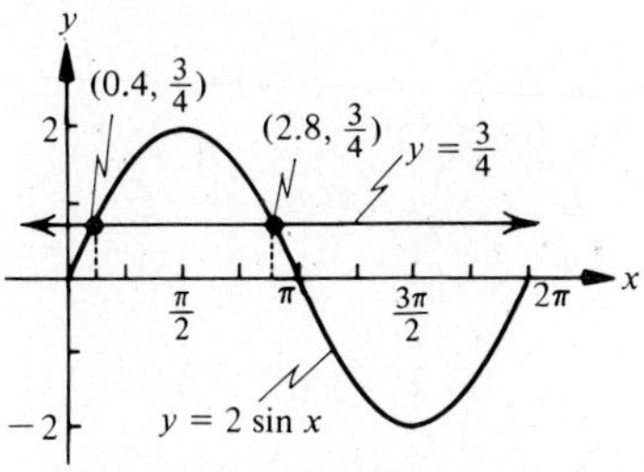

Solution set is $\{(0.4, \frac{3}{4}), (2.8, \frac{3}{4})\}$

15.

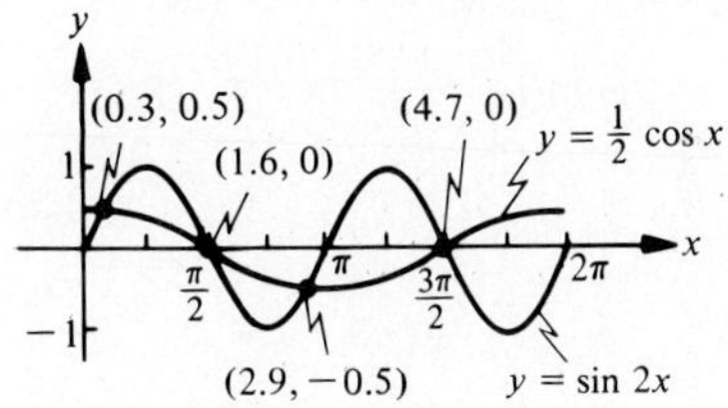

Solution set is
$\{(0.3, 0.5), (1.6, 0), (2.9, -0.5), (4.7, 0)\}$

16.

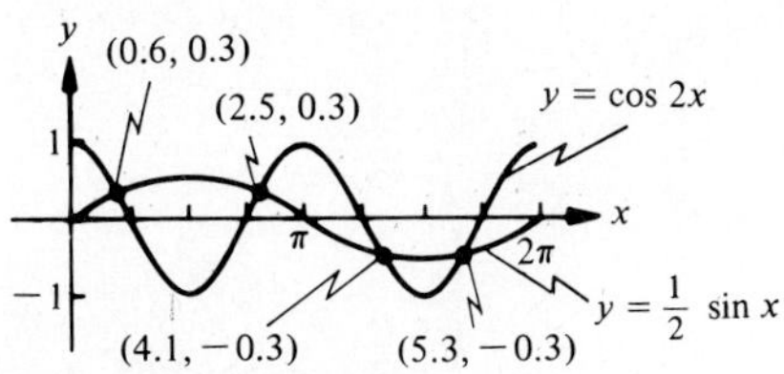

Solution set is
$\{(0.6, 0.3), (2.5, 0.3), (4.1, -0.3), (5.3, -0.3)\}$

17.

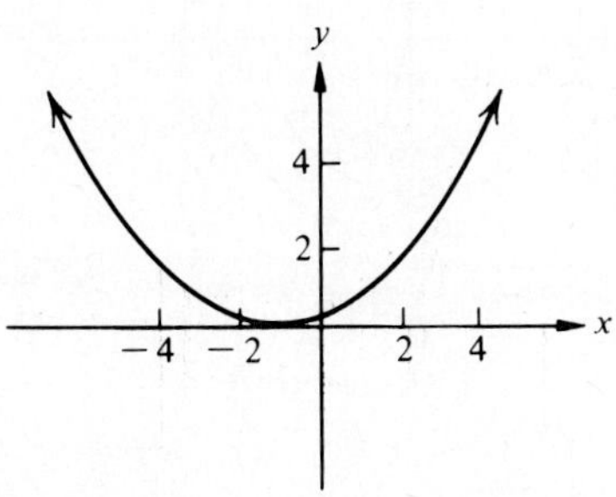

18.

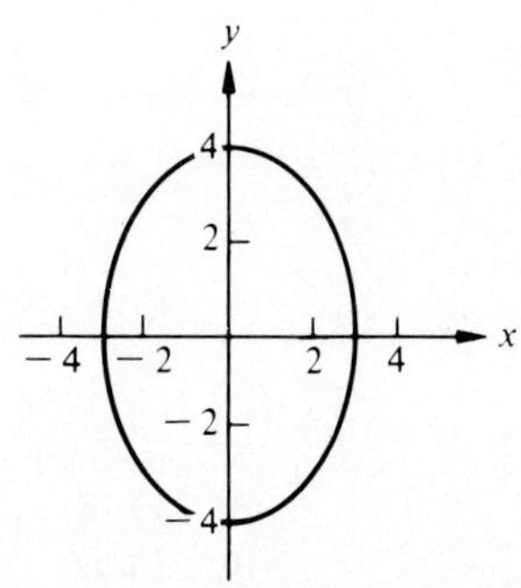

19. $x^2 + 2x - 8y + 1 = 0$ **20.** $\frac{x^2}{9} + \frac{y^2}{16} = 1$ **21.** $\cos y = 1/\sqrt{2}, \pi/4$
22. $\cot y = -1; -\pi/4$ **23.** $\csc y = -2; -\pi/6$ **24.** $\sec y = 2/\sqrt{3}; \pi/6$
25. $\pi/2$ **26.** $-\pi/6$ **27.** $3\pi/4$ **28.** $\pi/6$
29. 1.45 **30.** 0.95 **31.** 0.47 **32.** -1.02
33. $\sqrt{3}$ **34.** $\sqrt{3}/2$ **35.** $\pi/2$ **36.** $\pi/4$
37. 0.7945 **38.** 0.3169 **39.** 0.5459 **40.** 3.4106
41. $x = \text{Cos}^{-1}\frac{3}{5}$ **42.** $x = \frac{1}{2}\text{Sin}^{-1}\frac{1}{5}$ **43.** $x = \frac{1}{2}\text{Tan}^{-1}\frac{y}{3}$ **44.** $x = 2\,\text{Cos}^{-1}\frac{y}{2}$

Exercise Set 3.1

1. $m^\circ(\alpha) = \frac{180}{\pi} m^R(\alpha)$; $m^R(\alpha) = \frac{\pi}{180} m^\circ(\alpha)$

3. 60.0° **5.** 240.0° **7.** 288.0° **9.** -292.5° **11.** 10.3° **13.** -88.2°
15. 0.52^R **17.** 1.31^R **19.** 0.75^R **21.** -3.52^R **23.** 0.00029^R
25. 0.44^R

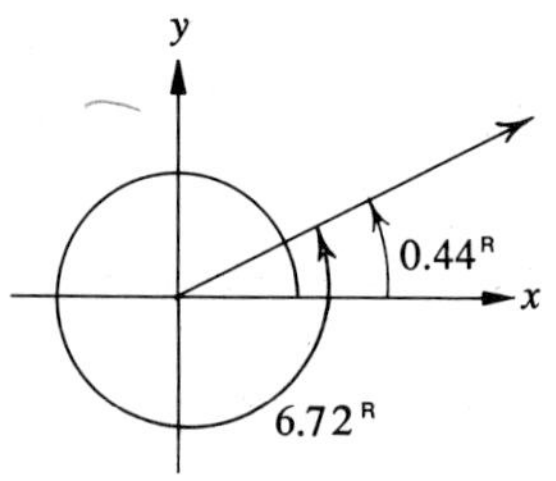

27. 19°

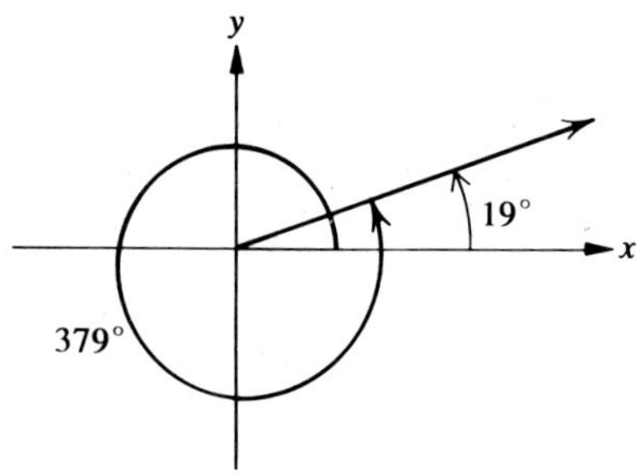

29. 4.83^R

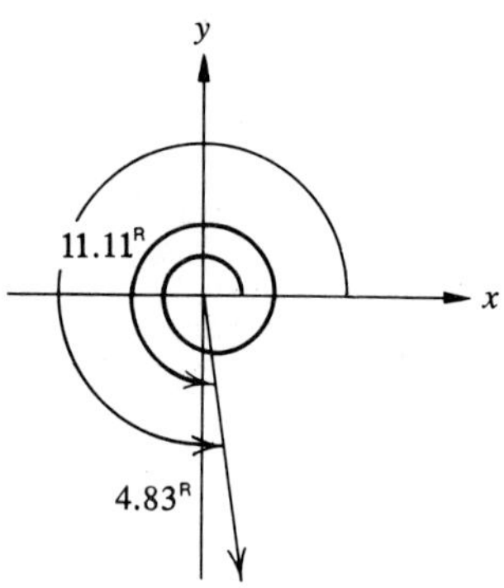

31. 288°

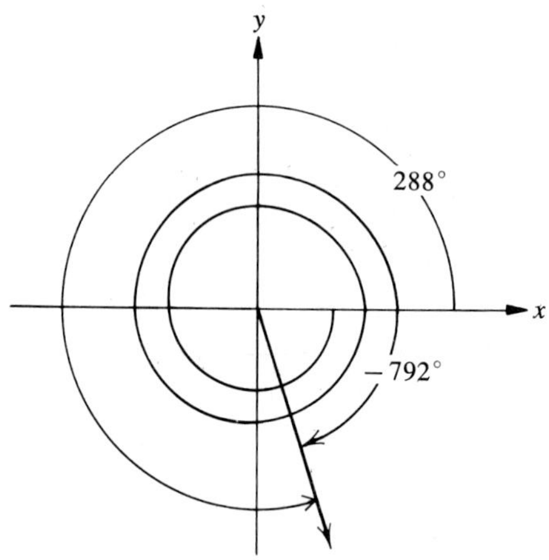

33. $(21 + k \cdot 360)^\circ$, $k \in J$ **35.** $(-5\pi/6 + k \cdot 2\pi)^R$, $k \in J$ **37.** $(242 + k \cdot 360)^\circ$, $k \in J$
39. $6\pi^R$; 1080° **41.** $12\pi^R$; 2160° **43.** $20\pi^R$; 3600°
45. 20.7° **47.** 63.7°

Exercise Set 3.2

1. 6.2 meters
3. $5\pi^2/6 \approx 8.2$ decimeters
5. $-9\pi/4 \approx -7.1$ feet
7. 126.9°
9. 280.7°
11. 3.6 meters per second
13. 3,000 radians per minute
15. $-54\pi \approx -169.7$ decimeters per second
17. $-12/\pi \approx -3.8$ degrees per second
19. 104.72 meters
21. 1,745 miles
23. 3.98 centimeters
25. 5024 meters per minute
27. 9.39 radians per minute
29. 7.17×10^{-4} radians per hour
31. 2.24×10^3 miles per hour
33. Area of the sector is to the area of the circle as the central angle θ is to 2π radians. Thus,

$$\frac{\mathscr{A}(\text{of sector})}{\pi r^2} = \frac{\theta^{\mathsf{R}}}{2\pi}, \qquad \text{from which} \qquad \mathscr{A} = \frac{\pi r^2(\theta^{\mathsf{R}})}{2\pi} = \tfrac{1}{2}r^2\theta^{\mathsf{R}}.$$

35. 1.92×10^{13} miles
37. 4690^{R}

Exercise Set 3.3

1. $\cos\alpha = 3/5$ $\quad\sec\alpha = 5/3$
   $\sin\alpha = 4/5$ $\quad\csc\alpha = 5/4$
   $\tan\alpha = 4/3$ $\quad\cot\alpha = 3/4$

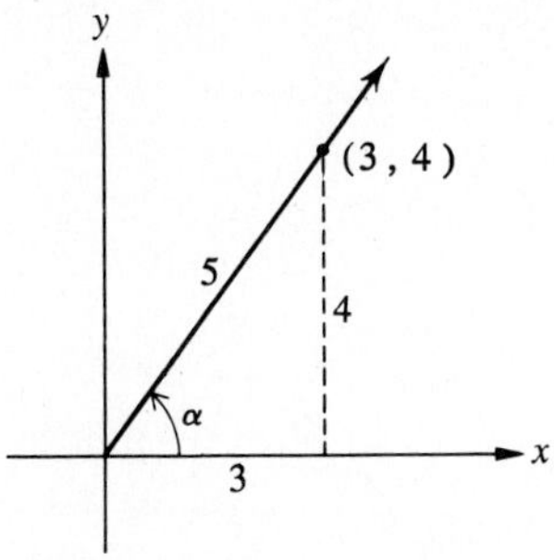

3. $\cos\alpha = -12/13$ $\quad\sec\alpha = -13/12$
   $\sin\alpha = -5/13$ $\quad\csc\alpha = -13/5$
   $\tan\alpha = 5/12$ $\quad\cot\alpha = 12/5$

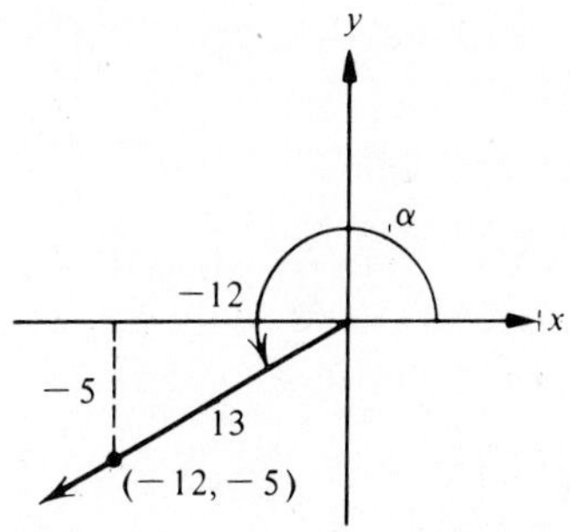

5. $\cos\alpha = 2/\sqrt{13}$ $\quad\sec\alpha = \sqrt{13}/2$
   $\sin\alpha = 3/\sqrt{13}$ $\quad\csc\alpha = \sqrt{13}/3$
   $\tan\alpha = 3/2$ $\quad\cot\alpha = 2/3$

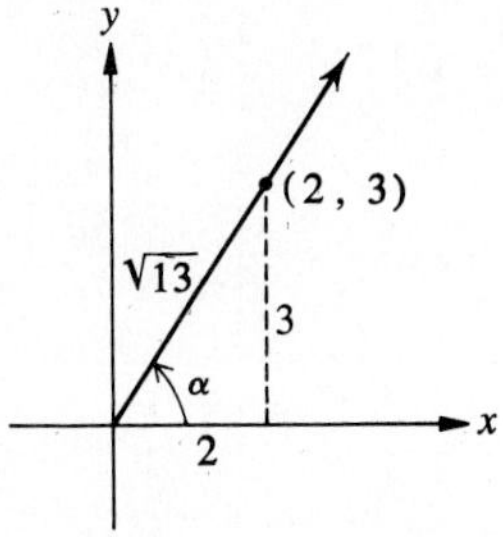

7. $\cos\alpha = -\sqrt{3}/2$ $\quad\sec\alpha = -2/\sqrt{3}$
   $\sin\alpha = -1/2$ $\quad\csc\alpha = -2$
   $\tan\alpha = 1/\sqrt{3}$ $\quad\cot\alpha = \sqrt{3}$

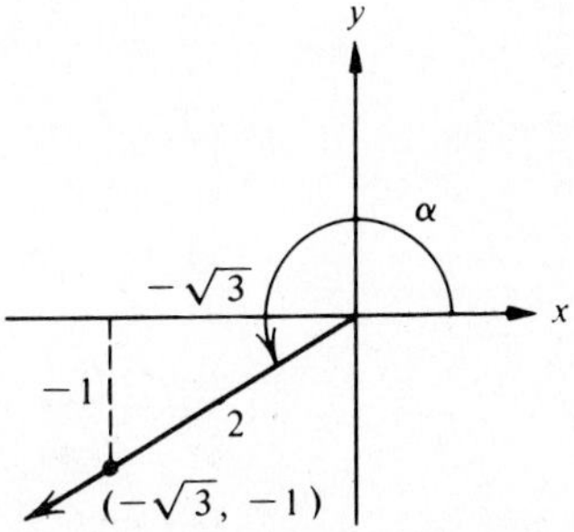

9. $\cos\alpha = \sqrt{3}/2$ $\quad \sec\alpha = 2/\sqrt{3}$
$\sin\alpha = 1/2$ $\quad \csc\alpha = 2$
$\tan\alpha = 1/\sqrt{3}$ $\quad \cot\alpha = \sqrt{3}$

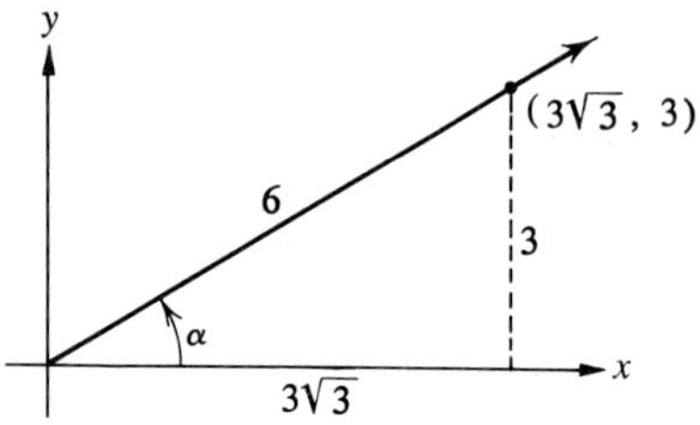

11. $\cos\alpha = 1$ $\quad \sec\alpha = 1$
$\sin\alpha = 0$ $\quad \csc\alpha$ undefined
$\tan\alpha = 0$ $\quad \cot\alpha$ undefined

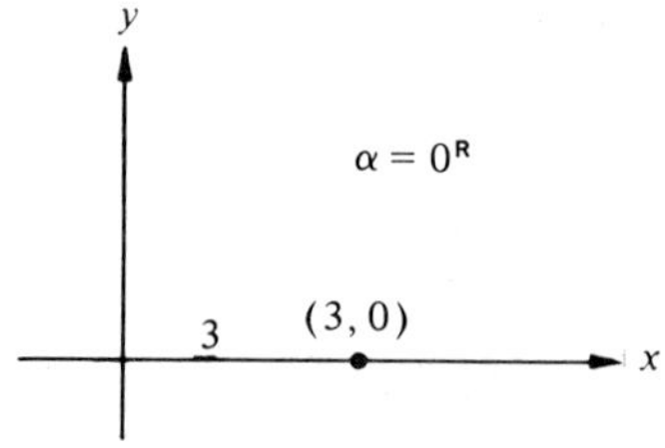

13. III **15.** I **17.** III **19.** I

21. $\cos\alpha = -3/\sqrt{13}$ $\quad \sec\alpha = -\sqrt{13}/3$
$\sin\alpha = 2/\sqrt{13}$ $\quad \csc\alpha = \sqrt{13}/2$
$(\tan\alpha = -2/3)$ $\quad \cot\alpha = -3/2$

23. $\cos\alpha = -4/5$ $\quad \sec\alpha = -5/4$
$(\sin\alpha = -3/5)$ $\quad \csc\alpha = -5/3$
$\tan\alpha = 3/4$ $\quad \cot\alpha = 4/3$

25. $\cos\alpha = \sqrt{15}/4$ $\quad \sec\alpha = 4/\sqrt{15}$
$\sin\alpha = -1/4$ $\quad (\csc\alpha = 4)$
$\tan\alpha = -1/\sqrt{15}$ $\quad \cot\alpha = -\sqrt{15}$

27. a. Not defined for $\alpha = (\pi/2 + k\cdot\pi)^R$, $k \in J$,
b. equal to 0 for $\alpha = k\cdot\pi^R$, $k \in J$

29. a. Not defined for $\alpha = (\pi/2 + k\cdot\pi)^R$, $k \in J$,
b. no values equal to 0.

31. $\cos\alpha = x/2$ $\quad \sec\alpha = 2/x$
$\sin\alpha = \sqrt{4 - x^2}/2$ $\quad \csc\alpha = 2/\sqrt{4 - x^2}$
$\tan\alpha = \sqrt{4 - x^2}/x$ $\quad \cot\alpha = x/\sqrt{4 - x^2}$

33. $\cos\alpha = \sqrt{16 - y^2}/4$ $\quad \sec\alpha = 4/\sqrt{16 - y^2}$
$\sin\alpha = y/4$ $\quad \csc\alpha = 4/y$
$\tan\alpha = y/\sqrt{16 - y^2}$ $\quad \cot\alpha = \sqrt{16 - y^2}/y$

35. $\cos\alpha = x/\sqrt{2x^2 - 4}$ $\quad \sec\alpha = \sqrt{2x^2 - 4}/x$
$\sin\alpha = \sqrt{x^2 - 4}/\sqrt{2x^2 - 4}$ $\quad \csc\alpha = \sqrt{2x^2 - 4}/\sqrt{x^2 - 4}$
$\tan\alpha = \sqrt{x^2 - 4}/x$ $\quad \cot\alpha = x/\sqrt{x^2 - 4}$

Exercise Set 3.4

1. 1/2 **3.** $\sqrt{3}$ **5.** $2/\sqrt{3}$ **7.** Undefined **9.** -1 **11.** -1
13. $-1/2$ **15.** $1/\sqrt{3}$ **17.** $-\sqrt{2}$ **19.** $-1/2$ **21.** $-1/2$ **23.** $-1/2$
25. 0.7379 **27.** 0.9872 **29.** 0.7774 **31.** 0.9857 **33.** 1.0575 **35.** 1.9458
37. 3.7849 **39.** 6.0405 **41.** 0.1132 **43.** 45° **45.** 30° **47.** $-30°$
49. 73.9° **51.** 30.7° **53.** $-52.5°$ **55.** 4/5 **57.** 2/5 **59.** 3/4
61. $\dfrac{\sqrt{9 - x^2}}{3}$ **63.** $\sqrt{1 - x^2}$

Exercise Set 3.5

1. sin 28° **3.** cot 10.7° **5.** sin 1° 48′ **7.** $\alpha = 48°$ **9.** $a \approx 1.78$

11. $\beta \approx 56.0°$ **13.** $c \approx 7.09$ **15.** $c \approx 7.57$ **17.** $b \approx 2.32$
19. $\alpha \approx 33.7°,\ \beta \approx 56.3°,\ c \approx 3.61$ **21.** $\alpha \approx 61.9°,\ \beta \approx 28.1°,\ b \approx 8.00$
23. $\alpha \approx 39.0°,\ a \approx 14.8,\ b \approx 18.3$ **25.** $\alpha \approx 67.8°,\ b \approx 4.49,\ c \approx 11.9$
27. $\beta \approx 39.7°,\ a \approx 18.1,\ c \approx 23.5$ **29.** $\alpha \approx 48.4°,\ a \approx 33.8,\ c \approx 45.2$
31. 360 meters **33.** 53.1° **35.** 27.1 decimeters **37.** 239 meters
39. 61.4 feet **41.** 12,300 feet **43.** 5.88 centimeters **45.** $4\sqrt{3}$
47. $\cos\alpha = \dfrac{l-h}{l}$, from which $l\cos\alpha = l - h$ and $h = l(1-\cos\alpha)$.

3

Exercise Set 3.6

1. $\gamma = 85°,\ b \approx 14.1,\ c \approx 14.6$ **3.** $\alpha = 42.8°,\ b \approx 52.0,\ c \approx 84.7$
5. $\beta = 52.9°,\ a \approx 0.307,\ c \approx 0.526$ **7.** $\gamma = 77.5°,\ a \approx 7.05,\ b \approx 13.3$
9. One triangle: $\beta \approx 20.9°,\ \gamma \approx 129.1°,\ c \approx 10.9$
11. Two triangles: $\gamma \approx 50.7°,\ \beta \approx 97.1°,\ b \approx 7.82$ or $\gamma' \approx 129.3°,\ \beta' \approx 18.5°,\ b' \approx 2.50$
13. One right triangle: $\alpha = 60°,\ \gamma = 90°,\ a \approx 27.7$
15. Two triangles: $\alpha \approx 60.4°,\ \gamma \approx 76.9°,\ c \approx 5.60$ or $\alpha' \approx 119.6°,\ \gamma' \approx 17.7°,\ c' \approx 1.75$
17. 1.51 **19.** 3049 feet
21. 7.90 centimeters; 14.8 centimeters **23.** 78.4 meters
25. 57.8 feet **27.** 64.3 meters

29. From the figure, $\tan\gamma = \dfrac{c}{AC}$, or $AC = \dfrac{c}{\tan\gamma}$.

Using the Law of Sines, $\dfrac{AC}{\sin\beta} = \dfrac{a}{\sin[180° - (\alpha+\beta)]} = \dfrac{a}{\sin(\alpha+\beta)}$.

Hence, $AC = \dfrac{a\sin\beta}{\sin(\alpha+\beta)}$, from which $\dfrac{c}{\tan\gamma} = \dfrac{a\sin\beta}{\sin(\alpha+\beta)}$ and $c = \dfrac{a\sin\beta\tan\gamma}{\sin(\alpha+\beta)}$

31. From the Law of Sines, $\dfrac{a}{b} = \dfrac{\sin\alpha}{\sin\beta}$. Adding 1 to both members gives

$$\frac{a}{b} + 1 = \frac{\sin\alpha}{\sin\beta} + 1, \quad \text{from which} \quad \frac{a+b}{b} = \frac{\sin\alpha + \sin\beta}{\sin\beta}.$$

33. From the results of Exercises 31 and 32 and dividing left members and right members, respectively, gives

$$\frac{\dfrac{a-b}{b}}{\dfrac{a+b}{b}} = \frac{\dfrac{\sin\alpha - \sin\beta}{\sin\beta}}{\dfrac{\sin\alpha + \sin\beta}{\sin\beta}}, \quad \text{from which} \quad \frac{a-b}{a+b} = \frac{\sin\alpha - \sin\beta}{\sin\alpha + \sin\beta}.$$

Exercise Set 3.7

1. $a \approx 4.37,\ \beta \approx 59.9°,\ \gamma \approx 49.1°$ **3.** $c \approx 3.39,\ \alpha \approx 65.9°,\ \beta \approx 38.8°$
5. $b \approx 1.42,\ \alpha \approx 17.9°,\ \gamma \approx 20.6°$ **7.** $a \approx 3.83,\ \beta \approx 24.3°,\ \gamma \approx 55.3°$
9. $a \approx 95.7°,\ \beta \approx 50.7°,\ \gamma \approx 33.6°$ **11.** $\alpha \approx 4.0°,\ \beta \approx 31.6°,\ \gamma \approx 144.4°$

13. $\alpha \approx 26.4°$, $\beta \approx 36.4°$, $\gamma \approx 117.2°$
15. $\alpha \approx 41.4°$, $\beta \approx 55.8°$, $\gamma \approx 82.8°$
17. 42.7°
19. 82.5°
21. 10.9
23. 5.33 kilometers
25. 11.7 centimeters
27. 47.8°
29. 1,240 feet
31. 875 kilometers

33. Adding 1 to both members of $\cos \alpha = \dfrac{b^2 + c^2 - a^2}{2bc}$ gives

$$1 + \cos \alpha = 1 + \frac{b^2 + c^2 - a^2}{2bc}$$

$$= \frac{2bc + b^2 + c^2 - a^2}{2bc} = \frac{b^2 + 2bc + c^2 - a^2}{2bc}$$

$$= \frac{(b + c)^2 - a^2}{2bc} = \frac{(b + c + a)(b + c - a)}{2bc}.$$

35. $a + b + c = 2s$ and $b + c - a = 2s - 2a = 2(s - a)$. Substituting $2s$ for $a + b + c$ and $2(s - a)$ for $b + c - a$ in the results of Exercise 33 gives

$$1 + \cos \alpha = \frac{(2s)(2)(s - a)}{2bc}, \qquad \text{from which} \qquad \frac{1 + \cos \alpha}{2} = \frac{s(s - a)}{bc}.$$

37. Multiplying left members and right members, respectively, of the result of Exercise 35 and 36 gives

$$\left(\frac{1 + \cos \alpha}{2}\right)\left(\frac{1 - \cos \alpha}{2}\right) = \left[\frac{s(s - a)}{bc}\right]\left[\frac{(s - b)(s - c)}{bc}\right],$$

$$\frac{1 - \cos^2 \alpha}{4} = \frac{s(s - a)(s - b)(s - c)}{b^2c^2},$$

$$\frac{\sin^2 \alpha}{4} = \frac{s(s - a)(s - b)(s - c)}{b^2c^2}.$$

Taking the square root of both members gives

$$\frac{1}{2} \sin \alpha = \frac{\sqrt{s(s - a)(s - b)(s - c)}}{bc},$$

$$\frac{1}{2} bc \sin \alpha = \sqrt{s(s - a)(s - b)(s - c)}.$$

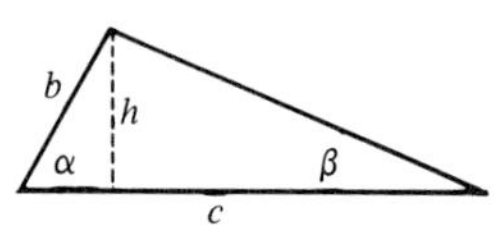

Because $h = b \sin \alpha$ (see the figure), the area $\mathscr{A}$ of a triangle is $\frac{1}{2}$(base × altitude), or $\frac{1}{2}c(b \sin \alpha)$. Thus,

$$\mathscr{A} = \sqrt{s(s - a)(s - b)(s - c)}.$$

39. 3.14
41. Noting in the figure that the radius r is the altitude of the three inner triangles, we have that the total area

is $\frac{1}{2}(ra + rb + rc) = \frac{1}{2}(a + b + c)(r) = sr$. From the results of Exercise 37, we have

$$\mathscr{A} = \sqrt{s(s-a)(s-b)(s-c)} = sr,$$

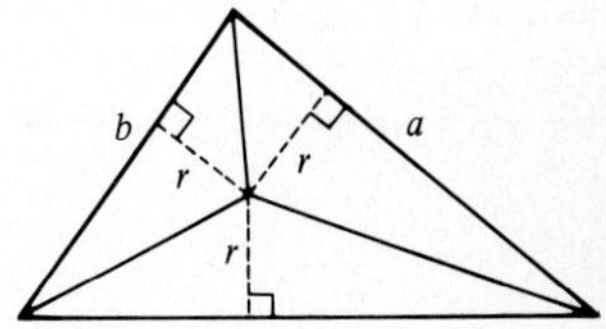

from which

$$r = \sqrt{\frac{s(s-a)(s-b)(s-c)}{s^2}} = \sqrt{\frac{(s-a)(s-b)(s-c)}{s}}.$$

Review Exercises

1. 72.0° **2.** 138.1° **3.** 1.31^R
4. 4.49^R **5.** 218° **6.** 0.95^R
7. $(131 + k \cdot 360^\circ),\ k \in J$ **8.** $(7\pi/4 + k \cdot 2\pi)^R,\ k \in J$ **9.** 2.86 meters
10. 0.194^R **11.** 12 decimeters per minute **12.** 0.394 radian per second

13. $\cos\alpha = 4/5$, $\sec\alpha = 5/4$, $\sin\alpha = 3/5$, $\csc\alpha = 5/3$, $\tan\alpha = 3/4$, $\cot\alpha = 4/3$

14. $\cos\alpha = 5/13$, $\sec\alpha = 13/5$, $\sin\alpha = 12/13$, $\csc\alpha = 13/12$, $\tan\alpha = 12/5$, $\cot\alpha = 5/12$

15. $\cos\alpha = -\sqrt{3}/2$, $\sec\alpha = -2/\sqrt{3}$, $\sin\alpha = 1/2$, $\csc\alpha = 2$, $\tan\alpha = -1/\sqrt{3}$, $\cot\alpha = -\sqrt{3}$

16. $\cos\alpha = -2/\sqrt{29}$, $\sec\alpha = -\sqrt{29}/2$, $\sin\alpha = 5/\sqrt{29}$, $\csc\alpha = \sqrt{29}/5$, $\tan\alpha = -5/2$, $\cot\alpha = -2/5$

17. IV **18.** IV **19.** III **20.** IV **21.** $1/\sqrt{2}$ **22.** $\sqrt{3}$
23. 0 **24.** -1 **25.** $-1/2$ **26.** $\sqrt{2}$ **27.** 0.8196 **28.** -1.8270
29. 0.0854 **30.** -1.0017 **31.** -0.5985 **32.** -1.0372 **33.** 30° **34.** 45°
35. 66.8° **36.** 122.3°
37. $\alpha = 39^\circ,\ a \approx 8.81,\ b \approx 10.9$
38. $\alpha \approx 53.7^\circ,\ \beta \approx 36.3^\circ,\ b \approx 3.97$
39. $\beta \approx 57.4^\circ,\ b \approx 20.2,\ c \approx 24.0$
40. $\beta \approx 18.7^\circ,\ a \approx 28.7,\ c \approx 30.3$
41. $\gamma \approx 77.5^\circ,\ a \approx 7.03,\ b \approx 13.3$
42. $\beta \approx 10.0^\circ,\ a \approx 1.73,\ b \approx 0.425$
43. Two triangles: $\alpha \approx 60.4^\circ,\ \gamma \approx 76.9^\circ,\ c \approx 5.60$ or $\alpha' \approx 119.6^\circ, \gamma' \approx 17.7^\circ, c' \approx 1.75$
44. One triangle: $\beta \approx 30.6^\circ,\ \gamma \approx 13.9^\circ,\ c \approx 17.5$
45. $\alpha \approx 51.3^\circ,\ \beta \approx 70.5^\circ,\ c \approx 2.07$
46. $\beta \approx 14.1^\circ,\ \gamma \approx 142.8^\circ,\ a \approx 3.7$
47. $\alpha \approx 35.2^\circ,\ \beta \approx 42.2^\circ,\ \gamma \approx 102.6^\circ$
48. $\alpha \approx 30.8^\circ,\ \beta \approx 56.0^\circ,\ \gamma \approx 93.2^\circ$

Exercises Set 4.1

1. $\sec^2\alpha$ **3.** $\sin^2\alpha$ **5.** $-\sin^2\alpha$ **7.** $\tan^3\alpha$

9. $\dfrac{1}{\sin\alpha}$ **11.** $\dfrac{\cos\alpha}{\sin^2\alpha}$ **13.** $\dfrac{\cos^2\alpha - \sin^2\alpha}{\cos^2\alpha}$ **15.** $-\dfrac{\sin^2\alpha}{\cos^2\alpha}$

17. $\dfrac{1}{\sin\beta\cos\beta}$ **19.** $\dfrac{\cos\beta}{\sin^3\beta}$ **21.** $\dfrac{1}{\sin\gamma}$ **23.** $\dfrac{\cos^2\gamma}{\sin\gamma}$

25. $\cos^2 x$ 27. $\dfrac{\sin^2 x}{\cos^4 x}$ 29. $\dfrac{1}{\sin y \cos y}$ 31. $\dfrac{\sin^4 \theta + \cos^4 \theta}{\cos^2 \theta \sin^2 \theta}$

33. $\dfrac{1}{\cos \theta \sin^2 \theta}$ 35. $\dfrac{\cos^2 \alpha}{\sin^3 \alpha}$ 37. $\dfrac{\sin^2 \alpha - \cos^2 \alpha}{\cos \alpha \sin^3 \alpha}$ 39. $\dfrac{-1}{\cos \alpha}$

41. $\dfrac{-1}{\cos \beta \sin \beta}$ 43. $\dfrac{-1}{\cos x}$ 45. $\dfrac{-1}{\sin y \cos y}$ 47. $\dfrac{1}{\cos \alpha - \sin \alpha}$

49. $\dfrac{\sin \alpha + \cos \alpha}{\sin \alpha}$ 51. Odd 53. Odd 55. Even

Exercise Set 4.2

51. If $x = \pi/4$, then $\sin \frac{\pi}{4} + \left(\cos \frac{\pi}{4}\right)\left(\tan \frac{\pi}{4}\right) = 2;\ \frac{1}{\sqrt{2}} + \left(\frac{1}{\sqrt{2}}\right) \cdot 1 \neq 2$

53. If $y = \pi/2$, then $2\left(\sin \frac{\pi}{2}\right)\left(\cos \frac{\pi}{2}\right) + \sin \frac{\pi}{2} = 0;\ 2(1)(0) + 1 \neq 0$

Exercise Set 4.3

1. $(\sqrt{6} - \sqrt{2})/4$ 3. $-(\sqrt{2} + \sqrt{6})/4$ 5. $(\sqrt{2} + \sqrt{6})/4$ 13. $\cos 9x$
15. $\cos 2x$ 17. $\cos x$ 19. $\cos 2x$ 21. $16/65$
23. $13/\sqrt{170}$ 25. $-\cos 60°$ 27. $-\cos 30°$ 29. $\cos 45°$
37. If $\alpha = 0°$ and $\beta = 60°$, then

$$\cos(\alpha + \beta) = \cos(0° + 60°) = \cos 60° = \tfrac{1}{2};$$
$$\cos \alpha + \cos \beta = \cos 0° + \cos 60° = 1 + \tfrac{1}{2} = \tfrac{3}{2} \neq \tfrac{1}{2}.$$

39. Using the difference formula and substituting $\pi/2$ for α and α for β, we obtain

$$\cos\left(\frac{\pi}{2} - \alpha\right) = \cos \frac{\pi}{2} \cos \alpha + \sin \frac{\pi}{2} \sin \alpha$$
$$= 0(\cos \alpha) + 1(\sin \alpha) = \sin \alpha.$$

49. 1 51. $(\sqrt{3} + 1)/2\sqrt{2}$ 53. $(\sqrt{3} - 1)/2\sqrt{2}$

Exercise Set 4.4

1. $(\sqrt{6} + \sqrt{2})/4$ 3. $(\sqrt{6} - \sqrt{2})/4$ 5. $(\sqrt{2} + \sqrt{6})/4$ 13. $\sin 3x$
15. $\sin 4x$ 17. $\sin 2x$ 19. $\sin 2x$ 21. $-44/125$
23. $5/\sqrt{754}$ 25. $-\sin 60°$ 27. $\sin 30°$ 29. $-\sin 30°$

37. If $\alpha = 90^\circ$ and $\beta = 60^\circ$, then

$$\sin(\alpha - \beta) = \sin(90^\circ - 60^\circ) = \sin 30^\circ = \tfrac{1}{2};$$

$$\sin\alpha - \sin\beta = \sin 90^\circ - \sin 60^\circ = 1 - \frac{\sqrt{3}}{2} \neq \frac{1}{2}.$$

39. 0 **41.** $-\frac{1}{2}$ **43.** $(\sqrt{3} - 1)/2\sqrt{2}$

Exercise Set 4.5

1. $(1 + \sqrt{3})/(1 - \sqrt{3})$ **3.** $(\sqrt{3} - 1)/(1 + \sqrt{3})$ **5.** $(\sqrt{3} - 1)/(1 + \sqrt{3})$
13. $\tan 9x$ **15.** $\tan 2x$ **17.** $\tan x$ **19.** $-\tan 60^\circ$ **21.** $\tan 30^\circ$ **23.** $-\tan 45^\circ$

Exercise Set 4.6

1. $\sin 2x$ **3.** $\cos 2x$ **5.** $\sin 4x$ **7.** $\frac{1}{2}\sin x$ **9.** $\cos 10\alpha$ **11.** $2\cos 12\alpha$
13. $\sin\alpha$ **15.** $\cos 2x$ **33.** $24/25$ **35.** $5/\sqrt{34}$ **37.** $-24/7$

Exercise Set 4.7

1. $\frac{1}{2}\sqrt{2 - \sqrt{3}}$ **3.** $2 - \sqrt{3}$ **5.** $\frac{1}{2}\sqrt{2 - \sqrt{3}}$ **7.** $\frac{1}{2}\sqrt{2 + \sqrt{2}}$
9. $2 + \sqrt{3}$ **11.** $-\frac{1}{2}\sqrt{2 - \sqrt{3}}$ **19.** $1/\sqrt{3}$ **21.** $\sqrt{2}/3$
23. $-\frac{1}{2}\sqrt{2 + \sqrt{2}}$ **31.** $2\cos 12^\circ \cos 8^\circ$ **33.** $2\sin 4\alpha \cos\alpha$ **35.** $2\cos 3\alpha \sin 2\alpha$

37.
$$\begin{aligned}\cos\alpha\cos\beta &= \tfrac{1}{2}[\cos(\alpha + \beta) + \cos(\alpha - \beta)]\\ &= \tfrac{1}{2}[\cos\alpha\cos\beta - \sin\alpha\sin\beta + \cos\alpha\cos\beta + \sin\alpha\sin\beta]\\ &= \tfrac{1}{2}[2\cos\alpha\cos\beta] = \cos\alpha\cos\beta\end{aligned}$$

39.
$$\begin{aligned}\sin\alpha\cos\beta &= \tfrac{1}{2}[\sin(\alpha + \beta) + \sin(\alpha - \beta)]\\ &= \tfrac{1}{2}(\sin\alpha\cos\beta + \cos\alpha\sin\beta + \sin\alpha\cos\beta - \cos\alpha\sin\beta)\\ &= \tfrac{1}{2}(2\sin\alpha\cos\beta) = \sin\alpha\cos\beta\end{aligned}$$

41. Using Equation (7) we have

$$\begin{aligned}2\cos\left(\frac{\alpha + \beta}{2}\right)\cos\left(\frac{\alpha - \beta}{2}\right) &= 2 \cdot \frac{1}{2}\left[\cos\left(\frac{\alpha + \beta}{2} + \frac{\alpha - \beta}{2}\right) + \cos\left(\frac{\alpha + \beta}{2} - \frac{\alpha - \beta}{2}\right)\right]\\ &= \cos\alpha + \cos\beta\end{aligned}$$

43. ... $\beta - \frac{\alpha - \beta}{2}\Big)\Big]$

Exercise Set 4.8

1a. $30°$ **b.** $\{30°, 330°\}$
3a. $150°$ **b.** $\{150°, 330°\}$
5a. $30°$ **b.** $\{30°, 210°\}$
7a. $45°$ **b.** $\{45°, 315°\}$
9a. $210°$ **b.** $\{210°, 330°\}$
11a. $210°$ **b.** $\{210°, 330°\}$
13a. $135°$ **b.** $\{135°, 225°\}$
15a. $45°$ **b.** $\{45°, 135°\}$
17a. $0°$ **b.** $\{0°, 180°\}$
19a. $150°$ **b.** $\{150°, 210°\}$
21a. $0°$ **b.** $\{0°, 180°\}$
23a. $0°$ **b.** $\{0°, 180°\}$
25a. $14°$ **b.** $\{14°, 166°\}$
27a. $111°$ **b.** $\{111°, 291°\}$
29a. $14.3°$ **b.** $\{14.3°, 345.7°\}$
31a. $155.3°$ **b.** $\{155.3°, 204.7°\}$
33a. $25.4°$ **b.** $\{25.4°, 205.4°\}$
35a. $231.4°$ **b.** $\{231.4°, 308.6°\}$
37a. 0.70^R **b.** $\{0.70^R, 2.44^R\}$
39a. 0.23^R **b.** $\{0.23^R, 3.37^R\}$
41a. 0.98^R **b.** $\{0.98^R, 5.30^R\}$
43a. 2.89^R **b.** $\{2.89^R, 3.39^R\}$
45a. 3.26^R **b.** $\{3.26^R, 6.16^R\}$
47a. 3.63^R **b.** $\{3.63^R, 5.79^R\}$
49. $\{\alpha \mid \alpha \approx 16.2° + k \cdot 360° \text{ or } \alpha \approx 163.8° + k \cdot 360°\},\ k \in J$
51. $\{\alpha \mid \alpha \approx 142.0° + k \cdot 180°\},\ k \in J$
53. $\{\alpha \mid \alpha \approx 231.3° + k \cdot 360° \text{ or } 308.7° + k \cdot 360°\},\ k \in J$
55. $\{\alpha \mid \alpha \approx 16.8° + k \cdot 180°\},\ k \in J$
57. $\{x \mid x = \pi/4 + k \cdot 2\pi\} \cup \{x \mid x = 7\pi/4 + k \cdot 2\pi\},\ k \in J$
59. $\{x \mid x = 4\pi/3 + k \cdot 2\pi\} \cup \{x \mid x = 5\pi/3 + k \cdot 2\pi\},\ k \in J$
61. $\{x \mid x = 3\pi/4 + k\pi\},\ k \in J$
63. $\{x \mid x \approx 0.3498 + k \cdot 2\pi\} \cup \{x \mid x \approx 2.7918 + k \cdot 2\pi\},\ k \in J$
65. $\{x \mid x \approx 2.0580 + k \cdot 2\pi\} \cup \{x \mid x \approx 4.2252 + k \cdot 2\pi\},\ k \in J$
67. $\{x \mid x \approx 1.0238 + k\pi\},\ k \in J$
69. $\{x \mid x \approx -0.6557 + k\pi\},\ k \in J$

Exercise Set 4.9

1. $\{\pi/6\}$
3. $\{\pi/4\}$
5. $\{\pi/2\}$
7. $\{\pi/4, \pi/2\}$
9. $\{0, \pi/2\}$
11. $\{\pi/4, \pi/2\}$
13. $\{0\}$
15. $\{0, 2\pi/3, 4\pi/3\}$
17. $\{1.91, 4.37\}$

19. **a.** $\{270°\}$ **b.** $\{x \mid x = 270° + k \cdot 360°\},\ k \in J$

21. **a.** $\{60°, 180°, 300°\}$
b. $\{x \mid x = 60° + k \cdot 360°\} \cup \{x \mid x = 180° + k \cdot 360°\} \cup \{x \mid x = 300° + k \cdot 360°\},\ k \in J$

23. **a.** $\{30°, 90°, 150°\}$
b. $\{x \mid x = 30° + k \cdot 360°\} \cup \{x \mid x = 90° + k \cdot 360°\} \cup \{x \mid x = 150° + k \cdot 360°\},\ k \in J$

25. $\{x \mid x = k \cdot \pi\} \cup \{x \mid x = \pi/6 + k \cdot 2\pi\} \cup \{x \mid x = 5\pi/6 + k \cdot 2\pi\},\ k \in J$
27. $\{x \mid x = k \cdot \pi\} \cup \{x = \pi/3 + k\pi\},\ k \in J$
29. $\{x \mid x = \pi/6 + k \cdot 2\pi\} \cup \{x \mid x = 5\pi/6 + k \cdot 2\pi\} \cup \{x \mid x = \pi + k \cdot 2\pi\}.\ k \in J$

31. **a.** $\{218.2°, 321.8°\}$
b. $\{x \mid x = 218.2° + k \cdot 360°\} \cup \{x \mid x = 321.8° + k \cdot 360°\},\ k \in J$

33. **a.** $\{39.9°, 115.7°, 244.3°, 320.1°\}$
b. $\{x \mid x = 39.9° + k \cdot 360°\} \cup \{x \mid x = 115.7° + k \cdot 360°\} \cup \{x \mid x = 244.3° + k \cdot 360°\}$
$\cup \{x \mid x = 320.1° + k \cdot 360°\},\ k \in J$

35. $\{\pi/3, \pi\}$
37. $\{3\pi/2\}$
39. $\{x \mid x = 3\pi/4 + k \cdot \pi\},\ k \in J$

Exercise Set 4.10

1. $\{0, \pi/3\}$ **3.** $\{\pi/6, 5\pi/12\}$ **5.** $\{\pi/15, 7\pi/15, \pi/3\}$
7. $\{\pi/6\}$ **9.** $\{0\}$ **11.** $\{\pi/8\}$
13. $\{0\}$ **15.** $\{10°, 50°, 130°, 170°, 250°, 290°\}$
17. $\{15°, 75°, 135°, 195°, 255°, 315°\}$ **19.** $\{45°, 135°, 225°, 315°\}$
21. $\{15°, 45°, 75°, 105°, 135°, 165°, 195°, 225°, 255°, 285°, 315°, 345°\}$
23. $\{180°\}$ **25.** $\{60°, 180°, 300°\}$
27. $\{x \mid x = \pi/2 + k \cdot 2\pi\} \cup \{x \mid x = \pi/6 + k \cdot 2\pi\} \cup \{x \mid x = 5\pi/6 + k \cdot 2\pi\},\ k \in J$
29. $\{x \mid x = k\pi\} \cup \{x \mid x = \pi/6 + k\pi/3\},\ k \in J$
31. $\{x \mid x \approx 1.28 + k\pi/2\} \cup \{x \mid x = \pi/8 + k\pi/2\},\ k \in J$
33. $\{x \mid x = \pi/12 + k(2\pi/3)\} \cup \{x \mid x = \pi/4 + k(2\pi/3)\},\ k \in J$
35. $\{0.5\}$

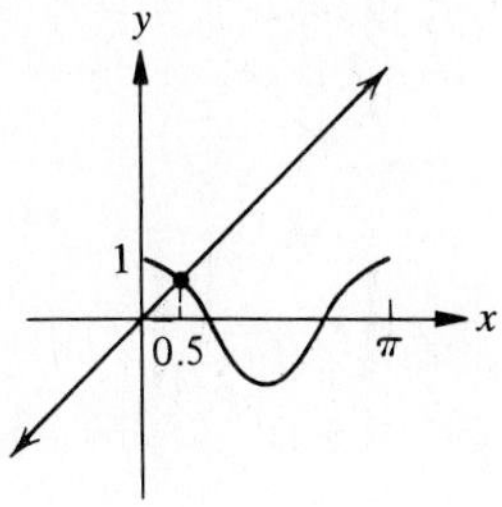

Review Exercises

1. $1/\cos\alpha$ **2.** $1/\sin^2\alpha$ **3.** $\cos^2\alpha$ **4.** $1/\sin\alpha$ **5.** $-\cos\alpha$ **6.** $1/\cos\alpha$

23. If $\alpha = \dfrac{\pi}{4}$, then $\sin\dfrac{\pi}{4}\sec\dfrac{\pi}{4} = 2\tan\dfrac{\pi}{4}$; $\dfrac{1}{\sqrt{2}} \cdot \sqrt{2} \neq 2 \cdot 1$

24. If $\alpha = \dfrac{\pi}{4}$, then $\dfrac{1 - \cos\pi/4}{\sin\pi/4} = \dfrac{1 + \sin\pi/4}{\cos\pi/4}$; $\dfrac{1 - 1/\sqrt{2}}{1/\sqrt{2}} \neq \dfrac{1 + 1/\sqrt{2}}{1/\sqrt{2}}$; $\sqrt{2} - 1 \neq \sqrt{2} + 1$

25. $-\sqrt{2 - \sqrt{3}}/2$ **26.** $\sqrt{3} - 2$ **29.** $\frac{1}{2}(\cos 6x + \cos 4x)$ **30.** $2\sin 5x\cos x$

31. a. $\{60°\}$
b. $\{\alpha \mid \alpha = 60° + k \cdot 360°\} \cup \{\alpha \mid \alpha = 120° + k \cdot 360°\},\ k \in J$

32. a. $\{135°\}$
b. $\{\alpha \mid \alpha = 135° + k \cdot 180°\},\ k \in J$

33. a. $\{118.0°\}$
b. $\{\alpha \mid \alpha \approx 118.0° + k \cdot 360°\} \cup \{\alpha \mid \alpha \approx 242.0° + k \cdot 360°\},\ k \in J$

34. a. $\{36.1°\}$
b. $\{\alpha \mid \alpha \approx 36.1° + k \cdot 360°\} \cup \{\alpha \mid \alpha \approx 143.9° + k \cdot 360°\},\ k \in J$

35. a. $\{0.369^{R}\}$
b. $\{\alpha \mid \alpha \approx 0.369^{R} + k \cdot 2\pi^{R}\} \cup \{\alpha \mid \alpha \approx 5.915^{R} + k \cdot 2\pi^{R}\},\ k \in J$

36. a. $\{1.810^{R}\}$
b. $\{\alpha \mid \alpha \approx 1.810^{R} + k \cdot \pi^{R}\},\ k \in J$

4

37. **a.** $\{0.316^{R}\}$
b. $\{\alpha \mid \alpha \approx 0.316^{R} + k \cdot 2\pi^{R}\} \cup \{\alpha \mid \alpha \approx 2.826^{R} + k \cdot 2\pi^{R}\},\ k \in J$

38. **a.** $\{0.459^{R}\}$
b. $\{\alpha \mid \alpha \approx 0.459^{R} + k \cdot 2\pi^{R}\} \cup \{\alpha \mid \alpha \approx 5.824^{R} + k \cdot 2\pi^{R}\},\ k \in J$

39. **a.** 0.8360
b. $\{x \mid x \approx 0.8360 + k \cdot 2\pi\} \cup \{x \mid x \approx -0.8360 + k \cdot 2\pi\},\ k \in J$

40. **a.** 0.2134
b. $\{x \mid x \approx 0.2134 + k \cdot 2\pi\} \cup \{x \mid x \approx 2.9282 + k \cdot 2\pi\},\ k \in J$

41. **a.** 2.3330
b. $\{x \mid x \approx -0.8086 + k\pi\},\ k \in J$

42. **a.** 1.1904
b. $\{x \mid x \approx 1.1904 + k\pi\},\ k \in J$

43. $\{\pi/3\}$ **44.** $\{\pi/6\}$ **45.** $\{\pi/4\}$ **46.** $\{\pi/2\}$
47. $\{30°, 150°, 210°, 330°\}$ **48.** $\{45°, 135°, 225°, 315°\}$
49. $\{x \mid x = \pi/3 + k\pi\} \cup \{x \mid x = 2\pi/3 + k\pi\},\ k \in J$ **50.** $\{x \mid x = \pi/4 + k\pi/2\},\ k \in J$
51. $\{15°, 105°, 135°, 225°, 255°, 345°\}$ **52.** $\{15°, 30°, 105°, 120°, 195°, 210°, 285°, 300°\}$
53. $\{0°, 45°, 135°, 180°, 225°, 315°\}$ **54.** $\{0°, 180°\}$
55. $\{x \mid x = \pi/24 + k\pi/4\},\ k \in J$
56. $\{x \mid x = \pi + k \cdot 2\pi\} \cup \{x \mid x = \pi/3 + k \cdot 2\pi\} \cup \{x \mid x = 5\pi/3 + k \cdot 2\pi\},\ k \in J$

Exercise Set 5.1

1. 2.3; 90° **3.** 3.5; −60° **5.** 2.4; −90°

7.
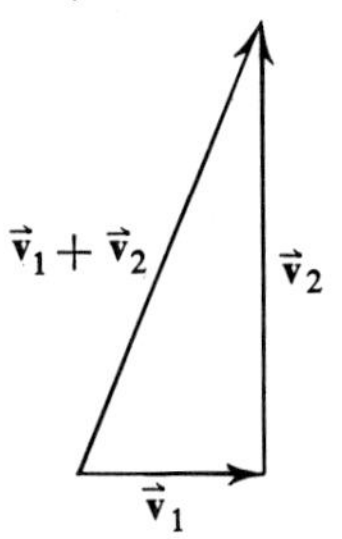

9.
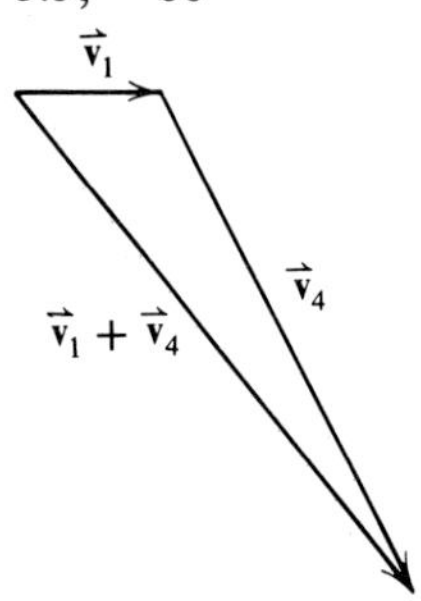

11.
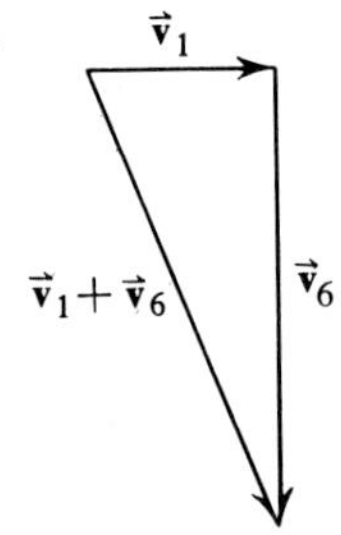

13.
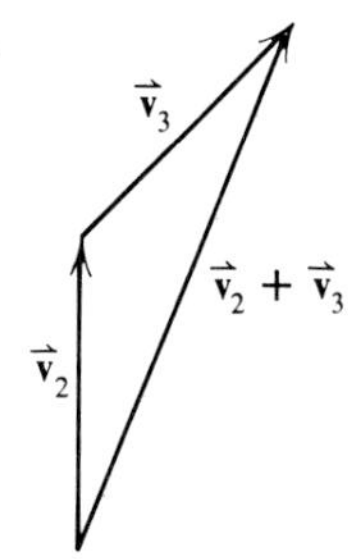

15.
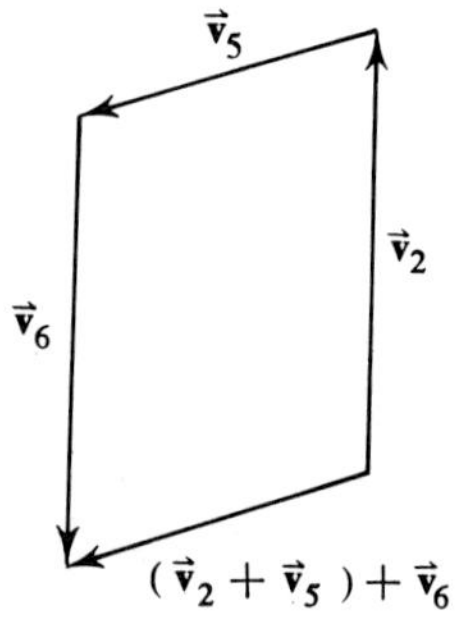

17.
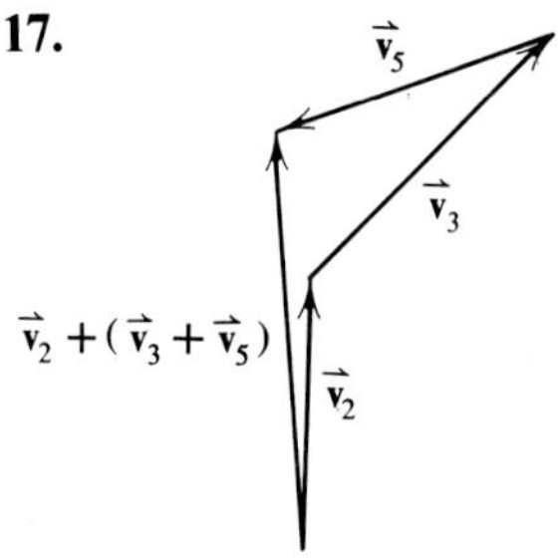

19.
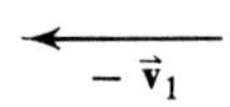

21.
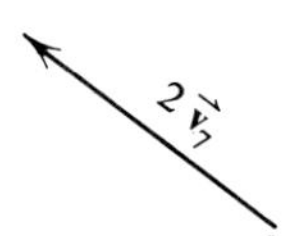

23.

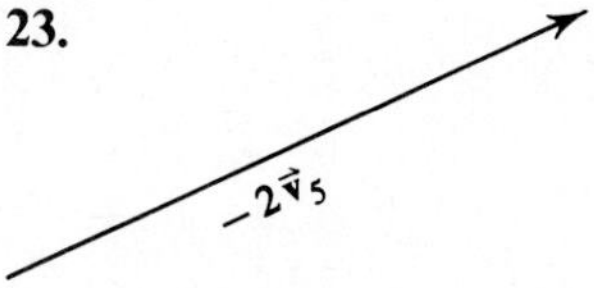

25.

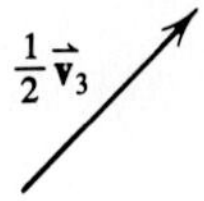

27.

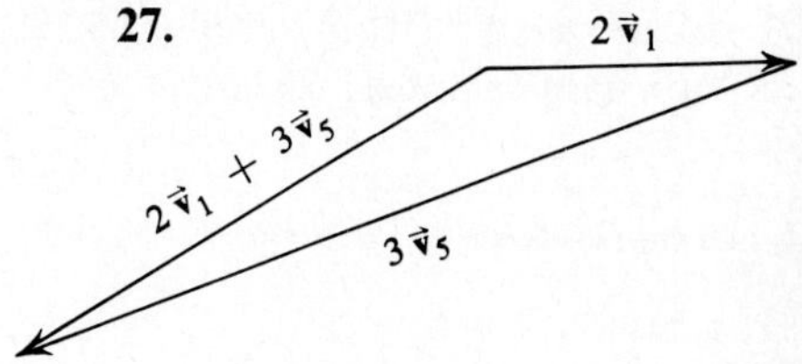

29.

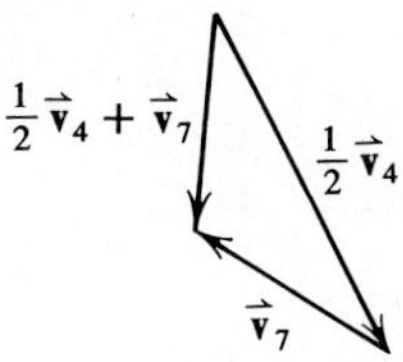

31.

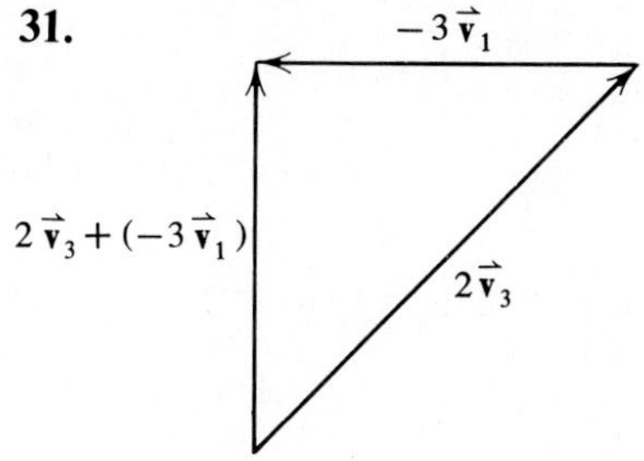

33.

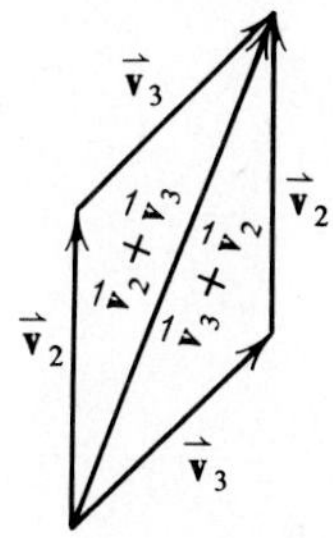

35.

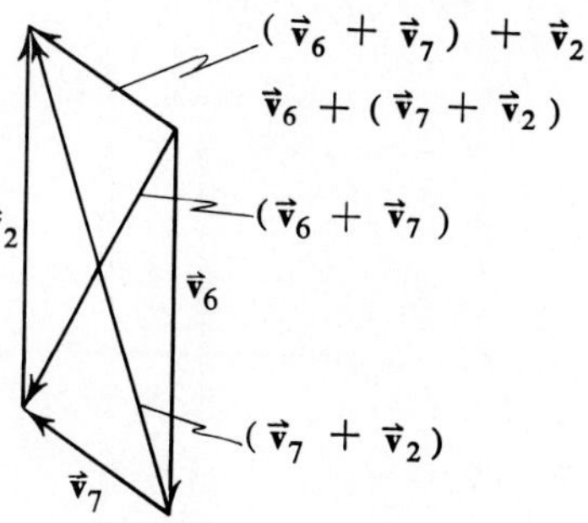

37.

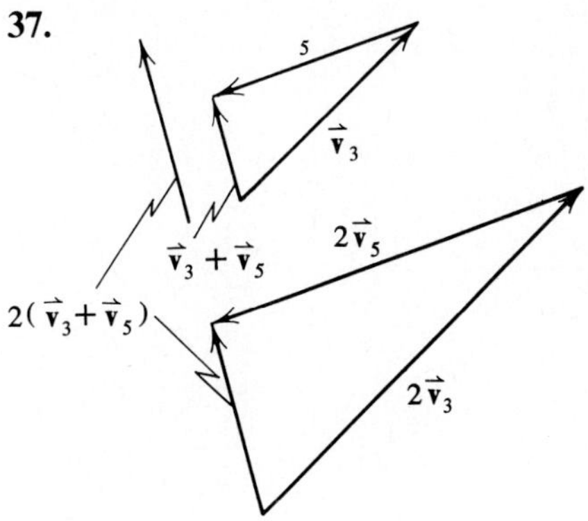

39.

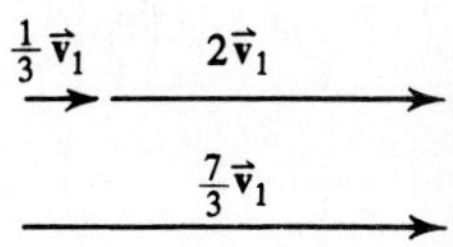

Exercise Set 5.2

1. $\|\vec{v}_x\| = 5, \; \|\vec{v}_y\| = 5$

3. $\|\vec{v}_x\| = 0, \; \|\vec{v}_y\| = 15$

5. $\|\vec{v}_x\| = 6\sqrt{3}, \; \|\vec{v}_y\| = 6$

7. $\|\vec{v}_x\| = 4, \; \|\vec{v}_y\| = 4\sqrt{3}$

9. $\|\vec{v}_3\| = 5$ pounds; 53.1° with 3-pound force, 36.9° with 4-pound force.

11. $\|\vec{v}_3\| = 27.9$ pounds; 43.8° with 15-pound force, 31.2° with 20-pound force.

13. 8.72 newtons, $\alpha \approx 36.6°$, $\beta \approx 83.4°$

15. 40 newtons, 36.9°

17. 19.7 pounds, 37.1 pounds

19. 511.3 pounds

21. 33.4°

23. $\|\vec{v}_x\| = 1.64$ meters/sec^2, $\|\vec{v}_y\| = 1.15$ meters/sec^2

25. 0.967 kilometer per hour

27. Drift angle is 11.3° to the right; ground speed is 612 miles per hour; course is 146.3°

29. Drift angle is 12° to the right; wind direction is 318.2° measured from true north; wind speed is 106.4 miles per hour

31. Heading is 85.2°; air speed is 602 kilometers per hour

33. 34.6 nautical miles, 265.3°

35. 70.7 pounds

Exercise Set 5.3

1. $x = -3,\ y = 6$

3. $x = -7/2,\ y = -7$

5. $x = -2,\ y = -2$

7. a.

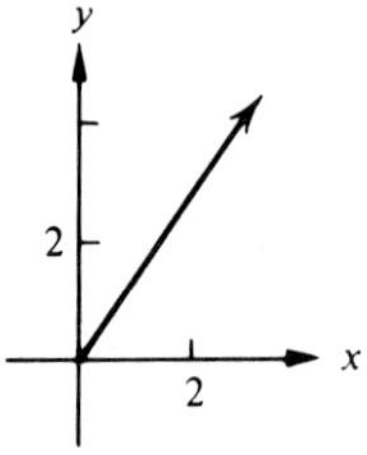

b. 5, 53.1°

9. a.

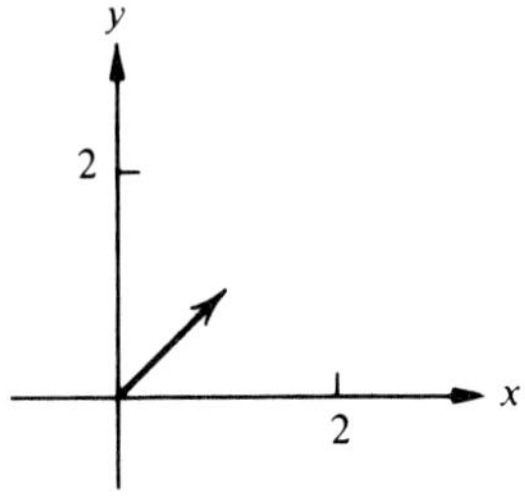

b. $\sqrt{2}$, 45°

11. a.

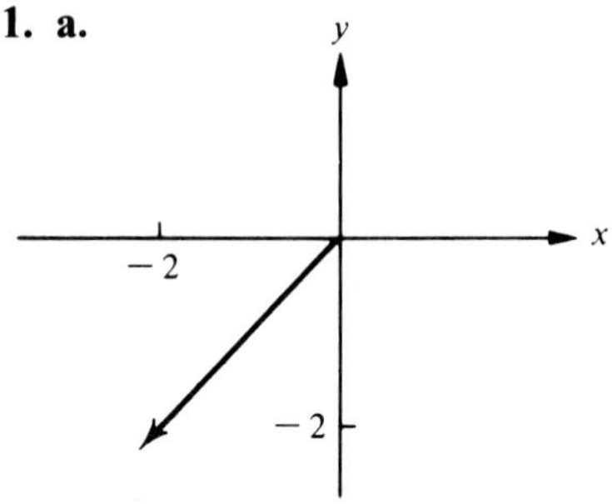

b. $3\sqrt{2}$, −135°

13. a.

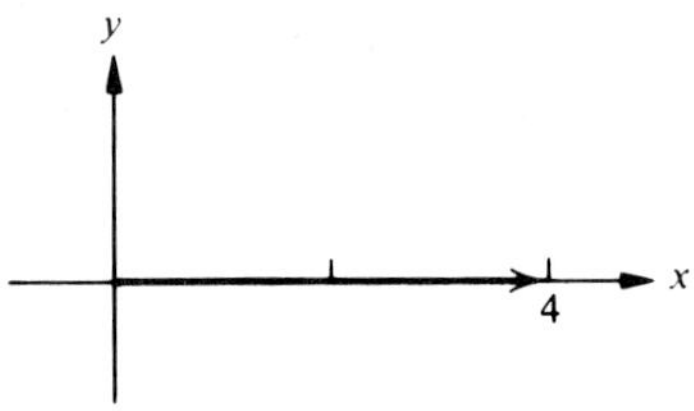

b. 4, 0°

15. (2, 2)

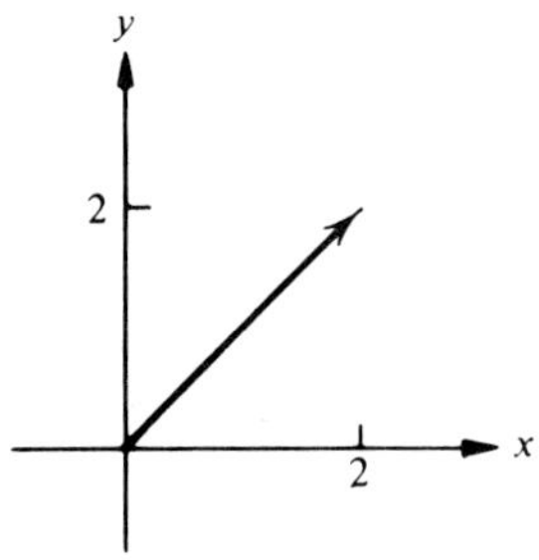

17. (0, 1)

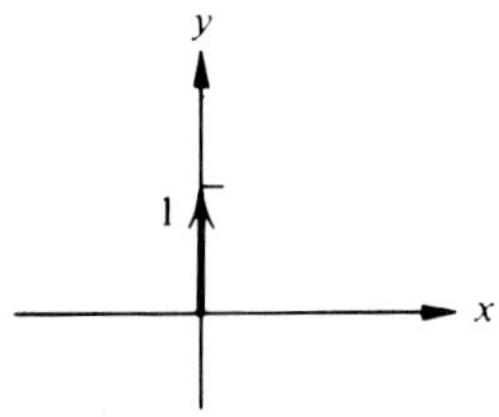

19. $(-\frac{5}{2}\sqrt{3}, -\frac{5}{2})$

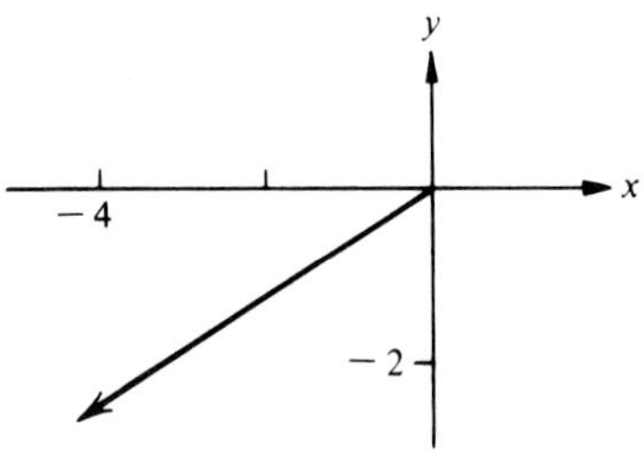

21. From Definition 5.11, the direction angles of (a, b) are given by

$$\sin\alpha = \frac{b}{\sqrt{a^2 + b^2}} \quad \text{and} \quad \cos\alpha = \frac{a}{\sqrt{a^2 + b^2}},$$

and of $(-b, a)$ by

$$\sin\beta = \frac{a}{\sqrt{a^2 + b^2}}, \quad \cos\beta = \frac{-b}{\sqrt{a^2 + b^2}}.$$

Hence,

$$\sin(\alpha + 90^\circ) = \cos\alpha = \sin\beta, \qquad \text{and} \qquad \cos(\alpha + 90^\circ) = -\sin\alpha = \cos\beta,$$

from which

$$\alpha + 90^\circ = \beta, \qquad \text{or} \qquad \beta - \alpha = 90^\circ.$$

Exercise Set 5.4

1. $(-4, 6)$

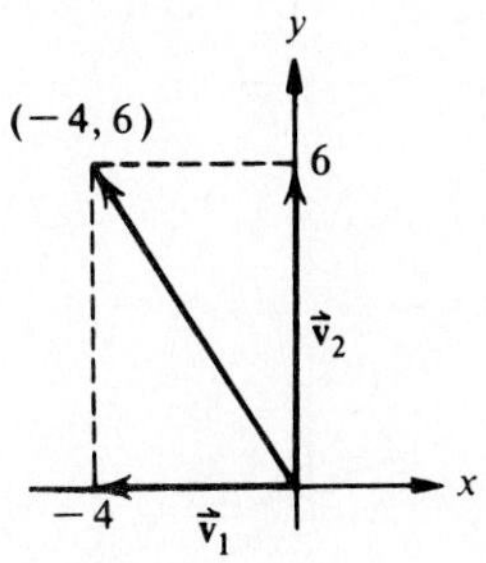

3. $(3, 1)$

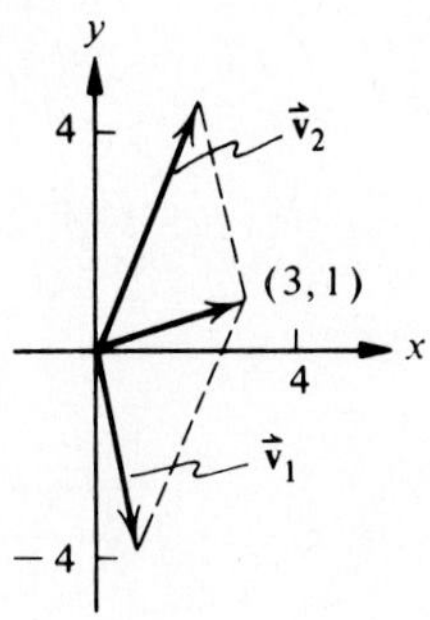

5. $(-4, 1)$

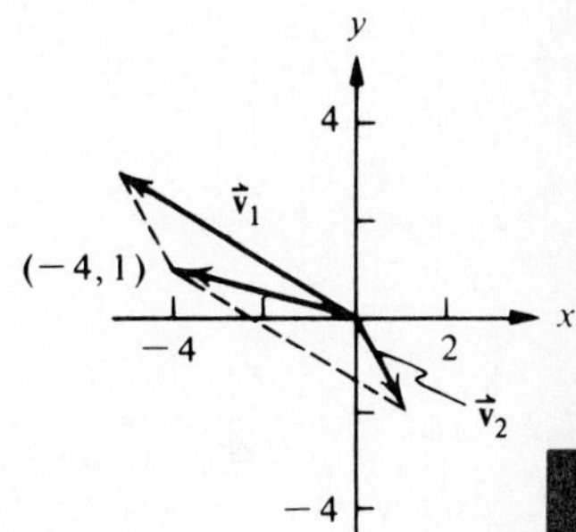

7. $(-3, -6)$

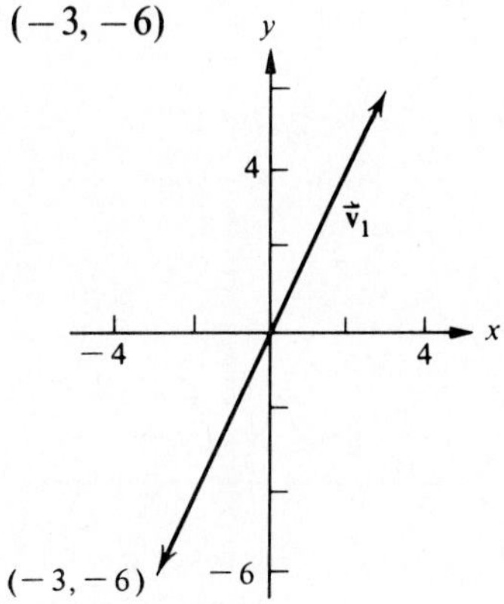

9. $(-4, 2)$

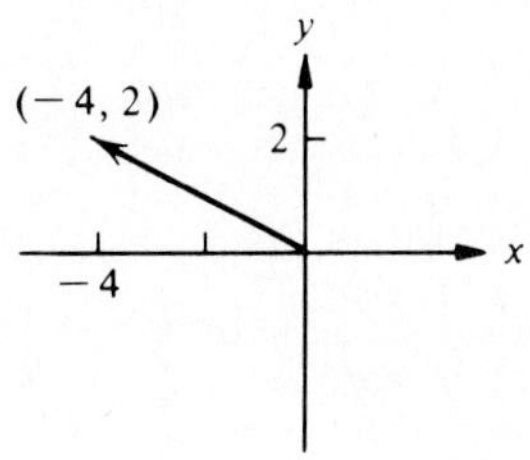

11. $(6, 12)$

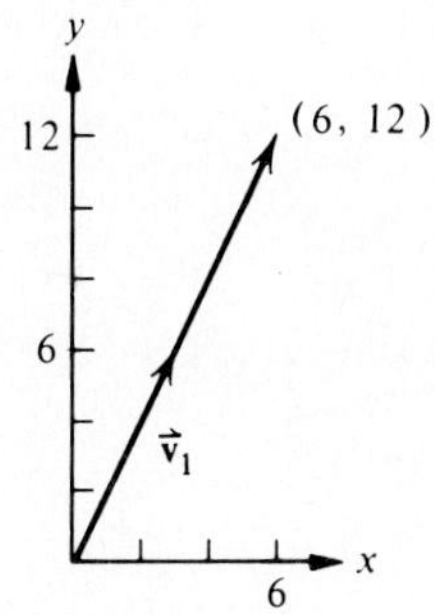

13. $(-13, -16)$

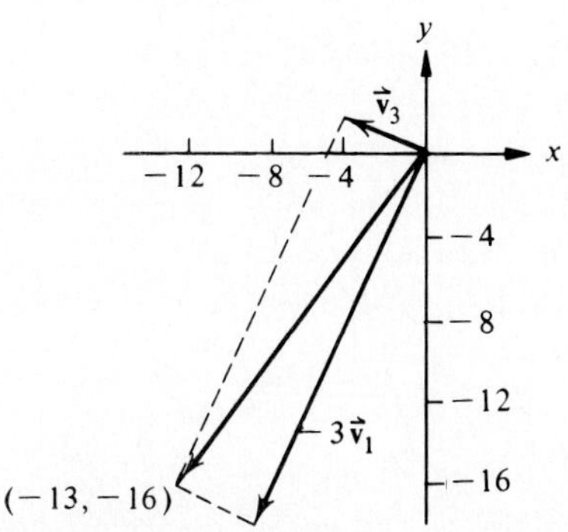

15. $(-9, 15)$

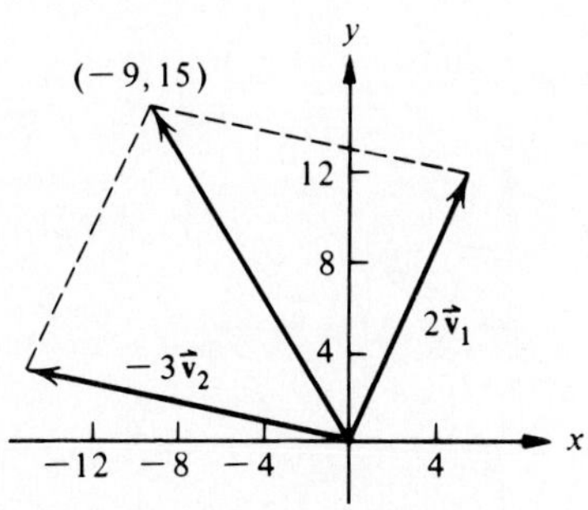

17. $(16, 10)$

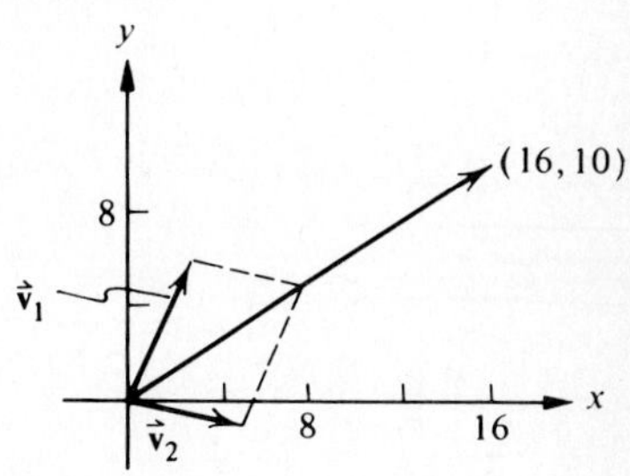

19. $(7, -8)$

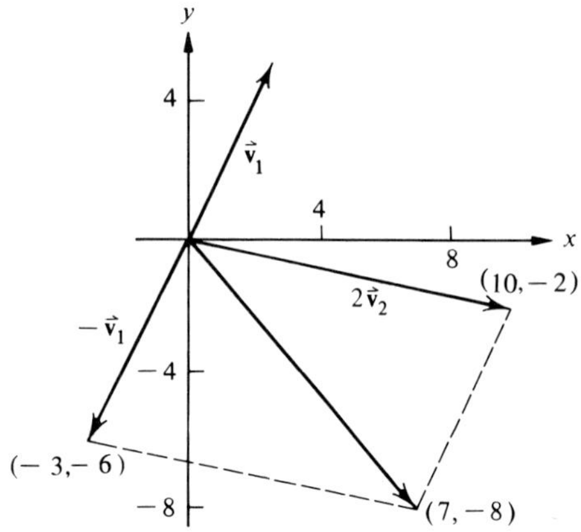

21. $(-7, -1)$

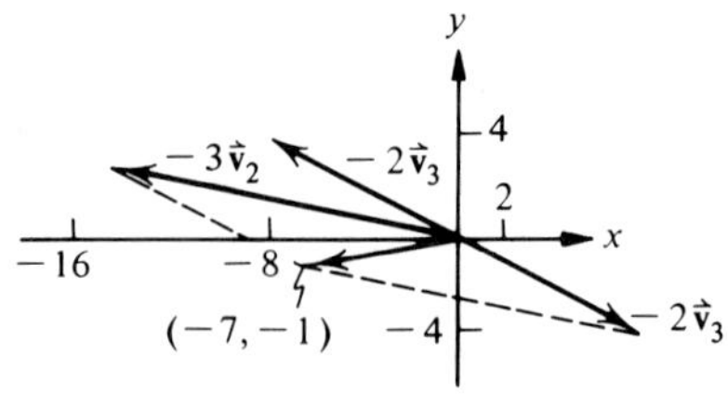

23. $(-27, 9)$

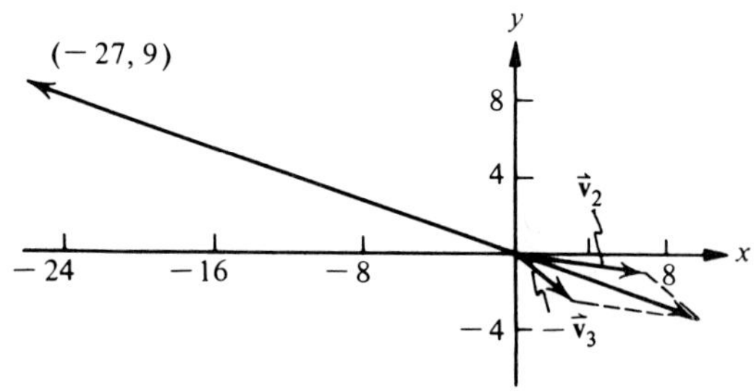

25. $(5/2, -1/2)$

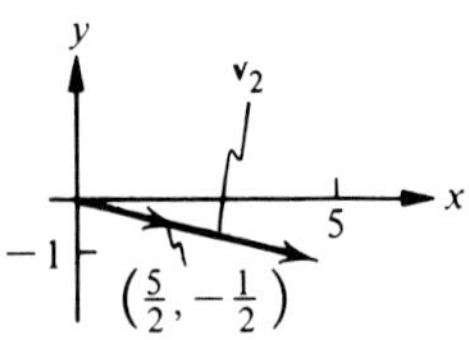

27. $(19/6, 8/3)$

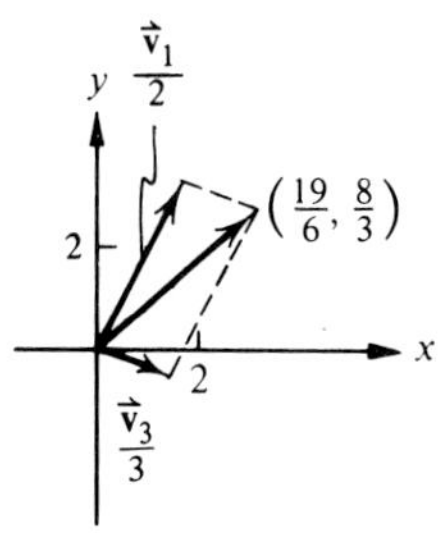

29. $(-5/2, 5)$

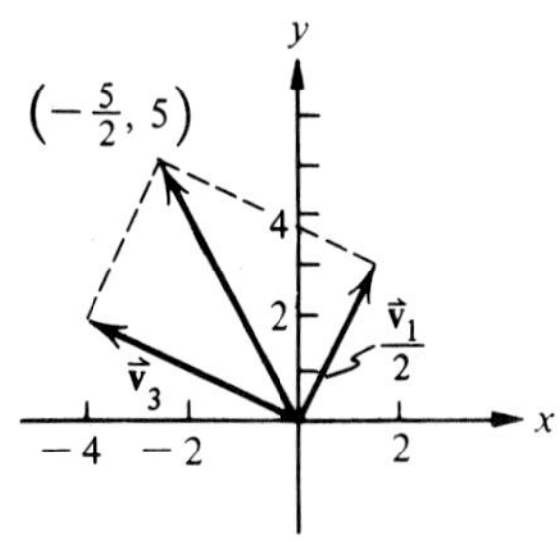

31. $3\mathbf{i} + 5\mathbf{j}$

33. $\mathbf{i} - 4\mathbf{j}$

35. $-3\mathbf{i}$

37. $4\sqrt{2}, -45°$

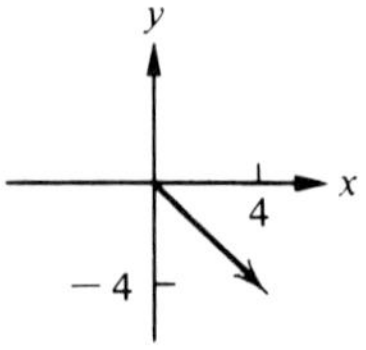

39. $\sqrt{5}, 26.6°$

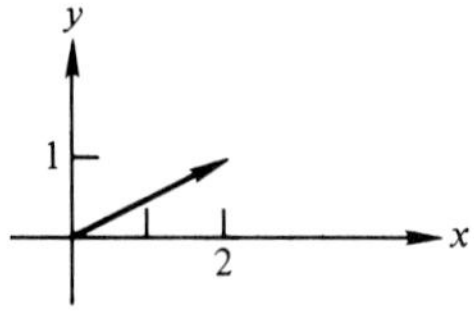

41. $\sqrt{10}, -108.4°$

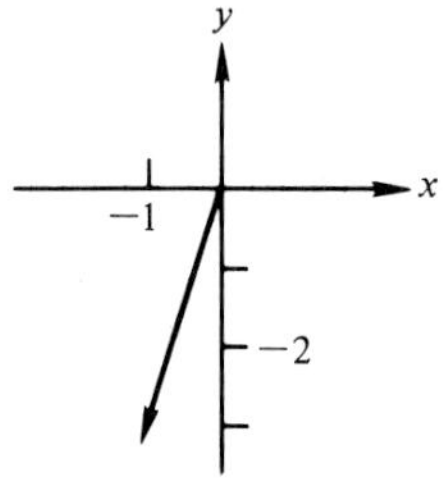

43. 2, 120°

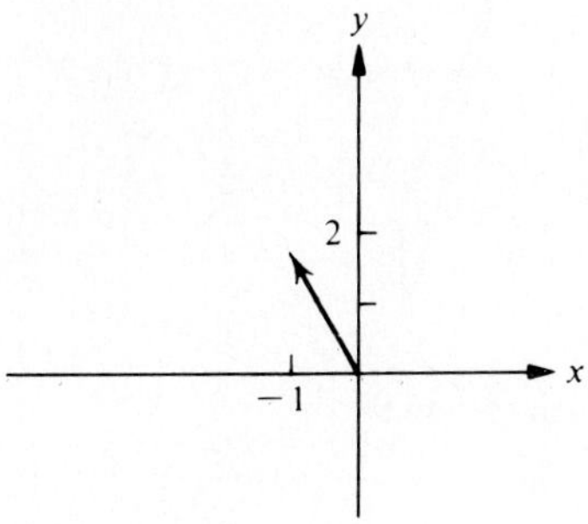

Exercise Set 5.5

1. 48 **3.** −2 **5.** 45 **7.** 13 **9.** $36\sqrt{2}$

11. 3/4 **13.** 156.8° **15.** 90° **17.** 24.0°

19. $20/\sqrt{29}$, 5 **21.** $7/\sqrt{41}$, $7/\sqrt{13}$ **23.** $-14/\sqrt{10}$, $-14/\sqrt{34}$

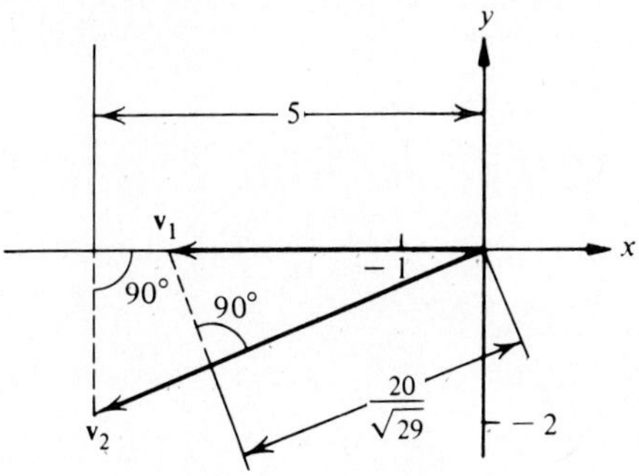

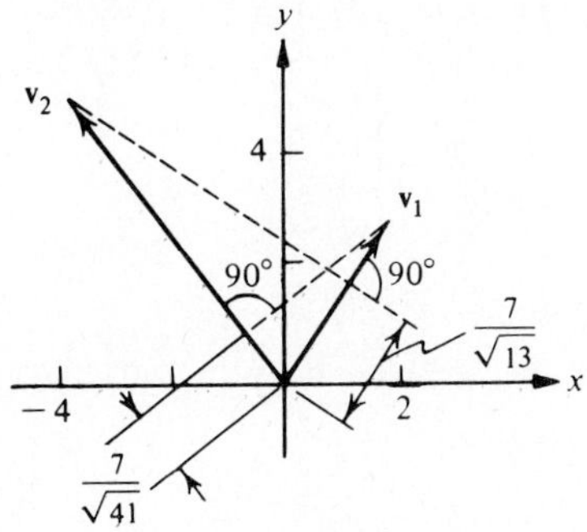

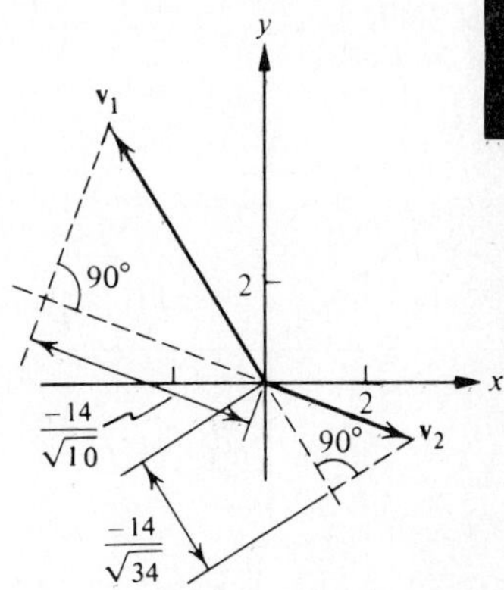

25. 0, 0 **33.** 24 **35.** 36

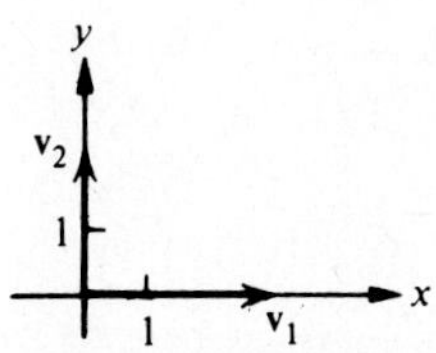

37. $1000\sqrt{3}$ meter-kilograms **39.** 1,000,000 meter-kilograms

41. By the property on page 236, $\mathbf{v}_1 \cdot \mathbf{v}_1 = \|\mathbf{v}_1\| \, \|\mathbf{v}_1\| \cos 0°$
$= \|\mathbf{v}_1\|^2$

Review Exercises

1. $(\vec{v}_1 + \vec{v}_2) + \vec{v}_3$

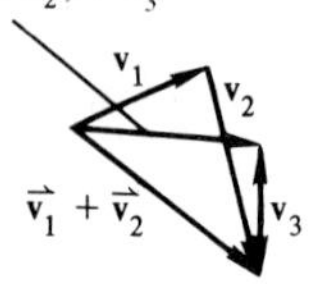

2.

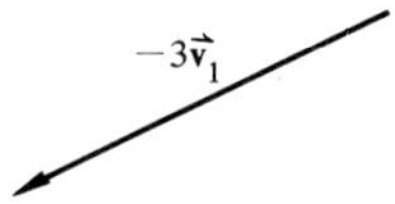

3.

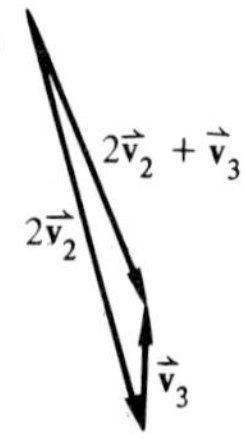

4.

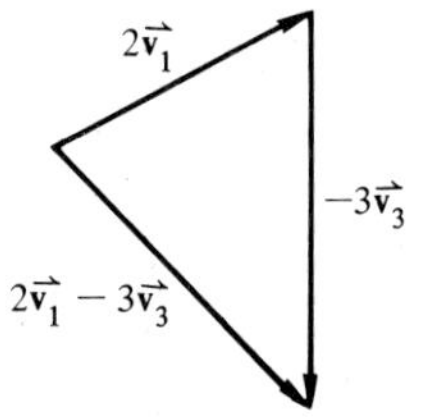

5. $6;\ 6\sqrt{3}$

6. $5;\ 36.9°$

7. 120 newtons; 22.6°

8. 6.71 miles per hour; drift of 63.4°

9. Wind bearing is 357.7°; wind speed is 183 kilometers per hour

10. 36.9°

11. $x = -1,\ y = 3$

12. $x = 4,\ y = 1$

13. 5, 143.1°

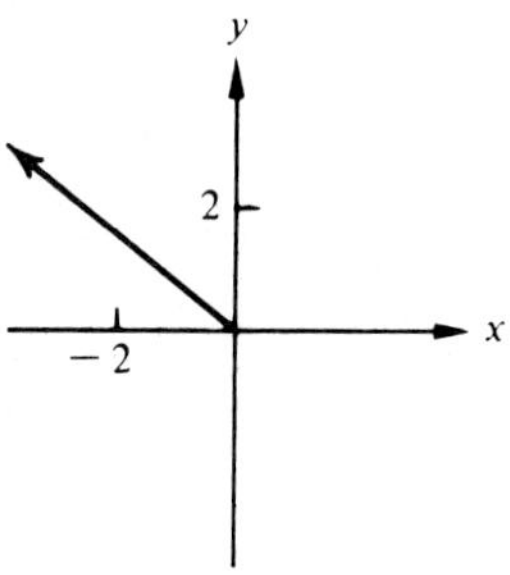

14. $5\sqrt{2},\ -45.0°$

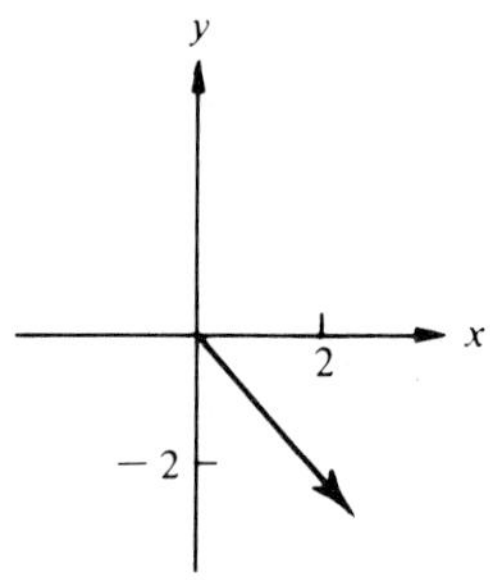

15. $(-5\sqrt{2}, -5\sqrt{2})$

16. $(-\sqrt{3}, 1)$

17. $(12, -4)$

18. $(-10, 10)$

19. $(8, 2)$

20. $(12, 2)$

21. $(-1, 3)$

22. $(5/2\sqrt{13}, 2/\sqrt{13})$

23. $7\mathbf{i} - 9\mathbf{j}$

24. $-8\mathbf{i} - \mathbf{j}$

25. $2\sqrt{29}$, 158.2°

26. $\sqrt{130}$, −52.1°

27. −54

28. 75

29. $16\sqrt{3}$

30. $25\sqrt{3}$

31. 136.8°

32. 56.5°

33. $3/\sqrt{101},\ 3/5\sqrt{2}$

34. $-82/\sqrt{89},\ -82/\sqrt{85}$

35. $(-4)(4) + (-5)(-16/5) = -16 + 16 = 0$

36. $(7)(5) + (6)(-35/6) = 35 - 35 = 0$

Exercise Set 6.1

1. $x = 1,\ y = -3$

3. $x = -\frac{1}{2},\ y = 8$

5. $1 + 10i$

7. $-10 - 11i$

9. $-1 + 6i$

11. $0 + 0i$

13. $12 - 12i$

15. $-36 - 12i$

17. $2 + 3i$

19. $-19 + 2i$

21. $62 - 39i$

23. $26 - 19i$

25. $a^2 + b^2$

27. i

29. -1

31. -1

33. $-i$

35. $5i$

37. $2\sqrt{3}\,i$

39. $1 + 3i$

41. $3 - 2\sqrt{5}\,i$

43. $8 + 2i$

45. $-8 + 8i$

47. $-10\sqrt{5} + 2\sqrt{10}\,i$

49. $23 + 2i$ **51.** $9 + \sqrt{3}\,i$

53. If $x = 2i$, then $(2i)^2 + 4 = 4i^2 + 4 = -4 + 4 = 0$;
if $x = -2i$, then $(-2i)^2 + 4 = 4i^2 + 4 = -4 + 4 = 0$

55. a. $\left(-\frac{1}{2} + i\frac{\sqrt{3}}{2}\right)^2 = \frac{1}{4} - i\frac{\sqrt{3}}{2} + i^2\left(\frac{3}{4}\right) = -\frac{1}{2} - i\frac{\sqrt{3}}{2}$;

$$\left(-\frac{1}{2} + i\frac{\sqrt{3}}{2}\right)^3 = \left(-\frac{1}{2} + i\frac{\sqrt{3}}{2}\right)^2\left(-\frac{1}{2} + i\frac{\sqrt{3}}{2}\right)$$
$$= \left(-\frac{1}{2} - i\frac{\sqrt{3}}{2}\right)\left(-\frac{1}{2} + i\frac{\sqrt{3}}{2}\right) = \frac{1}{4} - i^2\left(\frac{3}{4}\right) = 1$$

b. $\left(-\frac{1}{2} - i\frac{\sqrt{3}}{2}\right)^2 = \frac{1}{4} + i\frac{\sqrt{3}}{2} + i^2\left(\frac{3}{4}\right) = -\frac{1}{2} + i\frac{\sqrt{3}}{2}$;

$$\left(-\frac{1}{2} - i\frac{\sqrt{3}}{2}\right)^3 = \left(-\frac{1}{2} - i\frac{\sqrt{3}}{2}\right)^2\left(-\frac{1}{2} - i\frac{\sqrt{3}}{2}\right)$$
$$= \left(-\frac{1}{2} + i\frac{\sqrt{3}}{2}\right)\left(-\frac{1}{2} - i\frac{\sqrt{3}}{2}\right) = \frac{1}{4} - i^2\left(\frac{3}{4}\right) = 1$$

57. $a = 0,\ b = 1$ or $a = 0,\ b = -1$ **59.** $e^{\pi i} = \cos\pi + i\sin\pi = -1 + i(0) = -1$

61.
$$\frac{e^{ix} - e^{-ix}}{2i} = \frac{\cos x + i\sin x - [\cos(-x) + i\sin(-x)]}{2i}$$
$$= \frac{\cos x + i\sin x - \cos x + i\sin x}{2i} = \frac{2i\sin x}{2i} = \sin x$$

63. Using the results of Exercises 61 and 62, we obtain

$$\sin^2 x + \cos^2 x = \left(\frac{e^{ix} - e^{-ix}}{2i}\right)^2 + \left(\frac{e^{ix} + e^{-ix}}{2}\right)^2$$
$$= \frac{e^{2ix} - 2e^0 + e^{-2ix}}{4i^2} + \frac{e^{2ix} + 2e^0 + e^{-2ix}}{4}$$
$$= \frac{-e^{2ix} + 2 - e^{-2ix}}{4} + \frac{e^{2ix} + 2 + e^{-2ix}}{4} = \frac{4}{4} = 1.$$

6

Exercise Set 6.2

1. $13 - 4i$ **3.** $2 + 2i$ **5.** $5 - i$ **7.** $6 - 3i$ **9.** $-3 + 3i$

11. $-3i$ **13.** $\frac{3}{2} - \frac{9}{4}i$ **15.** $\frac{8}{85} + \frac{36}{85}i$ **17.** $\frac{9}{17} + \frac{2}{17}i$ **19.** $-\frac{6}{5} + \frac{11}{10}i$

21. $\frac{-i}{\sqrt{5}}$ **23.** $-\frac{2i}{\sqrt{2}}$ **25.** $\frac{4}{7} - \frac{\sqrt{5}}{7}i$ **27.** $\frac{18}{11} - \frac{6\sqrt{2}}{11}i$ **29.** $\frac{6\sqrt{2}}{7} - \frac{6\sqrt{5}}{7}i$

31. $\frac{2\sqrt{5}}{13} - \frac{4\sqrt{2}}{13}i$

Exercise Set 6.3

1. $(-1, 6)$ **3.** $(8, 1)$ **5.** $(0, 6)$ **7.** $-8 + 7i$ **9.** $-4 + 0i$ **11.** $7 - \sqrt{2}i$

13.

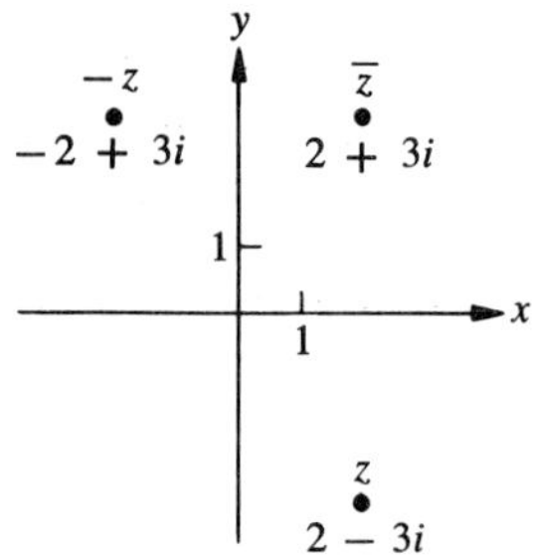

15.

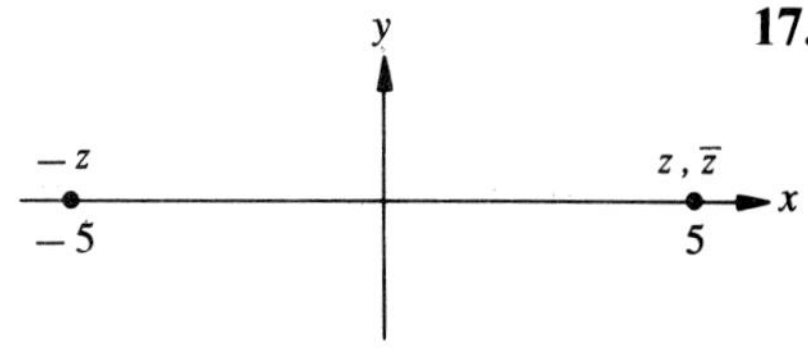

17.

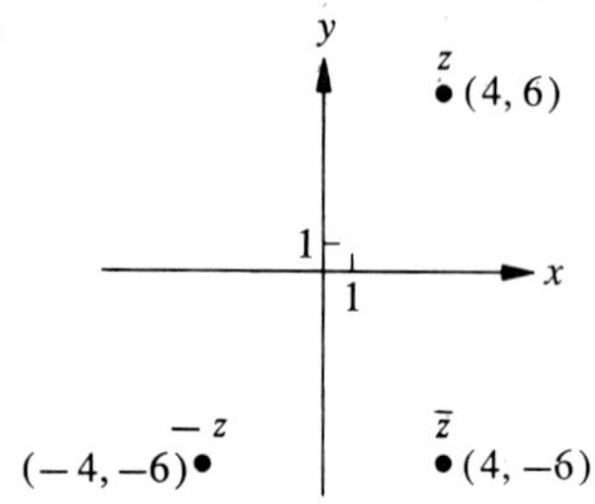

19.

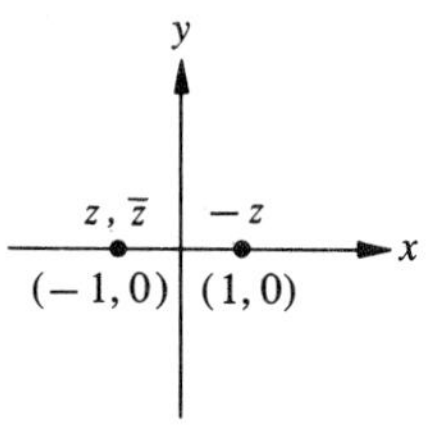

21. $-1 + 6i$

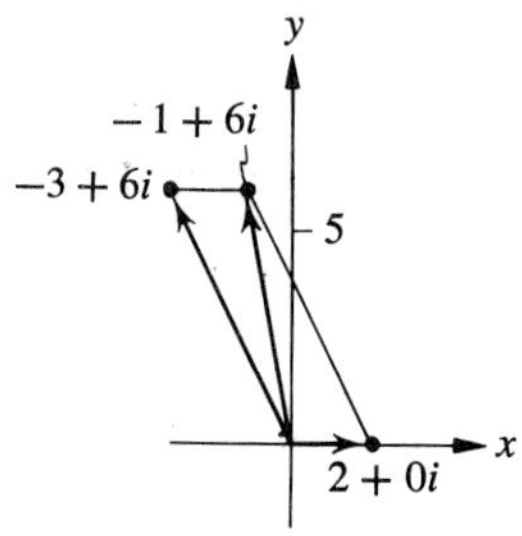

23. $-4 - 5i$

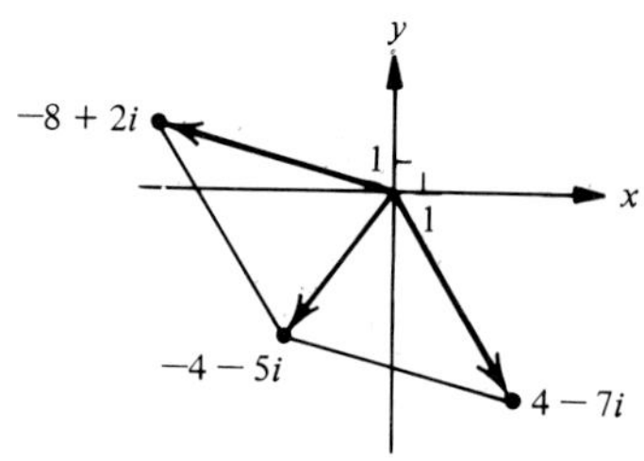

25. $-1 + 4i$

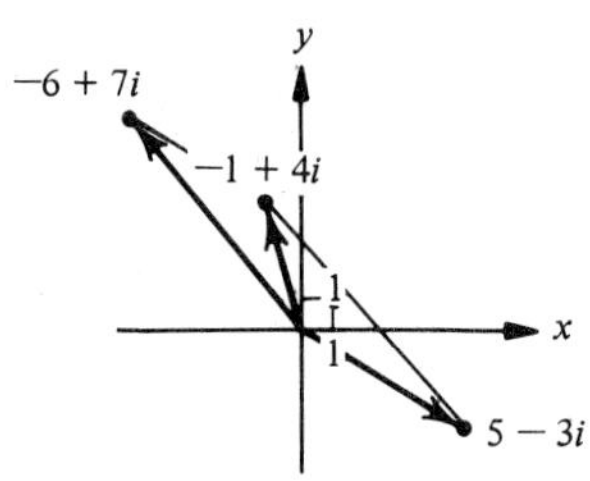

27. $\sqrt{29}$; 21.8° **29.** $2\sqrt{2}$; 225° **31.** 7; 0°

33. 5; 270° **35.** $3\sqrt{10}$; 288.4° **37.** 4; 0°

39. a. $a \neq 0,\ b = 0$ **b.** $a = 0,\ b \neq 0$ **c.** $b > 0$ **d.** $b < 0$

41.

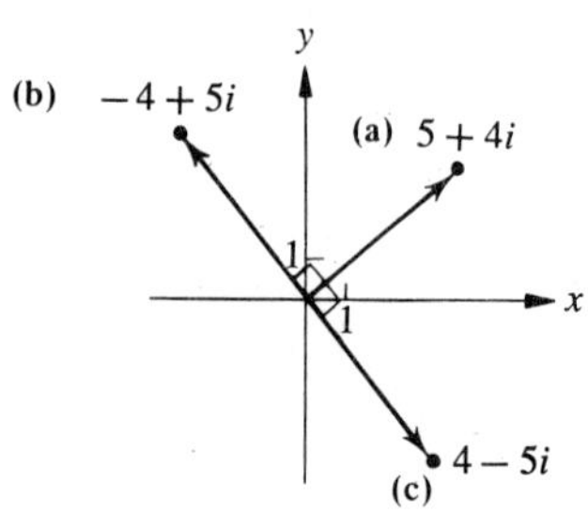

43.

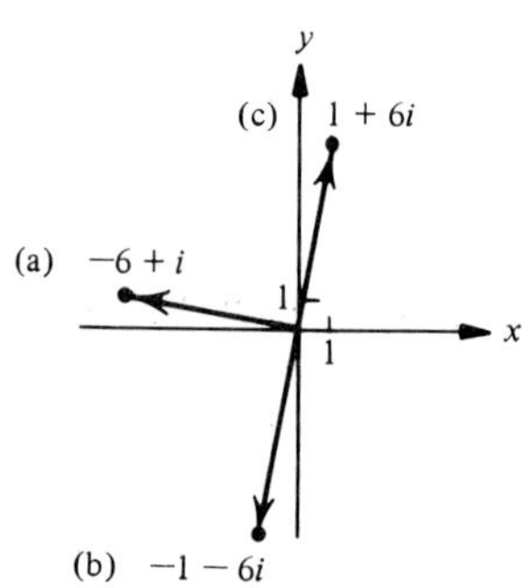

45.

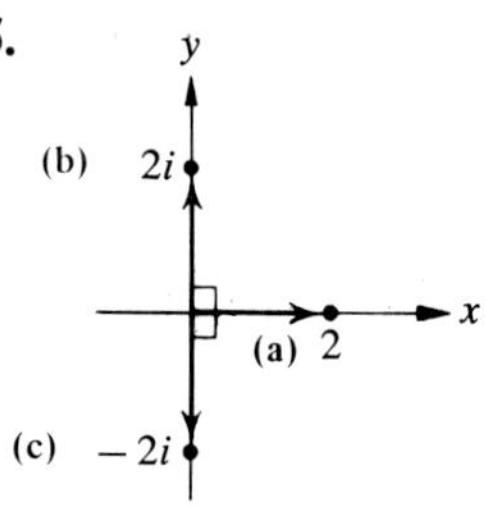

Exercise Set 6.4

1. $4(\cos 120° + i \sin 120°)$

3. $10(\cos 210° + i \sin 210°)$

5. $2\sqrt{2}(\cos 315° + i \sin 315°)$

7. $8(\cos 180° + i \sin 180°)$

9. $-1 - \sqrt{3}i$

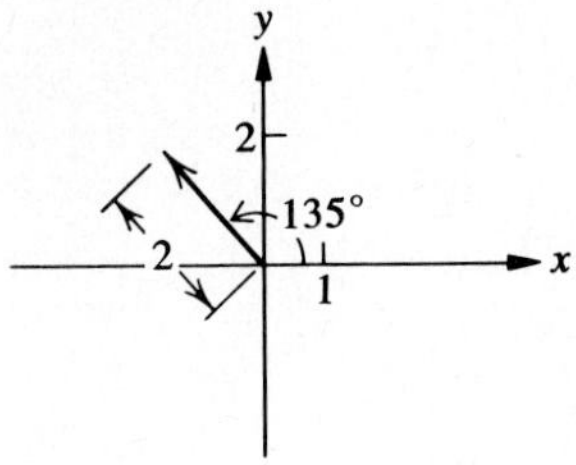

11. $-\sqrt{2} + \sqrt{2}i$

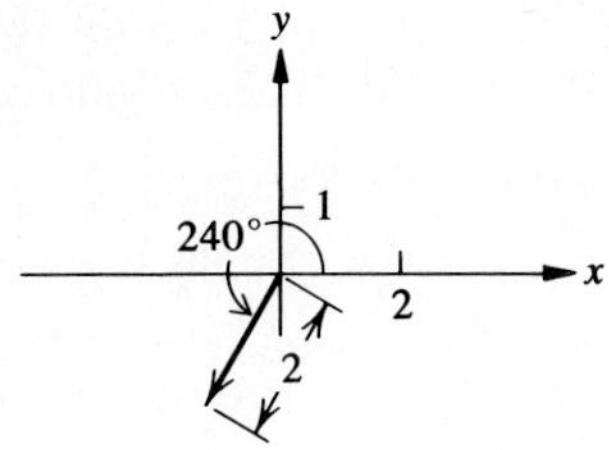

13. $\frac{1}{\sqrt{2}} - \frac{1}{\sqrt{2}}i$

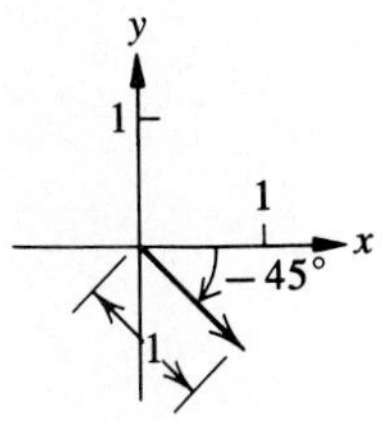

15. $-\frac{\sqrt{3}}{4} + \frac{1}{4}i$

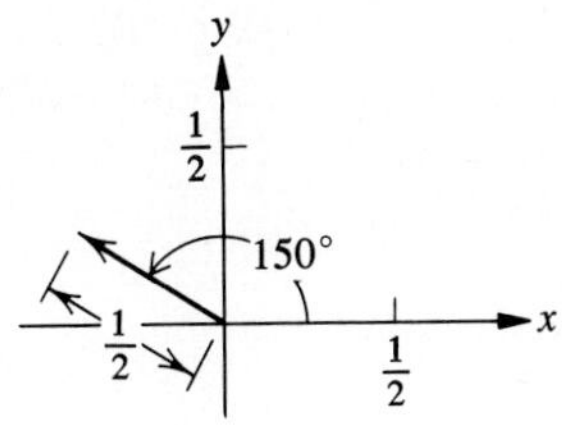

17. $-5\sqrt{2} + 5\sqrt{2}i$

19. $\frac{3}{2}\sqrt{3} - \frac{3}{2}i$

21. $-\sqrt{2} + \sqrt{2}i$

23. $-4 + 0i$

25. $-\frac{1}{5} - \frac{\sqrt{3}}{5}i$

27. $\frac{3}{2} - \frac{3\sqrt{3}}{2}i$

29. $4 + 4\sqrt{3}i$

31. $(a + bi)^2 = \rho(\cos\theta + i \sin\theta) \cdot \rho(\cos\theta + i \sin\theta)$; using Equation (1) in this section gives

$$\begin{aligned}(a + bi)^2 &= \rho \cdot \rho[\cos(\theta + \theta) + i \sin(\theta + \theta)] \\ &= \rho^2[\cos 2\theta + i \sin 2\theta].\end{aligned}$$

33. $(a + bi)^4 = (a + bi)^3(a + bi)$; using the results of Exercise 32 and Equation (1) in this section gives

$$\begin{aligned}\rho^3(\cos 3\theta + i \sin 3\theta) \cdot \rho(\cos\theta + i \sin\theta) &= \rho^3 \cdot \rho[\cos(3\theta + \theta) + i \sin(3\theta + \theta)] \\ &= \rho^4(\cos 4\theta + i \sin 4\theta).\end{aligned}$$

35. From Section 4.1, $\sin(-\theta) = -\sin\theta$ and $\cos(-\theta) = \cos\theta$.
Thus, $\cos(-\theta) + i \sin(-\theta) = \cos\theta - i \sin\theta$, the conjugate of $\cos\theta + i \sin\theta$.

Exercise Set 6.5

1. $\frac{27}{2} + \frac{27\sqrt{3}}{2}i$

3. $0 + 8i$

5. $-\frac{1}{8} + 0i$

7. $0 + \frac{1}{81} i$ **9.** $-\frac{1}{32} + \frac{1}{32} i$ **11.** $-\frac{1}{64} - \frac{\sqrt{3}}{64} i$

13. **a.** $\cos 40^\circ + i \sin 40^\circ$, $\cos 220^\circ + i \sin 220^\circ$
b. $\cos 30^\circ + i \sin 30^\circ$, $\cos 210^\circ + i \sin 210^\circ$

15. **a.** $2(\cos 40^\circ + i \sin 40^\circ)$, $2(\cos 160^\circ + i \sin 160^\circ)$, $2(\cos 280^\circ + i \sin 280^\circ)$
b. $\cos 20^\circ + i \sin 20^\circ$, $\cos 140^\circ + i \sin 140^\circ$, $\cos 260^\circ + i \sin 260^\circ$

17. **a.** $\cos 60^\circ + i \sin 60^\circ$, $\cos 150^\circ + i \sin 150^\circ$, $\cos 240^\circ + i \sin 240^\circ$, $\cos 330^\circ + i \sin 330^\circ$
b. $\cos 75^\circ + i \sin 75^\circ$, $\cos 165^\circ + i \sin 165^\circ$, $\cos 255^\circ + i \sin 255^\circ$, $\cos 345^\circ + i \sin 345^\circ$

19. $\left\{\frac{1}{2} + \frac{\sqrt{3}}{2} i, -1, \frac{1}{2} - \frac{\sqrt{3}}{2} i\right\}$ **21.** $\left\{1, -\frac{1}{2} + \frac{\sqrt{3}}{2} i, -\frac{1}{2} - \frac{\sqrt{3}}{2} i\right\}$

23. $\left\{\frac{1}{2^{1/3}} (\cos 110^\circ + i \sin 110^\circ), \frac{1}{2^{1/3}} (\cos 230^\circ + i \sin 230^\circ), \frac{1}{2^{1/3}} (\cos 350^\circ + i \sin 350^\circ)\right\}$.

25. In all cases, the graphs are spaced equally on the unit circle.

a.

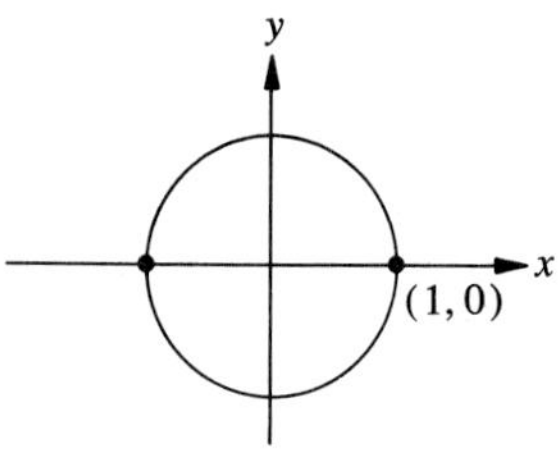

b.

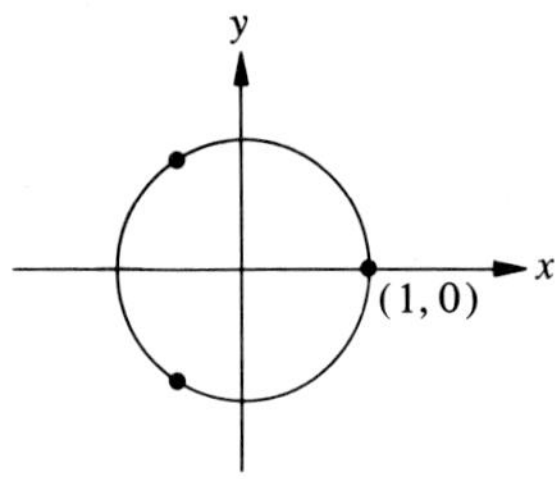

c.

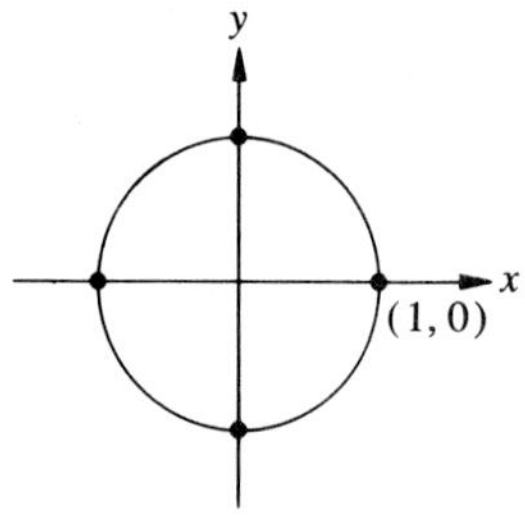

d.

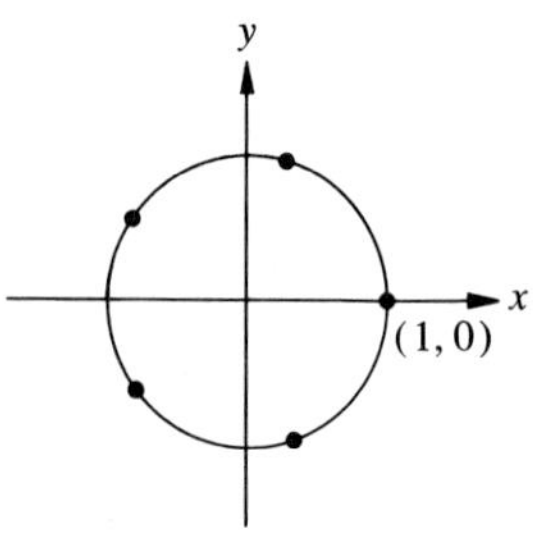

e.

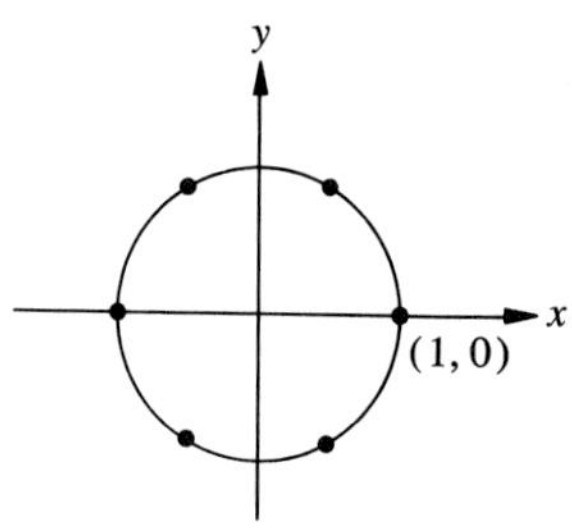

f.

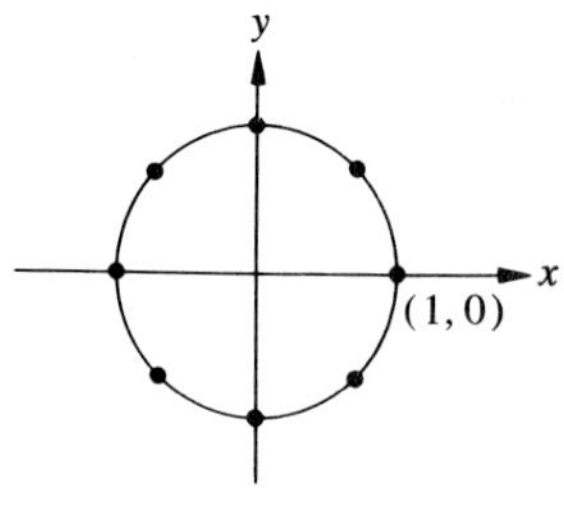

27. $\cos 3\theta = 4 \cos^3 \theta - 3 \cos \theta$

Exercise Set 6.6

1. $(3, 20^\circ), (-3, 200^\circ), (-3, -160^\circ), (3, -340^\circ)$ **3.** $(5, -90^\circ), (-5, -270^\circ), (-5, 90^\circ), (5, 270^\circ)$
5. $(-4, 60^\circ), (4, 240^\circ), (4, -120^\circ), (-4, -300^\circ)$
7. $(2\sqrt{3}, 2)$ **9.** $(-3/\sqrt{2}, -3/\sqrt{2})$ **11.** $(3\sqrt{3}/2, -3/2)$ **13.** $(4, 0^R)$
15. $\left(2\sqrt{2}, \frac{\pi}{4}^R\right)$ **17.** $\left(1, \frac{5\pi}{6}^R\right)$ **19.** $\rho \cos \theta = 4$ **21.** $\rho = \frac{8 \sin \theta}{\cos^2 \theta}$

23. $\rho = 4$ **25.** $\rho = 3 \sin \theta$ **27.** $x^2 + y^2 = 36$ **29.** $x^2 + y^2 = 4y$

31. $y = 2$ **33.** $x^2 - 8y - 16 = 0$

Exercise Set 6.7

1.

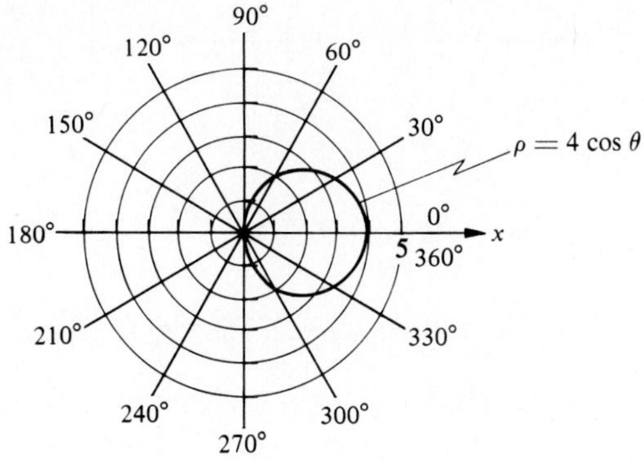

3.

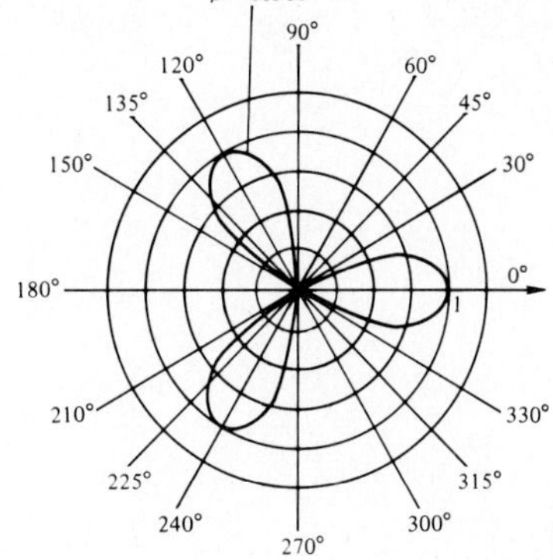

5.

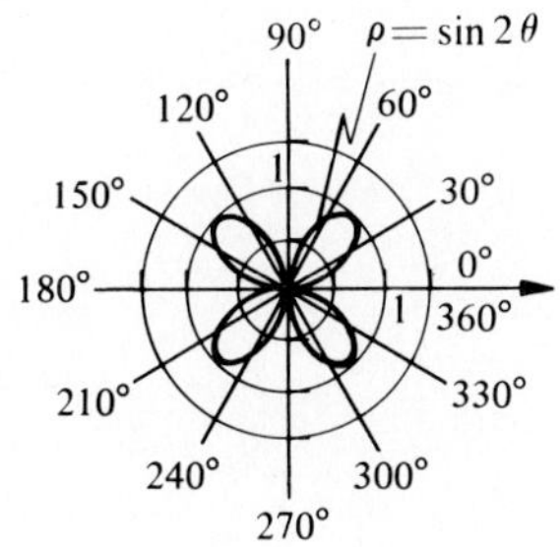

7.

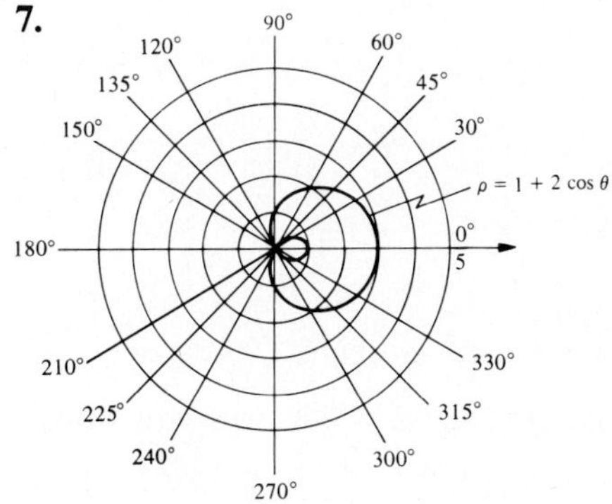

9.

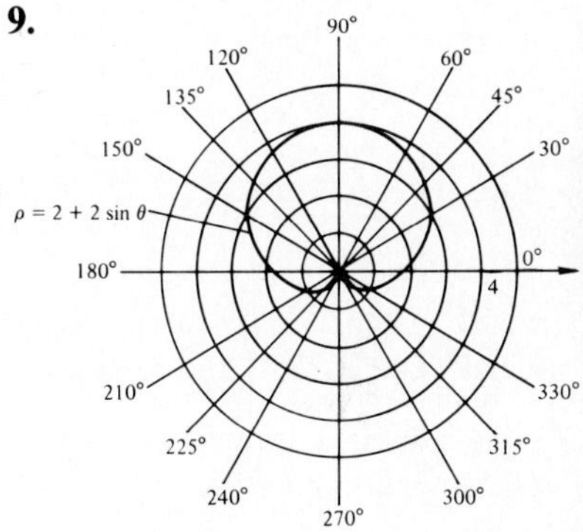

11.

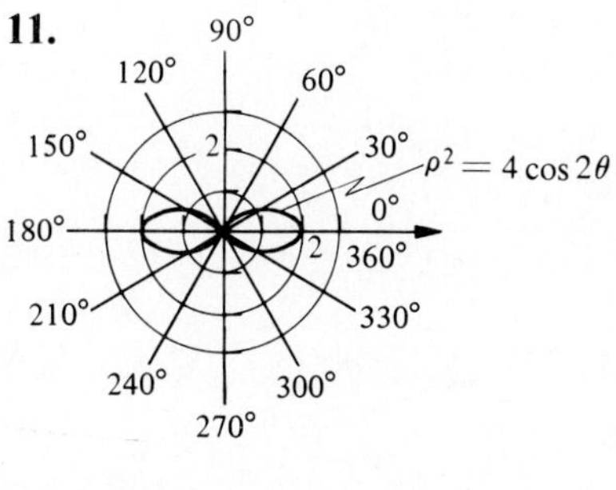

13.

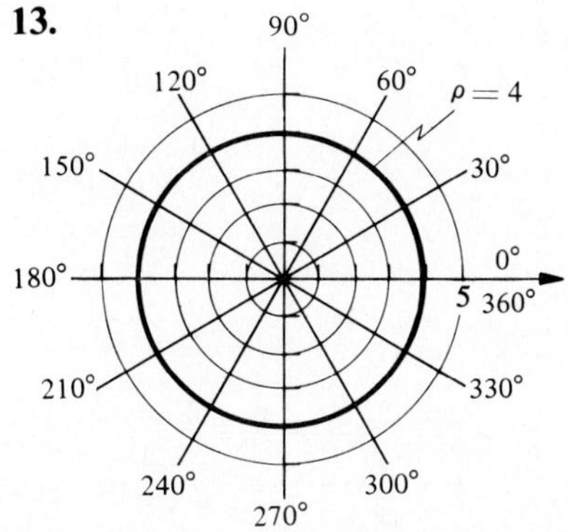

15.

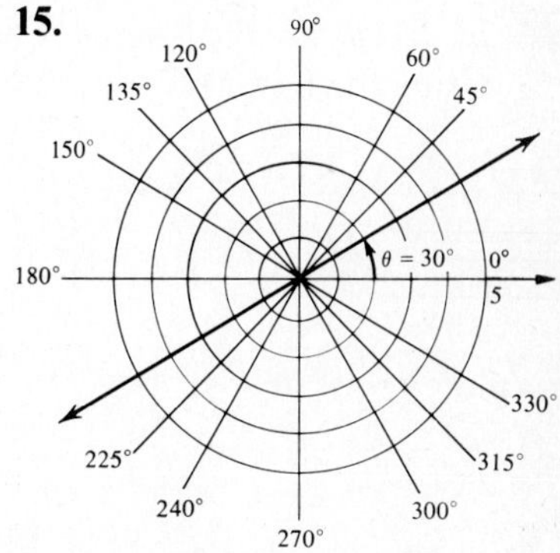

17.

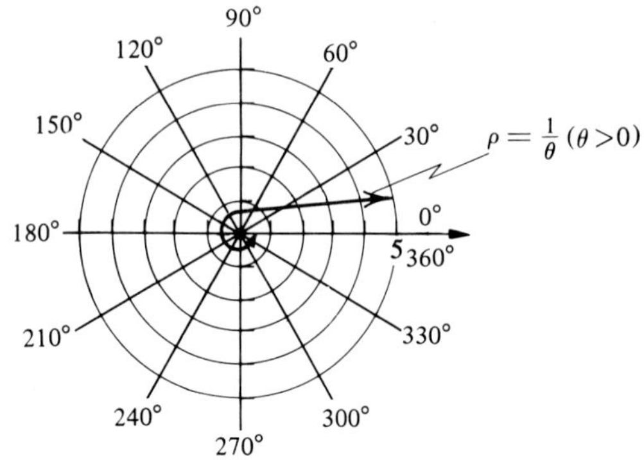

19.

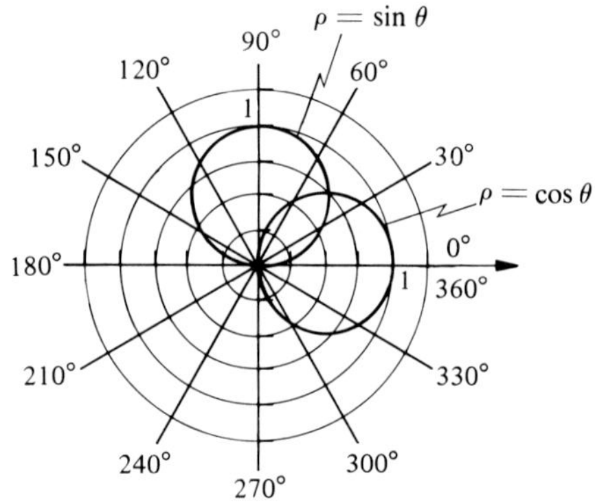

(0.7, 45°) and (0, θ) where θ has any measure

21.

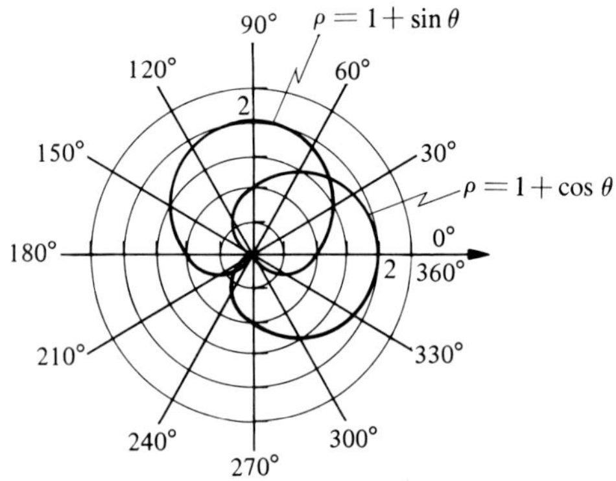

(1.7, 45°), (0.3, 225°) and (0, θ) where θ has any measure

23. $\{(1/\sqrt{2}, 45°)\}$; ρ may be zero for a different value of θ in each equation. In this case, $\rho = 0$ when $\theta = 0°$ in $\rho = \sin\theta$, and $\rho = 0$ when $\theta = 90°$ in $\rho = \cos\theta$.

25. $\left\{\left(\frac{\sqrt{2}+1}{\sqrt{2}}, 45°\right), \left(\frac{\sqrt{2}-1}{\sqrt{2}}, 225°\right)\right\}$

27. ρ_1 and ρ_2 are the lengths of sides with included angle $\theta_1 - \theta_2$.
Using the Law of Cosines, $d^2 = \rho_1{}^2 + \rho_2{}^2 - 2\rho_1\rho_2\cos(\theta_1 - \theta_2)$,

$$d = \sqrt{\rho_1{}^2 + \rho_2{}^2 - 2\rho_1\rho_2\cos(\theta_1 - \theta_2)}.$$

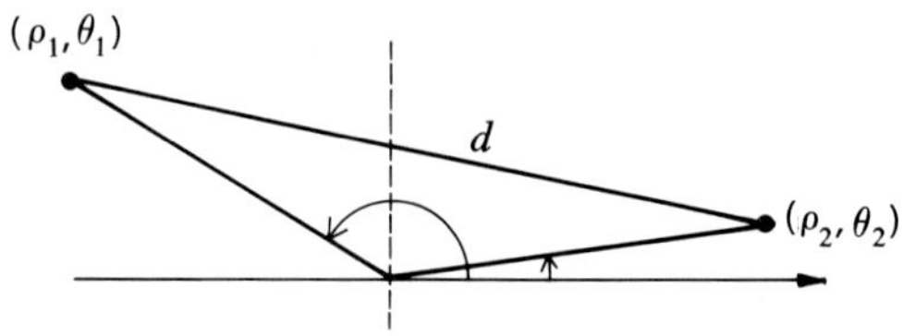

Review Exercises

1. $x = -1/3,\ y = -2/3$

2. $x = 1,\ y = 4;\quad x = -1,\ y = -4$

3. $-8 + 8i$

4. $0 - i$

5. $-46 + 28i$

6. $22 - 14i$

7. i

8. $-i$

9. $\sqrt{10} + 3\sqrt{5}i$

10. 14

11. If $x = 3i$, then $(3i)^2 + 9 = 9i^2 + 9 = -9 + 9 = 0$;
if $x = -3i$, then $(-3i)^2 + 9 = 9i^2 + 9 = -9 + 9 = 0$.

12. If $x = 1 + 3i$, then $(1 + 3i)^2 - 2(1 + 3i) + 10 = 1 + 6i + 9i^2 - 2 - 6i + 10 = 0$;
if $x = 1 - 3i$, then $(1 - 3i)^2 - 2(1 - 3i) + 10 = 1 - 6i + 9i^2 - 2 + 6i + 10 = 0$.

13. $-15 - i$

14. $3 - 5i$

15. $\frac{7}{5} - \frac{13}{10}i$

16. $-\frac{13}{34} - \frac{8}{17}i$

17. $\frac{2i}{\sqrt{6}}$

18. $\frac{8}{9} - \frac{4\sqrt{5}}{9}i$

19. $(-4, 1)$

20. $(2\sqrt{3}, -2)$

21. $9 - 6i$

22. $3 + 4i$

23.

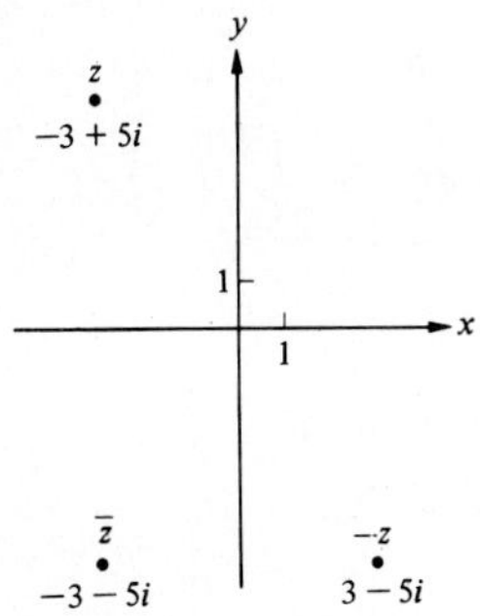

24.

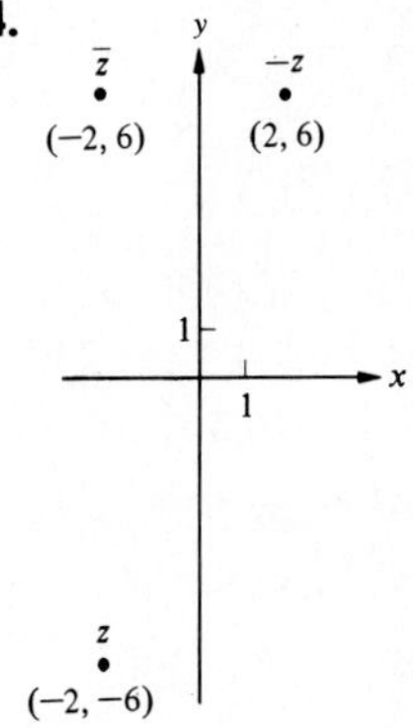

25. $4 + i$

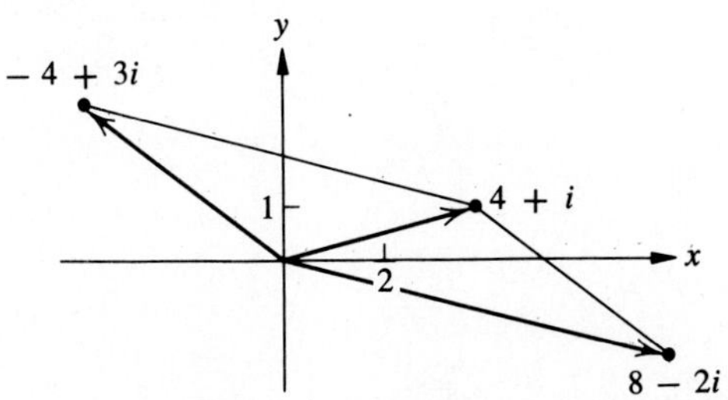

26. $(5, -1)$

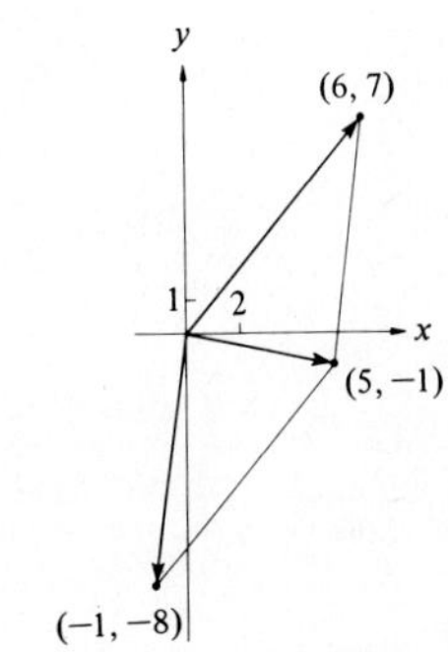

27. $2\sqrt{5}$, 26.5°

28. $\sqrt{29}$, 248.2°

29. $\sqrt{2}\,(\cos 45^\circ + i \sin 45^\circ)$

30. $6(\cos 120^\circ + i \sin 120^\circ)$

31. $-\frac{7}{2}+\frac{7\sqrt{3}}{2}i$

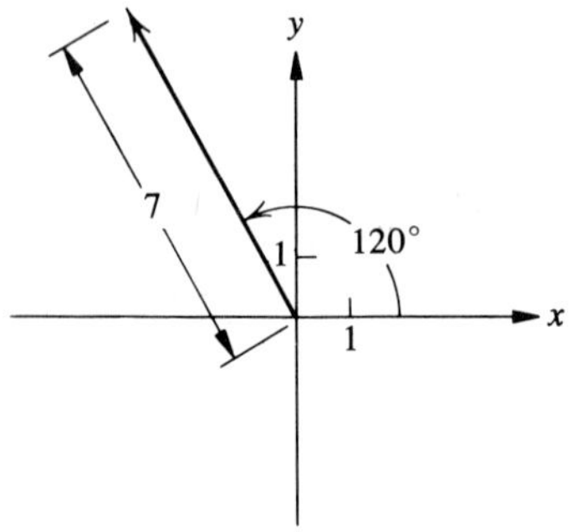

32. $\sqrt{2}+\sqrt{2}i$

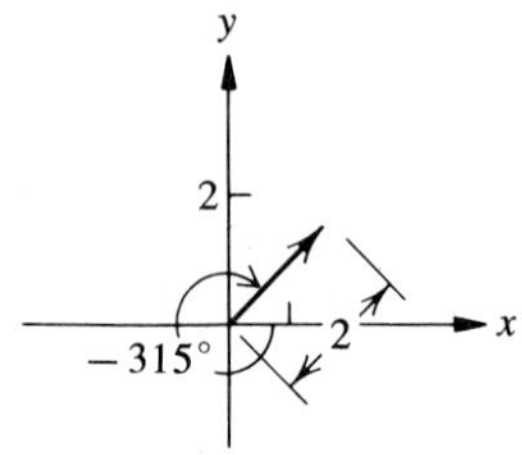

33. $-3\sqrt{2}-3\sqrt{2}i$ **34.** $3\sqrt{3}+3i$ **35.** $-\frac{8}{9}+0i$ **36.** $-2\sqrt{2}+2\sqrt{2}i$

37. $\frac{81}{2}+\frac{81\sqrt{3}}{2}i$ **38.** $0-8i$ **39.** $-\frac{1}{32}-\frac{\sqrt{3}}{32}i$ **40.** $-\frac{1}{256}+\frac{1}{256}i$

41. $\cos 15^\circ + i\sin 15^\circ$, $\cos 195^\circ + i\sin 195^\circ$

42. $5(\cos 110^\circ + i\sin 110^\circ)$, $5(\cos 230^\circ + i\sin 230^\circ)$, $5(\cos 350^\circ + i\sin 350^\circ)$

43. $\{\cos 45^\circ + i\sin 45^\circ, \cos 135^\circ + i\sin 135^\circ, \cos 225^\circ + i\sin 225^\circ, \cos 315^\circ + i\sin 315^\circ\}$

44. $\{10^{1/5}(\cos 48^\circ + i\sin 48^\circ), 10^{1/5}(\cos 120^\circ + i\sin 120^\circ), 10^{1/5}(\cos 192^\circ + i\sin 192^\circ),$
$10^{1/5}(\cos 264^\circ + i\sin 264^\circ), 10^{1/5}(\cos 336^\circ + i\sin 336^\circ)\}$

45. $(2, 250^\circ), (-2, 70^\circ), (2, -110^\circ), (-2, -290^\circ)$ **46.** $(-3, -330^\circ), (-3, 30^\circ), (3, 210^\circ), (3, -150^\circ)$

47. $\left(\frac{3}{2}, \frac{3\sqrt{3}}{2}\right)$ **48.** $(\sqrt{2}, -\sqrt{2})$ **49.** $\left(2\sqrt{2}, \frac{7\pi}{4}^{R}\right)$ **50.** $\left(2, \frac{7\pi}{6}^{R}\right)$

51. $\rho\cos\theta = 3$ **52.** $\rho^2(\cos^2\theta + 3\sin^2\theta) = 5$

53. $x^2 + y^2 = 5y$ **54.** $y^2 - 4x - 4 = 0$

55.

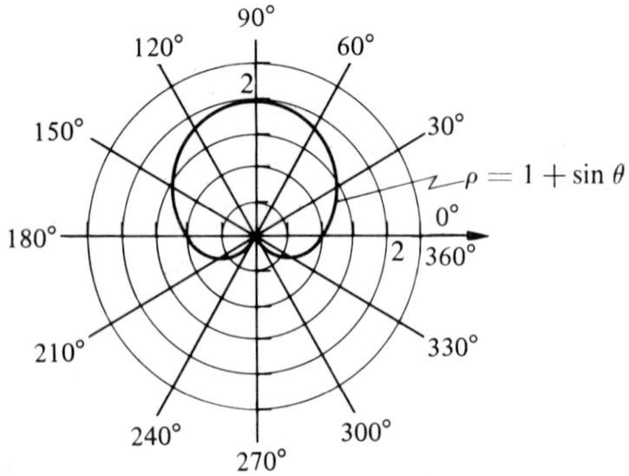

56.

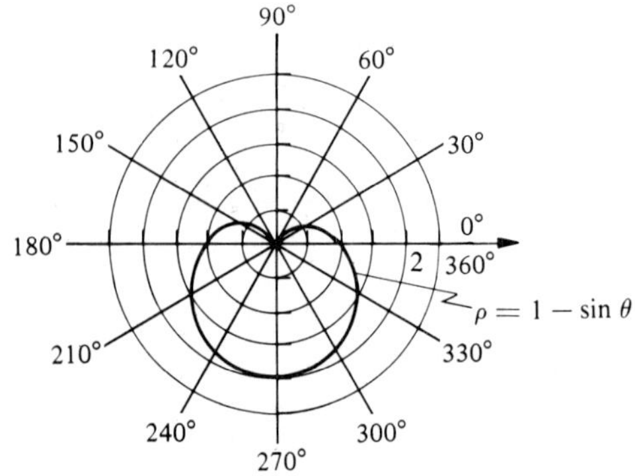

57.

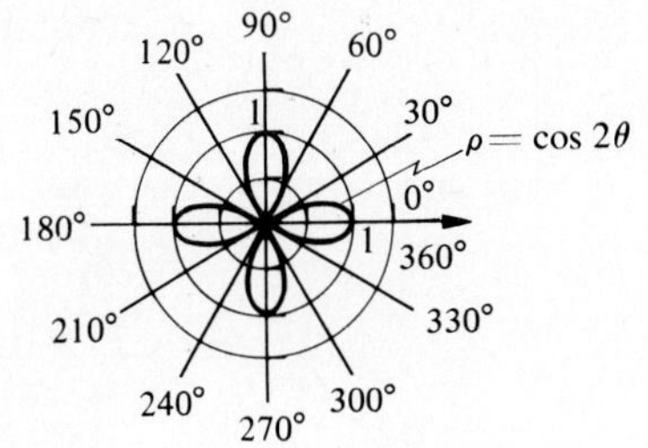
90°
120°
60°
150°
30°
$\rho = \cos 2\theta$
1
0°
180°
1
360°
210°
330°
240°
300°
270°

58.

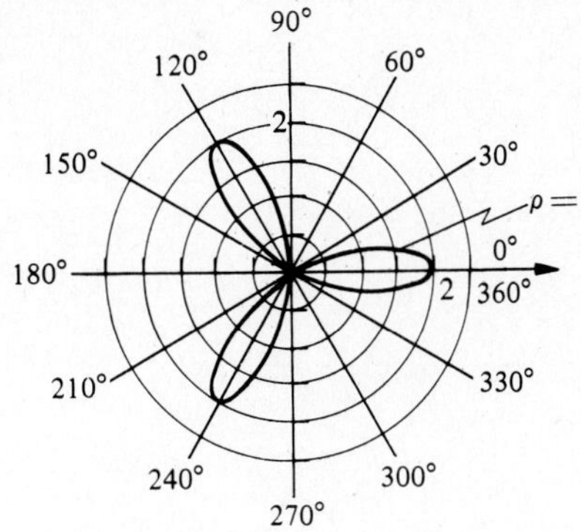
90°
120°
60°
2
150°
30°
$\rho = 2 \cos 3\theta$
0°
180°
2
360°
210°
330°
240°
300°
270°

6

INDEX

G

H

I

L

W

XYZ

SUMMARY

Chapter 1

Arc length s on a unit circle:
$\cos s = x$ and $\sin s = y$

Signs of function values:

Quad.	$\cos s$	$\sin s$	$\tan s$
I	+	+	+
II	−	+	−
III	−	−	+
IV	+	−	−

Special function values:

s	$\cos s$	$\sin s$	$\tan s$
0	1	0	0
$\frac{\pi}{6}$	$\frac{\sqrt{3}}{2}$	$\frac{1}{2}$	$\frac{1}{\sqrt{3}}$
$\frac{\pi}{4}$	$\frac{1}{\sqrt{2}}$	$\frac{1}{\sqrt{2}}$	1
$\frac{\pi}{3}$	$\frac{1}{2}$	$\frac{\sqrt{3}}{2}$	$\sqrt{3}$
$\frac{\pi}{2}$	0	1	undef.
π	−1	0	0
$\frac{3\pi}{2}$	0	−1	undef.

Reference arcs:

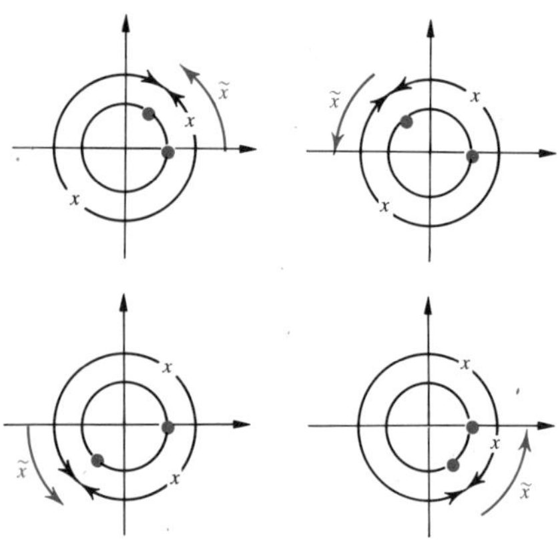

Basic identities:

$\sin^2 x + \cos^2 x = 1$

$\tan x = \dfrac{\sin x}{\cos x}$

$\cot x = \dfrac{\cos x}{\sin x} = \dfrac{1}{\tan x}$

$\sec x = \dfrac{1}{\cos x}$

$\csc x = \dfrac{1}{\sin x}$

Chapter 2

Graphs of circular functions:

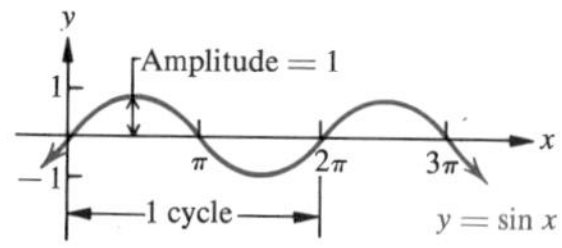

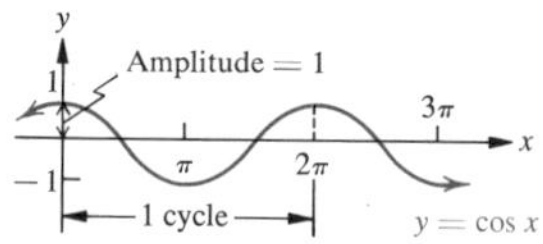

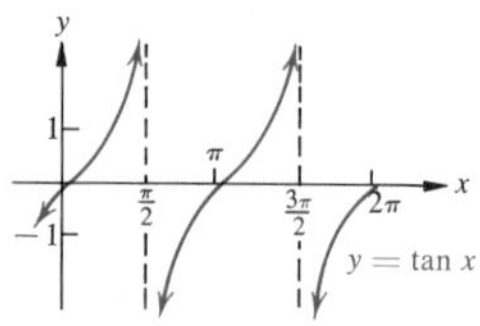

Inverse functions:

$y = \text{Sin}^{-1} x \leftrightarrow x = \sin y, \quad -\frac{\pi}{2} \le y \le \frac{\pi}{2}$

$y = \text{Cos}^{-1} x \leftrightarrow x = \cos y, \quad 0 \le y \le \pi$

$y = \text{Tan}^{-1} x \leftrightarrow x = \tan y, \quad -\frac{\pi}{2} < y < \frac{\pi}{2}$

$y = \text{Csc}^{-1} x \leftrightarrow x = \csc y, \quad -\frac{\pi}{2} \le y \le \frac{\pi}{2}, \; y \ne 0$

$y = \text{Sec}^{-1} x \leftrightarrow x = \sec y, \quad 0 \le y \le \pi, \; y \ne \frac{\pi}{2}$

$y = \text{Cot}^{-1} x \leftrightarrow x = \cot y, \quad -\frac{\pi}{2} \le y \le \frac{\pi}{2}, \; y \ne 0$

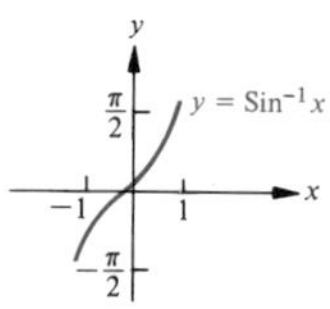

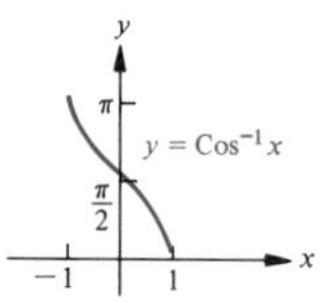

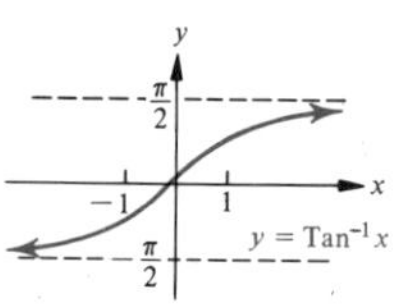

Chapter 3

Conversion formulas:

$\dfrac{m^\circ(\alpha)}{360} = \dfrac{m^R(\alpha)}{2\pi};$

$m^\circ(\alpha) = \dfrac{180}{\pi} m^R(\alpha); \quad m^R(\alpha) = \dfrac{\pi}{180} m^\circ(\alpha)$

Circular motion:

Length of arc: $s = r \cdot m^R(\alpha)$

Linear velocity: $v = \dfrac{s}{t} = \dfrac{r \cdot m^R(\alpha)}{t} = r\omega$

Angular velocity: $\omega = \dfrac{m^R(\alpha)}{t}$

Trigonometric ratios ($r = \sqrt{x^2 + y^2}, \quad r \ne 0$)

$\sin \alpha = \dfrac{y}{r} \qquad \csc \alpha = \dfrac{r}{y} \quad (y \ne 0)$

$\cos \alpha = \dfrac{x}{r} \qquad \sec \alpha = \dfrac{r}{x} \quad (x \ne 0)$

$\tan \alpha = \dfrac{y}{x} \quad (x \ne 0) \qquad \cot \alpha = \dfrac{x}{y} \quad (y \ne 0)$

Trigonometric ratios in a right triangle:

$\sin \alpha = \dfrac{\text{length of side opposite } \alpha}{\text{length of hypotenuse}}$

$\cos \alpha = \dfrac{\text{length of side adjacent to } \alpha}{\text{length of hypotenuse}}$

$\tan \alpha = \dfrac{\text{length of side opposite } \alpha}{\text{length of side adjacent to } \alpha}$

$\csc \alpha = \dfrac{\text{length of hypotenuse}}{\text{length of side opposite } \alpha}$

$\sec \alpha = \dfrac{\text{length of hypotenuse}}{\text{length of side adjacent to } \alpha}$

$\cot \alpha = \dfrac{\text{length of side adjacent to } \alpha}{\text{length of side opposite } \alpha}$

Solution of oblique triangles:

2∠'s and 1 side; 2 sides and an ∠ opposite one of the given sides — Law of sines:

$\dfrac{\sin \alpha}{a} = \dfrac{\sin \beta}{b} = \dfrac{\sin \gamma}{c}$

2 sides and the included ∠; 3 sides — Law of cosines:

$c^2 = a^2 + b^2 - 2ab \cos \gamma$

$b^2 = a^2 + c^2 - 2ac \cos \beta$

$a^2 = b^2 + c^2 - 2bc \cos \alpha$